Brewing

Related titles:

Brewing: Science and practice
(ISBN-13: 978-1-85573-490-6; ISBN-10: 1-85573-490-7)
Brewing: Science and practice updates and revises the previous work of this distinguished team of authors, producing what is the standard work in its field. The book covers all stages of brewing from raw materials, including the chemistry of hops and the biology of yeasts, through individual processes such as mashing and wort separation. Key quality issues are discussed such as flavour and the chemical and physical properties of finished beers.

Cereal biotechnology
(ISBN-13: 978-1-85573-498-2; ISBN-10: 1-85573-498-2)
This major work explains the techniques involved and their enormous potential for food producers and consumers. It also explains how this new technology is regulated, the methods for assessing its potential risks, and the ways in which cereal biotechnology can add value.

Details of these books and a complete list of Woodhead titles can be obtained by:

- visiting our web site at www.woodheadpublishing.com
- contacting Customer Services (email: sales@woodhead-publishing.com; fax: +44 (0) 1223 893694; tel.: +44 (0) 1223 891358 ext. 30; address: Woodhead Publishing Limited, Abington Hall, Abington, Cambridge CB1 6AH, England)

Brewing

New technologies

Edited by
C. W. Bamforth

BARTH-HAAS GROUP

FlavorActiV™

KERRY
Kerry Bio-Science

CRC

CRC Press
Boca Raton Boston New York Washington, DC

WOODHEAD PUBLISHING LIMITED
Cambridge England

Published by Woodhead Publishing Limited, Abington Hall, Abington, Cambridge CB1 6AH, England
www.woodheadpublishing.com

Published in North America by CRC Press LLC, 6000 Broken Sound Parkway, NW, Suite 300, Boca Raton, FL 33487, USA

First published 2006, Woodhead Publishing Limited and CRC Press LLC

British Library Cataloguing in Publication Data
A catalogue record for this book is available from the British Library.

Library of Congress Cataloging-in-Publication Data
A catalog record for this book is available from the Library of Congress.

Woodhead Publishing Limited ISBN-13: 978-1-84569-003-8 (book)
Woodhead Publishing Limited ISBN-10: 1-84569-003-6 (book)
Woodhead Publishing Limited ISBN-13: 978-1-84569-173-8 (e-book)
Woodhead Publishing Limited ISBN-10: 1-84569-173-3 (e-book)
CRC Press ISBN-13: 978-0-8493-9159-0
CRC Press ISBN-10: 0-8493-9159-8
CRC Press order number: WP9159

The publishers' policy is to use permanent paper from mills that operate a sustainable forestry policy, and which has been manufactured from pulp which is processed using acid-free and elementary chlorine-free practices. Furthermore, the publishers ensure that the text paper and cover board used have met acceptable environmental accreditation standards.

Project managed by Macfarlane Production Services, Dunstable, Bedfordshire, England (e-mail: macfarl@aol.com)
Typeset by Godiva Publishing Services Ltd, Coventry, West Midlands, England
Printed by TJ International Limited, Padstow, Cornwall, England

Contents

Contributor contact details

(* = main contact)

Chapters 1 and 21

Dr C. W. Bamforth
Chair and Professor
Department of Food Science and Technology
University of California
Davis, CA 95616
USA

E-mail: cwbamforth@ucdavis.edu

Chapter 2

Mr S. E. Heisel
American Malting Barley Association
740 N Plankinton Ave
Suite 830
Milwaukee, WI 53203-2403
USA

Chapter 3

Dr D. L. Goode*
Kerry Bio-Science BV
Veluwezoom 62
1327 AH Almere
The Netherlands

E-mail: declan.goode@kerrybioscience.com

Dr E. K. Arendt
Department of Food and Nutritional Sciences
University College Cork
Cork
Ireland

E-mail: e.arendt@ucc.ie

Chapter 4

Dr N. Davies
Technical Director
Muntons Plc
Cedars Maltings

Stowmarket
Suffolk IP14 2AG
UK

E-mail: nigel.davies@muntons.com

Chapter 5

Dr J. Henning
Research Hop Geneticist (plants)
USDA-ARS
3450 SW Campus Way
Corvallis, OR 97330
USA

E-mail: john.Henning@orst.edu

Chapter 6

Dr C. Schönberger
Joh. Barth & Sohn GmbH & Co. KG
Freiligrathstrasse 7/9
90482 Nuernberg
Germany

E-mail:
Christina.Schoenberger@johbarth.de

Chapters 7 and 8

Dr D. E. Quain
red.ts Ltd
9 Wheatfield Court
Willington
Derbyshire DE65 6PT
UK

E-mail: david.quain@red.ts.com

Chapter 9

M. Eumann
EUWA H.H. Eumann GmbH
Daimlerstrasse 2-10
D-71116 Gärtringen
Germany

E-mail: info@euwa.com

Chapter 10

Dr J. M. H. Andrews
Briggs of Burton Plc
Briggs House
Derby Street
Burton-on-Trent
Staffordshire DE14 2LH
UK

E-mail: jmha@briggsplc.co.uk

Chapter 11

Dr C. Boulton
Coors Brewers Ltd
Technical Centre
PO Box 12
Cross Street
Burton-on-Trent
Staffordshire DE14 1XH
UK

E-mail:
chris.boulton@coorsbrewers.com

Chapter 12

Dr I. Virkajärvi
Tuohimäki 3 B6
FIN-02130 Espoo
Finland

E-mail: ikka.virkajarvi@vtt.fi

Chapter 13

Mr G. Freeman
Brewing Research International
Coopers Hill Road
Nutfield
Redhill
Surrey RH1 4HY
UK

E-mail:
g.freeman@brewingresearch.co.uk

Chapter 14

Mr J. Browne
Technical Management Support Ltd
Claim Farm House
Manley Road
Kingswood
Frodsham
Cheshire WA6 6HT
UK

E-mail:
jeremy.browne@btinternet.com

Chapter 15

Mr D. Loeffler
Technical Director
Loeffler Chemical Corporation
5700 Bucknell Drive
Atlanta, GA 30336
USA

E-mail:
loeffler.dirk@loefflerchemical.com

Chapter 16

Dr R. Reed
Brackenbury House
Lodge Lane
Salfords
Redhill
Surrey RH1 5DH
UK

E-mail: reedr@bv.com

Chapter 17

Mr G. Jackson
Brewing Research International
Coopers Hill Road
Nutfield
Redhill
Surrey RH1 4HY
UK

E-mail:
g.jackson@brewingresearch.co.uk

Chapter 18

Dr K. J. Siebert
Food Science and Technology Department
Cornell University
Geneva, NY 14456
USA

E-mail: kjs3@cornell.edu

Chapter 19

Dr E. Storgårds*, A. Haikara and R. Juvonen
VTT Technical Research Centre of Finland
PO Box 1000
FI-02044 VTT
Finland

E-mail: erna.storgards@vtt.fi

Chapter 20

Dr W. J. Simpson
FlavorActiV Limited
Sanderum House
Oakley Road
Chinnor
Oxfordshire OX39 4TW
UK

E-mail: bill.simpson@flavoractiv.com

1

New brewing technologies: setting the scene

C. W. Bamforth, University of California, USA

1.1 Introduction

The aim of this book is not primarily to tackle the science underpinning malting and brewing. Rather, the focus is on practical issues. In this chapter I will set the scene with some underpinning information, but those seeking the necessary basic scientific descriptions of everything from barley to beer should consult a text such as Bamforth (2006) or, at a more advanced level, Briggs *et al.* (2005). For immediate purposes Table 1.1 offers basic summaries of the processes involved in malting and brewing and the importance of each unit operation.

Table 1.1 The essentials of malting and brewing

Process stage	Description
Selection of barley[1]	Malting barleys (moisture content <12%) of relatively low total N content (e.g. less than 1.7% N) with high viability and with endosperm of mealy texture that hydrates readily and possesses cell walls that are readily degraded
Storage of barley	Sometimes important to allow barley to free up from innate dormancy
Malting: steeping	Staged addition of water at 14–18°C separated by air rests to raise moisture content to 43–45%. Allows the embryo to start synthesising hormones (notably gibberellins); allows the aleurone to become receptive to the hormone action that will trigger enzyme synthesis; allows the starchy endosperm to become receptive to digestion ('modification')

Table 1.1 (continued)

Process stage	Description
Malting: germination	Controlled germination for 3–6 days at 16–20°C to degrade the endosperm cell walls and much of the protein. Enzymes synthesised by the aleurone migrate from proximal to distal end, digesting cell walls (β-glucanase family of enzymes, pentosanases) and protein (proteinase family). Causes softening of grain. Amylases also developed or activated, but limited in action on starch granules
Malting: kilning	Drying of malt at successively increasing temperatures (mainstream malts max. 105°C) to dry the grain (target <6% H_2O), whilst retaining much enzymic activity and developing colour and flavour through Maillard interactions between sugars and amino acids produced during modification
Malt: storage	2–4 weeks storage to avoid wort separation problems in the brewery
Milling and mashing	Generation of particles accessible to mashing water, mashing often starting at say 50°C (20 minutes) to allow remaining action of thermolabile β-glucanase, then passing through 65°C (e.g. for 1 h) for starch gelatinisation and action of amylase complex. Wort produced by conventional mashing has 20–25% starch left behind as non-fermentable digestion products called dextrins. Wort separated from grains
Boiling	Wort boiled with hops or hop preparations, typically for 1 hour. Isomerisation of bitter acids (main part of resin fraction) to increase their solubility and bitterness. Volatilisation of aroma components, including hop oils unless hops added late in the boil. Clarification stage follows to remove 'hot break' and residual hop material. Wort cooled and air or oxygen added
Fermentation	Ales typically fermented warmer (15–25°C) and therefore faster than lagers (6–15°C). Time range 3–14 days. Fermentation at a targeted rate of specific gravity drop and to a target 'attenuation'. Also diacetyl and pentanedione, which afford butterscotch/honey aromas, must be removed by prolonged contact of yeast with 'green beer'
Maturation, stabilisation and packaging	Minimum regime is −1°C, usually for 2–3 days. Some hold longer. Insolubilisation and settling of proteins and polyphenols. Filtration (kieselguhr or perlite-based). Removal of residual haze precursors by polyvinylpolypyrrolidone (polyphenols) and/or silica hydrogels or tannic acid or papain (proteins). Removal of any microbial contamination by pasteurisation or filtration. Adjustment of CO_2 content, then fill vessels

[1] *Process stages in italics occur in the maltings, prior to the brewery.*
Source: Bamforth, C.W. (2004) Opportunities for newer technologies in the oldest biotechnology, brewing. *Applied Biotechnology, Food Science and Policy*, 1, 213–222.

1.2 The materials used in brewing

The larger proportion of the world's beers is produced from *malted barley*. It has been this way for perhaps 8000 years. Barley has been retained as the primary cereal of choice, not least because it retains its husk on threshing and this traditionally form the filter bed through which wort is collected in the brew house. As we shall see in Chapter 10, the mash filter does not depend on the husk in this way, which might in future open up possibilities for huskless barleys for brewing. However, these handle in the maltings much as does *wheat*, with a tendency towards stickiness. Wheat is the second most employed cereal in brewing, notably for the production of weissbiers and weizenbiers in Germany.

Barley is malted prior to use, in order that the enzymes that degrade starch to fermentable sugars are synthesised. Additionally there is a synthesis of the enzymes that degrade the cell walls and much of the protein in the starchy endosperm, thereby softening the grain and making it more millable. There is an unavoidable development of embryo tissue (rootlets and acrospire), and the maltster seeks to balance the extent of this with the need for adequate 'modification'of the endosperm.

Although there are beers that are produced only from malted barley e.g. those produced under the terms of the five-centuries-old Bavarian 'purity' law, the *Reinheitsgebot*, many brewers employ various adjuncts for reasons of quality (different colours, better foams, interesting flavours) or cost. In fact there may not be quite the cost savings anticipated from use of an ostensibly cheaper starch source, because the brewer may have to employ more expensive processing procedures if problems are to be avoided. The major cost components in brewing are illustrated in Fig. 1.1. It will be apparent that the costs of packaging, processing, taxation and marketing vastly exceed those of raw materials. It is only when there are political reasons or taxation reasons – e.g. Happoshu, see Chapter 3) – that there are major justifications to be made on a cost basis for the use of adjuncts.

Of particularly low relative cost to the brewer are the *hops*, yet these make a huge contribution to product quality and stability: apart from the bitterness from the resins and aroma from the oils, the resins also afford anti-microbial properties and foam stabilisation but also comprise precursors of staling and of the skunky aromas that develop in beer exposed to light. Despite the low cost of hops relative to that of the beer, there is a plethora of hop preparations available to the brewer, including those that specifically protect against this light sensitivity (Chapter 6).

The bulk of most beers comprises *water*, hence the scrutiny which brewers devote to this product. Vastly more water is needed to make a pint of beer than actually finds its way into the beer, perhaps five times more for a well-run brewery and 20 times more for a badly operated facility.

Most brewers maintain their own *yeast* strains. As the alcohol concentration of most beers does not become too high during fermentation, the yeast that multiplies in fermentation remains healthy and suitable for re-pitching into

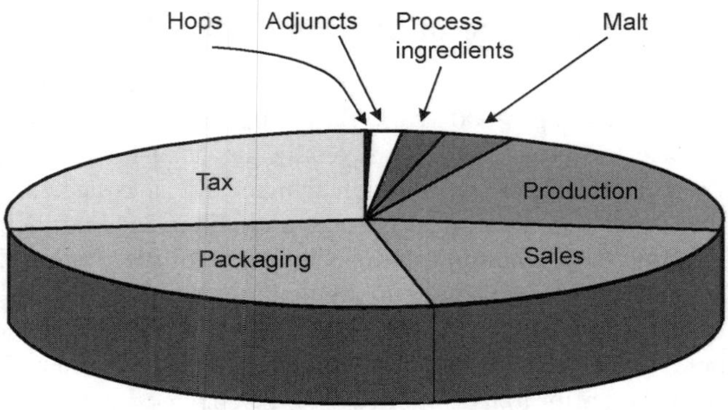

Fig. 1.1 A generalised breakdown of major costs associated with the brewing of beer.

subsequent fermentations. Some smaller brewers have done this in seeming perpetuity; however, the latter-day 'gold standard' is to re-pitch with newly propagated yeast every fourth or fifth fermentation. There is some interest in the use of dried yeast preparations of the type used *inter alia* in baking and wine making (Chapter 8).

As they will not be addressed elsewhere in this book, let us now consider miscellaneous process ingredients that may be employed in brewing. Not included in the discussion are the salts that may be added to adjust the ionic balance of the grist, e.g. calcium salts to increase the hardness ('Burtonisation'), and the zinc chloride or zinc sulphate that many brewers add to encourage yeast activity.

1.2.1 Exogenous enzymes

Various 'brewing enzymes' have been marketed over the years with the intent of satisfying various demands of the brewer (Table 1.2). The majority of these enzymes are not pure, insofar as they contain a range of enzymic activities. During the growth of the organism, different enzymes will be successively released into the medium in the order in which they are needed. So the precise balance of enzymes that is packaged and made available to the brewer will very much depend on where the 'cut' was taken. Nowadays most of the brewing enzymes are made using organisms that are genetically modified so as to optimise the level and type of enzymes in the broth.

1.2.2 Isinglass

One of the great beer genres (*viz.* the English cask ale) emerged on the backbone of a 'natural' clarification process rooted in a protein preparation called isinglass. Isinglass is a very pure form of collagen obtained from the dried swim bladders (some call them 'maws') of certain warm-water fish, amongst them the catfish, jewfish, threadfish and croaker. These fish are primarily caught for food use and the functional property of the maw represents added value.

Table 1.2 Exogenous enzymes used in brewing

Enzyme	Stage of use	Function
β-Glucanase[1]	Mashing	Eliminate glucans that cause wort separation, filtration and clarity problems
Pentosanase[1]	Mashing	Support glucanases in digestion of cell wall polymers from barley and wheat
Proteinase[1]	Mashing	Ensure generation of sufficient yeast-assimilable amino acids
α-Amylase[1]	Mashing	Starch digestion
Glucoamylase	Mashing or fermentation	Enhance starch digestion, to the extent of allowing conversion to totally fermentable sugars of value in the production of light and low carbohydrate beers
Pullulanase	Mashing	Promote digestion of branched-chain dextrins
Acetolactate decarboxylase	Fermentation	Accelerate elimination of vicinal diketones in beer maturation
Papain	Beer in storage	Eliminate haze-forming polypeptides
Prolyl endopeptidase	Beer in storage	Selectively remove haze-forming polypeptides; of potential value in producing beer for coeliacs

[1] Of especial significance if malt replaced by high levels of grain adjunct, e.g. unmalted barley.

The bladders are removed, washed and dried. At the smallest scale in a fishing village the maws are sun-dried, but modern fish processing plants use commercial dryers. Dried maws are ground up, washed and sterilised before being 'cut' by weak acids such as sulphurous acid to disrupt the structure of the collagen molecules so as to generate the correct balance and orientation of positively and negatively charged sites that are responsible for its functionality.

Worldwide, few brewers use isinglass, despite the fact that it is very effective in settling the contents of conditioning tanks, thereby minimising the loading of solids onto the filter without the expense of a centrifuge.

Various other fining agents are sometimes used in the brewing industry. These may be classified as kettle (or copper) finings, which are added during the boil, and auxiliary finings, which are used alongside isinglass to aid clarification of beer. The best-known kettle fining agent is the negatively charged carrageenan (Irish Moss), derived from seaweed, and the most frequently used auxiliary finings are silicates and alginates, the former derived from sand and the latter from seaweed. Again, both are negatively charged and complement the action of the positively charged isinglass.

1.2.3 Filter aids

There are fundamentally two types of filter aid: kieselguhr or perlite. Kieselguhr, or 'diatomaceous earth', comprises silica-based shells of ancient unicellular aquatic microscopic plants called diatoms. Its heat resistance means that it can be used as an insulator, but its abrasiveness means that it has also formed a component of toothpaste and metal polishes. Apart from being widely used as a filter aid to clarify syrups as well as alcoholic beverages, it is used as a filling material in paper, paints, ceramics, soap and detergents. Alfred Nobel found that it is a great absorbant of nitro-glycerine in the manufacture of dynamite.

Huge beds of kieselguhr, between 40 and 50 feet (12–15 metres) deep, are found in Virginia, but also in parts of Germany and in Aberdeenshire in Scotland. The microscopic appearance from different localities differs considerably. The deposits contain varying amounts of organic matter together with sand, clay, and iron oxide, and the raw material is first incinerated (calcined) to destroy organic matter. The successive process stages in rendering bags of kieselguhr in the form needed by the brewer are mining, crushing, drying, calcining, cooling, air classification and packaging.

The interests of purveyors of perlite are looked after by the Perlite Institute (go to http://www.perlite.org/). Perlite is a naturally occurring siliceous rock that, when heated, expands from four to 20 times its original volume. When heated to above 871°C, it pops like popcorn to produce many small bubbles, so perlite is very light and white.

There are many uses for perlite. Its insulating properties and lightness render it valuable as an insulator in masonry and cryogenic vessels. It is used as an aggregate in cement and plasters and for under-floor insulation, chimney linings, paint texturing, gypsum boards, ceiling tiles, and roof insulation boards. Perlite is used as a component of soil-less growing mixes, allowing aeration and moisture retention. It is also used as a carrier for fertiliser, herbicides and pesticides and for pelletising seed.

Apart from clarifying beer, perlite is also used for cleaning up pharmaceuticals, chemicals and water. Like kieselguhr, it can also be used as an abrasive.

1.2.4 Stabilisers

Silica hydrogels and xerogels have their origins in a pure form of sand. This is first converted into a soluble form by the action of alkali. Thereafter there is a controlled aggregation of sodium silicate from a sol form under acidic conditions. The washed aggregated particles are processed by techniques including micronisation, drying, milling and classifying to yield the desired balance of particle sizes and pore sizes, the range of which is taken advantage of by brewers (and others) to remove colloidal particles of various types from their products.

Polyvinylpolypyrrolidone (PVPP) is produced by a technique called 'popcorn polymerisation'. The monomer vinylpyrrolidone is heated with strong caustic and then cooled, in which phase the polymerisation takes place. Subsequently there are slurrying, filtering, hydrolysis (using phosphoric acid), washing, re-slurrying, and drying stages, in which any residual monomer and water are removed.

Alternative stabilisers are now in the market, including a PVP–silica gel composite.

1.2.5 Foam stabiliser

Propylene glycol alginate (PGA) is rather more relevant as a foam *protectant* than as a foam stabiliser. It helps prevent lipid and detergent damage at point of sale, without offering absolute protection. Alginates are polysaccharides extracted by alkali from brown seaweed. They have been known since their discovery by an English chemist in 1881. The seaweed is harvested on the coasts of North America, Scotland, Ireland, Norway, France, Japan, China, Korea, Chile, South Africa and Australia. Of these China easily generates the most, though the UK and the USA are the biggest processors, with one company dominating, at 70% of the market.

Insoluble alginate from the raw plant is solubilised by ion exchange, then filtered. The alginic acid obtained is partially dehydrated and then esterified by reaction with gaseous propylene oxide under pressure at 45–60°C. After washing with alcohol, the product is dried and milled.

Table 1.3 Issues of quality in brewing

Issue	Description
Foam	Most beers are expected to have an appealing, stable foam. However, increasing tendency for consumers to drink directly from the container. Impacting factors include amphipathic polypeptide survival from the grist, presence of bitter acids (reduced side-chain acids give more stable foams), levels of CO_2 and N_2, nucleation devices, absence of lipids and detergents.
Clarity	Most beers are bright (free from haze or sediment) – but not all. Insoluble materials can have diverse origins, including proteins cross-linking with oxidised polyphenols, residual grain polysaccharides, oxalate, micro-organisms
Colour	Derives from Maillard reaction in malt kilning and adjunct roasting and from polyphenol oxidation in the brew house
Absence of gushing	Principal cause is *Fusarium* infection of grain, with release of hydrophobin. But can also be cause by oxidised resins in hop bitterness preparations and insolubles in beer, e.g. oxalate
Package	Major determinant of impression customer gets of product – e.g. absence of scuffing on bottles, making returnable glass less appealing than one-trip glass
Flavour	Impacted by all the above. Pronounced marketing pressure to have beer in any coloured glass other than brown, but flint and green glass permit light passage that causes skunking of beer. Flavour impacted by all raw materials, demanding control of material selection and processing. Flavour instability the major unsolved problem facing brewers, in particular development of papery flavour. Stringent control of oxygen a major factor, but not alone. Cold shipping and stockholding of beer expensive but effective

Note: enhanced analytical capabilities (e.g. sensors) will permit increased control, with impacts on cost and quality.

Alginates, esterified or not, are very widely used in industry. About 50% go to textile printing, with 30% headed to food use. Ice cream, sherbets, milk shakes, yoghurts, icings, cake and pie fillings, meringues, glazes, salad dressings and non-carbonated fruit drinks may all contain alginates.

1.3 Brewing issues

Cost is only one of the drivers influencing business decisions in the brewing industry. Other major forces are quality (Table 1.3), product safety and

Table 1.4 Issues of product safety and wholesomeness in brewing

Issue	Detail
Absence of negatives	• Huge impact over the years of eliminating undesirables: nitrosamines, ethyl carbamate, monochloropropanols, pesticide residues, additives (e.g. sulphur dioxide) • 'Clean labelling' • Increased legislation
Presence of positives	• Alcohol – countering of atherosclerosis • Beer as a source of antioxidants, fibre, B vitamins, favourable potassium/sodium ratio, minerals • Perception issues – improved marketing compared with wine, which is perceived as healthier • Increased consumer awareness

Table 1.5 Issues of product innovation

- Traditional versus radical new approaches (*cf.* Happoshu/Third Category in Japan)
- Flavour chemistry – why products taste the way they do, and how to produce them consistently, whether by traditional or innovative approaches (e.g. by adding flavours to bland alcoholic bases)
- Alternative products from existing process streams – e.g. 'Malternatives'
- Global strategies – local production, licensing, control issues, matching
- New packaging modes – e.g. aluminium bottles, plastics

Table 1.6 Environmental issues

- Water availability and usage
- Carbon dioxide emissions
- Volatile organic compounds
- Energy demands
- Uses for co-products – spent grains, surplus yeast, spent filter aid, etc.
- Reduced inputs in agriculture
- Genetically modified organisms

wholesomeness (Table 1.4), product innovation to foster new global market opportunities (Table 1.5) and environment (Table 1.6).

1.4 Sources of further information

BAMFORTH CW, *Scientific Principles of Malting and Brewing*, St Paul, MN, American Society of Brewing Chemists, 2006.

BRIGGS DE, BROOKES PA, STEVENS R and BOULTON CA, *Brewing: Science and Practice*, Cambridge, UK, Woodhead, 2005.

2

Providing cereals for brewing

S. E. Heisel, American Malting Barley Association, USA

2.1 Introduction

Brewing grains and adjuncts are carefully selected for quality prior to their use. Quality factors vary widely among the various primary grains used for malting and those to be used for adjuncts. These quality factors influence how the grains destined for brewing are handled and processed. The selection of grain is a result of the genetics of the crop and the environmental factors such as soil, climate and farming practices. While environmental factors are critical to the suitability of grains for brewing, the current text will focus on the genetic or varietal specific traits.

In most conventional breweries, the malted grains are primarily barley and wheat, with barley use greatly outdistancing wheat. Rye and sorghum are also malting in some cases. Other grains are primarily used as adjuncts. Adjuncts are selected by the brewer to best fit their brewing process. Wheat and rye, either malt or not, are often used as significant contributors to the flavor profile of the final product. Other grains used as adjuncts, however, are typically included to provide a carbohydrate source for alcohol production, and in some cases to add body or foaming characters to the final product. Aside from imparting these positive attributes, adjunct brewers can also utilize these grains to attenuate the negative aspects of malt that may be more evident in lighter-style beers. There are breeding efforts to produce dual food and brewing varieties of rice, but worldwide, it is malting barley that breeders focus on to produce a cereal specifically adapted for use in brewing.

Barley has historically been the grain of choice as the main malt used in brewing. The use of barley in brewing is recorded on Sumerian clay tablets dating to 1800 BC (Katz and Maytag, 1991). Archaeological evidence suggests it was domesticated thousands of years prior to this (Harlan, 1968). Its early use in brewing may have resulted from its availability relative to other cereals (Burger and LaBerge, 1985), but several characteristics of barley make it well suited to the malting and brewing processes, which help explain its continued widespread use (Wiebe, 1968). Unlike other small grains, malting barley has a tightly adhering hull. This hull serves to protect the embryo from damage during grain handling and malting. It also plays a key role in brewing by providing a filter bed during the lautering process.

Early brewing relied on locally grown land races. These were often heterogeneous mixtures with the only pressure towards homogeneity applied by the producer of the barley. Quality evaluation was often limited to the visual examination of the crop (Hunter, 1926) and this did not always go very far, as evidenced in a letter of 6 October 1788 from Mr George Washington to Mr Clement Bidwell, where he writes 'My barley is raised from the seed you obtained for me from Mr. Haynes, and is, as I mentioned to you in a former letter, mixed in some degree with oats. This, I am told, is no disadvantage to the Malt tho' it is an objection to my sowing it again.' Other letters imply that Mr Washington was in contact with a nearby brewer who could use this barley. We have come a long way from such a quality perspective.

Attention to brewing quality has advanced over time. Chevalier was the first pure barley line and was developed by selecting seed from a single plant out of a land race in 1824. The first selections in North America were made out of a barley sample from Manchuria in 1873 (Weaver, 1950). Varieties released from these Manchurian lines represented the first real attempts to improve the adaptation of barley lines in the New World. It was not until the late 1800s that breeders began to purposefully cross varieties to produce superior barley varieties. Shortly after, European brewers began to assess the malting quality of barley (Burger and LaBerge, 1985). Burger and LaBerge (1985) identified the quality evaluation of new malting barley varieties, beginning in the 1930s with work in England (Bishop and Day, 1933), Canada (Harrison and Rowlands, 1932), and the United States (Dickson *et al.*, 1935). Many of these original quality measurements continue today. Prior to this period, most references to malting quality noted the nitrogen levels of the grain or the ratio of soluble nitrogen to the original barley grain.

2.2 Quality evaluations

There are many commonalities between the selection parameters of individual barley lots purchased for malting and those used to evaluate new breeder selections for malting and brewing quality. The focus here will be on methods currently used to differentiate between quality malting barley varieties and those that under the best environmental conditions would not produce suitable malt for today's brewers (feed varieties). At the same time, it is important that only barley with suitable quality be used for quality testing.

Most quality evaluation programs utilize internal checks or standards with which to compare new selections or varieties. Their use allows comparison among all the lines involved in testing where environmental influences may have caused one or more parameters to approach the limit of an accepted range. Quality programs still rely on traditional pilot and micro-malting evaluations and tend to treat all potential malting barley varieties (selections) the same. It may be true that some truly good varieties are lost along the way (Bamforth, 2003), an unfortunate result of the efficiencies and economies at all levels. First, there is the sheer number of lines to be evaluated. Micro-malting is labor intensive and quality analyses are expensive, though automation of analyses has slowed escalating manpower costs somewhat. The trend has been to increase the number of samples in a single pilot malting run and it is not practical to make multiple runs of samples under different malting conditions. This is particularly true during early generation testing when there is often limited seed available for testing. It is also where the greatest numbers of samples are discarded from breeding programs.

Varieties with superior agronomics or malting quality compared to existing varieties are also often lost in the marketplace. There is a limit to the number of varieties a market can handle and brewers incorporate new varieties into their malt blends slowly as they become accustomed to the processing nuances of each variety. They also want to be sure that a variety will be available in the quantity desired on a consistent basis. Further, grain handlers and maltsters within a region have the capacity to segregate only a limited number of varieties. This capacity gets quickly squeezed once the number of varieties are multiplied by crop year and growing district.

Malting barley breeders have been able to realize substantial yield and quality improvements despite the use of relatively narrow crosses (Wych and Rasmusson, 1983). This being the case, often there are only subtle differences among the progeny of malting barley crosses and some traditional quality measurements may be obsolete. At the same time, minor differences may limit the use of more cost-effective predictive tests such as Near Infrared Reflectance (NIR).

There has been considerable criticism of malt analyses as predictors of brewhouse performance (Axcell, 1998; MacGregor, 1997; Nischwitz *et al.*, 1999). While most of this criticism is valid, malt analysis is the best predictor that currently exists. Criticism often centers on the fact that the analyses do not account for the heterogeneity in the malt sample. Evaluations of new selections for malting and brewing are done on barley from relatively small field plots in comparison to varietal checks and in replicated nurseries. Compared to commercial lots, these barley samples are often relatively homogeneous and perhaps more credit can be given to the outcome of pilot malting evaluation trials.

Table 2.1 details ideal quality targets distributed to malting barley breeders by the American Malting Barley Association, Inc. (AMBA). The table is presented only to illustrate the type of information that may be distributed to malting barley development programs and not as a general guide for all breeders. Malting quality specifications are too diverse and any attempt to

Table 2.1 Ideal commercial malt criteria: malting barley breeding guidelines

	Two-row barley	Six-row barley
Barley factors		
Plump kernels 2.5 mm (on 6/64″)	>90%	>80%
Thin kernels thru 2.0 mm (5/64″)	<3%	<3%
Germination (4 ml 72 h GE)	>98%	>98%
Protein	11.0–13.0%	11.5–13.5%
Skinned and broken kernels	<5%	<5%
Malt factors		
Total protein (d.b.)	10.8–12.8%	11.3–13.3%
on 2.8 mm (7/64″) screen	>70%	>60%
Measures of malt modification		
β-Glucan (ppm)	<100	<120
F/C difference	<1.2	<1.2
Kolbach index (soluble/total protein)	40–47%	42–47%
Turbidity (NTU)	<10	<10
Viscosity (absolute cp)	<1.50	<1.50
Congress wort		
Soluble protein	4.4–5.6%	5.2–5.7%
Extract (FG db)	>81.0%	>79.0%
Color (°ASBC)	1.5–2.2	1.8–2.2
FAN	>180	>190
Malt enzymes		
Diastatic power (°L)	>120	>140
α-Amylase (DU)	>45	>45

General comments:

- Barley should mature rapidly, break dormancy quickly without pregermination and germinate uniformly.
- The hull should be thin, bright and adhere tightly during harvesting, cleaning and malting.
- Malted barley should exhibit a well-balanced modification in a conventional malting schedule with four-day germination.
- Malted barley must provide desired beer flavour.

Source: American Malting Barley Association, Inc., 2005.

specify analytical parameters, even for similar products, does not serve a purpose in these discussions. The key is communication between breeding programs, the barley research community, and the malting and brewing industry.

2.2.1 Barley evaluations

Barley evaluations are the easiest and most routine analyses. Most are utilized both at purchase points and in evaluation programs. Minimal barley quality standards should be determined before malting quality evaluations are performed. Check or standard varieties are extremely helpful in pilot malting and their use is widespread, if not universal. Checks should be carefully chosen so that they yield predictable malt quality profiles under the widest possible range of barley quality. Typically, these are established malting varieties that have demonstrated their ability to produce quality malt under a wide range of environmental conditions. Barley quality outside predetermined limits should not be evaluated.

Barley hull

The barley husk or hull is very important in regulating water uptake during steeping, protecting the developing embryo during germination, and as the major component of the filter bed during wort runoff in lautering. It also contains polyphenols that can contribute to astringent flavors in beer. The amount of hull is negatively correlated with malt extract (Roumeliotis *et al.*, 1999). Ultimately, a thin hull that tightly adheres to the kernel is desired for most brewing applications. Higher hull content was correlated with reduced germination rates, and Roumeliotis *et al.* (1999) speculate whether reductions could correlate with an increased propensity for pre-harvest sprouting, but this has not been studied.

Hull adherence is typically evaluated visually as a percentage of the total kernels in a sample missing a quantity of hull (ASBC Barley-2F). It is a routine measure of malting barley quality, but total husk content is not routinely analyzed although a method exists (EBC 3.9).

Kernel plumpness

Kernel plumpness is very important in the malting process. It is positively correlated with malt extract and negatively correlated with barley protein. All barley is cleaned and graded prior to malting. The thin barley is sold as livestock feed and the plumper fractions are malted separately according to size. Brewers request the larger-sized fractions in specified blends.

Barley selections are chosen not only for high kernel plumpness, but also for the uniformity of their plumpness. Two-row varieties are generally plumper and tend to be more uniformly plump than six-row varieties. Six-rows form three kernels at each node on the barley spike or head, whereas two-rows form a single kernel at each node. The outer two kernels (lateral kernels) of these triplets in six-row barley are squeezed aside resulting in kernels that are twisted and slightly smaller than the central kernels, as shown in Fig. 2.1. Gebhardt *et al.* (1993) showed that lateral and central kernels of six-row barley vary in malting quality. Using image analysis, it was shown that there are genotypic differences in the uniformity of kernel size within six-row varieties, suggesting that more uniformly plump six-rows could be developed if this trait were selected for.

Malting quality is influenced by both kernel size and shape (Fettell *et al.*, 1999). Perhaps the adoption of routine image analysis of barley kernels could address the importance of both kernel size and shape in evaluation programs (Armstrong *et al.*, 2003).

Germination and dormancy

Rapid germination after a short post-harvest period is a target of most breeders. The real key with any new variety is uniform germination. This is highly

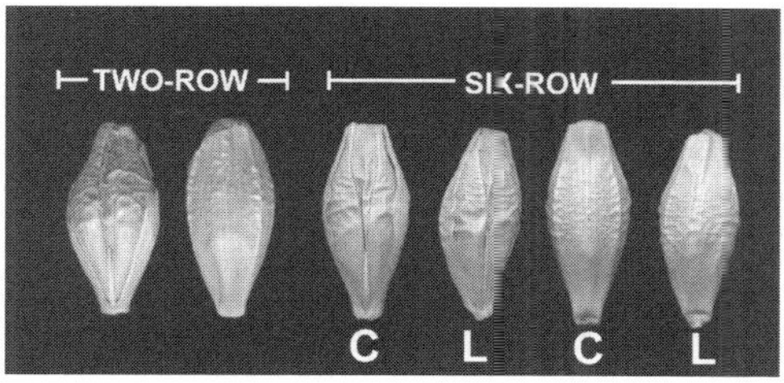

Fig. 2.1 Ventral and dorsal views of two-row (left) and six-row (right) barley kernels. The central kernels (C) of the six-row are symmetric like the kernels of two-row barley. The lateral kernels (L) of the six-rows are twisted and slightly smaller than the central kernel.

influenced by the physical attributes and environmental conditions under which the barley was grown. One significant attribute affecting germination that is inherent in the variety is dormancy.

Dormant varieties may tolerate prolonged storage periods, but can represent an obstacle to a rapid turnover in malting barley supplies. At the same time, a balance must be struck between dormancy and the capability of germination upon reaching physiological maturity. Some dormancy is important to prevent pre-harvest sprouting (Bamforth and Barclay, 1993). This is particularly important in environments where moderate to heavy rainfall could be expected once the barley has reached maturity, but has not yet been harvested. Many other environmental factors influence the level of dormancy expressed. Conditions that promote rapid maturation of the grain, such as high temperatures, reduce dormancy, once the kernel reaches physiological maturity, and cool-wet weather can be enough to break all dormancy (King, 1989).

Barley that has sprouted prior to harvest is unacceptable for malting. The germination potential of sprout-damaged barley can quickly and unexpectedly drop during storage, making it unusable for malting or for seed. Even if it retains its viability during storage, germination rates among the various kernels in a lot of sprout-damaged barley can vary widely, resulting in poorly modified malt (Pitz, 1991; Sole, 1994).

Dormancy is a complex quantitative trait governed by as many as 27 significant genes, according to Ullrich *et al.* (1992). Few, if any, malting barley development programs put forth an effort to maintain some degree of dormancy in their germplasm, even though techniques have been developed to select against pre-harvest sprouting (Mares, 1989). While not a direct measure of dormancy, pre-harvest sprouting is the greatest threat that a lack of dormancy presents. To the malting and brewing industry, significant dormancy is undesirable and over time dormancy has been reduced. In North America, it is likely that this reduction resulted not from a conscious effort to reduce dormancy, but through the selection of some other quality factors, like quick modification or possibly the production of higher levels of hydrolytic enzymes. The lack of significant dormancy in the North American barley crop can be demonstrated by widespread pre-harvest sprouting in the 2002 crop (Langrell and Edney, 2002; Heisel *et al.*, 2004).

Protein

Barley protein is one of the most important quality specifications in the malting barley trade. Kjeldahl procedures for the analysis of protein in barley are labor intensive and involve the use of harmful chemicals. They have been replaced by combustion and NIR methods (ASBC Barley-7; Buckee, 1994; Williams *et al.*, 1985).

High protein is one of the most limiting factors to achieving malting quality grades in North American (Edney *et al.*, 2005) and barley varieties with lower levels of total protein are selected for. Most of the malting barley produced in North America comes from semiarid regions and proteins tend to be higher than in many other regions of the world.

Barley and malt protein are negatively correlated with extract, and brewers prefer lower protein levels as long as there is plenty of soluble nitrogen for good yeast nutrition and beer foam potential. Solid adjunct brewers in North America prefer slightly higher protein levels as they result in higher levels of diastatic power. Moll (1979) provides a wide-ranging list of malt factors that are correlated to malt protein, soluble wort protein, and Kolbach index (see below). The Kolbach index is a ratio of soluble wort protein to the total malt protein expressed as a percentage (often noted as S/T).

2.2.2 Malt factors

No ideal malt specification exists and therefore no attempt will be made to present typical analytical parameters. Even within beer styles there is great variation among brewers. Targets must be set with the intent to satisfy the needs of end users and to reflect realistic expectations in the area in which the malting barley is grown.

The analysis of malt produced on a pilot scale is the most widely accepted method for evaluating new malting lines. Pilot brewing of new lines was once common, but is no longer mandatory in Australia (Healy, 2001) and its routine use in the US and Canada has been discontinued.

Measures of modification

Brewers want malt with a uniform modification profile. Measures of a balanced modification represent some of the important traits of new malting barley varieties. Many traditional measurements persist, while some older methods are being replaced. All are affected by malt variety (genetics) and environmental factors.

Procedures evaluate the mobilization and degradation of both protein and carbohydrate reserves during malting. Under-modified malt presents the brewer with a myriad of processing problems. Over-modification leads to reduced malt extract and high malt loss through respiration and excess rootlet and shoot growth.

Economics would dictate that a balanced modification be reached as quickly as possible provided the malt performs well in the brewery and yields a satisfactory beer. North American malting varieties are currently germinated for approximately four days, a significant improvement on the 8 to 12 days described by Hunter (1926) as used in floor malting systems at that time.

Fine–coarse (FC) extract difference

This traditional measure of modification has been given a back seat, at least in North American breeding programs. Elite germplasm exhibits only minor differences between fine-grind and coarse-grind extract differences (FC), and other methods are relied upon as measures of modification. In addition to its reduced value, there are considerable costs of duplicate extracts. Without FC only a single congress wort is needed for extract and other analytical measurements. The cost savings are highest in early generation evaluations where the number of samples is greatest. It then becomes a matter of deciding which extract measure better discriminates among the selections and which is most representative of plant-scale brewing. Fine-grind extract values are utilized for most quality analyses and therefore are routinely used in early generation testing in the US.

Kolbach index (S/T)

In the 1970s, the Kolbach index was the standard measure of malt modification (Seward, 1991) and it remains important today. It is a measure of the extent of proteolysis that has taken place during malting and mashing. The Kolbach index of North American varieties has risen in recent years. This is not likely a result of a conscious effort to select for varieties with a higher Kolbach index, but is probably an indirect result from selection of other traits such as extract or possibly α-amylase.

β-Glucan

β-Glucan is a major component of the cell walls of barley. It typically accounts for 4–7 percent of the weight of the grain (MacGregor and Fincher, 1993), and is found in the highest concentrations in the endosperm. It is analyzed in congress worts as one measure of carbohydrate modification. Excess β-glucans can reduce extract yields, increase wort and beer viscosities, and contribute to haze formation (Jin *et al.*, 2004).

β-Glucan levels show considerable variation among different barley varieties, but slight variations in malting conditions can also yield different results. Collaborative trials using several different laboratories may result in relatively large standard deviations. Experience has shown that while the numbers may vary, the relative rankings among the test selections are consistent and relationships to check varieties hold.

Wort turbidity

In the US, wort turbidity is noted by a simple visual rating (clear, slightly cloudy or cloudy) in early-generation trials. In later-generation trials by the industry, wort turbidity can be evaluated using methods commonly applied to beer (Gales,

2000; ASBC Beer-27; EBC 9.29). Turbidity is influenced by both variety and the environment. Turbid worts can result from poor modification and consist of proteins, polyphenols, undegraded starch, or polysaccharides.

Friability

Friability has long been an important analytical malt measurement in much of the world. It has not been an important parameter in the North American market, but is routinely used in most international markets. Friability is reported as the percentage of unmodified fragments of malt kernels (EBC 4.15).

Viscosity

Viscosity is a measure of modification in time-tested varieties, but also is a very heritable trait and thus serves as an evaluation trait of new selections. β-Glucans contribute much to viscosity but are not the only component. Arabinoxylans have also been shown to impact beer viscosity (Egi *et al.*, 2004; Sadosky *et al.*, 2002). Resistant starch and other malt constituents can contribute to increased wort viscosities.

2.2.3 Congress wort analysis

Congress worts are prepared for the measurement of a wide variety of analyses. These laboratory extracts are based upon traditional brewing practices and do not mimic the common commercial practice of high-gravity brewing (Budde and Jones, 2001). The American Society of Brewing Chemists (ASBC) is considering the evaluation of a high-gravity wort method (Budde *et al.*, 2005).

Extract

The percentage of the malt kernel that can be extracted and converted to beer has an important economic impact to the brewer. The goal is to increase the amount of fermentable carbohydrate and not to raise soluble protein to an excessive level. While not ideal, the AMBA evaluates carbohydrate extract as a simple difference between total fine grind extract and soluble protein. The impacts of barley protein, kernel plumpness and malt modification on extract were noted earlier in the chapter.

Soluble protein

The level of soluble protein in wort results from the amount of protein in the original barley, hydrolytic activity releasing the protein from protein bodies embedded in the endosperm, and proteolytic activity degrading protein to small peptides and amino acids.

Soluble protein is very important for proper yeast nutrition, and for foam production and stability (Bamforth and Kanauchi, 2003; Evans *et al.*, 2003). Excess wort protein levels can reduce beer stability by complexing with polyphenols and producing beer haze (Evans *et al.*, 2003; Sheehan *et al.*, 1999; Mikyška *et al.*, 2002). Greater levels of soluble protein are desired by adjunct brewers, as most of the common adjuncts have lower levels of protein than malt and dilute the level of soluble protein in wort.

Wort color

Wort color is affected by malt processing, growing environment and variety. Lines with higher levels of soluble protein and wort turbidity can be expected to produce higher levels of color. Color specifications are set by the brewer based upon the final product being produced. Malt processing can have a large impact on color development. Increasing modification or kilning temperatures can lead to higher wort color.

Free amino nitrogen

The malting and mashing processes yield insoluble protein, soluble protein and peptides, and free amino nitrogen (FAN). Peptides and amino acids are critically important to proper yeast nutrition (O'Conner-Cox and Ingledew, 1989). Predictive methods have been used to measure the amino acid composition of barley and malt (Williams *et al.*, 1985). The possible relationship between specific amino acids in wort and malting quality has also been explored (Edney *et al.*, 2005). Additional research elucidating a favorable balance of amino acid components in wort could lead to important screening methods in this area.

Fermentability

Evaluation programs that measure fermentability typically use procedures to test the apparent attenuation limit (AAL). These measures are conducted on advanced lines in European and Australian trials, but are not routine prior to commercial testing in North America. Edney and Langrell (2005) note the difficulties in measuring fermentability. More rapid methods for AAL determination that better reflect a broad spectrum of brewing processes could result in wider application of AAL evaluation.

Rather than directly analyzing AAL, it may be possible to measure the level of fermentable and nonfermentable sugars present in wort (Edney and Langrell, 2005). Evans *et al.* (2005) suggest measuring limit dextrinase and β-amylase activities in malt as predictors of fermentability.

α-Amylase

α-Amylase is produced during the malting process and is critical to the hydrolysis of starches during malting and mashing. It has been widely studied and routinely analyzed in malting barley breeding programs. At one time, six-row varieties had higher levels of α-amylase, but this is no longer true with present-day varieties (Schwarz and Horsley, 1996).

Diastatic power

The modification of barley starch to fermentable sugars is the result of four primary enzymes during malting and mashing. These are the limit detrinases, α-amylases, β-amylases, and α-glucosidases. The combination of these enzymes to break down starch to fermentable sugars is assayed (ASBC Malt-6; EBC 4.12) to yield the diastatic power. Thermostability, inhibitors, substrate, pH and other factors determine the contribution made by each enzyme involved in the determination of diastatic power.

Diastatic power is positively correlated to barley and malt protein while α-amylase is not. This is not surprising in that β-amylase exists in the mature seed (Allison and Swanston, 1974) and could be expected to be higher with higher grain protein. Nearly all α-amylase is produced during seed germination and regulated without regard to grain protein.

2.3 Commercialization of new malting varieties

Brewers need to produce a consistent, quality product that will retain the faithful and attract new customers. Generally, this results in a slow progression into new varieties, with malt blends dominated by one or two varieties. The remainder of the blend may consist of a lower percentage of a few varieties on their way in or out. In most regions of the world, a few varieties dominate acreage and production. The EU as a whole has a good number of different varieties, but only a few varieties dominate in most member countries.

Consolidation throughout the supply chain, from kernel to keg, applies pressures to winnow out minor varieties. Malting barley growers that have other crops not requiring varietal and crop year segregation are not going to bother with less popular malting varieties that may require extra marketing efforts. Grain handlers also have limited storage space, and as small rural elevators are being replaced by larger terminal facilities, the ability to segregate becomes more difficult. Similarly, the malting industry is replacing older plants with more efficient facilities that handle larger batch sizes. In the end, development programs releasing high yielding, high quality, widely adapted malting varieties will be seen as the most successful.

Success has its foundation in communication among brewers, maltsters, breeders, growers and grain handlers. All of the world's major malting barley regions have organizations that serve to encourage further interaction among these groups. Included are trade associations, government entities, scientific societies, and grower organizations. All have a responsibility to ensure that evaluation systems are developing varieties that will grow, handle, malt and brew better than the current varieties.

The tools used to evaluate new malting barley varieties are similar around the world. The institutions running the evaluations, and the process by which new barley varieties enter commercial production, vary significantly among countries and continents. The mix of public and private participation and the transition from one to the other occur at different points during the evaluation and commercialization processes. The following is a general summary of evaluation systems in Canada, the United States, Europe, and Australia.

2.3.1 Canada

Malting barley development programs are nearly all supported by federal or provincial governments in Canada. Development efforts shifted from spring six-rows, which dominated until the mid-1980s, to spring two-rows, as export markets for Canadian two-row malting barley grew along with a shift in domestic consumption from six-row to mostly two-row. Early-generation quality testing is done using a combination of micro-malting and predictive testing (Helms *et al.*, 2005). These are conducted either at the breeding institution or by Agriculture Agri-Food Canada (AAFC).

A barley variety must be registered with the government before seed for that variety can be sold in Canada. The variety registration system, which is administered by the Canadian Food Inspection Agency (CFIA), includes several years of 'merit testing' for agronomic performance, disease resistance and malting quality. Evaluation teams under the auspices of the Prairie Recommending Committee for Oat and Barley (PRCOB) review the trial data and make recommendations to the CFIA on the suitability of varieties for registration.

There are two stages to the malting quality evaluation trials: the cooperative trials and the collaborative trials. Malting quality evaluations of the cooperative trials are coordinated by the Grain Research Laboratory (GRL) of the Canadian Grain Commission, with micro-malting being done by the GRL and industry. The most promising malting selections in cooperative trials are advanced to collaborative testing. These trials are coordinated by the Brewing and Malting Barley Research Institute (BMBRI), a malting and brewing industry organization. The pilot malting trials are again conducted by industry members and the GRL. Upon successful completion of cooperative and collaborative trials, the PRCOB forwards a recommendation supporting registration to the CFIA.

The registration process is open to any company or institution that would like to put forward a malting variety, but a Canadian-based entity must be enlisted as a sponsor to enter a variety in the PRCOB trials. Varieties developed by public institutions are typically offered by license for marketing and distribution. This can take place prior to, during, or after PRCOB trials. Commercial-scale trials typically take place post-registration and are coordinated by the BMBRI. The Canadian Malting Barley Technical Center annually compiles and distributes a list of recommended malting varieties based on market projections.

2.3.2 United States

Malting barley is developed by a mixture of state universities, federal, and private entities. Development is driven primarily by domestic needs and attention is given to both six-row and two-row malting barley varieties. The current malting barley market is based on spring types, but there are efforts to develop winter malting barleys as well.

Early-generation quality analyses are conducted by the Cereal Crops Research Unit of the United States Department of Agriculture/Agricultural Research Service (USDA/ARS) for the public breeding programs. Anheuser-Busch, Inc.

and Molson Coors Brewing Company have breeding programs and run their own evaluation programs. Micro-malting analyses are conducted on all samples at the Cereal Crops Research Unit with little use of predictive testing (e.g. NIR).

Advanced-generation pilot malting evaluations are coordinated by the AMBA. Public programs as well as Anheuser-Busch, Inc. currently participate. Collaborators in these trials also pilot malt subsamples of commercial barleys, and analytical data are compared with those of malt from the same barley produced under plant-scale conditions. The aim is to adjust pilot conditions to mimic commercial-scale results as closely as possible. Commercial-scale trails may follow based upon pilot malting evaluations and agronomics of the line compared to current varieties. Again, these are coordinated by the AMBA.

Decisions to release barley varieties at public institutions are typically made by committees. These varieties are then made freely available to barley growers. Government oversight is limited to reviewing data establishing the variety as genetically unique and approving the name. AMBA releases an annual list of recommended malting barley varieties to barley growers.

2.3.3 Australia

Barley production is geographically diverse in Australia, with growing regions in the states of New South Wales, Queensland, South Australia, Tasmania, Victoria and Western Australia. Malting barley development is largely a public sector endeavor with strong programs in each of the states growing barley. The state-based breeding programs operate under a national body, entitled Barley Breeding Australia (BBA). BBA directs the focus of breeding targets and provides recommendations for strategic areas of barley research. Early and intermediate generation testing is conducted by the breeding programs. Advanced evaluation is run under a program called the National Variety Trials which is conducted in each state under a contract arrangement. Advanced stages of testing include industry trials on both pilot- and plant-scale levels. Varietal recommendations are made at the state level and varieties are released publicly, usually through a commercial partnership.

The last 25 years have seen the formation of a number of organizations related to malting barley research, development, evaluation and marketing. The industry body, the Malting and Brewing Industry Barley Technical Committee (MBIBTC), was formed in 1983 and provides guidelines for pilot malting evaluations (Healy, 2001). Final MBIBTC ratings are based upon weighting various quality parameters and result in a quality score ranging from 1 to 10. Separate ratings are established for liquid or solid adjunct brewing applications.

The Australian Malting Barley Centre (AMBC) was formed in 1990 with the intent to create a national malting barley breeding program and a center for quality evaluation (Inkerman *et al.*, 1999). The quality testing laboratory was set up in Toowoomba, Queensland in association with the barley quality laboratory of the Department of Primary Industries. Partial funding came from the Grains Research and Development Council.

2.3.4 Europe

Most barley development is done by private companies with public institutions providing ancillary support in the form of pathology, quality, genomic, and agronomic research. Malting varieties are then promoted by the companies developing them, or through an agent licensed for this purpose.

Quality evaluations of early-generation selections are conducted by developers or on a fee for service basis with public institutions. Testing of later generations is done according to national protocols which vary among the various European countries (Larsen, 2001). Once on a national list of a European Brewery Convention (EBC) member country, they may be accepted into EBC trials. These trials are broken down into four regions and are conducted to ascertain a malting variety's suitability throughout EBC member countries.

Europe is unique in that virtually all the malting barley development is done by private companies. Support for these programs is generated through seed sales, which can be challenging in an inbred species like barley. Profitability of European barley development programs likely results from a combination of faster turnover in varieties and a relatively low use of farm-saved seed for crop production.

2.4 Future trends

Some possible improvements or areas of further research have been mentioned in previous discussions. Future evaluation of barley and other cereals for the production of beer would obviously benefit from a better understanding of the constituents of beer. Even with a thorough understanding of the malting and brewing process, evaluation procedures of new varieties must be reasonably accurate and cost-effective and involve heritable traits.

New methods will evolve from additional research, but opportunities may also exist to rigorously statistically analyze data from current methods for relationships among various malting quality parameters, beer quality and brewing processes. Relationships between malting quality and milling energy (Swanston, 1990), hordein protein profile (Fox *et al.*, 2002), sedimentation tests (Palmer, 1975; Reeves *et al.*, 1978), kernel hydration (Davies, 1992), falling number (Best and Muller, 1991; Holmes, 1995), and many other parameters have been explored. Perhaps the most widely studied predictive test is NIR.

NIR could very well find a larger role in barley improvement. It is widely used to provide protein and moisture data at grain purchase points. Calibrations are made using current laboratory methods and theoretically NIR can not exceed these methods in accuracy, but it may provide a cost-effective means of screening large numbers of samples. NIR may also provide a screening tool that requires less training for laboratory staff. Newer equipment allows for analysis on whole grain and, in some cases, single kernels. Smaller sample sizes and nondestructive tests would allow for evaluations on earlier-generation selections. At a minimum, NIR and other predictive tests could be used to measure barley quality prior to pilot malting trials to insure that such evaluations are fair or to eliminate selections that fall well outside an accepted range for a quality parameter.

Image analysis has been used as well and may have a bright future. Not only could evaluations be based upon kernel shape and size, but also upon ventral crease depth and shape, kernel color, and presence of fungi or bacteria (Armstrong *et al.*, 2003). As with NIR, most of these evaluations could be done in a nondestructive manner.

Procedures will continue to be explored to evaluate malting barley and malt quality. New procedures will be adopted, some will replace or refine current procedures, and others will be relegated to being nice to know but not fitting the system for one reason or another. There will also be improvements to the efficiencies in how the methods are run. Many congress wort measurements have now been adapted to automated flow analyses resulting in increased efficiencies. Greatly reducing mash sizes (Schmitt *et al.*, 2006) would be another route allowing for evaluations when very little seed is available. It must be recognized, though, that there is a limit on how early quality evaluations can begin as the traits in the first filial stages are still segregating.

Limitations also exist in later-generation evaluations when plenty of seed is available for testing. Current malting evaluation procedures do not accurately portray brewers' needs relative to processing and flavor (Axcell, 1998; MacGregor, 1997; Nischwitz *et al.*, 1999) and little, if any, data is collected until at least pilot brewing trials begin. A lot of hard work and financial investment has been made by this point, and there is nothing harder for a malting barley breeder than to learn they did nothing wrong but their selection just doesn't taste or process right in commercial trials. Attempts to accurately mimic commercial brewing processes on a small scale have been tried (Ford *et al.*, 2001). The difficulty in scaling down brewing operations to a pilot level is further complicated by the wide range of processes employed and the myriad of beer styles.

Barley lines with novel traits have garnered considerable attention. Some have been discovered by testing barley from various collections worldwide and others have resulted from directed mutagenesis. Proanthocyanidins are major contributors to beer haze formation (von Wettstein *et al.*, 1985). Barley lines lacking them have been developed and examined for malting quality (Wesenberg *et al.*, 1989). Further, it has been demonstrated that beers produced from these malting barleys have greater haze stability (von Wettstein *et al.*, 1980).

Barley lines with low levels of phytate have also been developed (Bregitzer and Raboy, 2006). The phosphate in these lines is not bound in phytate, but is available as inorganic phosphate. Initially these barleys were developed for use as livestock feed, but they may have applications in brewing. Unfortunately, little research on brewing with these lines has been conducted to date.

The low-protein barley variety Karl (Wesenberg *et al.*, 1976) results from a reduction in storage protein (hordein) synthesis (Dailey *et al.*, 1988). Possible changes in the amino acid profile of wort produced from Karl have not been studied. More work in this area is needed to study possible impacts on fermentation, foam, haze, and other beer characteristics known to be influenced by protein. Karl has been shown to maintain low protein under varying nitrogen fertilizer rates (Weston *et al.*, 1993). It has been widely used as a parent in many US breeding programs, but little gain has resulted from its use (See *et al.*, 2002).

Barley starch is packaged in small and large starch granules and consists of both amylose and amylopectin. During mashing, the starch gelatinizes and becomes more available to degradation by carbohydrases. New malting barley varieties with altered ratios of small and large granules could improve quality. Similarly, altering the proportions of amylose and amylopectin could affect the fermentability of the wort, as waxy (high amylopectin) starch is more susceptible to degradation by α-amylase (MacGregor and Fincher, 1993). Unfortunately, existing waxy and high amylose varieties have higher gelatinization temperatures (MacGregor *et al.*, 2002) and lower levels of starch (Izydorczyk and MacGregor, 2001). The degradation of starch to fermentable sugars could be enhanced by increasing the levels of β-amylase (Evans *et al.*, 2005) or by producing a variety with a β-amylase with greater thermostability (Gómez *et al.*, 2005; Li *et al.*, 2003)

Genomic research has added greatly to our understanding of barley as a crop and has the potential to accelerate the development of new malting barley varieties. Genetic markers have been created that are capable of identifying the existence genes that contribute to important quality and agronomic traits. These markers facilitate the transfer of these genes to new barley selections without expensive and sometimes subjective field trials. The use of 'marker assisted selection' is becoming common, and the USDA/ARS is currently setting up genotyping laboratories to service small grain breeders throughout the US with the capability to do service work in this area. A USDA Cooperative State Research, Education, and Extension Service funded 'Barley Coordinated Agricultural Project' (Barley CAP) is aimed at developing new technologies and high throughput techniques to provide additional tools to barley breeders. Worldwide, genomics research is providing traditional breeders and researchers with additional tools that are distinct from those used to create genetically modified malting barley varieties.

2.5 References

ALLISON M J and SWANSTON J S (1974), 'Relationships between β-amylase polymorphisms in developing, mature and germinating grains of barley', *J Inst Brew* 80(3), 285–291.

ARMSTRONG B G, WEISS M, GRIEG R I, DINES J, GOODEN J and ALDRED G P (2003), 'Determining screening fractions and kernel roundness with digital image analysis', *Proc 11th Aust Barley Tech Symp*, http://www.cdesign.com.au/proceedings_abts2003/papers/064ArmstrongB.pdf.

ASBC Barley-2F, 'Skinned and/or broken kernels', in *Methods of Analysis* (8th edn), American Society of Brewing Chemists, St Paul, MN.

ASBC Barley-7, 'Protein', in *Methods of Analysis* (8th edn), American Society of Brewing Chemists, St. Paul, MN.

ASBC Malt-6, 'Diastatic power', in *Methods of Analysis* (8th edn), American Society of Brewing Chemists, St Paul, MN.

ASBC Beer-27, 'Physical stability', in *Methods of Analysis* (8th edn), American Society of Brewing Chemists, St Paul, MN.

AXCELL B (1998), 'Malt analysis prediction or predicament', *MBAA Tech Q*35(1), 28–30.

BAMFORTH C W (2003), 'Sourcing the soul of beer', *Brewers Guardian* 132(9), 22–25.

BAMFORTH C W and BARCLAY A H P (1993), 'Malting technology and the uses of malt', in MacGregor A W and Bhatty R S, *Barley Chemistry and Technology*, American Association of Cereal Chemists, St Paul, MN, 297–354.

BAMFORTH C W and KANAUCHI M (2003), 'Interactions between polypeptides derived from barley and other beer components in model foam systems', *J Sci Food Ag* 83(10), 1045–1050.

BEST S and MULLER R E (1991), 'Use of the Hagberg falling number apparatus to determining malt and barley quality', *J Inst Brew* 97(4), 273–278.

BISHOP L R and DAY F E (1933), 'Barley protein researches. Prediction of extract. II. The effect of variety on the relation between nitrogen content and extract', *J Inst Brew* 39(10), 545–551.

BREGITZER P and RABOY V (2006), 'Effects of four independent low-phytate mutations on barley (*Hordeum vulgare* L.) agronomic performance', *Crop Sci* in press.

BUDDE A, CHAN S, MAURICE M J and SCHWARZ P (2005), 'Malt extract under high-gravity conditions', *J Am Soc Brew Chem* 63(4), 225–227.

BUDDE A D and JONES B L (2001), 'Assessing difference in malting quality parameters when malts are extracted under standard ASBC and commercial conditions', *Brew Dig* 76(3), 40.

BUCKEE G K (1994), 'Determination of total nitrogen in barley, malt and beer by Kjeldahl procedure and the Dumas combustion method – collaborative trial', *J Inst Brew* 100(2), 57–64.

BURGER W C and LABERGE D E (1985), 'Malting and brewing quality', in Rasmusson D C, *Barley*, American Society of Agronomy, Madison, WI, 367–401.

DAILEY J E, PERTERSON D M and OSBORN T C (1988), 'Hordein gene expression in a low protein barley cultivar', *Plant Physiol* 88(2), 450–456.

DAVIES N L (1992), 'A new malting index: prediction of malting quality by endosperm hydration', *J Inst Brew* 98(1), 43–46.

DICKSON J G, SHANDS H L, DICKSON A D and BERKHART B A (1935), 'Barley and malt studies I. Developing new varieties of barley for malting and their properties', *Cereal Chem* 12(5), 596–609.

EBC method 3.9 'Husk content of barley', in *Analytica-EBC*.

EBC method 4.12 'Diastatic power', in *Analytica-EBC*.

EBC method 4.15 'Friability, glassy corns and unmodified grains of malt by friabilimeter (IM)', in *Analytica-EBC*.

EBC method 9.29 'Haze in beer', in *Analytica-EBC*.

EDNEY M J and LANGRELL D E (2005), 'Effect of fermentable sugars and amino acids on fermentability of malts made from four barley varieties', *MBAA Tech Q* 42(2), 101–106.

EDNEY M J, LEGGE W G and ROSSNAGEL B G (2005), 'Amino acid levels in wort and their significance in developing malting barley varieties', *Proc 18th N Amer Barley Res Workshop*, 104–107.

EGI A, SPEERS R A and SCHWARZ P B (2004), 'Arabinoxylans and their behavior during malting and brewing', *MBAA Tech Q* 41(3), 248–267.

EVANS D E, ROBINSON L, SHEEHAN M, TOLHURST R, HILL A, SKERRITT J S and BARR A (2003), 'Application of immunological methods to differentiate between foam-positive and haze-active proteins originating from malt', *J Am Soc Brew Chem* 61(2), 55–62.

EVANS D E, COLLINS H, EGLINTON J and WILHELMSON A (2005), 'Assessing the impact of the level of diastatic power enzymes and their thermostability on the hydrolysis of starch during wort production to predict malt fermentability', *J Am Soc Brew Chem* 63(4), 185–198.

FETTELL N A, MOODY D B, LONG N and FLOOD R G (1999), 'Determinants of grain size in malting barley', *Proc 9th Aust Barley Tech Symp*, http://www.cdesign.com.au/proceedings_abts1999/papers/N_A_Fettell.pdf.

FORD C, STEWART D, HEALY P and EVANS E (2001), 'Lautering: malt factors and their use in barley breeding', *Proc 10th Aust Barley Tech Symp*, http://www.cdesign.com.au/proceedings%5Fabts2001/papers/35%20Ford%20et al.pdf.

FOX G P, ONLEY-WATSON K and OSMAN A (2002), 'Multiple linear regression calibrations for barley and malt protein based on the spectra of hordein', *J Inst Brew* 108(2), 155–159.

GALES P W (2000), 'A comparison of visual turbidity with turbidity measured by commercially available instruments', *J Am Soc Brew Chem* 58(3), 101–107.

GEBHARDT D J, RASMUSSON D C and FULCHER R G (1993), 'Kernel morphology and malting quality variation in lateral and central kernels of six-row barley', *J Am Soc Brew Chem* 51(4), 145–148.

GÓMEZ B, ACEVEDO H and LÓPEZ A C (2005), 'Relationships among malt fermentability and malt quality parameters under the influence of barley β-glucan heat stable allele', *Proc 18th N Amer Barley Res Workshop*, 112.

HARLAN J R (1968), 'On the origin of barley', in *Barley: Origin, Botany, Culture, Winterhardiness, Genetics, Utilization, Pests*, United States Department of Agriculture, Agricultural Research Service, Washington, DC, 9–31.

HARRISON T J and ROWLANDS H (1932), 'Laboratory malting equipment', *J Inst Brew* 38(11), 502–508.

HEALY P (2001), 'Malting and brewing industry barley technical committee – future directions', *Proc 10th Aust Barley Tech Symp*, http://www.cdesign.com.au/proceedings_abts2001/papers/04.%20Healy.pdf.

HEISEL S, SCHWARZ P and BARR J (2004), 'Preharvest sprout damage as measured by pearling', *J Am Soc Brew Chem* 62(2), 59–62.

HELMS J H, OATWAY L and JUSKIW P (2005), 'Commercialization of near infrared reflectance spectroscopy (NIRS) for screening breeding lines in the breeding program and screening grain lots in a malt plant', *Proc 18th N Amer Barley Res Workshop*, 104–107.

HOLMES M G (1995), 'Studies on the malting potential of barley and malt with the falling number apparatus', *J Inst Brew* 101(3), 175–180.

HUNTER H (1926), *The Barley Crop*, Burleigh Press, Bristol, UK.

INKERMAN P A, CURTIS A, FERGUSON R and KAVANAGH T E (1999), 'Australian malting barley centre – current and future roles', *Proc 9th Aust Barley Tech Symp*, http://www.cdesign.com.au/proceedings%5Fabts1999/papers/PAInkerman.pdf.

IZYDORCZYK M S and MACGREGOR A W (2001), 'Effects of malting on phase transition behaviour of starch in barley cultivars with varying amylose content', *J Inst Brew* 107(2), 119–128.

JIN Y L, SPEERS R A, PAULSON A T and STEWART R J (2004), 'Barley ß-glucans and their degradation during malting and brewing', *MBAA Tech Q* 41(3), 231–240.

KATZ S and MAYTAG F (1991), 'Brewing an ancient beer', *Archaeology* 44(4), 24–33.

KING R W (1989), 'Physiology of sprouting resistance', in Derera N F, *Preharvest Field Sprouting in Cereals*, CRC Press, Boca Raton, FL, 27–60.

LANGRELL D E and EDNEY M J (2002), 'Quality of western Canadian malting barley 2002', Canadian Grain Commission, Winnipeg, Manitoba.

LARSEN J (2001), 'Evaluation of new European malting barley varieties', *Proc 10th Aust Barley Tech Symp*, http://www.cdesign.com.au/proceedings_abts2001/papers/22.%20Larsen.pdf.

LI C, LANCE R, TARR A, BROUGHTON S, HARASYMOW S, APPELS R, PARIS M and JONES M (2003), 'Can the thermostable β-amylase gene improve barley malting quality?', *Proc 11th Aust Barley Tech Symp*, http://www.cdesign.com.au/proceedings_abts2003/papers/059LiC.pdf.

MACGREGOR A W (1997), 'Dissatisfaction with current malt analysis – is there a solution?', *Proc 8th Aust Barley Tech Symp*, 2:2.4–2:2.7.

MACGREGOR A W and FINCHER G B (1993), 'Carbohydrates of the barley grain', in MacGregor A W and Bhatty R S, *Barley Chemistry and Technology*, American Association of Cereal Chemists, St Paul, MN, 73–130.

MACGREGOR A W, BAZIN S L and IZYDORCZYK M S (2002), 'Gelatinisation characteristics and enzyme susceptibility of different types of barley starch in the temperature ranges 48–72°C', *J Inst Brew*, 108(1), 43–47.

MARES D J (1989) 'Physiology of sprouting resistance', in Derera N F, *Preharvest Field Sprouting in Cereals*, CRC Press, Boca Raton, FL, 129–170.

MIKYŠKA A, HRABÁK M, HASÁKOVÁ D and Šrogl J (2002), 'The role of malt and hop polyphenols in beer quality, flavour and haze stability', *J Inst Brew* 108(1) 78–85.

MOLL M (1979), 'Composition of barley and malt', in Pollock J R A, *Brewing Science, Volume 1*, Academic Press, London, 2–143.

NISCHWITZ R, GOLDSMITH M, LEES M, ROGERS P and MACLEOD L (1999), 'Developing functional malt specifications for improved brewing performance', *Proc 9th Aust Barley Tech Symp*, http://www.cdesign.com.au/proceedings_abts1999/papers/Ralph_Nischwitz.pdf.

O'CONNER-COX E S C and INGLEDEW W M (1989), 'Wort nitrogenous sources their use by brewing yeasts: a review', *J Am Soc Brew Chem* 47(4), 102–108.

PALMER G H (1975), 'A rapid guide to endosperm malting potential of barleys using a sedimentation procedure', *J Inst Brew* 81(1), 71–73.

PITZ W J (1991), 'Rapid and objective methods for the estimation of pregermination and viability in barley', *J Am Soc Brew Chem* 49(3), 119–127.

REEVES S G, BAXTER E D, MARTIN H L and WAINWRIGHT T (1978), 'Prediction of the malting quality of barley by a modified zeleny sedimentation test', *J Inst Brew* 85(2), 141–143.

ROUMELIOTIS S, COLLINS H M, LOGUE S J, WILLSMORE K L, JEFFERIES S P and BARR A R (1999), 'Implication of thin husk in barley', *Proc 9th Aust Barley Tech Symp*, http://www.cdesign.com.au/proceedings_abts1999/papers/S_Roumeliotis.pdf.

SADOSKY P, SCHWARZ P B and HORSLEY R D (2002), 'Effect of arabinoxylans, β-glucans, and dextrins on the viscosity and membrane filterability of a beer model system', *J Am Soc Brew Chem* 60(4), 153–162.

SCHMITT M, BUDDE A D and MARINAC L A (2006), 'Research mash: a simple procedure for conducting congress mashes using reduced quantities of malt', *J Am Soc Brew Chem* in press.

SCHWARZ P and HORSLEY R (1996), 'A comparison of North American two-row and six-row malting barley', *Brewing Techniques* 4(6), 48–55.

SEE D, KANAZIN V, KEPHART K and BLAKE T (2002), 'Mapping genes controlling variation in barley grain protein concentration', *Crop Sci* 42(3), 680–685.

SEWARD B J (1991), 'Malt quality', *Ferment* 5, 275–278.

SHEEHAN M, EVANS E and SKERRITT (1999), 'Improved Methods for Determination of Beer Haze Protein Derived from Malt', *Proc 9th Aust Barley Tech Symp*, 2.9.1–2.9.4.

SOLE S M (1994), 'Effects of pregermination on germination properties of barley and resultant malt quality', *J Am Soc Brew Chem* 52(2), 76–83.

SWANSTON J S (1990), 'Early generation screening for malting quality', *Ferment*, 3(3), 170–172.

ULLRICH S E, HAYES P M, DYER W E, BLAKE T K and CLANCY J A (1992), 'Quantitative trait locus analysis of seed dormancy in Steptoe barley', in Walker-Simmons M K and Ried J L, *Proc 6th Int Symp Pre-Harvest Sprouting in Cereals*, American Association of Cereal Chemists, St Paul, MN, 136–145.

VON WETTSTEIN D, JENDE-STRID B, AHRENST-LARSEN B and SORENSEN J A (1980), 'Proanthocyanidin-free barley prevents the formation of beer haze', *MBAA Tech Q* 17(1), 16–23.

VON WETTSTEIN D, NILAN R A, AHRENST-LARSEN B, ERDAL F, INGVERSEN K, JENDE-STRID B, NYEGAARD KRISTIANSEN K, LARSEN J, OUTTRUP H and ULLRICH S E (1985), 'Proanthocyanidin-free barley for brewing: progress in breeding for high yield and research tool in polyphenol chemistry', *MBAA Tech Q* 22(2), 41–52.

WEAVER J C (1950), 'The distributional development of the American barley crop through the century, 1839–1939', in Weaver J C, *American Barley Production*, Burgess, Minneapolis, MN, 12–53.

WESENBERG D M, HAYES R M, STANDRIDGE N N, BURGER W C, GOPLIN E D and PETR F C (1976), 'Registration of Karl barley', *Crop Sci* 16(5) 737.

WESENBERG D M, JONES B L, ROBBINS G S and COHRAN J (1989), 'Malting quality and agronomic characteristis of selected proanthocyanidin-free barleys', *J Am Soc Brew Chem* 47(3), 82–86.

WESTON D T, HORSLEY R D, SCHWARZ P B and GOOS R J (1993), 'Nitrogen and planting date effects on low-protein spring barley', *Agron J* 85(6), 1170–1174.

WIEBE G A (1968), 'Introduction of barley into the new world', in *Barley: Origin, Botany, Culture, Winterhardiness, Genetics, Utilization, Pests*, United States Department of Agriculture, Agricultural Research Service, Washington, DC, 2–8.

WILLIAMS P C, NORRIS K H and SOBERING D C (1985), 'Determination of protein and moisture in wheat and barley by near-infrared transmission', *J Agric Food Chem* 33(2), 239–244.

WYCH R D and RASMUSSON D C (1983), 'Genetic improvement in malting in barley cultivars since 1920', *Crop Sci* 23(6), 1037–1040.

3

Developments in the supply of adjunct materials for brewing

D. L. Goode, Kerry Bio-Science, The Netherlands and E. K. Arendt, University College Cork, Ireland

3.1 Introduction

Brewing adjuncts are materials other than malted barley that bring additional sources of carbohydrate and protein into wort. This chapter will deal with developments in the supply of adjunct materials, focusing not only on current state-of-the-art technologies, but also on new and emerging technologies and products. Section 3.1 will cover the main reasons why brewers incorporate adjunct materials into brewhouse recipes. Section 3.2 will cover the traditional adjuncts barley, corn, rice, sorghum and wheat, focusing on their product attributes and processabilities and how they are currently incorporated into brewing recipes and brewhouse procedures. Section 3.3 will cover new potential adjunct sources with specific emphasis on triticale. Section 3.4 will cover adjunct manufacturing, new products and new processes. It will highlight different adjunct manufacturing procedures, namely micronisation, barley and malt fractionation, high pressure treatment, and advances in enzyme technology. Section 3.5 will cover new and future products, focusing on how high adjunct levels may be incorporated into these products.

3.2 Why use adjuncts?

It has been reported (Bamforth, 2003) that when the total cost of beer production is taken into consideration (from raw material purchase and processing through to packaging, sales and taxation), then malt costs in general have been estimated to represent just ~3.5% of the total cost. Therefore, it becomes apparent that grain costs represent only a relatively minor contribution to the total cost of beer production. Then, why replace malted barley with an unmodified substrate 'adjunct'? In less developed countries, malting facilities and malting conditions are quite often less than optimal. Therefore, because of its lower price, locally produced adjunct material can be used to supplement malted barley grain (Grujić, 1999). Apart from the direct cost benefits of using cheaper raw materials, indirect costs (much greater than the direct costs) can also influence raw material selection. In Kenya, for example, beer made from unmalted grain is taxed at 60% of the rate of beer made from malted grain (Cege *et al.*, 1999). Kenyan brewers are therefore encouraged to develop beer from exclusively non-malted grain (mainly raw barley). Likewise, in Japan a much lower rate of taxation is applied to products containing high adjunct levels (Happoshu) (Brewers Association of Japan; Shimizu *et al.*, 2002). Therefore, Japan's brewers have a great incentive to brew products from grists containing adjunct levels in excess of 50%. Likewise, in Nigeria a 1988 government economic decision to ban the importation of malted barley forced local brewers to develop alternative brewing procedures to utilise locally grown sorghum and maize crops (Hallgren, 1995; Little, 1994). Additionally, factors associated with product quality, tradition and consumer product expectations can be the decisive reason to use adjuncts, such as the impact that rice has on the flavour, colour and colloidal stability of an American pale lager, or the role that wheat plays in the taste and appearance of a Belgian or German style wheat beer (Delvaux *et al.*, 2001). Likewise, much of the distinct flavour profile of an Irish whiskey can be attributed to the traditional use of high proportions of raw barley in its manufacture (Booth *et al.*, 1989). Also the use of liquid adjunct materials in today's high gravity brewing culture can increase production output and significantly reduce production costs, whilst contributing to product character.

Table 3.1 gives an outline of the main adjunct types which are currently available to the international brewing industry. The type of adjunct available to an individual brewer largely depends on the geographical location of that brewery. Table 3.2 shows the 2004 production quantities of the world's top cultivated crops which are utilised by the brewing industry. Likewise the

Table 3.1 Common brewing adjuncts available

Whole cereal	Barley, wheat, sorghum, triticale, maize, millet, buckwheat
Grits	Maize, rice, sorghum, barley
Flaked	Corn, rice, barley, oats
Torrified/micronised	Corn, barley, wheat
Extrusion cooked	Maize, rice, sorghum, wheat
Flour/starch	Corn, wheat, rice, potato, cassava, soya, sorghum
Syrup	Corn, wheat, barley, potato, sucrose
Malted cereals	Wheat, oats, rye, sorghum
Malted pseudo-cereals	Buckwheat, quinoa

Table 3.2 Worldwide grain production (million tonnes) in 2004

	Africa	Asia	Europe	North and Central America	Oceania	South America	Total
Maize	41 610	182 753	88 222	329 783	574	6 235	705 293
Wheat	21 204	253 816	216 732	85 856	22 843	23 643	624 093
Rice	19 224	549 461	3 381	12 706	556	23 168	608 496
Barley	5 714	21 785	97 186	20 249	8 172	2 006	155 115
Sorghum	23 278	10 543	556	18 504	1 904	5 440	60 225
Millet	14 677	11 058	1 620	250	58	11	27 676
Oats	192	1 173	17 574	5 327	1 446	1 250	26 961
Rye	33	1 242	17 582	621	21	44	19 545
Triticale	1	1 150	11 885	62	641		13 739
Buckwheat	0.3	1 557	1 175	75		48	2 856
Quinoa						53	53

Source: Food and Agriculture Organisation of the United Nations (2004).

physicochemical properties of that adjunct (Table 3.3) will dictate its addition rates to a grist recipe, its time of addition and how it will be processed.

3.3 The range of traditional adjuncts

While adjunct material can be derived from any carbohydrate source, the five main cereals which are currently used as a base for brewing adjuncts are barley, maize, rice, sorghum and wheat. The following section will deal with these cereals, focusing on their product attributes and processabilities.

3.3.1 Barley

The use of barley over other cereal adjuncts offers significant advantages to the brewer. Since its starch has a similar gelatinisation temperature (53–58°C) to that of malted barley (61–65°C), it can be easily incorporated into conventional malted barley mashing procedures (O'Rourke, 1996). Its endogenous β-amylase (McCleary and Codd, 1989) ensures maltose production during mashing. Likewise, the presence of a husk can aid mash filtration through a traditional lauter tun (Cege *et al.*, 1999).

Careful selection of raw barley adjunct is a priority for users of this grain. The quality of grain supply to the industry is diverse. Barley grain supplies can differ greatly in terms of varietal content, harvest time, country of origin, source of supply, pre-brewery handling history and most importantly chemical and structural compositions. This can result in major processing difficulties, poor extraction problems and reduced alcohol yields (Goode *et al.*, 2005a). For processing purposes, it is wise for brewers to use malting-grade barleys. However, economically it can be attractive to use the cheaper feed-grade barley supplies.

Table 3.3 Physicochemical composition of some adjunct materials

Cereals	Moisture	Extract (% dry)	Gelatinisation temperature (°C)	Fat (% dry)	Proteins (% dry)	Starch (% dry wt of cereal)	Amylose	Starch granule sizes (μm)[1]
Maize grits	11–13	88–93	62–75	0.8–1.3	9–11	71–74	24–28	1–5, 10–20
Maize starch	8–12	101–106	62–74	<0.1	0.2–0.3	71–74	24–28	1–5, 10–20
Rice grits	10–13	89–94	61–78	0.2–0.7	6–9	57–88	14–32	2–10
Sorghum grits	10–12	75–82	68–75	0.5–0.8	6–10	70–74	24–28	0.8–10
Wheat starch	10–14	101–107	52–75	0.2–0.4	0.4–0.5	67–69	25–28	<10, 10–35
Barley	12–16	75–80	57–65	2–3	9–14	54–65	20–24	2–3, 12–32
Triticale	8–14	70–75	55–70	2–4	13–16	63–69	28–29	5, 22–36
Rye	10–15	76–80	55–70	1.5–2.0	8–16	58–62	23–25	2–3, 22-36
Oats	10–16	45–50	55-62	3–7	8–18	40–63	19–28	2–14
Millet	10–13	79–84	67–77	3–7	10–14	61–70	17–25	0.8–10
Potato starch	10–12	101–105	56–69	<0.1	0.05	65–85	20–23	
Manioc, cassava	8–11	87–97	52–70	0.3–0.6	9–12	85–87	15–17	

[1] Taken from Lindeboom *et al.* (2004).

Source: Moll (1994).

Therefore most brewers/distillers will seek a quality which is intermediate between feed grade and malting grade. Likewise, it is usual to select a malt of high diastatic power when brewing with high barley adjunct levels, particularly in the case of Scotch grain whisky production where the inclusion of exogenous enzymes is prohibited.

Raw barley grain is abrasive and difficult to mill, resulting in a high percentage of fine material which can give problems during lautering. In the past, hammer milling was preferred over roller-milling. However, new designs in roller milling such as wet roller milling systems can alleviate these problems by controlling the level and temperature of moisture addition to the grain prior to milling. Due to its low levels of essential enzymes (α-amylase, proteases and β-glucanases) together with a relatively inaccessible starchy endosperm, high inclusions of unmalted barley (>20%) in the mash (without the aid of commercial enzymes) can lead to problems such as low extract yields, high wort viscosities, decreased rate of lautering, fermentation problems and beer haze problems (Schwarz and Han, 1995; Viëtor *et al.*, 1991). In recent years an increased knowledge of the structural complexity of the barley starchy endosperm cell wall membranes together with their native enzyme inhibitors has enabled a more specific approach by enzyme producers to increase processability of raw barley adjunct.

When brewing with barley as adjunct together with malted barley, increases in the level of barley adjunct can result in decreases in extract recovery, wort α-amino nitrogen and fermentability, and increases in wort viscosity and β-glucan levels, if commercial enzyme levels are not optimised (Goode and Arendt, 2003b; Goode *et al.*, 2005d). Whilst increases in wort amino acid levels result from inclusions of higher levels of malted barley, the endogenous malt enzymes exhibit very poor raw barley protein and starch hydrolysing ability. Likewise, the endogenous malt amylases have been reported to exhibit very poor raw barley starch hydrolysing ability. As the level of malt is increased, their raw barley hydrolytic effects decrease (Goode *et al.*, 2005d).

Whilst hydrolysis of barley adjunct (~20%) can be achieved by using the enzyme capacity of malted barley, high adjunct levels may dilute the malt enzymes to a limiting level. These are then required to be augmented or replaced by commercial enzyme preparations. The enzyme preparations are usually available as single products or as part of mixed enzyme cocktails (Butcher, 1987; Goode *et al.*, 2005d; O'Rourke, 1996; Power, 1993). Exogenous proteolytic activity is required to modify endosperm structure and to facilitate saccharification, to release bound β-amylase, and to adjust the ratio of soluble nitrogen necessary for yeast growth (O'Rourke, 1996; Power, 1993). The most suitable preparation reported (O'Rourke, 1996; Power, 1993) is one containing only bacterial neutral protease from *Bacillus subtilis*. Additions yield increases in total soluble nitrogen, free amino nitrogen, wort colour and extract recovery. However, the hydrolysing efficiency of the protease decreases as its dosage level is increased (Goode and Arendt, 2003a; Goode *et al.*, 2005d). Addition of β-glucanase from *Bacillus*, *Aspergillus*, *Penicillium* or *Trichoderma* sources has been found to improve filtration, when undermodified malt or unmalted barley is used (Letters *et al.*, 1985; Oksanen *et al.*, 1985). When mashing with 100% raw barley substrate and commercial enzymes, exogenous β-glucanase (*Bacillus subtilis*) has little impact on mash filtration, but was found to reduce high molecular weight wort β-glucan levels (Goode and Arendt, 2003a; Goode *et al.*, 2005d).

High heat thermo-labile α-amylase derived from *Bacillus subtilis* is widely used for the degradation of gelatinised starch and high molecular weight dextrins to lower molecular weight dextrins and fermentable sugars (Butcher, 1987; Marshall *et al.*, 1982; O'Rourke, 1996; Power, 1993). An alternative is to use high heat thermo-stable bacterial α-amylase from *Bacillus licheniformis*, which is inactivated only at temperatures close to boiling. Thus it can hydrolyse a difficult starch even though swelling and gelatinisation occur at higher temperatures than usual (Marshall *et al.*, 1982). When mashing with 100% barley adjunct, exogenous α-amylase (*Bacillus subtilis*) addition has the greatest positive impact on mash separation (Goode and Arendt, 2003a; Goode *et al.*, 2005d). Increasing the level of exogenous α-amylase results in higher wort glucose and maltotriose levels and lower wort maltose levels (Goode *et al.*, 2005d). Optimal addition of an exogenous high heat stable α-amylase (*Bacillus licheniformis*) in combination with an exogenous α-amylase (*Bacillus subtilis*) is necessary for complete starch conversion and maximum extract recovery from the raw barley substrate.

Other pre-processed forms of barley are available for brewing purposes. These include pre-gelatinised barley flakes and dehusked barley. Pre-gelatinisation or partial gelatinisation of barley is performed by subjecting barley to mild pressure cooking or by steaming at atmospheric pressure followed by passage of the hot grits through rollers held at approximately 85°C, with a final reduction in moisture content to 8–10%. Pre-gelatinisation of barley allows easier extraction of β-glucans during mashing. Barley may be dehusked before brewing, to increase extract yields and decrease polyphenolic materials in the wort. However, the absence of husk material can lead to mash separation difficulties when a traditional lauter tun is used.

3.3.2 Maize

Whole grain maize corn consists of 76–80% carbohydrate, 9–12% protein and 4–5% oil. The oil fraction is located in the germ of the corn. Therefore maize is de-germed to limit beer foam damaging effects that would otherwise occur. During processing to grits or flakes the protein content is decreased to 7–9%. However, this protein remains largely undissolved during mashing and so free amino nitrogen (FAN) can be a limiting factor when brewing with high maize levels. The gelatinisation temperature of maize is reported to be 60–70°C, whilst the extract content is similar to that of malted barley at 77–78%. Maize is processed to make corn grits, maize flakes, refined grits and corn syrup. Maize flakes are pre-gelatinised and so can be mashed directly with malted barley.

Since the flakes do not contribute to the worts' protein pool, they can be added at the saccharification stage of mashing. Refined corn grits and corn syrups are starch grits that have been brought through a purification process, removing husks and protein through a series of washing steps, resulting in a product which gelatinises very easily and is therefore easy to pour. Essentially these products are pure starch. As a consequence yields from these products are often above 90%. They require cooking in order to achieve proper gelatinisation, though boil times are brief, typically 5–15 min (Hug and Pfenninger, 1980).

3.3.3 Rice

Brewers' rice is a by-product of the edible rice milling industry, the main business of which is the provision of whole grain rice for culinary use. The objective of rice milling is to completely remove the outer layers of bran, aleurone and germ, with a minimal amount of damage to the starchy endosperm, resulting in whole kernels for domestic consumption. Any kernels which may get fractured during the milling process (~30%) are considered undesirable for domestic use and are therefore sold to the brewing industry at a cheaper price. They are almost pure fragments of endosperm that contain starch exclusively. As a brewing adjunct, rice has a very neutral flavour and aroma, and when properly converted in the brewhouse yields a light clean-tasting beer.

The quality of brewers' rice can be judged by several factors, including cleanliness, particle size, gelatinisation temperature, mash viscosity, mash aroma, moisture, lipid, ash and protein content. Not all varieties of rice are acceptable brewing varieties (Teng *et al.*, 1983). Short-grain rice is preferred because medium and long-grain varieties can lead to viscosity problems (Bradee, 1977; Teng *et al.*, 1983). Owing to its relatively high gelatinisation temperature (61–78°C, Pierce, 1987) due to the presence of very small starch granules (2–10 μm, Lindeboom *et al.*, 2004), rice is extremely viscous prior to liquefaction in the cereal cooker. Thus the careful selection of varieties that liquefy well is important. Storage under unfavourable conditions such as high temperatures and high humidity can result in oil rancidity which can result in extraction problems. A high lipid content can cause increased yeast growth and reduced ester formation during fermentation (Ayraapaa and Lindstrom, 1973; Anderson and Kirsop, 1974), reduced foam stability, flavour problems and gelatinisation difficulties. Brewers' rice should therefore contain less than 1.5% lipid. In such concentrations lipids do not affect beer quality unless they become rancid. Extract yield differences depend greatly on the rice cultivar. It is important that brewers' rice is finely milled before brewing, otherwise gelatinisation problems will occur. Rice supplies little FAN, therefore the employment of a high-yielding FAN malt may be important to balance this deficiency. Flaked rice has the advantage of being pre-gelatinised and therefore does not need to be added to the cereal cooker.

The double-mashing system was developed in North America, to deal with grist containing large proportions of rice or maize grits (25–60%) and to use the nitrogen-and enzyme-rich malts that were available (Briggs, 1998). The adjunct mash containing the grits and a small proportion of enzyme-rich malt or bacterial α-amylase is mashed in at about 35°C in the cereal cooker. The stirred mash is heated to about 70°C and after remaining at this temperature for 20 min is brought to 85–100°C. It is held at this temperature for 45–60 min to ensure that any starch that has not been liquefied is gelatinised. Meanwhile, the malt mash has been mashed in at 35°C. After a stand of about an hour the adjunct mash is pumped in, with mixing, so that the final temperature of the mash is around 65°C. The whole process may take 3.5 h. It is important to remember that the actual time and temperature programme varies between breweries, depending on the ratio of adjuncts, the quality of adjuncts, modification of the malt, exogenous enzyme quality, process equipment, and the capabilities and capacities available to the brewer.

3.3.4 Sorghum

Sorghum, like barley, can be used in many forms, including malted grain sorghum (Agu and Palmer, 1998a; Owuama, 1999), sorghum grits (MacFadden and Clayton, 1989), extruded dehulled sorghum (Dale *et al.*, 1989; Delcour *et al.*, 1989) and unmalted whole grain sorghum (Goode, 2001; Goode and Arendt, 2003c; Goode *et al.*, 1999, 2002, 2003). Each has its own advantages and disadvantages.

Malted sorghum

Comprehensive reviews on brewing lager beer from sorghum, particularly malted sorghum, have been compiled by Ogbonna (1992), Adejemilua (1995), Agu and Palmer (1998b) and Owuama (1999). These show that remarkable progress has been made in investigating various factors which influence malting of sorghum. Therefore the sorghum malting process will not be the subject of this chapter. Limited endosperm cell wall degradation, low extract yields, poor wort separation and poor beer filtration are obstacles which have been widely reported when sorghum malt is used in lager production (Aisien, 1982, 1988; Aniche and Palmer, 1990a, b; Bajomo and Young, 1990; Etokakpan and Palmer, 1990a, b; Glennie *et al.*, 1984; Glennie, 1983; Morrall *et al.*, 1986; Okon and Uwaifo, 1985; Palmer, 1991).

Different mashing methods have been carried out using sorghum malt. These include a three-stage decoction method and a decantation method (Agu and Palmer, 1998b; Owuama, 1999). The three-stage decoction mashing procedure is an extensive mashing procedure designed to overcome the problem of low extract yield from sorghum malt (Skinner, 1976). In this process, about 70% of the mash is boiled to gelatinise the sorghum starch. Mashing at 65°C and 70°C for 30 min each at the second and third stages respectively provide wort with complete hydrolysis (Okafor and Aniche, 1980; Solomon *et al.*, 1994). High extract recovery yields of 82.7% and reasonable levels of attenuation (Dufour *et al.*, 1992) have shown that it is possible to produce malted sorghum worts of a quality similar to malted barley worts if the right sorghum cultivar is selected.

The decantation mashing procedure developed for extracting sorghum malt (Agu and Palmer, 1996) is a non-conventional method whereby active enzyme wort is decanted after mashing sorghum malt at 45°C for 30 min. The starchy grist residue is then gelatinised at 80–100°C before mixing with the decanted enzyme wort to achieve a saccharifying temperature of 65°C (Agu and Palmer, 1996). The extract recovery rates for this method are similar to or higher than those of well-modified barley malt. However, low levels of fermentable extract have been reported (Agu and Palmer, 1996).

Because of its lower levels of endogenous enzymes, it is quite common for brewers to add external enzymes during mashing of sorghum malt to increase extract yields and free amino nitrogen in the wort (Agu *et al.*, 1995; Bajomo and Young, 1992). A major advantage of using malted sorghum over unmalted sorghum is that the proteolytic enzymes of the malted grain produce sufficient free amino nitrogen (FAN) for efficient buffering capacity and optimal yeast performance (Bajomo and Young, 1993; Palmer, 1989). Less exogenous proteolytic enzymes are therefore required. This can be beneficial from a costing perspective and also from a beer quality perspective. Studies (Agu and Palmer, 1998a) show that the use of exogenous enzymes in mashing with raw sorghum reduced the foam head retention because commercial proteolytic enzymes destroyed the foam proteins.

Although beer has been brewed successfully at both a laboratory and a pilot brewery scale level from sorghum malt without the need to supplement endogenous enzymes with external heat-stable enzymes, there is yet no evidence available to suggest the existence of a commercial production plant based on such a process of lager beer brewing. These factors combined with the lack of sorghum malting capacity in native sorghum countries have led some brewers to favour the use of sorghum as an unmalted adjunct in combination with the necessary exogenous enzymes (Agu and Palmer, 1998a; Agu *et al.*, 1995; Bajomo and Young, 1992; Dale *et al.*, 1990; Goode, 2001; Goode and Arendt, 2003c; Goode *et al.*, 1999, 2002, 2003; Hallgren, 1995; Little, 1994; MacFadden and Clayton, 1989).

Sorghum grits

As with other cereals, the quality of sorghum grits depends on the quality of the sorghum grain, the milling method applied, the lipid and crude fibre content, and the extract yield. Sorghum brewers' grits are obtained from dry milling of the dehulled grain (Hallgren *et al.*, 1992). A high proportion of vitreous endosperm favours the yield of grits (Hallgren and Murty, 1983). One of the main problems identified in breweries using sorghum grits as adjunct is that of poor lautering runoff. This can be caused by insufficient separation in the milling system, leading to a high proportion of soft endosperm. Endosperm vitreousness (hardness) of sorghum brewer's grits is therefore a very important parameter to measure both at the grain level in selection of the raw material for the production of grits and on the final grits received at the brewery (Hallgren, 1995; Hallgren and Murty, 1983).

Extruded sorghum

The use of extruded cereal adjuncts in mashing offers a cheap and high yielding source of extract relative to malt. The technology of extrusion is relatively simple and is therefore suitable for use in developing countries for the processing of locally grown cereal crops (Dale *et al.*, 1989). The higher gelatinisation temperature of sorghum starch often means that brewers are limited in the amount of adjunct that they can include in their mash because of simple brewhouse capacity logistics. Extrusion has the ability to offer a solution to this problem as it is an efficient way of pre-gelatinising sorghum. Extruded products can therefore be directly mixed with malt during mashing. Extrusion of sorghum is carried out at temperatures of 165–190°C (Delcour *et al.*, 1989). Dale *et al.* (1989) found that, with regard to mash filterability, 175°C was the optimum extrusion temperature. Delcour *et al.* (1989) found that extrusion as a pre-gelatinisation step led to a significant improvement of the extract yield, but does not allow for complete saccharification and a good filtration of the mash. Increases in the addition of exogenous enzymes or a preliminary boil of the extruded material prior to mash-in gave increases in extract but also had no effect on the saccharification or filtration rate (Delcour *et al.*, 1989). The saccharification and filtration problems encountered when brewing with extruded sorghum are due not solely to the incomplete gelatinisation of the starch molecules but is also due to the formation of complexes that are not hydrolysable by the mash or exogenous enzymes (Delcour *et al.*, 1989). A possible explanation for this is the formation of an amylase–lipid complex upon extrusion (Delcour *et al.*, 1989). This impairs filtration and saccharification rates in beer production, since it is unhydrolysable by both exogenous and malt enzymes during gelatinisation.

Unmalted grain sorghum

Brewing beer with unmalted sorghum as adjunct involves many technical considerations such as the capacity of the cooker, energy costs and the high gelatinisation temperature of sorghum (71–80°C). A proper liquefaction step resulting in a low viscosity mash is suggested (Goode, 2001; Goode and Arendt, 2003c; Goode *et al.*, 1999, 2002, 2003; Lisbjerg and Nielsen, 1991; MacFadden and Clayton, 1989), regardless of the proportion of sorghum adjunct to barley malt. This can be achieved by heating the sorghum to 80–100°C in the presence of a thermostable bacterial α-amylase (Goode, 2001; Goode and Arendt, 2003c; Goode *et al.*, 1999, 2002, 2003; Hallgren, 1995).

Brewing beer with sorghum adjuncts (at levels $\leq$50%) has the same limitations as in producing beer containing rice or maize as adjunct. If the sorghum is of good quality, there should be no lautering problems, as the 50% content of malt has sufficient husk material to ensure the formation of the necessary filter bed in the lauter tun (Goode and Arendt, 2003c; Hallgren, 1995). In all cases when mashing with unmalted sorghum, efficient amylolytic hydrolysis of starch will occur only if the starch has been effectively gelatinised (Agu and Palmer, 1998a; Bajomo and Young, 1992; Delcour *et al.*, 1989; Palmer, 1989).

When brewing with low levels of unmalted sorghum (5–10%) as adjunct to barley malt, the endogenous enzymes of the malted grain can be sufficient to maintain adequate extract recovery, wort FAN and fermentability levels. However, when increasing the amount of sorghum adjunct, resultant decreases in wort filtration, colour, viscosity, attenuation limit and FAN and an increase in pH can be expected (Goode, 2001; Goode *et al.*, 1999, 2002). Addition of commercial enzymes can alleviate these problems. The inclusion of a heat-stable α-amylase is essential for efficient saccharification. The inclusion of a fungal α-amylase can improve filtration rates to that of 100% malted barley mashes, while the addition of a bacterial protease increases the amount of nitrogen solubilisation and peptide degradation (Goode, 2001; Goode and Arendt, 2003c; Goode *et al.*, 1999, 2003).

A typical mashing regime involves mashing in hammer-milled unmalted sorghum at 50°C with a liquor:sorghum ratio of 3:1. The pH can be adjusted to 6.5–7.0 by the addition of calcium hydroxide to give a calcium level of 50–150 mg/l. A blend of neutral protease (for FAN production), thermostable α-amylase (for liquefaction), and a range of β-glucanases (to open up endosperm cell walls) can be added at this point. After 30 min at 50°C (the protein rest period), the temperature can be slowly increased to 85°C and maintained there for 30 min in order to liquefy the starch. The malt mash can be prepared with water at 20°C. After 15 min, the cold mash is combined with the hot liquefied sorghum mash or part of this and maintained for a further 60 min at 50°C. More enzymes are then added for starch saccharification. A fungal α-amylase can be added which hydrolyses α-1,4 linkages of starch and dextrins, producing maltotriose, oligosaccharides and large amounts of maltose. At the same time a new enzyme mixture of neutral protease, cellulase and amyloglucosidase may be added. After 60 min at 50°C, the combined mash is mixed with the rest of the 85°C sorghum mash, and the temperature is raised to 75°C. After 20 min at this temperature, the mash can be transferred to the mash filter (Hallgren, 1995; Lisbjerg and Nielsen, 1991; MacFadden and Clayton, 1989; O'Rourke, 1996). It is important to remember that the actual time and temperature programme varies between breweries, depending on the ratio of sorghum to malt, sorghum grain quality, modification of the malt, exogenous enzyme quality, process equipment, and capabilities and capacities available to the brewer (Hallgren, 1995).

Goode (2001) and Goode and Arendt (2003c) and showed that when brewing with a South African red sorghum at an adjunct level of 50%, the addition of a heat-stable bacterial α-amylase, a bacterial neutral protease and a fungal α-amylase was necessary to maximise extractability and minimise processing difficulties. The sorghum mashes showed comparable lautering behaviour and green beer filtration performance to that of 100% malted barley brews. Sensory analysis indicated that no significant differences existed between the sorghum beer and the malted barley beer. However, the apparent degree of fermentation of the sorghum gyles was less than that of the 100% malted barley gyles (75.4% versus 86% respectively). Likewise the foam stability (NIBEM) of the 50% sorghum gyles was considerably reduced in comparison to the 100% malted barley control (118 s versus 266 s respectively).

Although it is more common for some African-based brewers to brew with 50% unmalted sorghum and 50% unmalted maize, brewing with 100% unmalted sorghum is possible. Due to the fact that unmalted sorghum contains no enzymes, a considerable amount of exogenous enzyme must be added (Hallgren, 1995). Typical mashing procedures for 100% unmalted sorghum have been reported by Goode (2001), Goode *et al.* (2003), Hallgren (1995), Lamidi and Burke (1995), Little (1994), MacFadden and Clayton (1989), and O'Rourke (1996).

Goode *et al.* (2003), when brewing with a grist containing 100% unmalted sorghum (Nigerian Fara Fara variety), suggested that the potential for brewing a high quality beer could be improved by the following procedures:

1. Adjusting the mash-in liquor to give a calcium content of 200 parts per million.
2. Adjusting the mash-in pH to 6.5.
3. Using a mashing programme with temperature/time stands of 50°C × 50 min, 80°C × 10 min, 95°C × 40 min and 60°C × 30 min.
4. Using a heat-stable α-amylase added at the end of the 50°C stand, a neutral protease added at mash-in and a fungal α-amylase added at the start of the 60°C stand.
5. Adjusting the pH to 5.5 prior to the 60°C stand.

The addition of calcium prevents the thermal inactivation of α-amylase by extending the pH range of the enzyme. Stabilisation of the added α-amylase can result in increased liquefaction and therefore in increased extraction of the grist. It can also mean that inclusion of calcium ions in the mashing liquor could allow the same amount of extract recovery, but with a lower proportion of added exogenous α-amylases in the mash (Goode, 2001; Goode *et al.*, 2003). With the incorporation of a 10 min stand at 80°C, significant increases in the levels of filterability, extract recovery and FAN were observed. The pH adjustment from pH 6.2 to pH 5.5 prior to the 60°C stand optimised pH conditions for the fungal amylase whilst increasing mash bed permeability as had been shown in malted barley mashes by Taylor (1990). The inclusion of an amyloglucosidase can significantly increase attenuation levels.

Of particular interest to brewers is the requirement for mash cooling after gelatinisation and before the addition of the saccharifying enzyme. This can be achieved with some difficulty by means of chilled water addition. This is suitable only when brewing beers of conventional gravities. With high gravity brews an external mash cooler is required, such as a plate and frame or a shell and tube heat exchanger (Little, 1994). Sorghum has no husk, therefore there are filtration problems when a traditional lauter tun is used in mash separation (Ogbonna, 1992). Okafor (1985) proposed the use of artificial husks manufactured from nylon materials or plant fibres as filter aids when a traditional lauter tun is used. However, there is no evidence to suggest that this idea has been commercially adopted. Many breweries in Nigeria have switched to conventional mash filters or thin bed filters (Hallgren, 1995; Little, 1994; Waesberghe, 1990). Little (1994) reported on mash filters being installed in Guinness breweries in Nigeria.

3.3.5 Wheat

Wheat is famous for its inclusion in Belgian- and German-style white beers. Traditional Belgian white beers are generally brewed with 60% barley malt and 40% unmalted wheat. German-style wheat beers are brewed with 50–80% malted wheat. The use of these cereals is based on tradition rather than economic reasons. However, wheat has important consequences on beer production and the quality of the final product (Delvaux *et al.*, 2001). The major difficulties in brewing with high levels of wheat include an increase in wort viscosity, slower wort separation (Bamforth, 1982) and lower wort fermentability (Koszyk and Lewis, 1976). On the other hand, it is well established that wheat enhances and/or stabilises foam, due to high molecular weight proteins (Bamforth, 1985; Leach, 1968), glycoproteins, or viscosity increasing compounds such as arabinoxylans and β-glucans (Kolbach and Kremkov, 1968). For this reason wheat is often incorporated at a 5–10% level in the grist recipes of lager beer.

Other beer characteristics such as aroma profile, flavour stability and colour may also be affected by inclusion of wheat. In comparison to lager beers, haze is a desirable quality characteristic of wheat beers. However, there are contradictory reports regarding the influence of wheat on beer colloidal stability. According to Bamforth (1999), barley and wheat based adjuncts increase haze formation due to haze-forming proteins, polyphenols, and in the case of wheat, pentosans. In earlier studies inclusions of unmalted wheat or wheat flour were shown to increase beer colloidal stability (Kolbach and Kremkov, 1968; Koszyk and Lewis, 1976) which in turn was attributed to the fact that wheat contains no haze-active polyphenols and provides less protein to the wort than barley malt. More recent studies (Delvaux *et al.*, 2001) showed that, at an unmalted wheat inclusion of 40%, unmalted wheat was found to have a strong positive influence on the haze stability. This effect was predominantly caused by wheat gluten proteins, most probably wheat gluten gliadins (Delvaux *et al.*, 2001). However, wheat gluten proteins were found to be haze-active since they interact with polyphenols and protein–polyphenol complexes. At low gluten levels haze is formed, but at high gluten levels these insoluble complexes are too large to stay in suspension and therefore precipitate (Delvaux *et al.*, 2003). Further studies (Delvaux *et al.*, 2004) revealed that an inclusion of malted wheat decreased colloidal stability. This could be attributed to the level of protein degradation in the malt, resulting in less precipitate being formed and hence a more stable haze.

3.4 Potential new adjunct sources

While the use of the traditional adjunct sources in brewing is already very well established, there are many other sources of carbohydrate Some of these are already used in commercial brewing, while others, from their physicochemical make-up and architecture, seem to have potential for incorporation as brewing adjuncts. These include carbohydrate-rich sources such as sugar beet, sugar cane, potato, millet, oats, rye, cassava, chick peas, mung beans, quinoa, buckwheat, amaranth, soya bean, banana, honey and the milk sugar lactose. The pseudo-cereal buckwheat is discussed extensively in Section 3.6.2 dealing with gluten-free raw materials. The following section will deal with unmalted triticale and the recent advances in its potential use as an adjunct.

3.4.1 Unmalted triticale

Triticale (*Triticosecale* ss. *Wittmack*) was the first manufactured cereal derived from an amphidiploid between wheat (*Triticum* spp.) and rye (*Secale* spp.). Recent studies have shown that unmalted triticale may be suitable as a brewing adjunct (Glatthar *et al.*, 2005). Most non-malt adjuncts do not contribute either enzyme activity or soluble nitrogen. However, triticale goes beyond this specification, since some triticale lines already contain high levels of amylolytic activity in their unmalted natural form, in conjunction with low levels of proteolytic activity (Flamme *et al.*, 2000; Jain and Khanna, 1991; Lorenz and Kulp, 1981; Madl and Tsen, 1974; Pomeranz, 1971; Ramanatha *et al.*, 1976; Senn, 2000; Senn and Pieper, 1996). Because of this and the low gelatinisation range of triticale starch (59–65°C) (Lorenz and Kulp, 1981), it is capable of degrading its own starch content with efficiencies equal to those of barley malt (Ande *et al.*, 1998). In addition some cultivars contribute considerable amounts of free amino nitrogen (FAN) to the wort accompanied by an arabinoxylan content similar to that of all malt worts (Glatthar *et al.*, 2002). It is therefore reasonable to assume that triticale could be used as a brewing adjunct at high adjunct ratios (>30–50%) without the need for the addition of commercial exogenous enzymes. Furthermore, because of its relatively low temperature of gelatinisation, triticale can be added directly to the mash tun, without the need for a cereal cooker or a second mashing vessel.

The most successful mashing regime reported by Glatthar *et al.* (2002) was a pre-liquefaction stage (64°C × 10 min, pH 5.9) with an adjunct:malt ratio of 9:1. After adding the remaining malted barley and water, the mash pH was adjusted to 5.5 followed by a 50 min rest at 50°C, a 60 min rest at 63°C, a 35 min rest at 70°C and a final 10 min rest at 77°C. Using this regime at a 50% adjunct level, FAN levels as high as 169 mg/l and attenuation limits of ~76% were achieved (Glatthar *et al.*, 2005). Careful selection of suitable genotypes is important. In one study the cultivar cv Trinidad was identified as the most suitable to serve as a brewing adjunct, due to its improved starch solubilisation properties and its ability to generate low wort viscosities (Glatthar *et al.*, 2005).

3.5 New developments to improve adjunct functionality

Many pre-processed forms of adjuncts are available to the brewer such as grits, flakes, torrefied grain, extruded grain, flour fractions and syrups. The following section deals with four distinctly different processing methods and possibilities for producing brewing adjunct materials, namely the already well-established heat-induced process of torrefication/micronisation, grain fractionation and its potential to improve beer quality, the application of high hydrostatic pressure as an emerging process for adjunct processing, and developments in enzyme technology and how it may influence adjunct production and processability.

3.5.1 Torrefication/micronisation – a well-established heat-induced process

Torrefied (or micronised) cereals, principally wheat and barley, are used in the brewing industry as a relatively low cost of extract (Lloyd, 1986). Torrefication occurs when a cereal is heated to gelatinise the starchy endosperm, creating expansion of the grains and a 'puffed' or 'popcorn' appearance. The process renders starch pre-gelatinised and thereby eliminates the cooking step in the brewhouse. Heating facilitates extract recovery during subsequent mashing and generates flavour, colour and aroma compounds. Traditionally, torrefication of barley was carried out by passing the grain through a stream of hot sand (Britnell, 1973; Brookes and Philliskirk, 1987). Other torrefication processes involve passing the grain in a stream of air at 260°C (Britnell, 1973). In the micronisation process, cereal grains are subjected to infrared radiation generated from burner-heated ceramic tiles. The grains are conveyed below the ceramic tiles, which exposes them to infrared radiation. This creates molecular vibrations within the endosperm at high frequencies and initiates starch gelatinisation at grain temperatures of approximately 140°C (Brookes and Philliskirk, 1987; South, 1991). The soft grains can be either flaked immediately and cooled, or cooled directly and used as whole grain in admixture with malt. Following torrefication the end product can be stored safely for many weeks (South and Ross, 1993).

Torrefication not only pre-gelatinises the cereal grain but also denatures a major portion of the protein in the kernel. This results in a wort soluble protein of only 10% of the total (~1.4% wort soluble protein). Use of such an adjunct may allow the use of higher protein malts of higher adjunct levels while maintaining soluble protein levels similar to those of worts produced with lower soluble protein adjuncts. From an economic viewpoint there is clearly an advantage in using the cheaper soft wheat varieties than the more expensive bread-making hardwheat varieties (Brookes and Philliskirk, 1987). There are no handling or dust problems associated with the use of torrefied cereals. It is common to include torrefied cereals at an adjunct level of 15% (South and Ross, 1993). Malt and torrefied cereals can be milled simultaneously and mashed in together. However, increased extract recovery is reported when the torrefied cereals are cooked separately at 71–77°C. Milled particle size is critical to control extraction and runoff time. A higher liquor to grist ratio is necessary, since torrefied cereals are reported to absorb more water than other grains during mashing. The puffed-up nature of this adjunct can cause an increase in the lauter-tun grain filter depth, which can lead to slight increases in the lauter runoff time. Likewise the formation of disulphide links between the grain proteins during the micronising process contributes to longer lautering times. The problem is reduced by heating the grain less, to give products with higher final moisture contents (South, 1991). Higher nitrogen wheats offer improved foam stabilising properties at the expense of lower extract yields (Brookes and Philliskirk, 1987). Importantly, torrefied products have low colours (2 EBC units) and yield extracts with bland flavours (South, 1991).

3.5.2 Grain fractionation – a physical means to improve beer quality

Physical techniques to fractionate a malt kernel into several components have been developed (Nishida *et al.*, 2005). The recently developed malt fractionation technique involves abrading and polishing the malt kernel from the outside using a stone grinder and then separating the fractions by using a sieve shaker. By using this technique a malt kernel is divided into three fractions: the inner fraction, the outer layer fraction and the husk fraction. The outer layer fraction is high in protein, amino acids and lipids, free fatty acids and polyphenols, with concentrations more than twice those of untreated malt. Lipids and amino acids are considered deterioration precursors of stale flavour substances in beer, whilst lipid materials can reduce foam stability (Nishida *et al.*, 2005). The inner fraction contains lower amounts of the deterioration precursors and astringent substances compared with the untreated malt. By brewing with this fraction, remarkable improvements in beer quality may be achieved in terms of beer foam, beer flavour, and taste and beer flavour stability. Likewise it is possible to fractionate milled barley (Sundberg and Aman, 1994) to render endosperm-enriched fractions. These enriched fractions can then be hydrolysed with suitable enzyme preparations to produce wort of good quality. Such a process may not involve the production of spent grains, and therefore the disposal problems that are part of every modern brewery's portfolio would be avoided (Bamforth, 2001).

With such continuous production methods the problems and costs associated with agriculture and malting, food safety issues such as nitrosamines (which have their precursors in the embryo), the high energy costs associated with kilning, and the emission of volatile organic compounds (Gibson *et al.*, 1995) could be reduced. Coupled with continuous fermentation plants (Linko *et al.*, 1998), such a method would embrace the downstream adjustment of quality parameters such as the addition of concentrated essences that are available for colour (Turner, 1986), bitterness (Westwood 1994), hop aroma (Murray *et al.*, 1987) and even foam (Bamforth and Cope, 1987).

The fractionation of cereals offers many possibilities in the area of product development for the brewer. The fractions can be used as mentioned above or they can be processed further to form sub-products such as flour fractions, B-starch fractions, A-starch fractions, barley syrups and starch syrups (Ahvenainen, 1989). These sub-fractions can be incorporated into the brewing process in a more traditional manner than was mentioned above. A- and B-starch fractions from barley starch can be added as a mashing adjunct, either in dry form or in water suspension. Barley syrups, which contain small amounts of proteins, can be added at the beginning of wort boiling. Highly purified starch syrups, such as maltose syrups, can be added at the end of wort boiling or even into fermentation. The use of such adjuncts could enable improvements in economy and quality and enable new opportunities in product development.

Moreover, barley and other cereals contain many components undesirable by the brewer but very desirable in other areas of food processing. Such components include arabinoxylans and β-glucans and other non-starch polysaccharides. With an increase in knowledge of the beneficial role of barley β-glucans and other non-starch polysaccharides in the human diet, there is a growing demand for the incorporation of barley into food systems (Izydorczyk *et al.*, 2000). For example, barley β-glucans bear the advantage over other sources of dietary fibre of being partially soluble in water (Izydorczyk *et al.*, 2000). Therefore in addition to physiological effects such as increase in faecal bulk, they also possess the biological benefits of soluble fibre such as reduction in plasma cholesterol and postprandial serum glucose levels in humans and animals (Jenkins *et al.*, 1995; Kahlon and Chow, 1997; McIntosh *et al.*, 1995; Newman and Newman, 1991; Wood *et al.*, 1994; Yokoyama *et al.*, 1997). The cell walls of the starchy endosperm contain about 75% β-glucan and 20% arabinoxylan, whereas the aleurone cell walls contain about 26% β-glucans and 71% arabinoxylans (Jadhav *et al.*, 1998). Therefore further processing of barley fractions to yield purified forms of such components could reduce brewing problems while supplying a valuable source of health-promoting compounds to the food industry.

3.5.3 High hydrostatic pressure treatment – an emerging process with much potential

High hydrostatic pressure (HHP) refers to the application of high pressure (100–1000 MPa) in the presence of an excess of water. The use of HHP to process foods is not a new concept and was investigated over 100 years ago (Ledward, 1995). There is increasing worldwide interest in the use of HHP because of the advantages it offers over other methods of processing and preservation of foods. HHP gives a homogeneous treatment at every point in the product, since the applied pressure is instantaneously and uniformly distributed within the HHP chamber (Mertens and Deplace, 1993). Therefore processing time is not a function of sample size. In addition, HHP offers significant energy savings in comparison to thermal stabilisation techniques, because once the desired pressure is reached it can be maintained without further need for energy input (Estrada-Giron *et al.*, 2005). Much has been reported on the effects of high pressure treatment in food processing (Barbosa-Cánovas *et al.*, 1997; Cheftel, 1995; Gould, 1995). HHP changes the conformation and coagulation of proteins by opening their native structures, inducing denaturisation and aggregation. It affects the melting properties of starches and the rearrangement of the polymorphic forms in lipids. It inactivates micro-organisms, and induces chemical changes at low temperatures (Ledward, 1995).

Many studies have shown that HHP will bring about gelatinisation of starch. Starch resistance to pressure depends largely on the size of the starch granules. The B-type starches are more pressure resistant than the A-types (Stute *et al.*, 1996). The A-type starches are the least pressure resistant, while the C-types show a resistance which is intermediate between the B- and A-type starches. For example, wheat starch granules gelatinise over a pressure range of 300–600 MPa, with complete gelatinisation at 600 MPa (Douzals *et al.*, 1996). On the other hand, potato starch needs 800–1000 MPa to reach total gelatinisation (Kudla and Tomasik, 1992). In addition, high pressure is known to modify a whole range of food enzymes, including amylases (Hayashi and Hayashida, 1989), polyphenoloxidases (Gomes and Ledward, 1996) and lipoxygenases (Estrada-Giron *et al.*, 2005). This property has been utilised in the inactivation of amylase in apple juice (Riahi and Ramaswamy, 2004). Under certain circumstances elevated pressures may bring about increased action of the enzyme. This is presumably due to some modification of the food, which enables the substrate and enzyme to work more effectively together. However, at sufficiently high pressures enzymes will invariably lose their activity as the active site is modified. This is possibly due to unfolding or partial unfolding of their structures (Gomes and Ledward, 1996) or perhaps oxidation of thiol group(s) (Gomes *et al.*, 1997).

HHP may also have application in the reduction of grain allergens. The consumption of rice is often associated with allergic disorders such as asthma and dermatitis (Baldo and Wrigley, 1984). These disorders are related to the ingestion of rice proteins, particularly 16 kDa albumin and 26 kDa α-globulin which have been identified as major rice allergens (Limas *et al.*, 1990; Shibasaki *et al.*, 1979). Another allergenic, 33 kDa globulin, was further identified as a new type of plant glyoxalase-I (Usui *et al.*, 2001). Application of HHP (100–400 MPa) to rice grains immersed in distilled water has been shown to cause the release of rice allergenic proteins with maximum amounts in the range of 300–400 MPa. However, these pressure released proteins were predominantly globulins Glb33, α-Glb and the albumin Alb16, therefore suggesting that as much as 80% of the allergenic proteins remained in the grain. To enhance the effects of HHP on the solubilisation and release of allergenic proteins, protease treatment has been employed. Since pressure promotes the permeation of protease solution into the endosperm cells through the cell walls and membranes, the combination of both treatments results in a more efficient release of allergenic protein (Estrada-Giron *et al.*, 2005).

A number of authors (Fischer *et al.*, 1998; Herdegen *et al.*, 1998; Pérez-Lamela *et al.*, 2002) have reported the effects of HHP on the brewing process.

However, its application in the preparation of malt and its effects on the brewing process are still very much in the early stages of investigation. HHP applied to milled malted barley induces gelatinisation of its starch (Ezaki and Rikimaru, 1992; Gomes, *et al.*, 1998). Gelatinisation of malted barley by high pressure treatment begins at 400 MPa and becomes more marked at 600 MPa (Pérez-Lamela *et al.*, 2002) when applied for 20 min at ambient temperature. Unlike in heat-induced gelatinisation, starch granules remain intact after pressure treatment and there is no leaking of amylose from the starch granules into the surrounding solution. While the birefringence of the starch is lost after 30 min at 450 MPa and within seconds at 600 MPa, the granular structure of starch is maintained even after a 50-min treatment at 600 MPa (Stolt *et al.*, 2001). Saccharification when mashing under high pressure treatment (400–600 MPa $\times$ 20 min) was shown to result in similar soluble sugar levels to those achieved when a conventional mashing temperature/time programme of 65°C $\times$ 90 min was employed (Pérez-Lamela *et al.*, 2002). Application of HHP to barley and wheat mashes has been shown to increase the apparent activity of the α and β amylases, with maximum effects observed at 500–600 MPa (Gomes *et al.*, 1998; Stolt *et al.*, 2001). Pressures in excess of 600 MPa caused inactivation of α- and β-amylases (Gomes *et al.*, 1998). β-Amylase is slightly more pressure sensitive than α-amylase. The decrease in enzyme activity is due to partial or total unfolding of the enzymes and depends on both time and pressure (Gomes and Ledward, 1996; Gomes *et al.*, 1998).

It is generally believed that HHP offers a new possibility of starch application in food products, for example as a fat substitute in low-energy food (Blaszczak *et al.*, 2005) or as a method of reducing toxic cereal allergens (Estrada-Giron *et al.*, 2005). Given the fact that high pressure induces the gelatinisation of cereal starches and increases the efficiency of starch hydrolysis by both α-and β-amylases, one can pose the question whether high pressure processing can be utilised in the development of adjunct materials of enhanced processability. Can high pressure be used to enhance processability of grain materials during brewhouse processing? Controlled application of heat is the usual method of bringing about a similar phenomenon and is routinely exploited in the manufacture and processing of brewing adjuncts. Therefore, the manipulation of such systems, possibly using both temperature and high pressure, would be of interest in producing brewing adjuncts with increased processability. Likewise, applications of high pressure treatment may exist in the preparation of malts or as a direct application during mashing of malts. However, the application of HHP in the brewing process is still very much in its infancy, so that many questions still remain unanswered. It is also important to emphasise that high pressure treatment is at present an expensive technology. Therefore along with research and development of products, feasibility studies concerning the cost of processing should be conducted to establish whether the successful application of high pressure technology can be implemented in the brewing adjunct processing industries.

3.5.4 Enzyme technology – new advances, new opportunities

When one discusses brewing adjuncts, it is virtually impossible not to mention commercial enzymes and their applications. Whilst increased use of adjunct materials inevitably means increased use of commercial enzymes, it should also be highlighted that developments in enzyme technology could also result in an increase in the use of adjuncts. Exogenous enzymes have been used safely in a wide variety of foods for centuries. Modern enzyme technology involving the use of exogenous enzymes was initially adopted for the hydrolysis of starch. Exogenous enzymes are tightly integrated into processes for starch hydrolysis and associated products (maltodextrins, glucose syrups, very high maltose syrups and high fructose syrups). With the advent of new technologies in enzyme liquefaction and enzyme hydrolysis, production of corn syrups of virtually any carbohydrate profile is possible (Birschbach *et al.*, 2004). Such syrups enable the brewer to introduce liquid adjuncts at any level without changing the carbohydrate profile of the wort. The routes to produce these products and the enzymes employed are already well established (Bently and Williams, 1996) and will not be the subject of this chapter

In much the same way as adjunct usage, much of the progress in using enzymes in food processes or in the manufacture of food ingredients has been evolutionary, involving catalytic activities, the ability to deal with a wider range of raw materials, wider variations in processing conditions, pH and temperatures, and importantly reduction in the cost of enzymes. Many of these improvements have been brought about by the introduction and use of enzymes from genetically modified micro-organisms (GMOs) (Birschbach *et al.*, 2004). In addition, there have been new developments that involve new enzyme activities and new applications for enzyme activities. Many of the recent and on-going developments for use of enzymes in starch hydrolysis are associated with improvements in enzyme activities (α-amylases, glucoamylases and debranching enzymes) and improvements in enzymatic hydrolysis of non-corn cereals such as wheat and barley (Birschbach *et al.*, 2004). One such example (Andersen *et al.*, 2005) of developments in protein engineering having benefited adjunct processing is that of α-amylases and their applications in syrup production. The introduction in 1973 of bacterial α-amylases that were capable of operating under industry-relevant conditions (>100°C) was revolutionary for starch processing. Before then acid hydrolysis of starch was the norm. Acid hydrolysis was environmentally unfriendly and led to significant levels of unwanted by-products.

However, two basic problems still existed with enzymatic hydrolysis. Firstly the pH of the starch slurry had to be adjusted from 4.0 to 6.0. Secondly calcium had to be added to stabilise the liquefying bacterial α-amylase (Andersen *et al.*, 2005). A breakthrough in the mid-1990s was the unveiling of the first three-dimensional structure of a *Bacillus*-derived α-amylase. This meant that through protein engineering new and improved commercial enzyme products could be produced (Bisgaard-Frantzen *et al.*, 1999). When the relationships between the structure of various *Bacillus* α-amylases and product specificities were obtained, the issue of panose formation in the process – when active α-amylase was present during saccharification – was resolved (Andersen *et al.*, 2005). The next breakthrough for the starch processing industries was the development of very efficient starch hydrolytic enzymes capable of acting below the gelatinisation temperature. This has meant that current processes can now be performed in a 'one step' starch hydrolysis process without the need for very high temperature, pH adjustment or calcium addition (Norman *et al.*, 2003). Developments such as this have significantly brought down the cost of starch processing and production of more specific liquid adjunct products.

The biodiversity of enzymes has provided the brewing industry with a wide range of functionalities. As biotechnology paves the way for making improvements to known enzyme functions as well as opening the door for designing new enzymes with new functionalities, this is likely to increase the possibility of increasing adjunct usage in brewing. Likewise it will enable the production of superior adjuncts with added benefits specially formulated to brewers' needs and may even decrease the cost of production of adjuncts, thereby decreasing the cost of beer production. The GMO issue may continue to be a source of controversy for the foreseeable future. The brewing industry has already benefited from several enzymes that are produced using genetically modified production hosts to reduce the cost or enhance the functionality of the enzyme. With increased knowledge of cereal grain structure such as barley endosperm cell walls, future biotechnology applications will focus on delivering the brewing industry with tailor-made enzymes with more specific hydrolysing capabilities.

Studies have also shown that biological acidification of mash and wort can result in improved mash and wort characteristics, ultimately resulting in a better beer. It has been shown that, when employed in high adjunct mashes, biologically acidified mash and wort can compensate for decreased endogenous grain enzyme activities (Lowe *et al.*, 2004, 2005; Ulmer *et al.*, 2003). It was recently shown that at barley adjunct levels of 20% (Lowe *et al.*, 2005; Ulmer *et al.*, 2003) and 50% (Lowe *et al.*, 2004) acidification of the mash with a biologically acidified (*Lactobacillus amylovorus*) stock wort resulted in improved extract, fermentability and FAN and reduced wort β-glucan levels. This could be attributed not only to a lowering of mash-in pH to 5.4, but also to the additional proteolytic and amylolytic enzyme activities that the biologically acidified stock brought into the mash. Thus biological acidification can offer the adjunct brewer an alternative natural way of bring additional enzyme activities into the mash.

3.6 New beverages based on high adjunct levels

In the past, the main drivers for the usage of brewing adjuncts have been cheaper cost of raw materials, together with opportunities of increasing product output capacity without the necessity of increasing brewhouse capacities (i.e. addition of syrups). In addition, usage of certain adjuncts has offered the brewer more control over product quality with regard to flavour, colour and colloidal stability. Likewise, governmental political decisions have encouraged the use of adjuncts, hence the manufacture of lager beer from unmalted sorghum and maize in Nigeria, the manufacture of barley beer in Kenya, and more recently the manufacture of happoshu in Japan (see below). There are other drivers which have the potential to increase adjunct usage. Very recent research efforts (Brauer *et al.*, 2005; NicPhiarais *et al.*, 2005; Wijngaard *et al.*, 2005a, b, c; Zarnkow *et al.*, 2005a, b) have concentrated on developing alternative beers and cereal-based beverages with the aim of fulfilling current consumer health needs and expectations. Two such beverage classes where both traditional and non-traditional adjunct materials will in the future play an important role in their recipe formulations are gluten-free beers and health-promoting functional beverages.

3.6.1 Happoshu

Whilst happoshu (a sparkling, low-malt beverage) is presented to the consumer as beer, it cannot be delineated as such (Shimizu *et al.*, 2002). The difference in definition between beer and happoshu, as defined by the Brewers' Association of Japan, depends on two conditions: (1) the raw materials used and (2) the ratio of malt used. To be classified as beer, ingredients are limited to malt, hops, rice, corn, koaling, potato, starch or sugar, and the ratio of malt should not exceed 67%. The classification of a liquor as happoshu occurs when an ingredient other than those previously mentioned is used or when the malt ratio is less than 67%. The liquor tax for happoshu is set in three stages corresponding to the ratio of malt contained, so that the tax for happoshu with a malt ratio of less than 50% becomes much less than the tax for beer. At malt levels greater than 50%, normal beer tax is imposed. At malt levels less than 50% and greater than 25%, taxation is at a rate of 69% relative to beer, whilst at malt levels less than 25%, taxation is at a rate of 47% relative to beer (Brewers' Association of Japan). Therefore, Japan's brewers have a great incentive to brew products, which very closely resemble beer, from high adjunct ratios and thereby considerably reduce their overall taxation costs.

3.6.2 Gluten-free beer

The possibility of purchasing a gluten-free beer is the desire of many coeliac sufferers. Coeliac disease is an auto-immune disease. It is a condition where the lining of the small intestine is damaged, when certain cereal proteins ('gluten') are consumed (Cooke and Asquith, 1974). The symptoms, which can develop at any age, include malabsorption, abdominal discomfort, weight loss, tiredness, anaemia and severe diarrhoea (Fasano and Catassi, 2001, 2005; Feighery, 1999; Murray, 1999). The non-classical symptoms only recently linked to coeliac disease are arthritis, constipation, dental enamel defects, dermatitis, hepatitis, iron-deficient anaemia, recurrent abdominal pain, short stature and vomiting (Fasano and Catassi, 2005).

Gluten is the general term relating to a protein fraction in wheat, of which the prolamin (alcohol-soluble protein) fraction is toxic to coeliac sufferers. Wheat, rye and barley are all members of the grass family (*Poaceae*) and taxonomically are closely related. All these cereals and their prolamins – wheat (gliadin), barley (hordein), rye (secalin) and possibly oats (avenin) – are toxic to coeliac suffers (Kasarda, 2001). The exact sequences of the peptides causing the reaction have not been established, but typically they have a high content of proline and glutamine residues (Arentz-Hansen *et al.*, 2002; Shan *et al.*, 2002). The only treatment for coeliac disease is a lifetime avoidance of wheat, rye and barley and all products derived from these cereals. Coeliac associations do not recommend the consumption of beer based on barley or wheat (Ciclitira *et al.*, 2005). Cereals that are not taxonomically closely related to wheat, rye and barley are likely to be safe. Such potentially safe grasses include sorghum, maize, brown rice, millet, teff, ragi and Job's tears, as well as pseudo-cereals such as buckwheat, amaranth and quinoa. Recent studies suggest that the majority of coeliac patients can tolerate a certain amount of oats in their diet (Ciclitira *et al.*, 2005).

Epidemiological studies suggest that there will be a significant increase in the incidence of coeliac disease, mainly due to improved diagnostic procedures and increased awareness (Balistreri, 2004). A major epidemiological study carried out by Fasano and Catassi (2001) suggests that an average of 1 in 266 worldwide are suffering from coeliac disease. The study clearly revealed that there is a large discrepancy between the prevalence of the disease based on clinical diagnosis and based on the screening data. For example, in the United States only 1 in 10 000 people are diagnosed with the disease. However, based on the screening data the prevalence of the disease could be as high as 1 in 111. This is expected to lead to an increased demand by the consumer for gluten-free products, such as gluten-free beer.

In recent years much research work has been carried out in the area of gluten-free bakery products. These results have been successfully transferred to industry and already represent a profitable market for bakers (Gallagher *et al.*, 2004). One way of producing gluten-free beverages is by using gluten-free raw materials. The second alternative would be to adjust the brewing process in such a manner that the offending protein fractions are removed. The second option is currently not possible. It is known that certain significantly cell-reactive areas of gliadin that are associated with the onset of coeliac disease are highly resistant to enzymatic degradation (Arentz-Hansen *et al.*, 2002; Shan *et al.*, 2002). In addition, gliadin-like epitopes have been detected in both beer and malt (Ellis *et al.*, 1990, 1994). Furthermore, there are no reliable or accurate methods for testing beer for gluten (hordein) fractions. Therefore it is not possible to provide adequate quality control to be able to conclusively categorise a beer as gluten-free. Current legislation also protects the consumer with ingredient-based labelling. Therefore the only conclusive way of producing a gluten-free beer is to brew it from exclusively gluten free raw materials and absolutely exclude any cross-contamination with gluten sources. This section will therefore focus on malting and brewing with gluten-free cereals.

The production of beer from gluten-free cereals is not a new technology. Some African-based brewers have for the past 20 years been producing beers based on sorghum and maize. In some cases these have been base products, where imported malted barley concentrated extracts are added for flavour and standardisation purposes. Such beers can therefore not be classified as gluten-free. Brewing with sorghum and maize has been previously discussed. Therefore the next section will focus on pseudo-cereals, such as buckwheat, which at present are largely unknown to brewers but have recently been shown (Brauer *et al.*, 2005; NicPhiarais *et al.*, 2005; Wijngaard *et al.*, 2005a, b, c; Zarnkow *et al.*, 2005a) to have great potential as raw materials for the production of gluten-free beers and cereal-based beverages.

Buckwheat as a brewing raw material

Buckwheat is a pseudo-cereal mainly grown in Central and Eastern Europe and Asia. It is a short-duration crop and requires a moist and temperate climate to grow (Mazza, 1993). Pseudo-cereals do not belong to the grass family, but like cereal grains they consist predominantly of starch, they are edible and they have a starchy endosperm and a non-starchy aleurone layer. Since buckwheat belongs to a different family than cereals such as barley, differences exist. For instance, the structure of the grain differs. Barley is a monocotyledonic and buckwheat a dicotyledonic plant (Kreft and Kreft, 2000). In recent years buckwheat has been associated with preventative nutrition and has been considered to be a health food (Li and Zhang, 2001). In addition to all its health benefits, buckwheat has also been reported to be the most important alternative crop, suitable for ecological growth, without the use of fertilisers or pesticides in Europe.

Malted buckwheat

Buckwheat can be obtained either hulled or unhulled. Wijngaard *et al.* (2005b, c) investigated the impact of hulled and unhulled buckwheat on the production of buckwheat malt and wort. They found that the use of hulled buckwheat has advantages over unhulled material, since the water uptake was slower and the enzymatic activity of the resulting malt was improved. In addition the hulls can be used as a filter aid in a lauter tun. Wijngaard *et al.* (2005a, c) reported that the optimal moisture content at the end of a 12 h steep at 10°C was 40–45%. Wijngaard *et al.* (2005b, c) revealed that optimal enzymatic activity in buckwheat malt can be obtained when buckwheat is germinated at 15°C × 4 days. A multistage drying process has been recommended (NicPhiarais *et al.*, 2005) since the enzymatic activities were affected by the kilning regimes employed. The authors also found that rutin levels were significantly increased during malting. Wijngaard *et al.* (unpublished data, 2005) also investigated the impact of germination on the ultra-structure of malted buckwheat compared to barley malt. Zarnkow *et al.* (2005a) recommended a steeping time of 4 days, degree of steeping 47% and germination conditions of 17°C × 5 days. An overview of the malt characteristics determined by Wijngaard (2005b, c) is given in Table 3.4.

Table 3.4 A comparison of malted buckwheat and malted barley

Parameter	Buckwheat malt	Barley malt
α-Amylase (IU g^{-1})	188.85	303.08
β-Amylase (U g^{-1})	37.61	1079.38
Protease (mg leucine h^{-1} g^{-1})	4.50	9.30
TN (%)	2.21	1.53
Extract (%)	65.31	79.88
FAN (mg l^{-1}) at 12°P	137.31	168.07
TSN (%)	0.062	0.055
Viscosity (mPas)	2.59	1.61
Fermentability (%)	60	81
Gelatinisation temperature (°C)	66–67	58–59

Source: adapted from Wijngaard *et al.* (2005a, b, c).

Brauer *et al.*, 2005 also carried out malting trials on a wide range of gluten-free cereals. They put specific emphasis on producing gluten-free crystal malts. Taste testing of the malts revealed that buckwheat crystal malt had a striking toffee, malty and nutty aroma. They suggested that this product could have the potential to be used as an ingredient in a traditional-style ale. Some of the other roasted products produced by this group included roasted soya flour and malted chickpeas. Some exhibited flavours that would not be regarded as typical for malt used in brewing, but the authors came to the conclusion that they would have potential as gluten-free brewing adjuncts.

Production of gluten-free beer

There has been very little work carried out specifically on the production of gluten-free beer. A patent by Maccagnan *et al.* (1999) exists, which concentrates on the production of gluten-free beer based on unmalted cereals with enzymes. A search of the Internet reveals that there are many home-brewer websites discussing gluten-free beer and how to produce it. A quick browse through some of their suggested recipes, however, reveals the presence of some gluten containing ingredients. These beers certainly would not be suitable for coeliac patients. One of the main problems when brewing with buckwheat malt is that the enzymatic contents of the malts that have been produced to date are significantly lower than that of barley malt (Table 3.4) (NicPhiarais *et al.*, 2005; Wijngaard *et al.*, 2005a, b, c; Zarnkow *et al.*, 2005a). In addition, buckwheat also contains polysaccharides, which result in a mash with a relatively high viscosity (Table 3.4). By combining mash rheological profiling tests (Goode *et al.*, 2005a, b, c, e, f) with traditional mashing experiments, it was found that an optimised infusion mashing procedure will result in a starch-free wort of good extract (Wijngaard *et al.*, 2005c). The grist:liquor ratio was of great importance, especially in relation to viscosity. Preliminary pilot-scale brewing experiments revealed that it is possible to produce gluten-free beer from buckwheat malt. The lautering performance of mash was good when unhulled buckwheat was used.

While brewing procedures with malted and unmalted sorghum, rice and corn are already very established, the search for other suitable malted gluten-free brewing materials is in its infancy. The studies performed so far have shown the potential that exists for producing gluten-free beer from pseudo-cereals such as buckwheat. While present studies have concentrated on producing malted pseudo-cereals, unmalted pseudo-cereals together with exogenous enzyme additions are also an option. With the expected surge in diagnosis of coeliac disease in the coming years, the market potential for such innovative products is potentially huge for the malting, adjunct, enzyme and beer producing industries. In addition, such products may not just appeal to the gluten intolerant but also attract a wider audience seeking new flavours, new tastes and alternative choices.

3.6.3 Cereal-based functional beverages

'*Americans are thinking healthier and taking unprecedented action to improve their health. This year, for the first time in food history, health will rival convenience as the most important new food product attribute*' (Sloan, 2004). This statement gives an indication of rising trends towards a more healthy food eating culture. This includes beverages of all categories, aimed at all ages. This is not just an American fad but a worldwide trend, and it is expected to stay. Consumers are seeking additional health benefits while consuming food or drinks. Surveys report that the functional drinks market is still growing. Dairy products represent the bulk of the current market, followed by energy-boosting beverages incorporating stimulants such as caffeine and guarine, gingko or ginseng. Terms such as functional beverages, neo-beverages, liquid meal replacers, sports drinks and breakfast drinks are becoming part of our everyday vocabulary. While in recent times there has been a move towards promoting the health benefits of moderate beer consumption (Walzl, 2005), in general brewers have surprisingly neglected the possibility of entering the functional beverages markets with products which could undisputedly truly claim health benefits.

Traditionally associated with high-fibre foods, a healthy digestive system has been linked to the consumption of probiotics and prebiotics. Probiotics are defined as bacteria, generally lactobacilli or bifidobacteria, which have beneficial effects beyond the nutrients that they claim. In general they favour the beneficial bacteria in the body while inhibiting harmful microbes. Prebiotics are defined as non-digestible food substances that stimulate the growth and activity of beneficial bacteria in the lower intestine. Non-digestible oligosaccharides and resistant starch are examples of prebiotics. Such beverages could be manufactured from cereal bases, incorporating malted cereals, unmalted cereals and pseudo-cereals and other adjunct materials. The dairy industry has for a long time embraced this challenge by leading the way with probiotic and prebiotic dairy based drinks. However, cereal-based probiotic and prebiotic beverages are virtually unknown, despite the fact that a large proportion of the world's population are lactose intolerant.

The benefits of producing beer with substantial levels of non-digestible soluble dietary fibre that can be used by the microflora in the lower intestine have been recognised. Soluble dietary fibre in the form of isomalto-oligosaccharides (and fructo-oligosaccharides), having a degree of polymerisation of 4 or more, can be produced during mashing by the action of D-glucosyltransferase. High gravity worts that have a higher ratio of maltose and maltotriose to glucose favour the formation of isomalto-oligosaccharides. The same effect can also be achieved by the direct addition of isomalto-oligosaccharide syrup directly to the fermenter (Brier *et al.*, 2002). When it comes to cereal processing, brewers are undoubtedly the industry experts in bringing cereal products into beverages. They have the necessary milling, conversion, fermentation and packaging technology capabilities to make cereal-based functional beverages without additional investment in processing equipment. Therefore, the transition from producing beer to producing both beer and functional cereal-based beverages with existing equipment is not difficult.

The main focus for selection of adjunct materials for these drinks should be on what bioactive substances these materials can bring into the beverage. For instance, while buckwheat is gluten-free, it also has many additional health benefits. Buckwheat protein has a high biological value, due to a high level of the amino acid lysine (Skrabanja *et al.*, 2000). Buckwheat contains phytosterols, such as rutin which has been reported to lower cholesterol levels as well as improving many chronic diseases (Li and Zhang, 2001). Advantageous effects in the treatment of diabetes II have been observed when buckwheat is consumed. This has been attributed to fagopyrins. Buckwheat also contains many trace elements (Ikeda and Yamashita, 1994) as well as high levels of both soluble and insoluble dietary fibre (Steadman *et al.*, 2001). Likewise malting processes have been shown to increase the level of the polyphenol rutin (Wijngaard *et al.*, 2005c). This prompted Wijngaard *et al.* (2005c) to develop a non-alcoholic carbonated functional drink using malted buckwheat wort as base together with various fruit juices and clinically proven probiotics. Likewise, complex carbohydrates such as β-glucan, arabinoxylan and starch in their unhydrolysed forms are undesirable in finished beer. However, due to their health-promoting benefits it may be the aim to retain as much of these compounds as possible in such functional beverages.

The main focus on the production methods would be to deliver extract containing adequate levels of fermentable substrate for probiotic fermentation whilst containing high levels of bioactive substances like fibre, polyphenols and Maillard products. Studies have shown that fermentation efficiency with probiotic strains such as *Lactobacillus* or *Enterococcus* is influenced by the substrate and by the presence of bioactive compounds such as polyphenols (Zarnkow *et al.*, 2005b). Therefore the careful selection of malted cereals and adjunct materials should complement the selection of probiotic cultures.

3.7 Future trends

Nobody envisages a dramatic shift in grist materials used in the current beer market. In recent years some brewers have shifted back from sizeable use of adjuncts to grists that are largely composed of premium malted barley. They are convinced that this offers genuine quality. However, there remains a clear justification for many brewers to use adjunct materials, since they offer unique product attributes such as flavour and colour. The quality attributes of some of the world's leading global beer brands are heavily based on the adjunct used in their formulation. Therefore the use of traditional adjuncts such as rice and corn will remain to the fore. Wheat is used only in low-volume speciality beers, so its incorporation as an adjunct in the brewing process is not expected to increase. It is only in the last 20 years that sorghum has been incorporated in sizeable quantities into the brewing recipes of clear lager beers. With the lifting of the Nigerian ban on the importation of malted barley, one might expect a significant decline in the use of sorghum. However, this is unlikely to occur, as some of the sorghum/maize beers have very specific product characteristics which have already been accepted by their loyal customer bases. With the expected increase in the diagnosis of coeliac disease, sorghum may make a comeback in terms of a brewing raw material in the US market. Indeed there may even be a demand for sorghum in non-sorghum-producing countries seeking to produce gluten-free beverages.

Pseudo-cereals such as buckwheat may make their entrance to the brewing marketplace, especially if brewers divest in the functional drinks markets. The recent and expected future advances in genetically modified crops may indirectly create new marketplaces for adjunct materials. Future GM malts with high enzymatic contents and thermo-stable enzymes would undoubtedly tempt the brewer to incorporate higher adjunct amounts. The expected advances in enzyme engineering and enzyme products can but mean increases in the use of adjunct materials. What is sure is that government directives will continue to play active roles in setting adjunct usage trends. The market potential for gluten-free beers and cereal-based functional drinks is huge. By the very nature of these products regarding health claims, the potential for incorporating sizeable proportions of adjuncts in these product recipes is massive.

3.8 Sources of further information

ABDEL-AAL E and WOOD P (2005), *Speciality Grains for Food and Feed*, American Association of Cereal Chemists, St Paul, MN, USA.

BELTON P and TAYLOR J (2004), *Pseudocereals and Less Common Cereal*, Springer Publications, Germany.

COURTIN C M, VERAVERBEKE W S and DELCOUR J A (2004), *Recent advances in enzymes in grain processing*, Katholieke Universiteit Leuven, Belgium, ACCO, Leuven, Belgium.

DENDY D A V (1995), *Sorghum and Millets – Chemistry and Technology*, American Association of Cereal Chemists, St Paul, MN, USA.

GODFREY T and WEST S (1996), *Industrial Enzymology*, 2nd edition, Macmillan, London.

3.9 References

ADEJEMILUA F (1995), 'Processing and profile of sorghum malt', *Technical Quarterly of the MBAA*, 32: 15–18.

AGU R C and PALMER G H (1996), 'Enzymic breakdown of endosperm proteins of sorghum at different malting temperatures', *Journal of the Institute of Brewing*, 102: 415–418.

AGU R C and PALMER G H (1998a), 'Effect of mashing with commercial enzymes on the properties of sorghum worts', *World Journal of Microbiology and Biotechnology*, 14: 43–48.

AGU R C and PALMER G H (1998b), 'A reassessment of sorghum for lager-beer brewing', *Bioresource Technology*, 66: 253–261.

AGU R C, OKENCHI M U, ANEKE G and ONWUMELU A H (1995), 'Brewing properties of Nigerian white and yellow malted sorghum varieties mashed with external enzymes', *World Journal of Microbiology and Biotechnology*, 11: 591–592.

AHVENAINEN J (1989), 'Total use of barley', *Proceedings of the 22nd European Brewery Convention Congress*, Zurich. IRL Press, Oxford, 101–115.

AISIEN A O (1982), 'Enzyme modification of sorghum endosperm during seedling growth and malting', *Journal of the Science of Food and Agriculture*, 33: 754–759.

AISIEN A O (1988), 'Sorghum: A suitable substitute for brewing beer?', *Brewing and Distilling International*, 3: 20–22.

ANDE A, PIEPER, H J and SENN T (1998), 'Production of glucose syrup by direct saccharification from triticale with high autoamylolytic activity', *Starch*, 50: 518–523.

ANDERSEN C, VIKSO-NIELSEN A, PEDERSEN S, BISGAARD-FRANTZEN H, SVENDSEN A and BARCHERT T V (2005), *Proceedings of the European Symposium on Enzymes in Grain Processing*, Nantes, France, in press.

ANDERSON R G and KIRSOP B M (1974), 'The control of volatile ester synthesis during the fermentation of wort of high specific gravity', *Journal of the Institute of Brewing*, 80: 48–55.

ANICHE G N and PALMER G H (1990a), 'Microscopic assessment of increasing moisture treatments on endosperm modification in sorghum', *Ferment*, 3: 378–380.

ANICHE G N and PALMER G H (1990b), 'Development of amylolytic activities in sorghum and barley malts', *Journal of the Institute of Brewing*, 96: 377–379.

ARENTZ-HANSEN H, MCADAM S N and MOLBERG O (2002), 'Celiac lesion T-cells recognize epitopes that cluster in regions rich in proline residues', *Gastroenterology*, 123: 803–809.

AYRAAPAA T and LINSTROM I (1973), 'Influence of long-chain fatty acids on the formation of esters by brewer's yeast', *Proceedings of the 14th European Brewery Convention Congress*, Salzburg, 271–285.

BAJOMO M F and YOUNG T W (1990), 'Measurement of diastatic power', *Journal of the Institute of Brewing*, 96: 373–375.

BAJOMO M F and YOUNG T W (1992), 'Development of a mashing profile for the use of microbial enzymes in brewing with raw sorghum (80%) and malted barley or sorghum malt (20%)', *Journal of the Institute of Brewing*, 98: 515–523.

BAJOMO M F and YOUNG T W (1993), 'The properties, composition and fermentabilities of worts made from 100% raw sorghum and commercial enzymes', *Journal of the Institute of Brewing*, 99:153–158.

BALDO, B A and WRIGLEY C M (1984), 'Allergies to cereals', *Advances in Cereal Science and Technology*, 6: 289–356.

BALISTRERI W F (2004), 'Celiac disease – not a rare disease after all', *Journal of Pediatrics*, 5: 632.

BAMFORTH C W (1982), 'Barley beta-glucans. Their role in malting and brewing', *Brewers Digest*, 22–27.

BAMFORTH C W (1985), 'The foaming properties of beer', *Journal of the Institute of Brewing*, 91: 370–383.

BAMFORTH C W (1999), 'Beer haze', *Journal of the American Society of Brewing Chemists*, 57: 81–90.

BAMFORTH C W (2001), 'Beer: a proud past and a promising future', *Brewer International*, 1(2): 26–29.

BAMFORTH C W (2003), 'Opportunities for newer technologies in the oldest biotechnology, brewing', *Applied Biotechnology, Food Science and Policy*, 1: 213–222.

BAMFORTH C W and COPE R (1987), 'Egg albumen as a source of foam polypeptide in beer', *Journal of the American Society of Brewing Chemists*, 45: 27–32.

BARBOSA-CÁNOVAS G V, POTHAKAMURY U R, PALOU E and SWANSON B G (1997), *Non Thermal Preservation of Foods*, Marcel Dekker, New York, 9–47.

BENTLY I S and WILLIAMS E C (1996), 'Starch conversion', in Godfrey T and West S, *Industrial Enzymology*, 2nd edition, Macmillan, London, 339–357.

BIRSCHBACH P, FISH N, HENDERSON W and WILLRETT D (2004), 'Enzymes: tools for creating healthier and safer foods', *Food Technology*, 58: 20–27.

BISGAARD-FRANTZEN H, SVENDSEN A, NORMAN B, PEDERSEN S, KJAERULFF S, OUTTRUP H and BORCHERT T V (1999), 'Development of industrially important alpha-amylases', *Journal of Applied Glycoscience*, 46: 199–206.

BLASZCZAK W, VALVERDE S and FORNAL J (2005), 'Effect of high pressure on the structure of potato starch', *Carbohydrate Polymers*, 59: 377–383.

BOOTH M, SHAW W and MORHALO L (1989), 'Blending and bottling', in Piggott J R, Sharp R and Duncan R E B, *The Science and Technology of Whiskies*, Longman Scientific and Technical, Harlow, UK, 295–326.

BRADEE L H (1977), 'Adjuncts', in Broderick H M, *The Practical Brewer*, MBAA, Madison, WI, USA.

BRAUER J, WALKER C and BOOER C (2005), 'Of pseudo-cereals and roasted rice. The quest for gluten-free brewing materials', *Brewer and Distiller*, 1: 24–26.

BREWERS ASSOCIATION OF JAPAN, home page http://www.brewers.or.jp (accessed November 2004).

BRIER M, SHETTY J K and KING J W (2002), 'High soluble dietary fibre fermented beverage and process for its production', PCT #WO 0220712A1.

BRIGGS D E (1998), *Malts and Malting*, Blackie Academic and Professional, London.

BRITNELL J (1973), 'Torrified cereals reassessment of their potential', *Technical Quarterly MBAA*, 10: 176–180.

BROOKES P A and PHILLISKIRK G (1987), 'Brewing properties of micronised cereals', *Proceedings of the European Brewery Convention Congress, Madrid*. IRL, Oxford, 337–344.

BUTCHER D M (1987), 'Enzymes for efficiency', *The Brewer*, 61–64.

CEGE P, SHAH S and KUBAI E (1999), 'Kenyan beer brewed with unmalted barley', *Ferment*, 12: 41–45.

CHEFTEL J C (1995), 'Review: high-pressure, microbial inactivation and food preservation', *Food Science and Technology*, 1: 75–90

CICLITIRA P J, ELLIS H J and LUNDIN K E A (2005), 'Gluten free diet – what is toxic? Best practice and research', *Clinical Gastroenterology*, 3: 359–371.

COOKE W T and ASQUITH P (1974), 'Introduction and definition', in *Clinics in Gastroenterology*, WB Saunders Co., London, 3–9.

DALE C J, YOUNG T W and MAKINDE A (1989), 'Extruded sorghum as a brewing material', *Journal of the Institute of Brewing*, 95: 157–164.

DALE C J, YOUNG T W and OMOLE T (1990), 'Small scale mashing experiments with grist containing high proportions of raw sorghum', *Journal of the Institute of Brewing*, 96: 403–409.

DELCOUR J A, HENNEBERT M E, VANCRAENENBROECK R and MOERMAN E (1989), 'Unmalted cereal products for beer brewing. Part 1. The use of high percentages of extruded or regular corn starch and sorghum', *Journal of the Institute of Brewing*, 95: 271–276.

DELVAUX F, GYS W, MICHIELS J, DELVAUX F R and DELCOUR J A (2001), 'Contribution of wheat and wheat protein fractions to the colloidal haze of wheat beers', *Journal of the American Society of Brewing Chemists*, 59: 135–140.

DELVAUX F, DEPRAETERE S A, DELVAUX F R and DELCOUR J A (2003), 'Ambiguous impact of wheat gluten proteins on the colloidal haze of wheat beers', *Journal of the American Society of Brewing Chemists*, 61: 63–68.

DELVAUX F, COMBES F J and DELVAUX F R (2004), 'The effect of wheat malting on the colloidal haze of white beers', *MBAA Technical Quarterly*, 41: 27–32.

DOUZALS J P, MARECHAL P A, COQUILLE J C and GERVAIS P (1996), 'Microscopic study of starch gelatinisation under high hydrostatic pressure', *Journal of Agriculture and Food Chemistry*, 44: 1403–1408.

DUFOUR J P, MELOTTE L and SREBNIK S (1992), 'Sorghum malts for the production of a lager beer', *Journal of the American Society of Brewing Chemists*, 50: 110–119.

ELLIS H J, FREEDMAN A R and CICLITIRA P J (1990), 'Detection and estimation of the barley prolamin content of beer and malt to assess their suitability for patients with coeliac disease', *Clinica Chimica Acta*, 189: 123–130.

ELLIS H J, DOYLE A P and DAY P (1994), 'Demonstration of the presence of coeliac-activating gliadin- like epitopes in malted barley', *International Archives of Allergy and Immunology*, 104: 308–310.

ESTRADA-GIRON Y, SWANSON B G and BARBOSA-CÁNOVAS G V (2005), 'Advances in the use of high hydrostatic pressure for processing cereal grains and legumes', *Trends in Food Science and Technology*, 16: 194–203.

ETOKAKPAN O U and PALMER G H (1990a), 'A simple diamylase procedure for the estimation of α-amylase and diastatic activity', *Journal of the Institute of Brewing*, 96: 89–91.

ETOKAKPAN O U and PALMER G H (1990b), 'Comparative studies of the development of endosperm degrading enzymes in malting sorghum and barley', *World Journal of Microbiology and Biotechnology*, 6: 408–417.

EZAKI S and RIKIMARU H (1992), 'High pressure effects on starch: structural change and retrogradation', in Balny C, Hayashi R, Heremans K and Masson P, *High Pressure and Biotechnology*, Vol 224, Colloque INSERM/John Libbey Eurotext, Montrouge, France, 163–165.

FASANO A and CATASSI C (2001), 'Current approaches to diagnosis and treatment of celiac disease: an evolving spectrum', *Gastroenterology*, 120: 636–651.

FASANO A and CATASSI C (2005), 'Coeliac disease in children', *Best Practice and Research Clinical Gastroenterology*, 19: 467–478.

FEIGHERY C (1999), 'Coeliac disease', *British Medical Journal*, 319: 236–239.

FISCHER S, SCHOBERT H, RUSS W and MEYER-PITTROFF R (1998), 'Effects of hydrostatic high pressure on the brewing process and on beer', *Monatsschrift für Brauwissenschaft*, 51: 120–123.

FLAMME W, SEDDIG S, JANSEN G and JÜRGENS H U (2000), 'Biochemische und rheologische Methoden zur Analyse des Auswuchsverhaltens von Triticale in Relation zu Roggen und Weizen', *Vortrage, Pflanzenzücht*, 49: 143–159.

FOOD AND AGRICULTURE ORGANISATION OF THE UNITED NATIONS (2004), FAOSTAT database: http://apps.fao.org.

GALLAGHER E, GORMLEY T R and ARENDT E K (2004), 'Recent advances in the formulation of gluten-free cereal based products', *Trends in Food Science and Technology*, 15: 143–152.

GIBSON N B, COSTIGAN G T, SWANNELL R P J and WOODFIELD M J (1995), 'Volatile organic compound (VOV) emissions during malting and beer manufacture', *Atmospheric Environment*, 29: 2661–2672.

GLATTHAR J, HEINISCH J J and SENN T (2002), 'A study on the suitability of unmalted triticale as a brewing adjunct', *Journal of the American Society of Brewing Chemists*, 60: 181–187.

GLATTHAR J, HEINISCH J J and SENN T (2005), 'Unmalted triticale cultivars as brewing adjuncts: effects of enzyme activities and composition on beer wort quality', *Journal of the Science of Food and Agriculture*, in press.

GLENNIE C W (1984), 'Endosperm cell wall modification in sorghum grain during germination', *Cereal Chemistry*, 61: 285–289.

GLENNIE C W, HARRIS J and LIEBENBERG N V W (1983), Endosperm modification in germinating sorghum grain, *Cereal Chemistry*, 60: 27–31.

GOMES M R A and LEDWARD D A (1996), 'Effect of high pressure treatment on the activity of some polyphenoloxidases', *Food Chemistry*, 56: 1–5.

GOMES M R A, SUMNER I G and LEDWARD D A (1997), 'Effect of high pressure on papain activity and structure', *Journal of the Science of Food and Agriculture*, 75: 67–72.

GOMES M R, CLARK R and LEDWARD D A (1998), 'Effects of high pressure on amylases and starch in wheat and barley flours', *Food Chemistry*, 63: 363–372.

GOODE D L (2001), 'Brewing with Unmalted Sorghum and Commercial Enzymes', MSc Thesis, University College, Cork, Ireland, L7a, 52-17786.

GOODE D L and ARENDT E K (2003a), 'Effects of addition of exogenous enzymes when mashing with 100% unmalted barley', in Courtin C M, Veraverbeke W S and Delcour J A, *Recent Advances in Enzymes in Grain Processing*, ACCO, Leuven, Belgium, 341–346.

GOODE D L and ARENDT E K (2003b), 'Effects of endogenous enzymes on final wort quality when mashing with increasing levels of unmalted barley as adjunct', in Courtin C M, Veraverbeke W S and Delcour J A, *Recent Advances in Enzymes in Grain Processing*, ACCO, Leuven, Belgium, 347–352.

GOODE D L and ARENDT E K (2003c), 'Pilot scale production of a lager beer from a grist containing 50% unmalted sorghum', *Journal of the Institute of Brewing*, 19: 208–217.

GOODE D L, HALBERT C and ARENDT E K (1999), 'The use of malted barley and commercial enzymes in unmalted sorghum brewing', in Simoinen T and Tenkanen M, *Proceedings of the 2nd European Symposium on Enzymes in Grain Processing ESEGP-2*, VTT Finland, 321–324.

GOODE D L, HALBERT C and ARENDT E K (2002), 'Mashing studies with unmalted sorghum and malted barley', *Journal of the Institute of Brewing*, 108: 465–473.

GOODE D L, HALBERT C and ARENDT E K (2003), 'Optimisation of mashing conditions when mashing with unmalted sorghum and commercial enzymes', *Journal of the American Society of Brewing Chemists*, 61: 69–78.

GOODE D L, BEAUMONT V H, QUINN D and ARENDT E K (2005a), 'Grouping of barley grain lots according to their basic compositions, wort qualities and rheological characteristics using multivariate data analysis', *Journal of the Science of Food and Agriculture*, under peer review.

GOODE D L, RAPP L, SCHOBER T, ULMER H M and ARENDT E K (2005b), 'Development of a new rheological laboratory method for mash systems – its application in the characterization of grain modification levels,' *Journal of the American Society of Brewing Chemists*, 63: 76–86.

GOODE D L, ULMER H M and ARENDT E K (2005c), 'Model studies to understand the effects of amylase additions and pH adjustment on the rheological behaviour of simulated brewery mashes', *Journal of the Institute of Brewing*, 111: 153–164.

GOODE D L, WIJNGAARD H H and ARENDT E K (2005d), 'Mashing with unmalted barley – impact of malted barley and commercial enzyme (*Bacillus* sp) additions', *Master Brewers Association of the Americas, Technical Quarterly (MBAA TQ)*, 42: 184–198.

GOODE D L, WILTSCHKO E A, BEAUMONT V, QUINN D, ULMER H and ARENDT E K (2005e), 'Mash rheological studies – the development and application of novel grain characterisation methods', *Proceedings of the 30th European Brewery Convention Congress*, Prague.

GOODE D L, WILTSCHKO E A, ULMER H M and ARENDT E K (2005f), 'Application of the Rapid Visco Analyser as a rheological tool for the characterization of mash viscosity as affected by the level of barley adjunct', *Journal of the Institute of Brewing*, 111: 165–175.

GOULD G W (1995), New Methods of Food Preservation, Blackie Academic and Professional, Glasgow, 151–175.

GRUJIĆ O (1999), 'Application of unconventional raw materials and procedures in wort production', *Journal of the Institute of Brewing*, 105: 275–278.

HALLGREN L (1995), 'Lager beers from sorghum', in Dendy D A V, *Sorghum and Millets, Chemistry and Technology*, American Association of Cereal Chemists, St Paul, MN, USA, 283–298.

HALLGREN L and MURTY D S (1983), 'A screening test for grain hardness in sorghum employing density grading in sodium nitrate solution', *Journal of Cereal Science*, 1: 265–274.

HALLGREN L, REXEN F, PETERSEN P B and MUNCK L (1992), 'Industrial utilisation of whole crop sorghum for food and industry'. in Gomez M I, House L R, Rooney L W and Dendt D A V, *Utilisation of Sorghum and Millets*, International Crops Research Institute for the Semi Arid Tropics (ICRISAT), Patancheru, India, 121–130.

HAYASHI R and HAYASHIDA A (1989), 'Increased amylase digestibility of pressure treated starch', *Agricultural and Biological Chemistry*, 53: 2543–2544.

HERDEGEN V, MEZGER R, RUSS W, VOGEL RF and MEYER-PITTROFF R (1998), 'A pressing matter. High pressure treatment new ways for the beverage industry', *Brauindustrie*, 83: 105–107.

HUG H and PFENNINGER H (1980), 'Utilisation of maize in brewing', in Inglett G E and Munck L, *Cereals for Foods and Beverages*, Academic Press, New York, 287–300.

IKEDA K and YAMASHITA Y (1994), 'Buckwheat as a dietary source of zinc, copper and manganese'. *Fagopyrum*, 14: 101–111.

IZYDORCZYK M S, STORSLEY J, LABOSSIERE D, MacGREGOR A W and ROSSNAGEL B G (2000), 'Variation in total and soluble β-glucan content in hulless barley: effects of thermal, physical and enzymic treatments', *Journal of Agricultural and Food Chemistry*, 48: 982–989.

JADHAV S J, LUTZ S E, GHORPADE V M and SALUNKHE D K (1998), 'Barley: chemistry and value added processing', *Critical Reviews in Food Science*, 2: 123–171.

JAIN J and KHANNA V K (1991), 'Enzyme activity in developing grains of triticale', *Indian Journal of Plant Physiology*, 34: 109–112.

JENKINS D J A, JENKINS A L, WOELVER T M S, VUKSAN V, VENKET RAO, THOMPSON L U and JOSSE R G (1995), 'Effect of reduced rate of carbohydrate absorption on carbohydrate and lipid metabolism', *European Journal of Clinical Nurtrition*, 49: S68–S73.

KAHLON T S and CHOW F I (1997), 'Hypocholesterolemic effects of oat, rice and barley dietary fibers and fractions', *Cereal Food World*, 42: 86–92.

KASARDA D D (2001), 'Grains in relation to celiac disease', *Cereal Foods World*, 46: 209–210.

KOLBACH P and KREMKOV C (1968), 'Zusammensetzung, Schaum und Haltberkeit der hellen Vollbiere. 1. Mitteilung. Chemische Zusammensetzung und Schaum', *Monatsschrift für Brauerei*, 21: 257–269.

KOSZYK P F and LEWIS M J (1976), 'Unmalted grains as malsters' adjuvant and brewers' adjunct', *Journal of the American Society of Brewing Chemists*, 35: 257–269.

KREFT S and KREFT M (2000), 'Localization and morphology of the buckwheat embryo', *Fagopyrum*, 17: 15–19.

KUDLA E and TOMASIK P (1992), 'Modification of starch by high pressure, part II: Compression of starch with additives', *Starch*, 44: 253–259.

LAMIDI A A and BURKE S (1995), 'Studies of the flavour stability of sorghum beer', *Ferment*, 8: 363–367.

LEACH A A (1968), 'Nitrogenous components of worts and beers brewed from all-malt and malt plus wheat flour grists', *Journal of the Institute of Brewing*, 74: 183.

LEDWARD D A (1995), 'High pressure processing – the potential', in Ledward D A, Earnshaw R G, Hastings A P M and Johnston D E, *High Pressure Treatment of Food*, Nottingham University Press, 1–5.

LETTERS R, BYRNE H and DOHERTY M (1985), 'The complexity of beer β-glucans', in Proceedings *of the Congress of the European Brewery Convention*, Helsinki, 20: 395–402.

LI S and ZHANG Q H (2001), 'Advances in the development of functional foods from buckwheat', *Critical Reviews in Food Science and Nutrition*, 41: 451–464.

LIMAS G G, SALINAS M, MONEO I, FISHER S, WITTMANN-LIEBOLD B and MENDEL E (1990), 'Purification and characterisation of ten new rice NaCl-soluble proteins: Identification of four protein synthesis inhibitors and two immunoglobulin-binding proteins', *Planta*, 181: 1–9.

LINDEBOOM N, CHANG P R and TYLER R T (2004), 'Analytical, biochemical and physicochemical aspects of starch granule size, with emphasis on small granule starches: a review', *Starch*, 56: 89–99.

LINKO M, HAIKARA A, RITALA A and PENTTILA M (1998), 'Recent advances in the malting and brewing industry', *Journal of Biotechnology*, 65: 85–98.

LISBJERG S and NIELSEN H (1991), 'Brewing with 80% or more unmalted cereal', in *Proceedings of the Scientific Technical Convention of the Institute of Brewing, Central and Southern African Section*, Victoria Falls, Zimbabwe.

LITTLE B T (1994), 'Alternative cereals for beer production', *Ferment*, 7: 163–168.

LLOYD W J W (1986), 'Adjuncts', *Journal of the Institute of Brewing*, 92: 336–345.

LORENZ K and KULP K (1981), 'Sprouting of cereal grains – effects on starch characteristics', *Starch*, 33: 183–187.

LOWE D P, ULMER H M, VAN SINDEREN D and ARENDT E K (2004), 'Application of biological acidification to improve the quality and processability of wort produced from 50% raw barley', *Journal of the Institute of Brewing*, 110: 133–140.

LOWE D P, ULMER H M, BARTA R C, GOODE D L and ARENDT E K (2005), 'Biological acidification of a mash containing 20% barley using *Lactobacillus amylovorus* FST 1,1: its effects on wort and beer quality', *Journal of the American Society of Brewing Chemists*, in press.

MACCAGNAN G, PAT A, COLLAVO F, RAGG G-L and BELLINI M P (1999), 'Gluten-free beer', European Patent, EP0949329.

MACFADDEN D P and CLAYTON M (1989), 'Brewing with sorghum – Use of exogenous enzymes', *Brewing and Beverage Industry International*, 1: 71–81.

MADL R L and TSEN C C (1974), 'The proteolytic enzyme system of triticale' in Tsen C, *Triticale – First Man-Made Cereal*, American Association of Cereal Chemists, St Paul, MN, USA, 157–167.

MARSHALL J J, ALLEN W G, DENAULT L J, GLENISTER P R and POWER J (1982), 'Enzymes in brewing', *Brewers Digest*, 58: 14–18.

MAZZA G (1993), 'Storage, processing, and quality aspects of buckwheat seed', in Janick J and Simon J E, *New Crops*, Wiley, New York, 251–254.

McCLEARY B V and CODD R (1989), 'Measurement of β-amylase in cereal flours and commercial enzyme preparations', *Journal of Cereal Science*, 9: 17–33.

McINTOSH G H, NEWMAN R K and NEWMAN C W (1995), 'Barley foods and their influence on cholesterol metabolism', *World Review of Nutrition and Dietetics*, 77: 89–108.

MERTENS B and DEPLACE G (1993), 'Engineering aspects of high pressure technology in the food industry', *Food Technology*, 47: 164–169.

MOLL M (1994), 'Beers and coolers, definition, manufacture, composition', Intercept Ltd, Andover, UK.

MORRALL P, BOYD H K, TAYLOR J R N and VAN DER-WALT W H (1986), 'Effect of germination time, temperature and moisture on malting of sorghum (*Sorghum bicolor*)', *Journal of the Institute of Brewing*, 92: 439–445.

MURRAY J A (1999), 'The widening spectrum of celiac disease', *American Journal of Clinical Nutrition*, 69: 354–365.

MURRAY J P, WESTWOOD K and DAOUD L (1987), 'Late hop flavour', *Proceedings of the 21st European Brewery Convention Congress*, Madrid. IRL Press, Oxford, 321–328.

NEWMAN R K and NEWMAN C W (1991), 'Barley as a food grain', *Cereal Foods World*, 36: 800–805.

NICPHIARAIS B P, WIJNGAARD H H and ARENDT E K (2005), 'Kilning of buckwheat', *Journal of the Institute of Brewing*, in review.

NISHIDA Y, TADA N, INUI T, KAGEYAMA N, FURUKUBO S, TAKAOKA S and KAWASAKI Y (2005), 'Innovative control technology of malt components by use of a malt fractionation technique', *Proceedings of the 30th European Brewery Convention Congress*, Prague, in press.

NORMAN B, VIKSO-NIELSEN A, OLSEN H and PEDERSEN S (2003), 'Starch process', International Patent Application, W003/068976.

O'ROURKE T (1996), 'Brewing', in Godfrey T and West S, *Industrial Enzymology*, 2nd edition, Macmillan, London, 105–131.

OGBONNA A C (1992), 'Developments in the malting and brewing trials with sorghum', *World Journal of Microbiology and Biotechnology*, 8: 87–91.

OKAFOR N (1985), 'Prospects and challenges of the use of local cereals and starchy substrates in brewing', in *National Symposium on the Brewing Industry, Now and the 21st Century*, Owerri, Nigeria, 22–25.

OKAFOR N and ANICHE G N (1980), 'Brewing a lager beer from Nigerian sorghum', *Brewing and Distilling International*, 10: 32–35.

OKON E U and UWAIFO A O (1985), 'Evaluation of malting sorghum 1: The malting potential of Nigerian varieties of sorghum', *Brewers Digest*, 60: 24–29.

OKSANEN J, AHVENAINEN J and HOME S (1985), 'Microbial cellulase for improving filterability of wort and beer', in *Proceedings of the Congress of the European Brewery Convention*, Helsinki. IRL Press, Oxford, 20: 419–425.

OWUAMA C I (1999), 'Brewing beer with sorghum', *Journal of the Institute of Brewing*, 105: 23–34.

PALMER G H (1989), 'Cereals in malting and brewing', in Palmer G H, *Cereal Science and Technology*, Aberdeen University Press, UK.

PALMER G H (1991), 'Enzymic degradation of the endosperm cell walls of germinated sorghum', *World Journal of Microbiology and Biotechnology*, 7:17–21.

PÉREZ-LAMELA C, LEDWARD D A, REED R J R and SIMAL-GÁNDARA J (2002), 'Application of high-pressure treatment in the mashing of white malt in the elaboration process of beer', *Journal of the Science of Food and Agriculture*, 82: 258–262.

PIERCE J S (1987), 'Adjuncts and the effect on beer quality', *Proceedings of the 21st European Brewery Convention Congress*, Madrid. IRL Press, Oxford, 49–60.

POMERANZ Y (1971), 'Functional characteristics of triticale, a man-made cereal', Wallerstein Laboratory Communications, 34(115): 175–186.

POWER J (1993), 'Enzymes in brewing', in *Enzymes in Food Processing*, 3rd Edition, Academic Press, 439–547.

RAMANATHA R V, SHANTI L M and MADHAV G J (1976) 'Peroxidase and amylase activity in developing grains of triticale, wheat and rye', *Phytochemistry*, 15: 893–895.

RIAHI E and RAMASWAMY H S (2004), 'High pressure inactivation kinetics of amylase in apple juice', *Journal of Food Engineering*, 64: 151–160.

SCHWARZ P B and HAN J-Y (1995), 'Arabinoxylan content of commercial beers', *Journal of the American Society of Brewing Chemists*, 53: 157–159.

SENN T (2000), 'Aspekte der technologischen Nutzung von Triticale', *Vortrage Pflanzenzücht*, 49: 179–186.

SENN T and PIEPER H J (1996), 'Ethanol – classical methods', in Rehm H J, Reed G and Roehr M, *Biotechnology, Products of Primary Metabolism*, VCH, Weinheim, Germany, 59–120.

SHAN L, MOLBERG O and PARROT I (2002), 'Structural basis for gluten intolerance in celiac sprue', *Science*, 297: 2275–2279.

SHIBASAKI M, SUZUKI S, NEMOTO H and KURUOME T (1979), 'Allergenicity and lymphocyte-stimulating property of rice protein', *Journal of Allergy and Clinical Immunology*, 64: 259–265.

SHIMIZU C, OHNO M, ARAKI S, FURUSHO S, WATARI J and TAKASHIO M (2002), 'Effect of reduction of carbonyl compounds by yeast on flavor stability of Happoshu', *Journal of the American Society of Brewing Chemists*, 60: 122–129.

SKINNER R (1976), 'Tropical lager beer brewing with sorghum malt', *Brewing and Distilling International*, 26–27.

SKRABANJA V, NYGAARD L H and KREFT I (2000), 'Protein–polyphenol interactions and *in vivo* digestibility of buckwheat groat proteins', *Pflugers Archiv*, 440: R129–R131.

SLOAN A E (2004), 'The top 10 functional food trends 2004', *Food Technology*, 58(4): 28–51.

SOLOMON B O, LAYOKUN S K, IDOWU A O and ILORI M O (1994), 'Prospects for the utilisation of the endogenous enzymes in sorghum malts in the hydrolysis of starch – Case study with utilisation of breadfruit starch for ethanol production', *Food Biotechnology*, 8: 243–255.

SOUTH J B (1991), 'Improvements in micronised cereal production', *Proceedings of the European Brewery Convention Congress*, Lisbon, 177–184.

SOUTH J B and ROSS R J (1993), 'Evaluation of cereal quality for micronising', *Aspects of Applied Biology*, 36: 433–442.

STEADMAN K J, BURGOON M S, LEWIS B A, EDWARDSON S E and OBENDORF R L (2001), 'Buckwheat seed milling fractions: description, macro-nutrient composition and dietary fibre', *Journal of Cereal Science*, 33: 271–278.

STOLT M, OINONEN S and AUTIO K (2001), 'Effect of high pressure on the physical properties of barley starch', *Innovative Food Science and Emerging Technologies*, 1: 167–175.

STUTE R, KLINGLER R W, BOGUSLAWSKI S, ESHTIAGHI M N and KNORR D (1996), 'Effects of high pressures treatment on starches', *Starch*, 48: 399–408.

SUNDBERG B and AMAN P (1994), 'Fractionation of different types of barley by roller milling and sieving', *Journal of Cereal Science*, 19: 179–184.

TAYLOR D G (1990), 'The importance of pH control during brewing', *Technical Quarterly Master Brewers Association of the Americas*, 27: 131–136.

TENG J, STUBITS M and LIN E (1983), 'The importance of rice variety selection for optimum brewhouse operation', *Proceedings of the 19th European Brewery Convention Congress*, London. IRL Press, Oxford, 47–54.

TURNER B K (1986), 'Malty coloured: use of malt as a natural ingredient for taste, colour, etc., to exclude additives', *Food, Flavourings, Ingredients, Packaging and Processing*, 8: 63–65.

ULMER H M, HOULIHAN J F, BARTA R C, GOODE D L, VAN SINDEREN D and ARENDT E K (2003), 'Proteolytic activity of lactic acid bacteria and their effect on beer quality', in Courtin C M, Veraverbeke W S and Delcour J A, *Recent Advances in Enzymes in Grain Processing*, ACCO, Leuven, Belgium, 353–358.

USUI Y, NAKASE M, HOTTA H, URISU A, AOKI N, KITAJIMA K, *et al.* (2001), 'A 33-kDa allergen from rice (*Oryza sativa* L. Japonica). cDNA cloning expression and identification as a novel glyoxalase I,' *Journal of Biological Chemistry*, 276: 11376–11381.

VIËTOR R J, VORAGEN A G J, ANGELINO S A G F and PILNIK W (1991), 'Non-starch polysaccharides in barley and malt: a mass balance of flour fractionation', *Journal of Cereal Science*, 14: 73–83.

WAESBERGHE J W M (1990), 'Sorghum mash separation on thin bed filter', *Brauwelt International*, 125–126.

WALZL M (2005), 'Beer – a factor of health', *Proceedings of the 30th European Brewery Convention Congress*, Prague.

WESTWOOD K (1994), 'Hop products and their effect on bittering quality', *Brew Guard*, 123: 31–34.

WIJNGAARD H H, ULMER H H and ARENDT E K (2005a), 'The effect of different moisture percentages on the malting quality of buckwheat', *Journal of the Institute of Brewing*, in review.

WIJNGAARD H H, ULMER H M and ARENDT E K (2005b), 'The effect of germination temperature on malt quality of buckwheat', *Journal of the American Society of Brewing Chemists*, 63: 31–36.

WIJNGAARD H H, ULMER H M, GOODE D L and ARENDT E K (2005c), 'Gluten-free beverages based on buckwheat', *Proceedings of the 30th European Brewery Convention Congress*, Prague.

WOOD P J, BRAATEN J T, SCOTT F W, RIEDEL K D, WOLYNETZ M S and COLLINS M W (1994), 'Effect of dose and modification of viscous properties of oat gum on plasma glucose and insulin following an oral glucose load', *British Journal of Nutrition*, 72: 731–743.

YOKOYAMA W H, HUDSON C A, KNUCKLES B E, CHIU M C M, SAYRE R N, TURNLUND J R and SCHNEEMAM B O (1997), 'Effect of barley β-glucan in durum wheat pasta on human glycemic response', *Cereal Chemistry*, 74: 293–296.

ZARNKOW M, KESSLER M, BURBERG F, KREISZ S and BACK W (2005a), 'Gluten-free beer from malted cereals and pseudocereals', *Proceedings of the 30th European Brewery Convention Congress*, Prague.

ZARNKOW M, KREISZ S, BACK W, SEIDEL B and KURZ T (2005b), 'Innovative drinks based on malted cereals and pseudocereals', *Poster Presentation, International Cereal Chemistry Conference*, Vienna.

4

Malt and malt products

N. Davies, Muntons Plc, UK

4.1 Introduction – factors driving change

As we progress into the twenty-first century, the main drivers for change in malting are food safety, energy conservation and maintaining a benign environmental impact. Developments in malt processing equipment appear to have reached a plateau and attention is now focused on assurance of quality, whilst at the same time using industry best practice to minimise cost and generate resources for investment. Particularly in the past decade there has been a cascade of accreditation frameworks designed to protect the human and animal food supply chains, which have been readily adopted by maltsters. These systems have created a very structured control of quality and risk and have driven malting practice from agricultural to food factory standards. There are many very good accounts of the biochemistry of malting and brewing (Briggs, 1998; Briggs *et al.*, 1995). In contrast this chapter aims to present a practical and business perspective of malt and malted ingredients production and illustrate our challenges and opportunities.

Supply chain profit sharing is a major factor affecting the sustainability of maltings to invest in improvements that achieve the three main aims stated at the outset. There is a predicted steady rise of around 2% per annum in beer consumption figures over the next few years and malting capacity is being planned to meet that need (Braks, 2005). There continues to be the expectation from customers that maltsters can pass on savings in production costs even in a market that is currently out of balance with oversupply, which limits the opportunities to develop a sustainable profit margin. This situation becomes untenable if there is to be a programme of investment in malting to maintain standards commensurate with the brand image of the products for which malt forms an integral part. Most of the malting capacity is within the European Union, which produces around 60% of the world's malting barley and 50% of the world's malt (Fig. 4.1). The difficulty is that new malting capacity coming on line suddenly creates an oversupply in the market, with a consequent depression on malt margins. This depression in profit stalls investment programmes, resulting in a stepwise rather than gradual increase in malting capacity. Investment in new malting capacity can also be quickly hampered if certain parts of the economy suddenly develop free trade agreements, thus giving an unfair advantage for imports to one part of the market until the process of getting the free trade arrangement extended or changed across the entire supply market takes effect, if indeed it is changed.

Currently there is a commodity price structure for malting barley that creates a challenge to balance cost and quality. Barley suitable for beer must generally have a low protein content and thus attracts a premium. The level of the premium is affected by the size of the malting-barley crop at harvest. In a good year the supply of malting barley is plentiful and the premium is depressed; in a poor harvest year the premium tends to increase. Crop failures in other parts of the world can at times give an unexpected rise in the premium. Not only does this make planning for the farmer difficult, it also complicates the forward planning of malt prices to generate sufficient profit for reinvestment. Further up the supply chain, brewers aim to reduce their exposure to price volatility by entering into long-term agreements or contracts with maltsters. This presents the

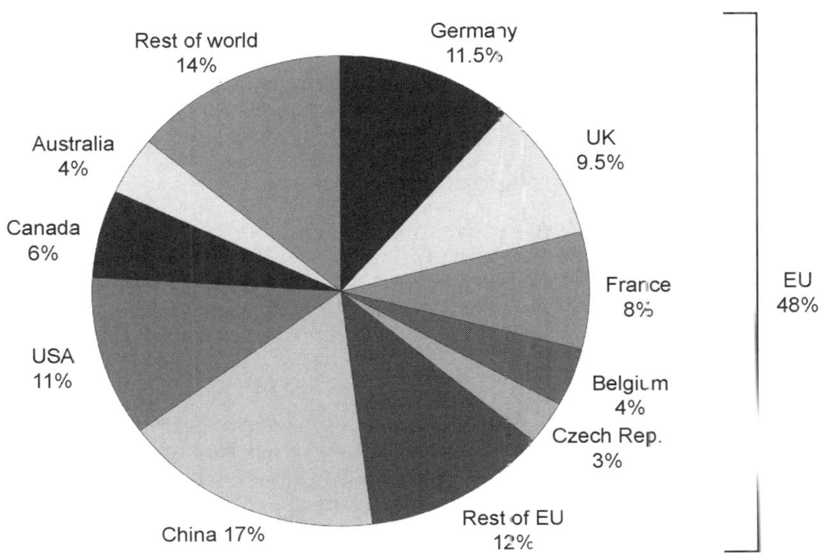

Fig. 4.1 World malt production.
A great proportion of the world's malt production comes from Europe, but the largest single country production is within China. However, much of the EU production is destined for export around the world.

maltster with a challenge – whether to fix barley price at the time of the contract to limit their risk to price volatility or to wait and see. The latter approach will result in either a financial windfall or a financial disaster. Some mechanisms of smoothing the risk (hedging the price) of barley going forward have been proposed, such as the European Barley Office of Trade (EBOT) and the Commodity Exchange Hannover (WTB Warenterminbörse). The aim is to trade malting barley contracts on the same basis as futures markets. These markets are not yet well developed and the profit margin on malt remains very small, financial success often being determined by volume rather than margin. This pressure will undoubtedly continue the drive to consolidation in the malting industry among the non-independent malting groups, but place increasing pressure on the smaller independent malt companies.

Interest is increasing in the use of processed malt in brewing as it has been already established in food manufacture. In contrast to the market for malt as a commodity product, malted ingredients can attract a much better margin, since they are perceived as niche products adding value. The starting point for malted ingredients can be either white malts or any of the range of coloured or speciality (roast) malts available. These malts are most often used to create a malt extract. Here malt is milled and mashed as in a normal brewing process, but 80% of the water is evaporated to leave a viscous flavour-rich extract. Malt extract can then be added in relatively small amounts to the brewing process or in food manufacture to create flavour or enhance existing flavours. Extracts can have a wide range of other ingredients added prior to packaging and then be dried in various ways such as spray drying or vacuum band drying. The possibilities for innovation in this area are enormous and could easily have a major impact on malt usage in the next 20 years.

4.2 Malting barley development

4.2.1 Barley breeding – the conventional norm

There is a very strong and successful history of breeding good quality malting barley around the world. However, the cost involved in bringing a new variety to acceptance by maltsters and brewers is substantial. Estimated costs for maintaining a breeding programme approach £1.5 million (€2 million) a year. This cost has to be offset by collecting royalties when the seed of each variety is sown – a system called plant breeders' rights. From an initial cross of two parental lines through to acceptance of a new barley as a malting variety for brewing can take 11 years. The development of new varieties progresses from the first parental or filial cross (F1) through to the final variety accepted for malting and brewing, likely at F11. In the early stages of the breeding programme barley varieties are assessed for distinctiveness, uniformity and stability (DUS) and the variety's value for cultivation and use (VCU). This assessment is through National List trials.

Breeders have a vast array of possibilities for breeding malting barley varieties, since the barley genome contains around 40 000 genes. For the initial crosses, selection is generally on the basis of appearance alone. The first involvement of the maltster or brewer in this programme is usually at stage F8 when sufficient grain is available for micromalting.

Looking back 10 to 15 years the number of new varieties in barley breeding programmes had proliferated. The result was a large number of varieties in a market that appeared to lack focus on the positive attributes of new varieties, but was influenced more by tradition – a reluctance to change from the old and trusted varieties. It took time for new varieties to gain acceptance as brewing varieties – time the breeders didn't have if they wished to recoup their development costs. The time frame for a variety to remain on the recommended list for brewing could then be as little as three to five years, often insufficient to make breeding new barley varieties attractive. Quite often barley breeders expressed the concern that they didn't have a clear picture of the brewer's requirements. In practice the aims were simple, if not fully elucidated: farmers wanted good harvest yield, maltsters wanted ease of processing and flexibility to make a variety of malt types, and brewers wanted good extract, fast runoff and good flavour attributes.

The key factors for a good malting variety are now much more widely accepted and discussed, such that the requirements of the various supply chain partners are incorporated into decisions in developing and evaluating new varieties. In many countries across the world there are formal systems in place to evaluate the suitability of barley for malting. There is, however, a difference in the degree of rigour of the assessment systems and the numbers of varieties screened, but the basic principles are the same: new varieties must be better than existing ones and be proven to be suitable by pilot malting and brewing.

Table 4.1 outlines the similarity of malting barley evaluation schemes across the world. The result is not always an official recommendation, but is always designed to compare varieties against existing varieties of known malting and brewing performance. The UK and French systems are very systematic in approach and have very structured committee-based evaluation. Membership of these committees comprises breeders, maltsters and brewers at all stages. The Canadian system is similarly merit-based but with fewer committees. For Australia and the USA the evaluation is rigorous and process driven, but is reviewed by their brewing membership rather than a series of committees. At the micromalting stage the key parameter for all the systems is malt quality. Brewers' hot water extract is universally of prime importance in determining whether a variety is improved over existing good quality control varieties. However, there is also a distinct emphasis in having low β-glucan and high diastatic power in new varieties, because both of these have a direct impact on brewing performance. These same parameters apply as a variety progresses to the pilot malting and brewing stages, but always with an eye on yield on the farm and disease resistance. It can be extremely frustrating for a variety to fulfil all the malt quality parameters only to fail because farmers don't view it as economic to grow because of variable agronomic performance. A new variety needs to have a distinct financial benefit to farmers, because the premium for

Table 4.1 Evaluation of barley varieties for malting quality

Country/region	Evaluation procedure	Controlling committees	Physical parameters considered	Quality analysis	Partners involved	Official recommendation
UK	Micromalting; National List Trials; Recommended List Trials; pilot malting and brewing; specific brewing recommendation	EMMG/SMMG[1] EWP/SWP[2] IoB MBC[3]	Agronomy, disease resistance, yield	Hot water extract, colour, diastatic power, α-amylase, soluble and total protein, FAN, viscosity, friability, homogeneity, β-glucan, glassy corns	Growers, breeders, maltsters, brewers, research laboratories (BRi[5]), commercial	Institute of Brewing HGCA recommended list (via CEL[4])
France	Micromalting; National List Trials; Recommended List Trials; pilot malting and brewing	CBMO[6] CTPS[7]	Agronomy, disease resistance, yield	Hot water extract, Kolbach, diastatic power, viscosity, final attenuation, filtration index	Breeders, maltsters, brewers, IFBM[8]	Malteurs de France recommended list
Europe	Micromalting of varieties on National Lists, 4 geographical areas[10]	EBC[9] Barley and Malt Committee	Agronomy, disease resistance, yield	Hot water extract, Kolbach, diastatic power, total and soluble protein, viscosity, final attenuation, β-glucan, friability	Breeders, maltsters, brewers	EBC list and results published by region
Australia	Micromalting; pilot malting and brewing	MBIBTC[11]	Agronomy, disease resistance, yield	Hot water extract, diastatic power, viscosity, apparent attenuation limit, β glucan; different ratings for home and export	Breeders, brewers, maltsters, AMBC[12]	Recommendations used by brewers supporting the consortium

USA	Micromalting; pilot malting and brewing	AMBA[13]	Disease resistance, yield	Hot water extract, total and soluble protein, screenings, wort turbidity, viscosity, colour, diastatic power, α-amylase (DU)	Maltsters and brewers	Recommendations used by brewers supporting the consortium
Canada	Micromalting; pilot malting and brewing	CFIA[14] PRRCG[15]	Agronomy, disease resistance, yield	Hot water extract, soluble and total protein, soluble nitrogen ratio, viscosity, diastatic power, α-amylase (DU), β glucan	Growers, breeders, researchers, maltsters, CMBTC[16]	Canadian Malting Barley Industry Group issue a final merit-based rating
Denmark	Micromalting and brewing	None	None directly	Hot water extract, colour, soluble protein, friability, modification, homogeneity, diastatic power, α-amylase (DU)	Private trials	No official recommendation; brewers' own list

[1] EMMG: English Micromalting Group; SMMG: Scottish Micromalting Group.
[2] EWP/SWP: English/Scottish Working Party.
[3] IoBMBC: Institute of Brewing Malting Barley Committee.
[4] CEL: Crop Evaluation Limited (HGCA: Home Grown Cereals Authority are sole shareholder).
[5] BRi: Brewing Research International, Surrey, UK.
[6] CBMO: Comité Biére Malt Orge (Malt, Barley and Beer Committee).
[7] CTPS: Comité Technique Permanent de la Sélection (The Permanent Technical Committee for plant breeding).
[8] IFBM: Institut Français des Boissons de la Brasserie et de la Malterie (French Brewing and Malting Institute).
[9] EBC: European Brewery Convention.
[10] North (Finland, Sweden, Denmark, Estonia); West (United Kingdom, Ireland, France, Netherlands, Belgium); Central (Hungary, Germany, Slovak Rep., Austria, Slovenia, Czech Rep.); South (Portugal, Spain, Italy, Bulgaria).
[11] MBIBTC: Malting and Brewing Industry Barley Technical Committee.
[12] AMBC: Australian Malting Barley Centre.
[13] AMBA: American Malting Barley Association Inc.
[14] CFIA: Canadian Food Inspection Agency.
[15] PRRCG: Prairie Registration Recommending Committee for Grain.
[16] CMBTC: Canadian Malting Barley Technical Centre.

malting quality barley over feed grade has been severely restricted by virtue of cost squeezing further up the supply chain.

Specifications for β-glucan have been progressively lowering over the past 15 years since this parameter became more common on specifications. Diastatic enzyme potential is also now more important, particularly where maltsters export to those brewers who may use significant amounts of starchy adjunct that require amylolytic breakdown via malt diastatic enzymes. The modern brewmaster, however, recognises that a balanced profile of malting characteristics is vital to give reproducible brewing performance at all times. The complex interactions between the various malting analyses means that only those varieties that have passed through the review and assessment panels and proven acceptable for brewing will be specified in brewing malt contracts.

Barleys can be bred to grow in different seasons of the year. For example, there is the choice of winter- or spring-sown malting barleys. Although brewing tradition can result in a rejection of winter varieties, there is acknowledgement from some brewers that the differences in brewing performance are not well established and they readily accept winter varieties for both lager and ale production. From the farmers' perspective a winter variety is very economically sound. For example, in the Northern hemisphere the farmer can start to sow the winter varieties in the autumn, follow on with a drilling of spring varieties in March and harvest both during July and August.

Currently breeding programmes reflect the more traditional view of supporting spring-sown barleys. The balance of new varieties coming through for assessment in the breeding programmes is predominantly spring. This could signal a halt to breeding of winter varieties. However, since most businesses now establish a risk-averse strategy for success, the relatively poor support for winter varieties may lose the supply chain a degree of protection from crop failure. As such it is vital that we continue to support breeders in developing winter varieties at least as good as if not better than those now available. Although there is a healthy succession of spring malting varieties coming through the stairway assessment system, there are now relatively fewer varieties from which to choose. This is more attractive to breeders who can gain a potentially greater share of the market for their variety and thus recoup the development costs. That said, there is still a tendency for a variety to dominate, particularly if it is suited to the farmer and flexible for the maltster. The spring malting variety Optic is such a variety and achieved a significant proportion of the hectarage of spring malting barley grown for many years since it was introduced. Even this variety is now in sharp decline as new varieties take over.

4.2.2 Where next for barley breeding?

With the vast number of genetic combinations possible in a barley breeding cross, it would be advantageous to have a greater degree of certainty that the desired gene or genes would indeed be present in the new variety. Current barley breeding practice cannot guarantee that a cross will generate a variety suitable for malting. Although in theory it is possible to use genetic modification (GM) to manipulate genetic sequences into new varieties, there is not yet public acceptance of such technology. Even if GM crops become acceptable in the future, the technology is still subject to a requirement to know precisely which genetic sequence should be inserted and into which part of the genome. Perhaps surprisingly, the interest in GM technology has actually generated something of more acceptable and immediate use. For GM to be possible, more detailed maps of the grass (*Triticeae*) family genome were required with barley and wheat as the primary targets to map. Back in 1997 the International Triticeae Mapping Initiative (ITMI) was set up to identify these genetic sequences. In 1998 there were only 80 expressed sequence tags (ESTs) for the entire grass family in the ITMI database. By 2002 this had risen to nearly 400 000 (National Center for Biotechnology Information (NCBI), June 2005). Such a resource enables an innovative approach to barley breeding and has sophisticated database manipulation to predict gene combinations (Close *et al.*, 2004). By linking the genetic map to highly specific malting and brewing traits, it is possible to target quite complex groups of genes in a barley variety very accurately and to track their incorporation into a barley cross made in the traditional manner. This field of breeding is referred to as *genomics* and is a very exciting breakthrough.

The technology of importance to genes is already proven. For example, the *sdw* allele (genetic sequence) has been identified as controlling grain size (Fettel *et al.*, 1999); three alleles, B*my*-Sd1, B*my*-Sd2L and B*my*-Sd2H, control thermostability of β-amylase (Eglington *et al.*, 1999). The useful genes are identified and tagged with a piece of DNA that is easy to detect once the sequence is incorporated into a new variety. Frequently there will be groups of genes acting together that need to be inserted all together for a trait to be transferred into a new variety. There may not always be markers available for all of the genes, so regions of DNA that are closely associated with the start and end of the group of genes, called quantitative trait loci (QTL), are used, for which there are genetic markers. A simple detection of the QTLs determines whether the required set of genes is present in the new variety. Similar approaches have been tried already in Australia (Dennis, 2001; Paris *et al.*, 2001), but work at the Scottish Crop Research Institute (see Section 4.7.3) is now set to bring a powerful collaborative group of supply chain partners together to create a very specific barley breeding programme. The aim is to reduce the breeding time from 11 years to perhaps five years, thus making it possible to react more quickly to changing demands and reducing breeding costs.

While this work is progressing, there is also the possibility of more traditional methods providing better barley. In conventional breeding, barleys from the same genetic line are crossed. A variant of this is to render the female line incapable of self-pollinating and crossing barleys from different genetic lines. The varieties produced are found to be more vigorous and are termed hybrid barleys. In the UK the feed barley variety Colossus is the first hybrid barley to be included on the Home Grown Cereals Authority (HGCA) recommended list (Syngenta, 2004). As with most plant F1 hybrids, the vigour is enhanced and claimed by the breeder to give greater harvest yield than traditionally crossed varieties.

The drive to improve malting barley will continue to be from within the industry. In a recent survey examining the potential for developing cereal varieties with specific end user traits, it was concluded that in the mind of the consumer there was little benefit of an improvement in malting variety for malt and beer production (Farmers Fund, 2004). This was quite different from inclusion of cereal in milled products where consumers readily accepted that a more wholesome product with a health advantage was worth a premium. The reason? Cereals are a more obvious part of a baked product, but are not marketed as such within beer that is heavily brand-led. In addition the malting industry is very much a commodity market where price and cost dominate the strategy and this short-term pressure does not allow maltsters to be diverted into niche markets. For distillers their lead-time to product launch is also not suited to rapid development of niche products, which by their very nature are rather ephemeral.

4.3 Malt processing

4.3.1 Malthouse process equipment

The technology of malting equipment has remained largely unchanged for a number of years because it is has proved reliable and cost-effective. Process flow is similar across the world. Grain selected for size to assure even water uptake is steeped for 48 hours using two wet (immersion) periods, resulting in chitted grain at 44–46% moisture. The most common steep vessels are cylindroconical because they are easy to clean. They do, however, have large bed depths which require very efficient aeration systems to keep the grain moving and avoid hot pockets of grain which would subsequently have poorer germination. Many older plants have a central aeration pipe that generates a central surge mixing of the barley. It is more common now to have many smaller injection points arranged in rings around the lower part of the vessel to give an even and diffuse aeration during the immersion periods. Whichever vessel is used, it is important to achieve sufficient aeration and pulse the circulation throughout the immersion period to minimise anoxic conditions. Flat-bottomed steeps, although having the advantage of small bed depths and more even chitting, have always been associated with difficulties in cleaning the plenum to maintain hygiene. In a recent innovation a flat-bottomed steep without a plenum chamber is being trialled. The first proven installation of such a system in a commercial maltings is awaited, but if successful it will no doubt become a standard installation for new maltings. To achieve a vigorous start to germination it is important to have excellent removal of carbon dioxide generated by the barley. Inefficient removal of carbon dioxide can significantly reduce germination vigour and impair malting yield. If the fans are not properly regulated and are overused, the temperature within the steep vessel can rise and adversely affect grain modification. Therefore pulsed removal of carbon dioxide by drawing air through the bed is preferred.

Germination can take place in a wide variety of systems with either box or circular geometry and will typically take 4–5 days. For new plants circular vessels are the preferred method of construction. Because many malting plants have restricted space, new plants often use the tower method of construction. It is possible to have all vessels in one tower utilising gravity transfer, or separate steeps and kilning buildings. If land is not an issue, it is more cost-effective to construct a box germination vessel, but conveying systems to empty and fill these vessels are more complex than the simple screw loading and emptying system found in circular vessels. In general, bed depths for any germination vessel will not exceed 1 metre. The most important controls in germination are temperature and humidity. This is achieved by powerful fans and injection of water into the air stream using either compressed air or spinning discs. The most significant change in vessel construction is batch size. Some new circular germination vessels are up to 35 m in diameter with a batch size of 600 tonnes. It is not inconceivable that a batch size of 1000 tonnes will be commissioned within 10 years. The reasoning behind increased batch size is economy of scale and reduced movement per tonne of product. It is estimated that for a 50 000 tonne malting plant there is little cost differential for capital expenditure between box and circular construction. However, for a 100 000 tonne plant it is possible to reduce capital costs by up to €50 per tonne of malt by using circular vessels and tower construction rather than boxes.

Kiln construction still has essentially two variants: direct and indirect firing. The direct-fired systems pass air heated directly by burners through the kiln bed. Indirect-fired kilns heat the air by heat exchangers which are generally stainless steel pipes with circulating water inside that is heated by a gas-or oil-fired boiler. Concern over many years about potential carcinogenic compounds (e.g. N-nitroso dimethylamine – NDMA) in malt using direct-fired systems has resulted in a switch to indirect firing. For a normal brewing malt the kilning process starts with air being passed through the grain (air on) at around 65°C with a high fan speed to remove the surface water. Once the surface is dry it is more difficult to remove the water inside the malt, so the air on temperature is increased and the fan speed gradually reduced. During the second phase and the final curing phase of kilning the temperature is increased to around 80°C for lager malts and 90°C for ale malts. The process is monitored by a probe placed in the air flow above the bed (air off probe). Once the surface wetness is removed the temperature of the grain bed begins to increase and the air off temperature rises, this flexion point being termed the 'break'. A double-deck or two linked single-deck kilns can operate more efficiently by utilising the relatively hot dry air coming off a kiln in the final curing phase to dry a second kiln in the first phase of kilning. The recirculated air requires less heating and is drier than ambient air. Even with just one kiln it is good practice to recirculate up to 100% of the air during the second phase.

4.3.2 Food safety and hygiene

Legislative and audit structures

Food safety and assurance across all stages of processing is the most important customer requirement next to price and is maintained by good traceability. In the

UK, since the Food Safety Act was introduced in 1990, all food processing factories have targeted improved hygiene levels. This approach has been applied in many other countries via legislation, registration or best practice protocols. Malting has thus moved a long way from being an agricultural process to become food factory oriented, even though malt will undergo further processing before reaching the consumer. The scope of the definition of a consumer has also changed over the past three years to incorporate both human and animal feed.

The most common and useful structure for ensuring that a malting plant has been properly assessed and is in control of food safety is to have a Hazard Analysis and Critical Control Points (HACCP) system in place. This is a prerequisite of a number of other hygiene standards. In many cases where maltsters supply into the large food supermarkets either directly or via another processor, there is a requirement for a higher level standard such as the British Retail Consortium – Global Standard Food (BRC). This is similar in scope to the International Food Standard (IFS) adopted in other parts of Europe, but BRC is much more rigorous in the standards of hygiene, documentation, audit and training.

Hygiene standards that are now essential in and around the malting plant are included as prerequisites in a good HACCP programme. Table 4.2 lists some of the more important controls that should be in place.

Any point in the malting plant at which the operator comes directly into contact with the malt is treated as a food grade area. The surrounding areas are still required to be kept clean, but not to the same food grade standard. Where possible, plant that is in contact with the raw material is now constructed of stainless steel and all lubricants and greases should be food grade and non-toxic.

The HACCP risk assessment is carried out by a multi-disciplinary team representing a cross-section of the staff, e.g. operators, engineers, micro-biologists and managers, and led by a certificated HACCP lead auditor or equally qualified person. The team looks at every aspect of the malting operation and establishes which points are critical control points (CCP), i.e. if the control fails at that point, there will be no other control in place to prevent the malt becoming dangerous as a food product. An example of a CCP could be a magnet in a conveyor just before loading into a truck for dispatch. If the magnet fails to collect metal it could go directly into the delivered malt. Whilst it would be removed by intake controls at the brewery, it has to be a CCP at the maltings because for that operation it is the last point at which any control is possible. It is very important to ensure that operators are well aware of each CCP. The best practice is to have a formal written description of each CCP at the point where operators can see them. It describes what the CCP is, why it is needed and the impact if the control fails.

It is a great strength of the UK malting industry that a HACCP protocol for malting has been created and published by the Maltsters Association of Great Britain (MAGB) that is adopted by all UK maltsters (MAGB, HACCP protocol, version 3, 2006). This document is under constant review as new legislation is introduced or incidents in other industries create concern across the entire food supply chain.

Table 4.2 Examples of prerequisite control measures that form a basis for a HACCP protocol
This table lists some of the key prerequisites that must be in place for a suitable HACCP system for malting. It is not intended to be exhaustive, but demonstrates the foundation on which a HACCP study can build.

Area for control	Methods of control
Grain intake and storage	Strict control of reception and storage of grain according to best practice guidelines, e.g. moisture. Strict terms and conditions for supply and supplier evaluation procedures. Use of an assured grain supply scheme. Due diligence sampling to assure controls effective
Housekeeping	Must be of a high standard throughout the maltings. For example, different coloured brooms should be used for areas with different risks, e.g. green for outside roadways and paths, blue for inside the plant and red for germination boxes, thus keeping cleaning of food and pedestrian areas separated
Jewellery	No loose jewellery, watches, spectacles or anything else that could fall into the product
Glass	Generally minimal glass allowed on site. All glass or brittle plastics must be on an audited register and risk assessed
Smoking	If allowed on site it must be restricted to a defined area away from production
Personal hygiene	Hand washing or sanitising stations must be provided at entrances to food areas, e.g. germination boxes
Cleaning	Chemicals must be food grade. Cleaning schedules must be documented
Pest control	Must be extremely thorough and any actions identified must be rectified immediately
Visitors and contractors	Brief induction in food safety at reception to site
Personnel hygiene	Basic hygiene rules are mandatory and staff are restricted from work if they or close family members are il

A vital area with all food safety audits is to effectively communicate and train staff in the importance and operation of the controls required. It can be advantageous to develop a specific food safety training course based on the type used for those handling food products. Companies may well develop these in house to address issues of specific importance to malting such as mycotoxins, traceability, due diligence analysis and food safe packaging. This type of course engenders a change in attitude among the workforce and lifts the standard of hygiene by education rather than enforcement.

Traceability and due diligence
Traceability is a key area for legislative control at present. In practice the maltsters' supply chain already has much better traceability than is currently

required, but these laws will undoubtedly continue to tighten. Legislation has been introduced in the EU recently that begins to define what is meant by traceability (EC Regulation 178/2002). The US Food and Drugs Authority also introduced the US Bioterrorism Preparedness and Response Act 2002, which requires good traceability, albeit for protection against criminal attack rather than adulteration.

Traceability is important not just for individual maltsters in minimising potential damages, but to protect the perception of the entire industry upstream and downstream (Golan *et al.*, 2003). Customer requirements currently are well ahead of legislation in describing the extent of traceability. For malting, traceability involves both physical tracing of documents relating to processing and assurance of food safety in that supply chain.

Speed would be of the essence if a food safety issue was ever identified, and maltsters have to run simulations to test their traceability and product recall procedures annually. An example of the speed of response is given in Fig. 4.2. Although hypothetical, it demonstrates the speed at which it is possible to trace upstream and downstream. Food safety audit schemes also demand that traceability procedures are tested at least annually, both backwards down the chain from barley at intake to the farm it was grown on and upwards to all customers supplied from a specific barley silo. Modern Enterprise Resource Planning (ERP) systems such as SAP enable such a seamless, rapid and traceable linkage from raw material to customer. However, the resource required to trace back to the farm is considerable. Therefore, in the UK there are established grain assurance schemes such as the Assured Combinable Crops Scheme (ACCS – currently changing its name to AFS, the Assured Farm Scheme) and Scottish Quality Crops (SQL), and there are a number of others

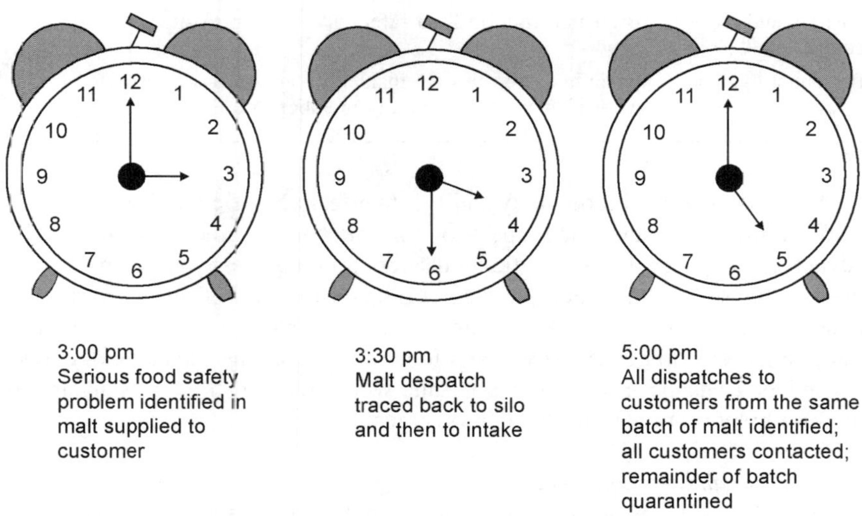

Fig. 4.2 Scenario showing speed of response to a potential food safety issue.

(see Section 4.7.2). These schemes audit the farms on which malting barley is grown to ensure they are working to the correct standards regarding pesticide application and grain storage. Farms satisfying the rigorous checks become certificated and receive authentication stickers to attach to documentation sent with deliveries to maltings. Generally certification is renewed annually. Maltings in the UK will now only accept grains via an assurance scheme that greatly protects the supply chain.

Whereas the assurance schemes are excellent gatekeepers of good growing practice, it is not acceptable as a business risk to assume that the barley being supplied at intake is indeed food safe. Therefore samples are taken according to schedules designed to reflect the risk to health of pesticide residue levels, mycotoxins and heavy metals. The sampling ratio, according to British Standards for sampling in the UK, is one sample for every 6000 tonnes of barley and for every 4000 tonnes of malt (BS6001-1:1999; BS6002-4.1:1996). These are minimum standards and many maltings choose to carry out spot checks more frequently. Analysis is carried out by accredited laboratories and confirms, or validates, that our assurance schemes are indeed effective. Other countries adopt similar sampling systems, but all based on product risk.

Pesticides and mycotoxins – hot issues, but in control

It is a non-negotiable contractual term that barley coming into a maltings must have no chemical residues on it nor any fungal growth or secretion that could be harmful to human health. To determine what is harmful or not, there is national and international legislation in place setting maximum residue limits (MRLs). There is often variation globally between acceptable levels of pesticide residues for malting barley. However, there is an international standard for pesticide residues published by the Codex Committee on Food Additives and Contaminants (CCFAC, 'CODEX'). In some cases national limits can be set lower than CODEX. As an additional control for malting barley, it can only be treated with pesticides that have been proven to have no effect on brewing performance and are not harmful to human health. Using the UK as an example, the only chemicals that may be applied to barley intended for malting are published by the British Beer and Pub Association (BBPA) and Brewing Research International (BRi) at least annually (BBPA/BRi, 2005). Every chemical on the list has been applied to barley at much higher rates than normal, then trialled by malting and brewing with analysis for chemical residues and sensory attributes. If there are no differences in performance or analysis compared to a control, the chemical passes the test. To be included on the BBPA/BRi list, however, an additional thorough search of key databases worldwide by BRi is made to ensure there are no reported health dangers. Many of the chemicals that are allowed to be used on malting barley are applied long before grain emergence, thus reducing even further those that could persist on grain at harvest. Generally no pesticide application is allowed onto the grain after harvest, but if any application is found to be necessary, only chemicals on the BBPA/BRi list are allowed and a declaration must be made on the post-harvest pesticide certificate that is required at grain intake. To validate these controls, it is conventional to take samples for testing on a random basis to confirm residues are well below the MRL, if present at all. The malting process will further dilute any residues, thus making a very safe control of the process.

For mycotoxins the story is a little different because geographical area and agronomic practice have a greater influence. There has been a marked increase in those countries having mycotoxin regulations, to almost 100 countries representing 90% of the world's population by 2003 (van Egmond and Jonker, 2004). In countries where maize is grown that is particularly susceptible to fungal growth the problem is more acute. In the USA it has been estimated that economic losses due to contamination by mycotoxins amount to over \$900 million per annum. With the added concern in the USA that mycotoxins could be used in bioterrorism, there is a need to develop rapid detection and detoxification procedures (Bhatnagar *et al.*, 2003). Rapid testing is also on the agenda in other countries as part of an ever-improving aim of better traceability. Mycotoxins are secondary metabolites produced by certain fungi in certain conditions – normally referred to as stressful conditions. Fungi can be found on barley grown in the field, e.g. *Fusarium*, which can produce mycotoxins such as deoxynivalenol (DON), zearalenone and fumonisin. In storage, if the moisture is not controlled, it is possible to find species of *Aspergillus* and *Penicillium*, which can produce the mycotoxin Ochratoxin A (OTA). There is no correlation between the level of fungus on a grain and the amount of mycotoxin produced, thus vigilance at intake for any fungal growth in essential. Operators at barley intake are trained to visually assess and sniff the grain. Pink grains signify infection by *Fusarium* and could signal presence of mycotoxins, thus causing immediate rejection of the barley. A slightly musty or mouldy aroma could indicate that the grain was damp or that there was a very low level of fungal infection. This check is actually a Critical Control Point (CCP) in the HACCP system because it is currently the last check before grain is dried and put into silo for storage.

Fungi that could potentially produce mycotoxins do not grow at grain moisture contents below 14.5%. Therefore good drying practice rapidly brings all grain at intake down to below 14.5% (HGCA, The Grain Storage Guide, version 2, 2003). In practice a level of 13% moisture is reached because this also ensures good germination later in the process.

As a validation check to confirm that grain is being effectively screened at intake, samples are drawn from silos of dried barley throughout the year. The first samples are taken just after harvest and a second set are taken after six months in storage. These two tests enable a check to be made on control of field and storage fungi. The degree of concern over mycotoxins can be attributed by country based on climatic conditions and the crops grown. Taking the mycotoxin DON as an example, Table 4.3 illustrates the differences in levels in five major areas of the world. The safe maximum level for DON has been set in the EU at 1250 ppb, but there is a drive from brewers to reduce this level as low as 300 ppb and even to be non-detectable. Historically the UK has had very low

Table 4.3 Mycotoxin risk around the world illustrated by levels of deoxynivalenol (DON) The mycotoxin deoxynivalenol (DON) is produced by the fungus *Fusarium* in the field. It is thus a good indicator of the influence of geographical location and agronomy.

Country	Mycotoxin risk for barley	Typical DON levels
Australia	Low incidence of DON due to dry climate	Negligible
UK	Low country risk due to agronomy: no maize grown. Grain at harvest in England usually less than 14%, quickly dried to <14.5%	<25 ppb (μg/kg)
France	Climate similar to UK, but greater amount of maize grown	30–150 ppb (μg/kg)
USA	Greater amount of maize and agronomy increase incidence of causative fungi. DON found in up to 80% of barleys, but endemic *Fusarium* strains produce low levels of mycotoxin	700–900 ppb (μg/kg)
Canada	Similar risk to USA at up to 73% incidence of *Fusarium*	150–500 ppb (μg/kg)

levels of mycotoxin on malting barley, which makes the country as a whole low risk (Baxter, 2004; Edwards, 2005; MAGB, 2005). Although the UK has reasonable rainfall, there is little cultivation of maize, which is a prime crop for harbouring fungi that can produce mycotoxins. The average levels of the key mycotoxins detected in UK malts over the past five years are shown in Table 4.4. Taking the most recent data illustrates the effectiveness of mycotoxin control, with mean OTA levels being just 10% of the MRL (mean 0.3 ppb, MRL is 3 ppb max.) and for DON 1.2% (mean 6 ppb, guideline MRL is 1250 ppb,

Table 4.4 Levels of key mycotoxins in UK malts
Results of testing for mycotoxins in UK malts over a five-year period indicate that there is a very low country risk for all the major toxins. Levels detected represent a tiny fraction of the permitted maximum level to assure food safety.

Mycotoxin	Maximum level	1999	2000	2001	2002	2003
Ochratoxin	3 ppb	0.27	0.07	0.15	0.30	0.23
Deoxynivalenol (DON)	1250 ppb[1]	3.5	3.0	7.0	8.0	6.0
T-2	None set[2]	–	–	–	2.5	2.5
HT-2	None set[2]	–	–	–	2.5	2.5
Zearalenone	<5 ppb	0	0	0.5	1.0	1.4

[1] Maltsters in the UK work to a maximum limit of 500 ppb for DON according to the requirements of most customers.
[2] Limit suggested of 200 ppb, but yet to be ratified.

Source: MAGB and Brewing Research International. Results represent samples taken as part of a due diligence testing scheme via the MAGB.

maltsters target 500 ppb) (MAGB, 2005). The data shows that not only is the country risk low, but storage is well controlled as evidenced by the OTA levels. Across the world there is wide variation in permitted levels for Ochratoxin A, from 1 to 50 ppb, with the EU taking a firm lead in setting levels of 5 ppb for unprocessed cereals and 3 ppb for malt.

Since the detection of a mouldy aroma at intake is such a critical step, it is important to determine the suitability of intake operators to detect such an aroma. Using a synthetic analogue of the mouldy aroma (2,4,6-trichloroanisole), the author has devised a method to test an operator's ability to smell in two ways. To test their detection capability and threshold, operators can be presented with samples spiked with trichloroanisole at three different levels. They need to detect a level of aroma equivalent to a sample of mouldy grain that would be rejected at intake. Those capable of properly detecting the aroma can be routinely tested using three samples either spiked or not spiked with the chemical, and asked to detect the odd one out. Operators achieving a satisfactory success rate in these routine tests remain on the assessor list for intake samples.

Once barley is dried and in a silo, it is imperative that the temperature of the grain bulk is regularly measured. The absolute temperature of the grain is not always the critical determinant for control, because it will vary depending on whether the silo has just been filled and to some extent on the external temperature. Most important is that once the silo has reached its desired storage temperature any rise in temperature is recognised and appropriate action taken. A rise in temperature could signal an infestation within the silo that if treated can preserve the safety of the whole bulk. There must be a documented action level set for a specified rise in temperature in a specified time period, For example, a rise of more than 3°C over two weeks. The silo manager should be automatically prompted to check the silo and record the decision to treat by cooling or other means to demonstrate a rapid response to minimise risk.

A plethora of audits – what next?

For maltings managers it is not unusual to have audits every month of every year. This has become a key feature of malting over the past 10 years. Most businesses saw the benefit of the ISO9000 quality systems introduced many years ago and the recent upgrade to ISO9001:2000 which established the need to report key performance indicators (KPI). These KPIs reflect important production parameters such as malt yield, energy usage, production volume and plant utilisation. These clearly focus businesses on cost savings and best practice. However, to continue supplying malt we now need to have our laboratories, storage facilities and haulage fleet operated and audited to such standards as the Trade Assurance Scheme for Combinable Crops (TASCC). Our co-products also receive scrutiny, since materials to be supplied as feed for animals now have dedicated assurance bodies such as the Feed Materials Assurance Scheme (FEMAS). The aims of food safety audit schemes are very similar, being based on a rigorous HACCP system with appropriate verification of the systems and validation that our controls really are appropriate and effective in protecting food safety. It would be desirable to combine some if not all of these audits into one, because operationally the standards are in practice the same. Indeed it is much easier to operate the entire malting plant and supply chain to high food safety standards than to have areas of differing standards.

As a collaborative group, UK maltsters have successfully devised a standard that addresses many of the areas of the individual audits described above, and have collected them into one single standard – Assured UK Malt (AUKM) – that is externally accredited (Murrell, 2003; MAGB, 2004). The scheme is comprehensive, but has five main features:

- Implementation of the existing MAGB HACCP code
- Grain sourced via an auditable assurance scheme
- Excellent supply chain traceability and product recall
- Effective quality management system
- Effective control of factory environment, standards, product, process and personnel.

The AUKM standard demonstrates the surge of interest and drive for food safety that is such a prominent feature of today's malting business. Setting high standards not only demonstrates our determination to continually improve food safety, but also protects the upward-chain brands of our customers and emphasises to the downward-chain suppliers of grain the importance of food safety from 'farm to fork' (see, e.g., European Food Standards Agency (EFSA), www.europa.eu.int/comm/food/intro_en.htm).

Malt analysis

Malt analysis has been well established for many years and new methods are not easily incorporated into a traditional specification. Table 4.5 gives a description of some key analytical parameters and what they indicate for the brewer. For most international contracts the specification is based on the EBC Congress method of wort production. However, because brewing is very traditional there is still a requirement to report by other methods such as the Institute of Brewing (IoB) or American Society of Brewing Chemists (ASBC). These methods are gradually merging into accepted international methods and over the next few years it is likely that there will be just one agreed set of recommended analyses. It can be confusing to those setting and reading specifications when different standards are used, since there can be considerable differences essentially due to enzyme activity during each mashing protocol.

New methods are regularly described in brewing journals whereby a better mimic of production can be achieved in the laboratory. The most prominent of these methods relate to runoff or filtration to augment viscosity and β-glucan measurements, and suitability of malt for milling by imaging internal structure to improve indices of friability or homogeneity. The slow take-up of new methods is likely because they are not necessarily improvements over existing ones and in themselves have other inadequacies. Also, if a method is not recommended by one of the major bodies, it cannot easily be verified and

Table 4.5 Key malt analytical parameters

Analysis	Brief description of scope	Typical values
Extract IOB[1] Fine grind 0.2 mm Coarse grind 0.7 mm	Mashing at 65°C Quickly inactivates all enzymes except α-amylase	Good 315 L°/kg Poor 290 L°/kg Fine:coarse difference Good <4
Extract EBC[1] Fine grind 0.2 mm Coarse grind 1.0 mm	Mashing at 45°C rising gradually to 70°C Gradually inactivates amylase and allows proteolysis	Good >82% Poor <79% Fine:coarse difference Good <2
Extract Hartong (H_x) (x = mash temperature)	Shows the effect of enzyme deactivation Expressed at Hartong at different temperatures: H_{45}, 45°C: α- and β-amylases and proteases active H_{55}, 55°C: α- and β-amylases active H_{65}, 65°C: α-amylase active	 H_{45} = 35–40 Lower values indicate undermodification Higher values indicate overmodification
Cold water extract[2]	Ammonia is added to deactivate malt enzymes to indicate base level of modification present	18–22% Good malt <18% undermodified; >22% overmodified
Viscosity	Due to β-glucan in cell walls; also small starch granules that only partially gelatinise in mashing	<1.60 cP (mPa.s)
Friability/homogeneity	Malt ground in a Friabilimeter. Measures the evenness of modification	Friability >80% Homogeneity >95%
 Dextrinising units (DU) Diastatic power (DP) Saccharification time	Measures amylase (diastatic[3]) activity in malt: DU is a measure of just α-amylase DP measures both α- and β-amylase and limit dextrinase Saccharification measures starch breakdown due to amylases	e.g. for lager malt DU >40 DP[5] >60° IoB = DPWK >215 Saccharification time <10 minutes

Colour	Wort can be measured visually using colour comparator discs or using a spectrophotometer set at 430 nm Varies significantly depending on mashing protocol and measurement type Can measure the red/green/blue proportions in a wort by tristimulus using either a spectrophotometer or a custom-made device, but not generally specified for malt	e.g. for ale malt EBC 3–4 IoB 2.5–3.5 ASBC 1.6–2.0
Fermentability	Reflects the suitability of the wort for yeast to ferment. Endogenous enzymes are inactivated first before yeast is added and fermentation started	>80% (different yeasts give different fermentabilities for the same wort)
Nitrogen/protein (= N × 6.25) Total nitrogen (TN) Total soluble nitrogen (TSN) Free amino nitrogen (FAN) Soluble nitrogen ratio/ Kolbach	Important for beer foams, to determine protein breakdown for good yeast nutrition (fermentation), related to enzyme levels since they are proteins. Higher nitrogen barleys are more difficult to malt, are good for foam, but can contribute to beer haze TSN shows the amount of protein modification in malting FAN indicates the gross level of amino acids available for yeast nutrition	e.g. for lager malt TN <1.6% TSN 0.6–0.7% FAN >140 mg/litre Kolbach[4] 38–44

[1] Laboratory mashes are more liquid than in a brewery and will therefore inactivate more enzymes than in a commercial mash. Highly coloured malts of known analysis need to have a pale malt added to supply sufficient enzyme activity to make an extract. The analysis of the coloured malt is determined by difference from the pale malt.
[2] Hot water extracts create extracts of 78–82%. Cold water extracts create extracts of 15–28%. Therefore enzyme action in mashing generates more than 50% greater extract.
[3] Diastatic activity is the ability to break down starch. Starch is a branched polymer of glucose and has different types of linkage: α-amylase attacks α-1,4 links within the polymer (endolytic); β-amylase attacks from the ends of the polymer (exolytic) to create maltose; the enzyme limit dextrinase attacks the α-1,6 linkages and thus further assists the actions of the amylases.
[4] Soluble:total nitrogen ratio measured on EBC wort. Where IoB wort method used it is termed SNR.
[5] DP can be expressed in three ways: IoB, Windisch-Kolbach (WK), Lintner (L). The values can be converted using the following formulae. DP°IoB = DP°L ÷ 1.1; DP°IoB = (DP WK + 16) : 3.85; DP WK = (DP°IoB × 3.85) − 16.

Table 4.6 Typical malt specifications

Analysis[1]	Lager malt	Ale malt
Moisture	4.5% max.	3.5% max.
Fine extract	80% min.	82% min. (>312 L°/kg)
Fine:coarse extract difference	3–4%	1–3%
Colour	3–4°EBC	5–7°EBC
Total nitrogen	1.65% max.	1.60% max.
Total soluble nitrogen	0.63–0.73	0.64–0.72
Soluble nitrogen ratio (Kolbach for lager with EBC mash)	38–44	40–45
Diastatic power	>200 WK >60°IoB	>150 WK >45°IoB
Friability	>80%	>85%
Homogeneity	>95%	>95%
Viscosity	<1.6	<1.55
β-Glucan	<200 ppm	<200 ppm
DMSP (dimethyl sulphide precursor)	5 ppm	<1 ppm

[1] Analytical results can differ significantly for analytes measured in wort because of different mashing protocols. EBC (Congress) mashes allow for more proteolysis and amylolytic activity than IoB, thus specifications will be proportionately higher.

therefore is of little use in setting a standard for malt purchase. For the foreseeable future malt specifications seem set to retain the traditional analyses. Table 4.6 lists some of the most common items on a malt specification for lager and ale malts. There are a host of company-specific specifications, such as higher sulphur precursor for some lagers and food safety checks such as pesticide residues, mycotoxins, NDMA and heavy metals, lipoxygenase activity (linked to beer staling), and for distilling malts analyses such as predicted spirit yield (PSY) and measurable cyanide (MC) or glycosidic nitrile (GN).

Food safety has also generated significant requirements for a range of other items to be certified. For roasted speciality malts interest has been generated in compounds formed during heating, such as 3-monochloropropane 1,2-diol (3MCPD), 1,3-dichloropropanol (1,3-DCP), acrylamide, polycyclic aromatic hydrocarbons (PAH) and furans. Whilst all these chemicals can be found in roasted malts, they are not considered a hazard to human health because of the very small amounts of malt that are incorporated into foods and beverages. It is also quite common to certify that the malt is sourced from a non-genetically modified source and to list its potential allergenicity status. The most common allergic possibility is for those sensitive to gluten when the hordein in barley can have a similar effect.

4.3.3 Environmental management

Malting plants are considered to have a relatively low impact on the environment in terms of the waste materials that go to land, air or watercourses. The major effluent source is of course water from the steeping process, with many sites having their own dedicated wastewater treatment plant. For plants producing the usual white malt for ales and lagers, unless situated adjacent to domestic dwellings, there is little impact from noise and dust. Where a maltings produces peated malt or roast (coloured/speciality) malt, the odour emitted is naturally different and can create a nuisance if not managed by phasing operations at the correct time of day or by fitting odour reduction mechanisms to the exhaust emission flues.

As a whole the malting industry has very simple and common goals to protect the environment. These include keeping waste and packaging to a minimum, selecting environmentally safe cleaning and process chemicals where possible, and minimising the use of water, energy and the emissions of noise and dust. A prime structure to achieve these aims is an environmental management system such as ISO14001.

When any control system is introduced there is an implied cost to achieve the standard. This is true of environmental standards. On the other hand, companies that introduce environmental control report on average a 10% reduction in waste created, which is a valuable saving to achieve.

Energy

Energy prices are one of the steepest rising costs for the malting industry, with rises of the order of 25% annually over the past few years. Currently energy utilisation is in the range 830–1010 kWh/tonne, comprising gas (thermal) energy of 750–850 kWh/tonne and electricity 80–160 kWh/tonne, representing approximately 20% of the overhead costs of malt production.

As a result of the Kyoto Protocol, countries that have agreed to cooperate are committed to reducing emissions of greenhouse gases, principally carbon dioxide, by around 5.2% below their 1990 levels by 2010. The overall target was set by the UK government at 12.5% to be achieved by 2008–2012. Responsibility in the UK for monitoring progress lies with the National Climate Change Policy Division (NCCP) within the Department for Environment, Food and Rural Affairs (DEFRA).

Two key strategies have been developed in the UK to focus industry on energy saving and reduction in emissions. Firstly a Climate Change Levy (CCL) was introduced that set stringent targets to reduce energy usage per unit of product. Country targets are sub-divided into targets for industry groups where they exist. The MAGB acted for the UK malting industry and agreed a reduction in primary energy usage of 8% in a series of stepped 'milestone' targets by 2010. Primary energy is an important measure, because for electricity primary energy is in fact 2.6 times the energy actually metered on site due to the inefficient nature of electricity generation and transmission. UK maltsters have to achieve the overall reduction as an industry collaborative sector, which allows for flexibility within sites that have varying possibilities for improvement. There is a real incentive to achieving the target on an on-going annual basis, because interim energy usage audits are conducted to determine progress. Failure to meet any of the milestone targets leads to a substantial financial penalty. Achieve the target and an 80% reduction in the levy is applied, a very significant saving of perhaps £300 000 for a 100 000 tonne malting plant over a two-year period in which the levy is either granted or declined, depending on energy targets being met.

A further control across Europe is a scheme designated the EU Emissions Trading Scheme (EU ETS); countries are required by the European Union to develop a National Allocation Plan (NAP) of carbon dioxide emissions from combustion plants, based on plant capacity. Both schemes have introduced the system of trading carbon dioxide allowances. If a company has achieved significant energy savings better than the target, those 'carbon credits' can be sold to another company that has used too much energy. Carbon credits can only be traded within the scheme through which they were generated, i.e. the CCL scheme or the EU ETS scheme, but not across both. Of course to achieve the energy saving targets, very considerable investment in more efficient plant and process control has to be made. The impact of the investment may be offset to some degree by energy savings achieved, but represents a considerable burden in an industry currently trading on a marginal commodity basis.

Water

On average the malting process uses 4–5.5 m^3 water per tonne of malt produced, with approximately 90% of that water being process water. The wastewater to be treated will be 80–85% of the process water entering the steep vessels. Thus for a 100 000 tonne malting plant the volume of water to be treated will approach 400 000 m^3 per annum. The waste is suitable for treatment aerobically on site if a dedicated plant is installed. It requires little pre-treatment because the pH is neutral to slightly acidic (pH 6.8–7.0). Various restrictions are applied to the water quality after treatment. These are referred to as consents and are frequently part of a combined site strategy licensed by the local environmental authority. In the UK Integrated Pollution Prevention and Control (IPPC) consents were introduced from 2005 and look at all environmental impacts on site and in particular, for maltings, at the impact of water release on local watercourses, the major environmental emission source.

Maltings effluent typically would have restrictions in place for Biological Oxygen Demand (BOD), Chemical Oxygen Demand (COD), suspended solids, temperature, ammonia and most recently phosphate. Depending on the position in the steeping cycle, the effluent quality varies, but typical average results going to the treatment plant could be:

COD (Chemical Oxygen Demand)	3000 mg/litre
BOD (Biological Oxygen Demand)	1950 mg/litre
Suspended solids	150 mg/litre

The wastewater from a plant producing malted ingredients such as liquid or dried malt extract and materials produced from dark or roasted cereals is entirely different. This is due to the increased sugar and colour of the washings from plant-cleaning operations. It would not be uncommon to find COD values of 30 000 mg/litre and very high suspended solids. Additionally, process vessels have caustic and acidic cleaning-in-place regimes, which cause a wide variation in pH. It is common to find extremely low pH values of 3.0–4.0, which require significant adjustment before treatment. Therefore balancing tanks are required to smooth out variations in the feed to the treatment plant, in order to avoid damage to living microorganisms in the sludge which are responsible for degrading the effluent constituents that could compete for oxygen in the water and harm the aquatic environment. This rather specialised effluent is more akin to brewery effluent and requires an intermediate anaerobic treatment prior to the aerobic plant. Due to the cleaning-in-place methods routinely adopted using phosphoric acid, the newly proposed consents for phosphate represent a challenge. Phosphate can be effectively removed using ferric chloride, but at a high cost. It is likely that cleaning-in-place systems will be the target of new treatment strategies to offset this additional cost of treatment.

Water reuse
The cost of wastewater treatment is high and consents for discharge will undoubtedly continue to be restricted, therefore reuse of water is on every malting manager's agenda. Maltings may be situated on aquifers and have dedicated boreholes with a plentiful supply of relatively cheap water compared to those on mains water supplies. Whatever the source of water it will still be attractive to consider the possibilities of reuse for either cleaning operations or process. Recycling water for cleaning operations where chlorine-releasing reagents like sodium hypochlorite may be added is a very straightforward operation with no impact on process performance. Since this represents only 10% of an average maltings water use, it is not as attractive as recycling water for process use. However, this is an area where food safety concerns become absolutely paramount. For a maltings with certification such as the BRC Global Food Standard, it is commonplace to carry out risk assessment of the water being used for steeping. Water from boreholes or town water supplies will most often be stored in large tanks prior to use in steeping. Measurements of *Salmonella* and *E.coli* are often made on water in these tanks at regular intervals and chlorination levels are checked daily. Levels of other organic and inorganic ions are often also monitored by a local authority at regular intervals. The water is of potable (drinking quality) standard going into the tanks, but positive confirmation of the water quality is the risk-averse strategy for a food grade process.

There would be real benefit to recycling water treated sufficiently to render it of potable quality suitable for reuse in processing. Trials of various recycling techniques over a number of years have provided water suitable for washing, but not yet of a quality suitable for re-steeping. It has been suggested that germination may be inhibited in recycled water even though it is apparently potable. Information about potential inhibition is scarce and a solution could save the malting industry considerable sums of money. Thus a consortium of British and French maltsters are currently engaged on a project with European and UK funding to track down the 'holy grail' of steep purification. Code-named project SWAN, it is driven by the best available techniques (BAT) principles governing all food processors who target water reuse as highly important. The project aims to develop novel treatment methods for steep water that will render it suitable for malting with no penalty in process yields, and also of a quality that will give confidence to consumers that it is entirely safe for use (*Global Watch*, March 2005).

4.4 Malted ingredients

Malt produced in a conventional malting process can be further processed in a number of ways to make products that are readily incorporated into beverages or foods to create or enhance flavour, or improve processing parameters.

A vast range of malts with higher colours and flavours is available (Table 4.7). Malts that have been conventionally kilned can be further roasted to create a range of colours from 110 through to 1500 EBC colour units (at 10% dilution). If the malt is heated slowly and kept for longer in the wetter part of the kilning process, it allows stewing of the grain and produces sweeter, more caramel flavours and colours in the range 2–400 EBC colour units (Bemment, 1985; Blenkinsop, 1991; Jupp, 1994; Gretenhart, 1997). The manner in which these different malts are blended together in a grist is very much a part of the brewer's art. The coloured malts offer the possibility of simply adjusting or enhancing colour by using a relatively small amount in the grist. Small proportions of crystal malt in the grist can enhance other malty flavours or create a more rounded mouthfeel. Ales incorporate crystal and cara malts. Porters utilise darker crystal and chocolate malts. Stouts use the very dark chocolate and black malts and roasted barley. The possibilities are almost endless.

For brewing applications malt extract is the most commonly used malted ingredient, generally as a brew extender. Extract production involves mashing and wort separation in the same manner as at the start of the brewing process, but the final stage evaporates water to create a viscous extract at 80% solids. Due to the viscous nature of malt extract, it is only sparingly soluble below 20ºC and is best handled above 40ºC when it is easy to pour or pump. Malt extract cannot, however, be stored at elevated temperature because this accelerates Maillard reactions with consequent colour increase and generation of carbon dioxide (Nursten, 2005). Extract is best added into the kettle at the end of the boil before cooling where it is easily dissolved and provides a cost-effective solution where increasing the size of a mash vessel is not possible, or where there are sudden peaks in product demand. Malt extract is also rich in carbohydrate and can be used in a similar manner to sugar syrups to balance out a wort that predominates in nitrogenous compounds to give a better profile for yeast growth.

The raw material for extract production can be barley rather than malt. A source of amylolytic enzymes can be provided either by a mixture of regular malt at up to 20% addition, or by the use of commercially available enzymes.

Table 4.7 Malt flavours and colours produced from varying kiln schedules

To make normal low colour malt, the malt is dried in two phases. In the first (drying) phase, moisture is removed from the outside of the malt at low temperatures. A second (curing) phase to remove moisture within the grain requires higher temperatures. The moist malt direct from germination is termed green (uncooked) malt. If temperatures are raised during the initial drying phases, the internal sugars and amino acids of the malt can 'stew' and create colour and flavours that are essentially sweet. Once the malt is dried it is termed white malt. It can be used in a further roasting process at varying high temperatures to create very high colour and bitter smoky flavours.

Starting material	Malt type	Key flavours	Colour (°EBC)	Kiln schedule temperature (°C)	
				Initial drying phase	Final curing phase
Green malt	White	Cereal, green (uncooked), sulphury	2–3	65	80
	Lager (pils)	Cereal, sulphury	2.5–4	65	85
	Pale ale	Biscuit, toast	5–7	65	90
	Mild	Nutty, caramel	6–8	70	95
	Vienna	Nutty, toffee	5–10	75	110
	Munich	Nutty, biscuity	10–20	75	120
	Caramalt	Caramel, sweet	25–60	50	100
	Crystal	Sweet fruit, toffee	100–200	50	130
		Burnt fruit, caramel, treacle	200–400	50	160
	Brown	Biscuity	90–150	130	130
	Peated	Phenolic[1]	2–4	65	80
White malt	Amber	Biscuity, baked, nutty, dry	40–100	N/A	160
	Chocolate	Dark chocolate, smoky, burnt, coffee, bitter, caramel	1000–1300	N/A	80 → 150 → 230
	Black	Smoky, bitter, coffee, acrid	1100–1500	N/A	80 → 230

[1] Peated malts have differing levels of phenols that give flavour, for example <10 ppm (low), 10–20 ppm (medium), >20 ppm (high).

The sugar profile and final flavour are essentially similar to those of malt extract produced using white malt, but in declarations on food products the material becomes barley rather than malt. Barley syrup finds applications in the soft drinks industry and in recent years in Japan in the production of 'happoshu' beer. In happoshu beer the malt ratio is less than 67% and, when introduced, the tax was 40% less than that of beer (Malone, 2001). The reduction in the amount of malt is achieved either by using more barley in the grist and adding food grade enzymes, or by diluting a conventional brew line with barley syrup in the kettle prior to fermentation.

A considerable range of colours and flavours is achievable in malt extract production depending on the composition of the grist, therefore introduction of coloured malts can create a range of worts that match those of any brewhouse. The main advantage in extract use lies in the concentration of the extract due to evaporation. Relatively small volumes of extract can be added to create beer with different colours or flavours. This greatly liberates the possibilities for new product development (NPD) because a range of quite different beers can be produced from the same wort without changing the grist. In NPD trials it is possible to add very small quantities of extract to the final beer to judge the impact of various malt extracts. In the author's experience the most dramatic change has been to create a sweet porter from a lager brew line by addition of just two malt extracts: 2% (w/v) crystal malt extract plus 1% (w/v) chocolate malt extract. The crystal malt extract gave a sweet fruity taste and smooth mouthfeel, while the chocolate extract provided colour and a degree of bitterness.

4.5 Sensory evaluation of malt and malted ingredients

In the pursuit of quality and new brand development, the impact of raw materials used in brewing is becoming more important. Some companies are already well known for having implemented a system of tasting all raw materials used in brewing, from liquor to kieselguhr. A rigorous system for malt and malt products flavour profiling is available, though it is not widely used at present apart from as a tool in new product development.

4.5.1 Malt

Currently malt tasting is not part of any malt specification, but offers possibilities to differentiate both positive and negative flavours. A malt analysis can indicate that malts are identical, yet the processing parameters could be different and the flavour profiles slightly different (Chandra *et al.*, 1997). It is thus important to have a sufficiently discriminatory method to flavour profile malts that reflects the range of malt flavours available. Some customers require tasting of worts for the correct flavour profile. This doesn't reflect contribution of the husk and the sweetness of worts predominates, making full profiling impractical. A method that fully describes the profile of malt has been developed

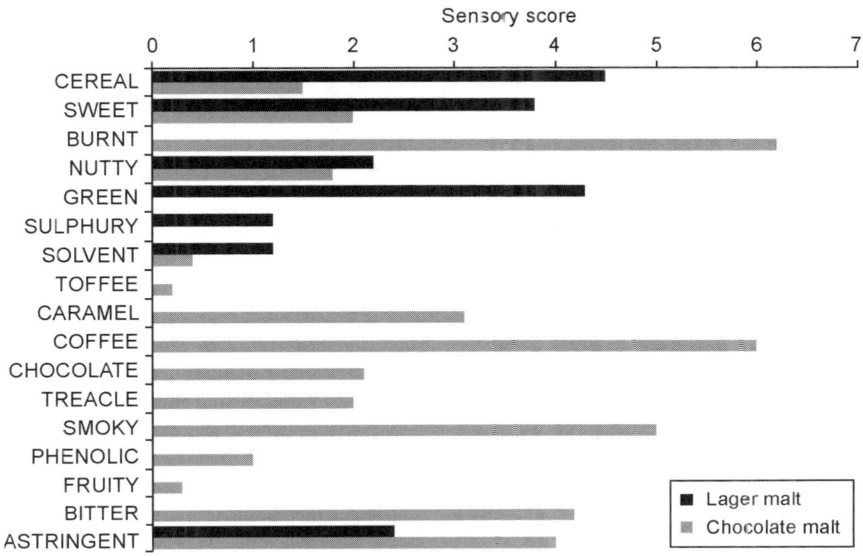

Fig. 4.3 Malt sensory analysis – example of good differentiation in profiles. A full range of sensory attributes for malt can be achieved by coarsely grinding malt and mixing with water to form a paste or porridge. Taste profiles can then be created with clear distinctions between malt types. The figure shows two malts at the extremes of colour: lager and chocolate. The lager malt has distinct cereal, sweet, nutty and green (uncooked) notes. The chocolate malt has pronounced burnt, coffee, smoky and bitter components.

that creates a 'porridge' by grinding the whole malt and mixing with a small amount of water. This avoids the problems for the taster of drying out the palate if tasting whole malt or dry ground samples. When given free rein to describe malt flavours, tasters tend to use food products as descriptors. It has proved possible to group these into a smaller number of sensory terms to create a unique profile for the entire range of malts (Murray *et al.*, 1999). An example of the clear difference in profile achieved with this method is given in Fig. 4.3. The flavour descriptors are fully described in the papers referenced above.

Malt sensory analysis can also assist in detecting flavour problems, as described in the following brief case histories that the author has been involved in as an expert witness.

- Case history 1: A brewery purchased malts from four maltsters that were all in specification by conventional analysis. Sensory profiling of the beer indicated a flavour off-note in beer brewed with one of the malts. It proved possible to identify a very different profile and particularly an increased solvent/wet note in this malt (Fig. 4.4). Subsequent investigation of that maltings indicated poor attention to hygiene
- Case history 2: An unusual oily note was detected in beer produced using malt that met the specification. Tasting the malt used, it was immediately apparent

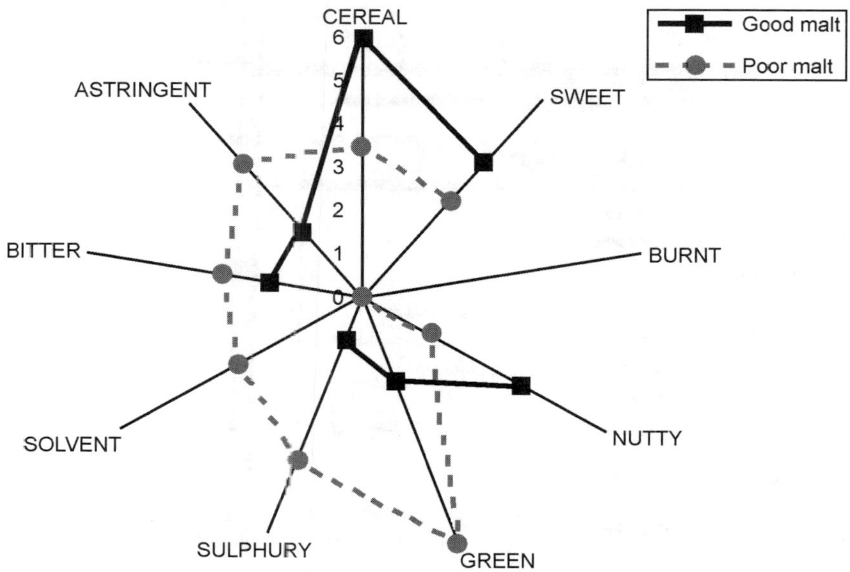

Fig. 4.4 Difference in profile of good and poor malt by sensory analysis. Taste profiles from two malts that were within specification by conventional analysis. The brewer found that one of the malts created an unusual flavour in the beer. By sensory profiling, the undesirable flavours were associated with a very different malt profile, which had distinct green (uncooked), solvent notes and reduced positive malty flavours. Although it is not yet possible to link specific processing methods to malt off-flavours, ultimately the maltings was found to have poor hygiene standards and would have been rejected as a supplier on that basis as well as by sensory analysis.

that there was a note variously described as goaty, oily or diesel. How could diesel notes have got into the malt? After investigation the maltster found that on the day the malt was produced there was a queue of trucks on site outside the kiln air intake. It was resolved that these fumes had tainted the husk of the malt. The solution was to ban trucks from queuing in this area.

Neither case history could have been resolved by conventional analysis. Thus malt sensory analysis can be a useful tool in pinpointing problems not easily resolved by conventional analytical methods.

4.5.2 Malted ingredients

Flavour profiling for malted ingredients utilises exactly the same profile terms as malt. The only difference is in the preparation. It is quite possible to taste malted ingredients, apart from flour, without preparation because they are all readily soluble in the mouth. If an extended tasting session is planned, it can be useful to dilute the malt extracts so that there is no build-up of sweetness on the palate.

Sensory profiling has also proved a useful tool in determining which speciality malt is most appropriate rather than relying on the conventional

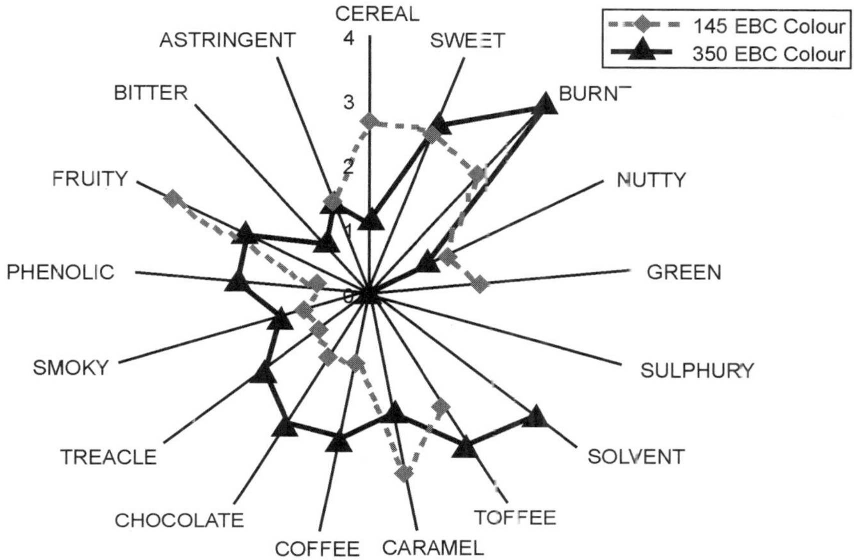

Fig. 4.5 Flavour profiles of different coloured crystal malts.
Sensory analysis of different malts designated as crystal malt. The flavour profiles indicate that the different colour of these malts is associated with very different flavour. The lower colour crystal has a sweet, fruity caramel profile, whereas the higher colour crystal has a burnt fruity and chocolate/treacle profile. These profiles show why it is important to carefully consider the required specification for crystal malt, not just the colour.

analysis of colour. This is particularly important in the rather broad description of crystal malt. This category has a colour range of 145–400 EBC colour units (10% dilution), but flavour characteristics are extremely different (Fig. 4.5). This is particularly important when specifying and using crystal malt. The lower colour crystal malts have a very sweet fruitiness, whereas the higher colour crystal malts have a burnt fruit note and some bitterness. It is therefore possible to detect the blending range of malts that have been used to create the crystal malt. For conventional colour analysis it is acceptable to blend colours from either end of the crystal specification to achieve an intermediate. If, however, a specific flavour is required, there is a very marked difference between higher and lower colour crystal malts and the flavour difference is simple to detect if the colours, and hence flavours blended, are too wide apart.

4.6 Future trends

The current climate of improved traceability, food hygiene and environmental concern will undoubtedly predominate over the next five years. Of these it is most likely that legislation on traceability will strengthen. It is acknowledged that where a number of deliveries of grain are bulked into a silo, traceability of individual deliveries is partially lost. However, since the whole silo becomes a unique batch, traceability is maintained, although potentially requiring a large bulk of grain to be written off in a real safety scenario. Therefore it is likely that more attention will be given to silo size to minimise this business risk.

Control of grain food safety in silo is also set to become higher profile. At present management of silos is achieved by monitoring and controlling temperature, by drying and cooling grain to keep it in a safe zone to prevent fungal growth, and by taking samples to validate that these controls are effective. A potential method is being developed whereby the Available Water (or water activity, A_w) content of the grain bulk can be monitored to give an earlier indication of contamination. Insect activity and mould activity are strongly related to moisture present inside and outside the grain, and moisture increase can be detected before the temperature rises, provided a probe is sufficiently close to detect the change. The absolute water content of grain is less important than the available water in the air around the grains. This is best measured by equilibrium relative humidity (ERH) or water activity (A_w). ERH and A_w are effectively the same, For example, if ERH $= 70\%$, $A_w = 0.70$ (Caddick, CSIRO website). Moulds cannot develop when A_w is below 0.68, but there are specific A_w limits for different fungi (WHO Technical Report, 2001). A number of companies and research institutes are developing software that links the recording of silo temperature and moisture together with an expert database of potential pests and treatments made to give further controls to the store manager and evaluate the overall risk from all potential storage pests.

The effect of environmental legislation will certainly require malting technology to use ever more energy-efficient equipment and to minimise the production of effluent. Reuse of steep water to reduce water abstraction will require the supply chain to be confident of the safety of the water but could provide a new marketing opportunity to herald environmentally friendly beverages. Novel breeding technology such as genomics may well provide new varieties of barley that require less energy in production and allow the maltster to create a wider range of specifications with less energy input: for example, production of ale malts or low lipoxygenase malts currently requires higher temperatures. New varieties could also be higher yielding, be easier to modify and require less water for satisfactory germination. All these are beneficial to the maltster, but ultimately new varieties will be dominated by the requirements for problem-free brewing performance.

Whether flavour specifications for malt will become the vogue is not clear. Certainly sensory analysis gives added insight in troubleshooting, can improve raw materials purchasing for maltster and brewer and channel new product development. One of its main advantages is the ability of the human palate to assess the interaction of the many chemical components in one operation. This requirement highlights the practicality of developing better methods of analysis. Novel methods are published every year, but in general find little favour with those who have grown up on the flawed but well-peddled malt analyses. Harmonisation of the various methods of analysis would certainly be prudent, whilst new methods will still be benchmarked against existing recommended methods and the ultimate analysis of brewing to test the vagaries of individual plants.

The possibilities for new product development using malted ingredients, and specifically malt extract, are exciting. It has to be attractive to brewers to be able to test a market with a new flavour profile without changing their basic buying portfolio for malt substantially until the market is proven. It is also a very easy way to support niche markets for specific products without the need to invest substantial amounts of capital in new plant.

4.7 Sources of further information

4.7.1 General information on barley, malt, malted ingredients and food safety

Maltsters Association of Great Britain: an excellent source of information on malting practice, food safety and due diligence, HACCP: www.ukmalt.org

Muntons plc website: information on malt and malted ingredients production, specification and technical: www.muntons.com

Home-Grown Cereals Authority: UK-based levy-funded body website with vast array of information on cereals: www.hgca.com

British Beer and Pub Association: represents the interests of the UK's vibrant beer and pub sectors – general brewing and legislative interest: www.beerandpub.com

UK Food Standards Agency: legislation and advice on food safety: www.food.gov.uk

4.7.2 Cereal assurance schemes

The Assured Combinable Crops Scheme (ACCS): www.assuredcrops.co.uk/accs2/

Scottish Quality Cereals: www.sqcereals.co.uk/

Farm Assured British Beef and Lamb (FABBL): assures beef, lamb and combinable crops: www.fabbl.co.uk/

Agricultural Industries Confederation: general trade assurance scheme information: www.agindustries.org.uk/issues/tradeassurance/default.asp

4.7.3 Analytical research centres for malted cereals and foods

Brewing Research International: www.brewingresearch.co.uk

Campden and Chorleywood Food Research Association: www.campden.co.uk

Scottish Crop Research Institute: www.scri.sari.ac.uk/SCRI/Web/Site/home/ResearchAreas

4.8 References

BAXTER, E.D. (2004), Review of food safety issues relating to the supply and market acceptability of UK malting barley and UK malt, HGCA project report 2804, www.hgca.com/document.aspx?fn=load&media_id=1706&publicationId=1537

BBPA/BRI (2005), The agrochemicals approved for use on malting barley 2005 (via MAGB website), www.ukmalt.com/downloads/Tech396Agrochem.pdf

BEMMENT, D.W. (1985), Speciality malts, *The Brewer*, December, 457–460.

BHATNAGAR, D., PAYNE, G.A., CLEVELAND, T.E. and ROBENS, J.F. (2003), Mycotoxins – current issues in the USA, *Second World Mycotoxin Forum*.

BLENKINSOP, P. (1991), The manufacture, characteristics and uses of speciality malts, *Tech. Q. Master Brew. Assoc. Am.*, 28(4), 145–149.

BRAKS, P. (2005), Development in the malting industry. Report by Braks, P. and Leijh, T. for Rabobank presented at the World Barley, Malt and Beer Conference. See also www.rabobank.com/far

BRIGGS, D.E. (1998), *Malts and Malting*, Blackie Academic and Professional.

BRIGGS, D.E., HOUGH, J.S., STEVENS, R. and YOUNG T.W. (1995), *Malting and Brewing Science* (2nd edn), *Volume 1 Malt and Sweet Wort*, Chapman & Hall.

BS6001-1:1999, Sampling procedures for inspection by attributes. Sampling schemes indexed by acceptance quality limits (AQL) for lot-by-lot inspection.

BS6002-4.1:1996, Sampling procedures for inspection by variables. Specification for sequential sampling plans for percentage nonconforming. Known standard deviation.

CADDICK, L., Water activity and equilibrium relative humidity, CSIRO website http://sgrl.csiro.au/storage/moisture/erh.html

CHANDRA, G.S., BOOER, C.D., DAVIES, N.L. and PROUDLOVE, M.O. (1997), Effect of air-recirculation during kilning on malt and beer flavour, *Proc. Conv. Inst. Brew. (Cent. & South African Sect.), Durban*, 6, 23–28.

CLOSE, T.J. *et al.* (2004), A new resource for cereal genomics: 22K barley GeneChip comes of age, *Plant Physiol.*, 134, 960–968.

DENNIS, E.S. (2001), Genomics and its relevance to barley, in *Proceedings of the 10th Australian Barley Technical Symposium*.

EC REGULATION 178/2002, Laying down the general principles and requirements of food law, establishing the European Food Safety Authority and laying down procedures in matters of food safety, Article 18 Traceability. http://europa.eu.int/eur-lex/pri/en/oj/dat/2002/l_031/l_03120020201en00010024.pdf

EDWARDS, S. (2005), Investigation of fusarium mycotoxins in UK barley and oat production www.hgca.com/document.aspx?fn=load&media_id=1803&publicationId=2028

EGLINGTON, J.K. *et al.* (1999), Application of β amylase thermostability analysis in a barley improvement program, in *Proceedings of the 9th Australian Barley Technical Symposium*.

FARMERS FUND (2004), Evaluating the potential markets for UK cereal varieties with specific end user traits. Report to Farmers Fund via Institute of Grocery Distribution (IGD): www.igd.com

FETTEL, M.N. *et al.* (1999), Determination of grain size in malting barley, in *Proceedings of the 9th Australian Barley Technical Symposium.*

GLOBAL WATCH (2005), Technology is grist to a perfect pint, *Global Watch* (Dti magazine), March, 4–5, UK Department of Trade and Industry or online at: www.globalwatchservice.com/magazines/globalwatch/pdfs/Issues/GWIssue_46.pdf

GOLAN, E., KRISSOFF, B., KUCHLER, F., NELSON, K., PRICE, G. and CALVIN, L. (2003), Traceability in the US food supply: dead end or superhighway?, *Choices*, 2nd quarter.

GRETENHART, K.E. (1997), Speciality malts, *Tech. Q. Master Brew. Assoc. Am.*, 34(2), 102–106.

HGCA: THE GRAIN STORAGE GUIDE, VERSION 2 (2003), www.hgca.com/document.aspx?fn=load&media_id=408 (Chapter 10: Fungi), www.hgca.com/document.aspx?fn=load&media_id=955&publicationId=820 (full publication).

JUPP, D.H. (1994), Coloured malts, *The Brewer*, November, 80(961), 466–469.

MAGB (2004), Assured UK Malt Standard website: www.assuredukmalt.com/

MAGB (2005), Checking for due diligence on mycotoxins (2005 update): www.ukmalt.com/foodsafety/UKM/UKMmycotoxin.html

MAGB (2006), The MAGB HACCP Protocol for Malting, version 3.0, www.ukmalt.com/downloads/MAGBHACCPprotocol3.pdf

MALONE, R. (2001), Japanese brewing: the 'Ji-buru' phenomenon, *Brew. Guardian*, 130(9), September, 24–26, 28–30.

MURRAY, J.P., BENNETT, S.J.E., CHANDRA, G.S., DAVIES, N.L. and PICKLES, J.L. (1999), Sensory analysis of malt, *Tech. Q. Master Brew. Assoc. Am.*, 36(1), 15–19.

MURRELL, I.R. (2003), UK maltsters assurance to its customers is a world first, *Brewer International*, 3(11), November, 47–48.

NATIONAL CENTER FOR BIOTECHNOLOGY INFORMATION (NCBI) (June 2005), Expressed Sequence Tags www.ncbi.nlm.nih.gov/dbEST/dbEST_summary.html

NURSTEN, H.E. (2005), *The Maillard Reaction: Chemistry, Biochemistry and Implications*, Royal Society of Chemistry Publishing.

PARIS, M., POTTER, R., LANCE, R. and JONES, M. (2001), Development and implementation of PCR markers in Western Australia's Barley Breeding Program, in *Proceedings of the 10th Australian Barley Technical Symposium*.

SYNGENTA (2004), Hybrid barley Colossus, www.syngenta-crop.co.uk/NR/exeres/DC48FE43-F848-465F-A450-FA788480785E,frameless.htm

VAN EGMOND, H.P. and JONKER, M. (2004), Worldwide and European regulations for mycotoxins and analytical methods, http://mycoglobe.ispa.cnr.it/Brussel/vanEgmond.pdf

WHO TECHNICAL REPORT SERIES 906 (2001), Evaluation of certain mycotoxins on foods, 56th report of the Joint FAO/WHO Expert Committee on Food Additives, http://whqlibdoc.who.int/trs/WHO_TRS_906.pdf

Note: The opinions expressed in this article are those of the author and do not necessarily reflect those held by Muntons Plc.

5

The breeding of hop

J. Henning, Oregon State University, USA

5.1 Introduction

Hop breeding is in the infancy stage in sophistication and application of classical and molecular techniques when compared to other cultivated crops. Much of this state is attributed to the minor crop status of this crop coupled with the specialized nature of use for flavoring and the need to maintain end-product consistency from year to year. Nevertheless, given the small number of hop breeding programs around the world, much can be said regarding the successful adaptation of new breeding techniques, particularly new molecular breeding techniques that are being applied to hop. As will become obvious, most of the early breeding efforts in hop used mass selection followed later by the use of simple pedigree breeding focused upon the use of a few select cultivars as parents. Hop production and breeding requires large plot sizes for individual genotypes and this fact alone has precluded breeders from observing large numbers of individuals or offspring for selection purposes. It has only been within the past 15 years that hop breeders have expanded beyond simple pedigree breeding technique to include recurrent selection or use of multiple designed crosses incorporating several traits into single genetic backgrounds.

5.1.1 Early hop breeding – mass selection

Beer brewing has been documented in Babylon from about 7000 BC. Nevertheless, hops were not used for brewing purposes until thousands of years later in Europe. Much controversy exists regarding the first use of hops in beer (Barth *et al.*, 1994; DeLyser and Kasper, 1994). Because the center of origin for hop is proposed as central China (Neve, 1991), it is theorized that migrations from central Mongolia brought wild hops west to central Europe and east to the North American continent. Concrete evidence exists that hop was first cultivated in 768 AD in the Halletauer region of Germany (Linke and Rebl, 1950; Neve, 1991; Moir, 2000) and then specifically mentioned for use in beer brewing in 822 AD (DeLyser and Kasper, 1994). Neve (1991) suggested that in Bohemia, Slovenia and Bavaria, hop cultivation was first established sometime during the seventh century to the ninth century AD and was firmly established by the end of the ninth century.

Early attempts at developing hop varieties using mass selection date from prior to the fifteenth century with the use of feral hop lines collected and replanted in cultivated yards. These early attempts almost certainly involved cultivation of mixed populations from which superior genotypes were kept and inferior individuals were culled or by selecting fortuitous seedlings (Fore and Sather, 1947; Neve, 1991; Moir, 2000). Interestingly, the famous English variety 'Fuggle' is thought to have arisen by means of a chance seed thrown out with the refuse of a hop-picker's lunch at Horsmonden in Kent and later cultivated and introduced as a commercial variety by Mr Fuggle in 1875 (Moir, 2000). Other examples of early mass selection involved the isolation of specific individuals in local regions and the asexual reproduction of these select individuals to such an extent that all growers in a specific region eventually came to grow the same genotype. This is how the cultivars 'Saazer', 'Hallertauer', 'Spalter' and 'Hersbrucker' arose – eventually acquiring the name of the locality from which they were derived as the cultivar name. Moir (2000) and Neve (1991) proposed that selection was initially based upon prolific and hardy growth in the particular soil and environment from which these varieties were grown. Neve (1991) went on to suggest that secondary selection was based upon suitability for brewing and that the hop line in greatest demand became the predominant variety for the region.

5.1.2 Early attempts at hybridization

Much of the early history of hop breeding has been adequately covered by Neve (1991). Nevertheless, a short summary of highlights is in order. It wasn't until the early 1900s that focused efforts utilizing hybridization followed by selection were performed for the development of new hop varieties (Fore and Sather, 1947; Neve, 1991; Moir, 2000). Prior to that point, the only variety that could be claimed as a genetic improvement over wild hops was the American 'Cluster' line. Neve (1991) stated that 'Cluster' was the result of hybridization between a European cultivar, brought over with the Massachusetts Company in 1629, and a wild American hop line.

Breeding programs utilizing hybridization techniques were initiated in Germany by Stambach in 1894 and Remy in 1898 and also in the USA by Fairchild in 1904 (Neve, 1991). None of these three programs survived for long and it wasn't until E.S. Salmon began a hop breeding program that success was obtained. It is generally believed that Salmon pioneered hop breeding, and the program he started at Wye College, England, in 1906 was the first successful program in the world (Moir, 2000). Salmon was a plant pathologist with specific interest in powdery mildew (*Podosphaera* spp.) and it has been suggested that his interest in hop powdery mildew *Podosphaera macularis* (Braun), formerly called *Sphaerotheca humuli*, caused Salmon to initiate hop breeding work (Neve, 1991). With the identification of downy mildew *Pseudoperonaspora humuli* (Miy. & Tak.) in 1905 (Miyabe and Takahashi, 1906), its rapid spread to the USA in 1909 (Royle and Kremheller, 1981) and its eventual spread to England and Europe in the early 1920s (Salmon and Ware, 1925; Lang, 1925), successful breeding programs were initiated in Germany in 1926 and the United States in 1930.

5.2 Developments in hop breeding

In almost all cases, breeding programs utilized simple phenotypic selection protocols to identify superior offspring and these individuals were released as varieties. Many early variety releases from the first part of the twentieth century, such as 'Brewers Gold' and 'Bullion', were the result of open pollinated offspring selected for superior alpha levels and high yields. Early attempts at breeding by Wye College involved the development of varieties with resistance to *Verticillium* wilt, resulting in open pollinated seedlings from a wild North American hop var. *neomexicanus* (Salmon, 1949). It wasn't until the 1930s and later that specific recurrent breeding programs were initiated by Zattler at the Huell in Germany and by Keyworth located at Wye. Zattler focused on the development of lines that were resistant to downy mildew while Keyworth focused on wilt resistance. The objective of US hop breeding was developing high yielding aroma lines that performed well under Pacific Northwest climatic conditions. Readers are directed to Neve (1991) for an excellent and extensive review of hop breeding for pest resistance.

5.2.1 Procedures used in hybridization of hops

As previously mentioned, one of the simplest procedures for developing hop varieties via hybridization involves open pollination of known female cultivars. Early hop breeding programs typically grew the desired female cultivar surrounded by unimproved male hop pollinators. After pollination and seed set, seeds were collected and planted in the field for phenotypic selection. Superior offspring were selected, cloned and subsequently grown in larger-scale plots. If a selection proved successful, it was propagated in commercial-scale plots and harvested for pilot brewing tests. If the experimental line proved acceptable in brewing trials the line was released as a cultivar. It is reasonable to assume that brewers have a vested interest in maintaining a consistent flavor in their beer and that flavor is highly influenced by the type and amounts of hop used during brewing. As a consequence, brewers have insisted upon using the cultivars that provide that consistent flavor and have been reluctant to switch hop cultivars.

The use of open pollination breeding schemes allowed control of only one-half the genome of the resulting offspring because contributions from males were not controlled. It became obvious to early breeders that undesirable traits were being transferred from unimproved male lines and that this process was limiting or slowing the success of new cultivars in brewery acceptance. Soon, breeders at all three locations began utilizing controlled crosses using pollen from specific males, some of which were seedlings from open pollination on known female cultivars. This process was the beginning of what is called 'pedigree breeding' and is quite analogous to methods used in animal breeding.

5.2.2 Classical breeding techniques in hop cultivar development

Further progress in breeding technique was the use of male test-crosses to help determine male breeding value. Breeding value is a term used to describe the potential genetic contributions that a particular male or female parent possesses. Much, if not most, of the early estimations of breeding value for males was performed on a trial and error basis, with a cumulative estimate of overall value arrived at via mental reasoning rather than statistical experimentation. As an example, certain male lines were noted for producing female offspring that were higher than average in alpha acid or had better storage ability than other crosses made with different males. Breeders took note of these breeding values for specific males and used these males for specific breeding goals. Fore and Sather (1947) noted that certain phenotypic traits in the male plant were useful for selection of potential male parents, including general plant vigor, insect and disease resistance, leaf and stem color (an aesthetic property rather than one affecting production), length and arrangement of flowers along primary branches, flower cluster arrangement, date when flowering initiates and length of 'pollen-shedding period'. Fore and Sather (1947) discussed this last characteristic, the length of pollen-shedding period, as a highly desirable trait in male phenotype. They reasoned that male plants with extended pollen shedding periods, in some cases up to 52 days, were the most vigorous plants. They also reasoned that male plants with a long pollen-shedding capability would allow for crosses onto female plants that differ in time of pollen receptivity.

Most brewers have specific hop cultivars, particularly the 'noble hop cultivars', as primary hops used in brewing. In the US and Germany the choice of female parent was almost always made by brewers, while the male parent determination was based solely upon the specific needs for improvement in the female line. If a particular brewer primarily used the noble hop line 'Saazer' for brewing but disliked the low yields produced by this cultivar in the US, they requested the breeder to develop a new 'Saazer-like' cultivar with many of the quality characteristics of the original 'Saazer' mother but with superior yields and characteristics that benefited growers and brewers. Crosses were subsequently made using male lines with proven breeding value for yield or aroma characteristics. The resulting offspring were screened for 'Saazer-like' aroma characteristics coupled with higher yields.

Recent work by the USDA-ARS (Henning *et al.* 1997a, b; Henning and Townsend, 2005) and others (Murakami, 1999), along with previous research by Keller and Likens (1955), Roberts *et al.* (1980) and Haunold *et al.* (1983), demonstrated the genetic potential for gain due to selection for traits such as yield and for quality traits such as bittering acid levels and essential oils. These reports validated breeders' observations that male parents can be selected for crossing based upon the performance of their respective daughters. Furthermore, these studies illustrated that in the case of some traits, such as alpha acid levels, selection of male parents was highly accurate using appropriate test-crosses grown out in multiple locations (Henning and Townsend, 2005).

Currently, most hop breeding programs utilize a combination of pedigree breeding coupled with male test-crosses for selection of male parental lines. Several other classical breeding techniques commonly used in other crops, including mutation breeding and recurrent selection, have been attempted but not utilized to any great extent.

One report using mutation breeding has been published by Nesvadba and Krofta (2002) describing the cultivar 'Agnus', but it is not entirely clear what mutation treatment contributed towards the development of this new cultivar. Nevertheless, mutation breeding remains a potentially useful and virtually unexplored means of obtaining new genes for traits with little or no genetic variation across hop germplasm collections.

At the same time, some breeding groups are beginning to use genotypic recurrent selection coupled with backcrossing techniques in order to incorporate multiple traits of economic interest into a single genetic background (P. Darby, personal communication, 2001; J.A. Henning, unpublished data). Much of the work from these two groups is designed to incorporate multiple genes for resistance to powdery mildew into a single genetic background – a technique called 'Cascading Resistance Genes'. Other areas of work utilizing this technique lie in the realm of improving traits that are controlled by multiple loci in a single genetic background or group of genotypes under the guise of population improvement. Indeed, many believe these activities will become the primary goal of most public breeding programs in the future as private companies make advances in variety development.

5.3 Molecular techniques in hop cultivar development

Most hop breeding research during the past decade has focused upon catching up with the level of genetic knowledge found in other crop species. The specialized nature of the hop industry coupled with its minor crop status and limited geographical growth have contributed towards the lack of major funding necessary to make significant advances in understanding the hop genome. Indeed, the work that has been accomplished is a tribute towards those doing research in this field, given the reduced level of funding and limited equipment experienced worldwide by hop scientists! Most research performed with hops has involved adaptation of approaches used with crop species. One difficulty in adapting molecular techniques to hop research is the presence of large quantities of polyphenolic compounds which bind DNA during purification (Townsend *et al.*, 2000). As a result, additional purification steps have been developed for use in hop. Another difficulty encountered in applying molecular technologies to hop research has been the high level of methylation of hop genomic DNA which has limited the number of vectors that can be used for cloning DNA for sequencing purposes (J.A. Henning and D.L. Moore, 2002, unpublished data).

5.3.1 Molecular genetic diversity studies

Many of the molecular studies in hop have focused on characterizing the genetic diversity present in hop, including germplasm in local breeding programs and in worldwide collections. In some cases researchers have evaluated the genetic diversity of female hop collections to differentiate between closely related lines or to establish strongly related groups or clusters of hop accessions. Other research attempted to identify heterotic crossing groups between male and female lines (Townsend and Henning, 2005) or to identify specific males that are distantly or closely related to important hop varieties (Henning *et al.*, 2004). Work in progress by the author's research group will detail genetic diversity between specific male and female *Humulus lupulus* var. *lupulus* accessions and wild American (*H. lupuloides* and *H. neomexicanus*) accessions.

Multiple techniques have been used to estimate genetic diversity between accessions; each has nuances and powers of differentiation. Henning *et al.* (2004) published recent work using chemical and morphological traits to differentiate hop accessions. This research characterized 129 accessions including wild North American female accessions along with cultivated hop lines from around the world and suggested classification based on three primary groups: wild North American, European and Hybrids. The European group was further subdivided into 'English' and 'Continental European' groups, while the Hybrids group was subdivided into five different groups corresponding with geographic origin. The first report using molecular means to discriminate hop lines was published by Abbott and Fedele (1994) using RAPDs, although no estimates of genetic diversity were reported. Initial diversity studies in hop utilized RAPD analyses of 24 hop accessions (Pillay and Kenny, 1996a). These authors reported that nine out of 60 primers produced polymorphic bands and observed little genetic diversity between the selected hop accessions in their study. Three primers, A11, A17 and C9, studied for segregation in five different families provided segregation ratios that followed Mendelian patterns. Šustar-Vozlic and Javornik (1999) used the same PCR technique on 65 cultivars and found two primary groups corresponding to European and American hop cultivars. The European group was further sub-divided into five sub-clusters corresponding to regions of geographic adaptation. This study also utilized essential oil composition in a separate cluster analysis, stating that the RAPD clustering analysis agreed well with the essential oil analysis. Murakami (2000) later assayed 51 world cultivars using RAPD analysis and observed six groups among the cultivated hops in his study.

Pillay and Kenny (1996b) reported the only use of RFLP analysis of rDNA from cultivated and wild hop accessions and identified three different genotypes (A–C) with *XhoI* restriction endonuclease. Genotype A was observed in all accessions, genotype B was found in all North American cultivars, and genotype C was found only in wild North American lines. They suggested that genotype C could be used to classify wild North American lines. No fine-detail differentiation was possibly using this technique with the number of molecular markers used.

Initial AFLP studies (Hartl and Seefelder, 1998) characterized eight hop accessions and found low genetic diversity (GD) among the eight accessions. Three of them, 'Saazer', 'Tettnanger' and 'Spalter', could not be discriminated. The greatest genetic diversity was observed between cultivars 'Hallertauer Magnum' and 'Wye Target' with a GD value of 0.11. Later work by this same research group (Seefelder *et al.*, 2000) analyzed genetic diversity among 90 hop accessions (86 male, 4 female) and observed greater polymorphism than was reported in their previous work. Again, the cultivars 'Saazer', 'Tettnanger' and 'Spalter' were indistinguishable under the conditions of this study, while the GD for some pairs, e.g. 'Columbus' vs. 'Saazer', was estimated at up to 0.83. This study revealed two main clusters with an average GD between clusters of 0.55. The first cluster was made up entirely of European-derived aroma hop lines, while the second cluster consisted of European germplasm infused with genes from American hop accessions. Within each main cluster they observed several sub-groups. The authors claimed that the resulting groups derived by cluster analysis were consistent with known pedigrees using, apparently, qualitative judgment to determine accuracy.

Although GD between female hop accessions is important in germplasm collection, it has little direct value in terms of practical breeding. Hop is a dioecious crop species requiring pollination from male hop accessions for the production of offspring. Henning *et al.* (2004) conducted a study that included a large number of male accessions in addition to female accessions and characterized 19 of the primary hop cultivars grown or used for breeding in the USA and 82 male accessions representing male hop lines in the USDA-ARS hop collection. There was no attempt to identify genetic clusters in this publication, but specific male lines were documented as genetically similar or genetically distant from each female cultivar based upon AFLP fingerprints and coefficient of coancestry estimated from pedigrees. They found a significant correlation between coefficients of coancestry values and the genetic distance for extreme pairs – both closely related male–female pairs and distantly related or unrelated male–female pairs. However, this relationship was not true across all possible male–female pairs. The authors stated that the benefit of this research was the identification of male lines to potentially maximize heterosis. Townsend and Henning (2005) later analyzed 80 males and 26 females and identified potentially heterotic clusters between male and female lines. Their research showed high levels of polymorphism for AFLP markers (490/550 polymorphic loci) and concluded that this level of polymorphism was likely due to the inclusion of a broad range of male and female genotypes. Consistent with other molecular diversity studies of hop, Townsend and Henning (2005) observed two primary clusters – the so-called European-types and the wild American–European hybrids. The two primary clusters were further distinguished into 13 sub-groups consisting of two completely female groups, nine completely male groups and two mixed sex groups.

Several groups have reported the use of microsatellites to differentiate hop genotypes, estimate genetic diversity and classify genotypes into similarity groups. The first study compared microsatellites to RAPDs for distinguishing hop cultivars and reported that the most common dinucleotide repeats in hops were $(GA)_n$ and $(GT)_n$ (Brady *et al.*, 1996). They did not define genetic diversity levels in hops. Patzak (2001) compared the use of microsatellites (ISSR) to AFLP, RAPD and STS. The AFLP technique appeared to provide the best discrimination capabilities and was the only technique that differentiated three closely related clonal selections. Both AFLP and RAPD markers provided greater levels of polymorphism than did ISSR or STS. Only 10 genotypes were used in this study, so estimates of genetic diversity were rather limited. Jakse *et al.* (2001) reported the first large-scale microsatellite analysis of genetic diversity using 41 genotypes that represented female accessions from around the world. They compared discrimination results between microsatellites and AFLP and found that microsatellites were useful for grouping genetically related genotypes, but AFLP effectively clustered closely related genotypes and separated two geographically distinct hop clusters.

Other applications for microsatellites focused on identification of specific hop cultivars from unknown samples. Three publications (Jakse *et al.*, 2002; Hadonou *et al.*, 2004; Stajner *et al.*, 2004) reported on new microsatellites, and means to develop them, that were useful for varietal differentiation. Jakse *et al.* (2004) reported an extensive evaluation of microsatellites among 124 wild and cultivated female hop accessions using four microsatellite loci. This analysis identified 63 different alleles with an average of 15.7 alleles per locus. The highest number of alleles per locus in the various geographical groups was observed in the wild hop accessions from North America and Europe. Wild North American accessions presented the highest numbers of unique alleles, which the authors suggest was indicative of high levels of genetic diversity in this germplasm. Cerenak *et al.* (2004) used a two-stage microsatellite process to differentiate 63 hop accessions, some of which were clonal selections from other lines included in the study while others represented colchicine-induced tetraploids of female accessions included in the study. In this work, microsatellites did not differentiate between clonally-related lines and between diploid–tetraploid pairs of accessions. Recently Peredo *et al.* (2005) reported on the relative advantages of capillary electrophoresis over autoradiography for evaluation and detection of microsatellites. The authors observed higher resolution of bands using fluorescence-labeled microsatellites with capillary electrophoresis and fluorescence detection than what was observed with radioactively labeled microsatellites and visual inspection of autoradiographs.

Varieties and accessions published in four of the above diversity studies are listed in Table 5.1. These studies permit some general comments about the relative merits of microsatellites and AFLP markers in genetic diversity studies. Contrary to some reports (Peredo *et al.*, 2005; Cerenak *et al.*, 2004; Stajner *et al.*, 2004), microsatellites are not universally superior for use in hop diversity studies. Indeed, a number of papers utilizing AFLP in addition to microsatellites observed higher resolution with AFLP markers (Jakse *et al.*, 2001; Patzak, 2001, 2003; Peredo *et al.*, 2005). While claiming the superiority of microsatellites over other techniques, Peredo *et al.* (2005) reported that the number of microsatellite loci used in their study (four) was inadequate to differentiate between 'Hersbrucker' and 'Hallertauer' while AFLP proved sufficient in the task. It is entirely possible that insufficient loci have been used in microsatellite studies for superior discrimination of clonally related material and that inclusion of greater numbers of loci would result in resolutions similar to that provided by AFLP. Nevertheless, Jakse *et al.* (2001) reported that microsatellites and RFLPs did not differentiate mother plants from meristem-derived clones while AFLP was amenable to the task. AFLP markers also were more effective than microsatellites for differentiating naturally occurring clonal selections 'Oswalds Clones #31, 72, 72M and 114'. Microsatellites may have greater utility in molecular mapping studies or for the identification of molecular markers linked to traits of interest rather than use in molecular diversity studies or discrimination of cultivars.

5.3.2 Molecular markers and genetic mapping studies

The first report of molecular markers in hop was designed to identify male hop seedlings (Polley *et al.*, 1997). Using 900 RAPD primers on pooled male or pooled female DNA from a single cross, this group identified 32 bands associated solely with male DNA. These 32 primers were subsequently used to screen individual male and female lines and the authors identified three primers consistently associated with the male trait. STS markers were developed from one of these primers, providing what the authors claimed was a seedling selectable marker for male hop lines. Henning (unpublished results) was unable to replicate the specificity of this marker, and it appeared that the marker reported by Polley *et al.* (1997) was specific for European-derived males but not for males derived from wild North American parentage. This was later verified by Seefelder *et al.* (2000) and Patzak *et al.* (2002). Danilova and Henning (2005) reported on the use of three different STS markers to detect male seedlings, including one RAPD marker reported by Polley *et al.* (1997) and two ISSR markers. Individually, none were 100% effective for identifying male lines, although the combination of all three markers discriminated all male hop genotypes. Male plants that had a genotype where all three markers were observed were presumed to have European origins. The authors concluded that

Table 5.1 Varieties and accessions published in four diversity studies (Cerenak *et al.*, 2004; Seefelder *et al.*, 2000; Sustar-Vozlic and Javornik, 1999; Townsend and Henning, 2005). Abbreviations used for some of the more common lines are: Bullion (Bu), Brewers Gold (BG), Early Green (EG), East Kent Golding (EKG), Fuggle (Fu), Hallertauer (Hall.), Late Cluster (LC), Whitbred Golding variety (WGV), Saazer (Saaz.), seedling (-s) and Zattler-seedling (ZS)

81/8/13	Wye Target × 75/5/46 (Hall., Saaz.- type, wild hops)	Germany
82/39/37	Hall. Mfr, Spalter, Saaz., wild hops, NB, Perle, USDA 21055	Germany
85/54/15M	Male (Hall. Mittelfrue, Spalter, Saaz., wild hops, NB, Yeoman)	Germany
88/55/501M	Male 82/39/37 × 85/54/15M	Germany
88/55/502M	Male 82/39/37 x85/54/15M	Germany
91/44/39M	Male (Hall. Magnum × [Orion × (USDA 21055 × 66/2/10)])	Germany
Ahil	BG × 3/3	Slovenia
Ahil	4n Colchicine-derived tetraploid	Slovenia
Alliance	WGV × *Verticillium*-resistant breeding line from East Malling	England
Apolon	BG × 3/3	Slovenia
Apolon	4n Colchicine-derived tetraploid	Slovenia
Atlas	BG × 3/3	Slovenia
Atlas 4n	Colchicine-derived tetraploid	Slovenia
Aurora	NB × TG	Slovenia
Backa	Mass selection of Bavarian hop	Yugoslavia
Belg. Spalter	Origin unknown	Belgium
Blisk	Atlas ($2n = 4x = 40$) × 1/9	Slovenia
Bobek	NB × TG	Slovenia
Braustern	Clonal selection of NB	Germany
Brewers Gold	BB1 × OP	England
Buket	NB × 2/137	Slovenia
Bu	BB1 × OP	England
Callicross	LC × Fu seedling	New Zealand
Cascade	(Fu × [Serebrianka × Fu seedling]) × OP	USA
Cascade 4n	Colchicine-derived tetraploid	USA
Cekin	Aurora × 3/3 ($2n = 4x = 40$)	Slovenia
Celeia	Savinjski Golding ($2n = 4x = 40$) × 105/58	Slovenia
Cerera	Savinjski Golding ($2n = 4x = 40$) × 105/58	Slovenia
Chang Bei 1	Unknown	China
Chang Bei 2	Unknown	China
Changbai 0	Origin unknown	China
Chinook	Peltham Golding × 63012M (BG × wild Utah hop)	USA
Cicero	Aurora × 3/3 ($2n = 4x = 40$)	Slovenia
Cluster	Old American variety	USA
College Cluster	Y90 × OP	England
Columbus	Unknown	USA
Comet	Sunshine-s × Utah 524-2	USA
Coobs	Clonal selection of Golding	England
Crystal	Hall. Mittelfrue (4X) × 21381M (2X)	USA
Density	Bu × OP	England
Dunav	Northern Brewer × OP	Yugoslavia
Early Bird Golding	Clonal selection of Golding	England
Early Choice	Golding × male breeding line	England
East Kent Golding	Old English cultivar	England
Eastwell Golding	Old variety	England
Emerald	Northern Brewer × 63/5/27M	Germany
Estera	Unknown	England
Eurohop-T	Seedling of Hall. × seedling of Tettnanger	New Zealand
First Choice	LC × Fu-seedling	England

Table 5.1 Continued

First Gold	WGV × 23/77/56M	England
Fuggle	Old English variety	England
Fuggle	37 Clonal selection from Fu	England
Fuggle N	Clonal selection from Fu	USA
Galena	BG × OP	USA
Galena VF	Meristem-tip culture from Galena	USA
Ging Dao Do Hua	Origin unknown	China
Ging Dao Do Hua 791	Origin unknown	China
Golden Star	Mutant from Shinshuwase	Japan
Hall. Magnum	Galena × 75/5/3M (23% Hall., 8% Saaz., 5% Spalter, 28% wild hops)	Germany
Hall. Mittelfrue	Mass selection of Bavarian hop	Germany
Hall. Tradition	Hall. Gold × 75/17/106M (47% Hall., 15% Saaz., 9% Spalter, 28% wild)	Germany
Hall. Gold	(Huller Anfang × 8lt9) × (NB × 54/33/11M)	Germany
Herald	34/73/11 × 23/77/56 (large contributions from Omega-dwarf variety)	England
Hersbrucker Spalt	German landrace	Germany
Hersbrucker Pure	73/10/16 × 78/18/145M (42% Hall., 13% Saaz., 8% Spalter, 35% wild)	Germany
Horizon	Nugget × 64035M (25% BG, 25% German aroma hop, 12.5% EG)	USA
Huller Anfang	23gl × 21199 (46% Hall., 23% Saaz. type, 31% wild hops)	Germany
Huller Aroma	50/1/152 × 58/14/92	Germany
Huller Bitterer	Northern Brewer × 11/0/96M ($2n = 4x = 40$, tetraploid Huller breed)	Germany
Huller Fortschritt	Huller Anfang × (8lt9 3°45′x 7kM7)	Germany
Huller Start	47t34 (21199 × 2°29′)	Germany
Iwanovecki	Parentage unknown	Russia
Japan 826	Origin unknown	Japan
Japan C-730	Origin unknown	Japan
Japan C-845	C79-27-01 × C76-64-110	Japan
Japan C-966	Origin unknown	Japan
Keyworth Midseason	Y90 × OF	England
Kirin 1	Shinshuwase's clonal selection	Japan
Kirin II	Saaz. × White Bine OP	Japan
Kitamidori	C79-27-01 × C79-64-110 (related to Kirin II and OB79)	Japan
Liberty	Tetraploid Hall. Mittelfrue × 64035M (ZS male)	USA
Lubelski	Origin unknown (Saaz. ancestry)	Poland
M19005	LC-s	USA
M19007	Brewer's Favorite-s	USA
M19009	Fu × Fu-s	USA
M19036	LC × Fu-s	USA
M19037	Fu-s × Fu-s	USA
M19041	EG × OP	USA
M19046	LC-s × Fu-s	USA
M19047	Elsasser × Fu-s	USA
M19060	EKG × Bavarian-s	USA
M19061	Late Grape × Fu-s	USA
M19172	Cat's Tail × (Fu × Fu-s)	USA
M21009	Sunshine-s × (Utah-523-4 × (EG × OP))	USA
M21058	Fu × (Striesselspalt × LC-s)	USA
M21072	BG × Arizona-1-2	USA

Table 5.1 Continued

M21076	Comet × (BG × (Fu × Colorado-2-1))	USA
M21087	Yugoslavia Selection 3/3	Yugoslavia
M21089	Yugoslavia Selection 5/10	Yugoslavia
M21090	Yugoslavia Selection 12/17	Yugoslavia
M21109	[BG × (EG × OP)] × ZS	USA
M21110	Bu × ZS	USA
M21129	[Late Grape-s × (Fu × Fu-s)] × (BG × Utah-526-4)	USA
M21132	Yakima Cluster × ZS	USA
M21135	[BG × (EG × OP)] × ZS	USA
M21184	Unknown	USA
M21268	NB × [(BG × (EG × OP)) × ZS]	USA
M21272	NB × (Bu × ZS)	USA
M21273	Comet × [BG × (Fu × Colorado-2-1)]	USA
M21300	[BG × (EG × OP)] × ZS	USA
M21303	Bu × ZS	USA
M21306	Complex pedigree: Comet, BG, Fu, EG, Colorado 2-1, Bu, ZS, unknown	USA
M21313	Comet × [(BG × (EG × OP) × ZS)]	USA
M21329	Comet × (Bu × ZS)	USA
M21335	NB × (Bu × ZS)	USA
M21336	NB × (Bu × ZS)	USA
M21339	[Comet × (BG × (Fu × Colorado 2-1))] × [(BG × (EG × OP)) × ZS]	USA
M21345	[Comet × (BG × (Fu × Colorado 2-1))] × [(BG × (EG × OP)) × ZS]	USA
M21351	[Comet × (BG × (Fu × Colorado 2-1))] × (Bu × ZS)	USA
M21358	[Comet × (BG × (Fu × Colorado 2-1))] × [BG × (EKG × Bavarian-s)]	USA
M21360	Cascade × [(BG × (EG × OP)) × ZS]	USA
M21398	Native Yugoslavian Male 01P04	Yugoslavia
M21400	Native Yugoslavian Male 20P09	Yugoslavia
M21415	[BG × (EG × OP)] × (LC-s × Fu-s)	USA
M21416	Bu × ZS	USA
M21417	Complex pedigree: Comet, BG, Fu, EG, Colorado 2-1, Bu, ZS unknown	USA
M21420	Complex pedigree: Comet, BG, Fu, EG, Colorado 2-1, Bu ZS unknown	USA
M21424	Cascade × LC-s	USA
M21425	Cascade × (Semsch-s × 8-2B yrd)	USA
M21426	Cascade × (Fu × Fu-s)	USA
M21427	Cascade × (Red Vine × Fu-s)	USA
M21428	Cascade × (Fu-s × Fu-s)	USA
M21432	Cascade × [(Late Grape-s × (Fu × Fu-s)) × (EG × OP)]	USA
M21435	Cascade × Colorado-1-1	USA
M21437	Fu × OP	USA
M21444	Comet × [(BG × (EG × OP)) × ZS]	USA
M21446	NB × [BG × (EKG × Bavarian-s)]	USA
M21448	Cascade × [(BG × (EG × OP)) × ZS]	USA
M21461	Complex pedigree: Comet, BG, Fu, EG, Colorado 2-1, Bu, ZS unknown	USA
M21462	Cascade × (Fu × Fu-s)	USA
M21463	Cascade × Yugoslavian 3/3	USA
M21465	Comet × [(BG × (EG × OP)) × ZS]	USA
M21466	Comet × [(BG × (EG × OP)) × ZS]	USA
M21487	Cascade × (EKG × Bavarian-s)	USA
M21488	Cascade × [(BG × (EG × OP)) × ZS]	USA
M21603	Cascade × (Semsch-s × 8-3B yrd)	USA

Table 5.1 Continued

M21690	Complex pedigree: Late Grape, Fu, LC, unknown	USA
M21692	[(Late Grape-s × (Fu × Fu-s)) × (LC-s × Fu-s)] × (LC-s × Fu-s)	USA
M51114	[Landhopfen-s × (Golden Cluster × Fu-s)] × (Semsch-s × 8-2B Yrd)	USA
M52042	(Late Grape × Fu-s) × (Late Grape × Fu-s)	USA
M52047	[Striesselspalt × (EG × OP)] × (Striesselspalt × LC-s)	USA
M58111	[BG × (Belgian-31-s × Belgian-31)] × (Late Grape × Fu-s)	USA
M63011	(Late Grape × Fu-s) × (EG × OP)	USA
M63015	BG × (EKG × Bavarian-s)	USA
M64033	ZS	Germany
M64034	ZS	Germany
M64035	ZS	Germany
M64036	ZS	Germany
M64037	ZS	Germany
M64101	Unknown	Germany
M64102	Wild American × OP	USA
M64105	Fu × (Wild American × OP)	USA
Malling	Origin unknown (author's note: perhaps a Golding)	England
Mount Hood	Tetraploid Hall. Mittelfrue × USDA 19058M (EG seedling)	USA
Neoplanta	Northern Brewer × Sx-502 (Savinski Golding × Yugo. wild hop)	Yugoslavia
Newport	Magnum × {[(BG × (Belg. 31-s × Belg. 31)) × (Late Grape × Fu)] × OP}	USA
Nordgard 1478	Nordgard 978 × seedling of Bramling Cross	Denmark
Nordgard-978	Spalter × wild Danish hop	Denmark
Northern Brewer	Canterbury Golding × OB21 (BG × California male)	Germany
Nugget	[BG × (EG × OP)] × [BG × (EKG × Bavarian-s)]	USA
Olympic	Complex pedigree: 3/8 BG, 3/16 Fu, 1/8 EKG, 1/16 Bavarian, 1/4 unknown	USA
Omega Wye	Challenger × English male	England
Orion	Perle × German 70/10/15	Germany
Osvald clone 126	Clonal selection of Fu hop	Czech Republic
Osvald clone 72	Clonal selection of Saaz. hop	Czech Republic
OT 48 Bramling	Golding clone × OL34 (male seedling related to BB0)	England
Perle	NB × 63/5/27M (Hall. Mittelfrue, Spalter, Saaz.)	Germany
Phoenix	Yeoman × 38/77/14M (related to OB79)	England
Pioneer	34/73/11 × 23/77/56 (large contributions from Omega-dwarf variety)	England
Polish clone 12	Parentage unknown	Poland
Pride of Ringwood	(Pride of Kent × OP) × OP	Australia
Progress	WGV × OB79	England
Ringwood Special	Fu × OP	Australia
Saaz.	Old Czech landrace	Czech Republic
Saaz. 36	Clonal selection from Saaz.	Germany
Saladin	Clonal selection of Saaz. hop	Germany
Sara	Parentage unknown	Belgium
Savinjski Golding	Fu's clonal selection, introduced in Slovenia in 1872	Slovenia
Savinjski Golding 4n	Colchicine-derived tetraploid	Slovenia
Serebrianka	Wild Siberian hop selection	Russia

Table 5.1 Continued

Sirem	Mass selection of Saaz. hop	Czech Republic
Smooth Cone	LC × Fu's seedling	New Zealand
Southern Brewer	Fu × Fu's seedling	South Africa
Spalter	Mass selection of Saaz. hop	Germany
Spalter Select	76/18/80 × 71/16/7 (43% Hall., 12% Saaz., 9% Spalter 34% wild)	Germany
Strisselspalter	Reportedly derived from Hersbrucker Spalt	France
Styrian	Yugoslavian selection from Fu	Yugoslavia
Sudafrika	Origin unknown	South Africa
Tardif de Bourgogne	France landrace	France
Taurus	82/39/37 × 85/54/15M	Germany
Tettnanger	Old German landrace	Germany
Toyomidori	NB × OB79	Japan
Tutsham	Clonal selection of Golding	England
U.S. Tettnanger	Fu	USA
Ultra	(Hall. Mittelfrue; $2n = 4x = 40$ tetraploid) × 21373M	USA
Urozani	Origin unknown	Russia
USDA 21055	Comet × 65009M [BG × (Fu × Colo 2-1)]	USA
Vojvodina	NB × SX-502	Yugoslavia
Vojvodina	NB × SX-502	Yugoslavia
W45	Wild hop from Thüringen	Germany
WGV	Whitbread Golding Variety	England
Willamette	Fu ($2n = 4x = 40$) × seedling of Fu	USA
Wurtemberger	Mass selection of Bavarian hop	Germany
Wye Challenger	(Zattler × OP) × (NB × Wye 22/56)	England
Wye Northdown	NB × OP England	
Wye Saxon	Svalof × (Bramling Cross × Wye 1/63/42)	England
Wye Target	(NB × Wye 22/56) × (Eastwell Golding × OB79)	England
Wye Viking	Svalof × (Bramling Cross × Wye 1/63/42)	England
Wye Yeoman	Wye 43/69/17 × Wye 25/68/173	England
Wye Zenith	Wye 25/68/22 × OP	England
Zitomir clone 18	Clonal selection of Volinsky hop	Ukraine
Zlatan	Clonal selection of Saaz. hop	Czech Republic

there appeared to be differences in the distribution of simple repeats in the Y-chromosome between the North American male hops and the European male hops. One other publication, by Seigner *et al.* (2005), discussed attempts to identify AFLP molecular markers to identify powdery mildew resistant hop germplasm. At this point, one marker (R2-N3_212; Seigner *et al.*, 2005) appears to be approximately 1.5 cM from the R2 gene from the resistant cultivar 'Target'. The authors are continuing to pursue additional markers that eventually would be converted to sequence characterized amplified regions (SCARs).

Although hop geneticists have discussed the possibility of combining forces to develop a saturated genomic map, the funding for such an effort has not been identified. Seefelder *et al.* (2000) published the only genetic map to date in hops. Using a limited population consisting of 60 females and eight males, this group identified eight linkage groups in the map generated for female offspring and nine linkage groups for the map generated for male offspring. The authors used 217 AFLP markers, three male-associated RAPD markers, one male-associated STS marker and three microsatellites in their study for a total of 224 polymorphic markers. Out of the 224 loci, the authors found 121 loci segregating in a 1:1 ratio presumed to be 'ab × aa' (using the authors' nomenclature) or 'maternally informative', 49 loci segregating in a 1:1 fashion presumed to be 'aa × ab' or 'paternally informative', and 53 loci apparently resulting from an 'ab × ab' cross segregating in a 3:1 fashion. Interestingly, one of the microsatellites, '5-2' (Brady *et al.*, 1996), segregated in a 1:1:1:1 ratio, indicating a multi-allelic locus with four alleles. Seefelder *et al.* (2000) identified a specific linkage group that was tightly linked to males. Whether eight individual male genotypes is representative of males for a mapping population is debatable. Nonetheless, this excellent study provides a starting point from which to pursue further map studies.

Koie *et al.* (2005) described a mapping effort using AFLP and QTL analysis to identify hop chemical constituents. They screened 130 female offspring from the cross between 'Chinook' and 'SaM' using 212 AFLP molecular markers for a 'Chinook'-based map and a 'SaM' (Saazer male)-based map. The total length of the 'Chinook'-based map was 272 cM which mapped to 10 linkage groups, while the SaM-based map was 102 cM consisting of seven linkage groups. The authors deduced that the number of markers specific to SaM was insufficient to provide enough information to construct 10 linkage groups. QTLs for seven chemical components mapped to four different linkage regions. No information was presented regarding the identification of specific QTL markers.

5.3.3 Gene identification and expression

Few genes or promoter regions have been identified and characterized from hop. The first set of genes cloned and identified were the 7SL RNA genes (Matoušek *et al.*, 1999), four of which were cloned and subsequently characterized by *in vitro* heterologous expression system (HeLa Extract). Their work identified differences in transcription rates among four clones in the expression system and the authors suggested this was indicative of the complexity required for 7SL RNA. These same authors conducted Southern-blot analysis and HindIII restriction endonuclease studies which suggested that some of the 7SL RNA genes were arranged in genomic clusters. During the same year, Henning and Moore (1999) cloned and sequenced the first non-conserved gene in hop presumed to be involved in plant resistance to fungal pathogens – an endochitinase precursor gene (HCH1). The authors used a heterologous probe from pea (*Pisum sativum* L.) to screen genomic libraries made from the powdery mildew resistant cultivar 'Zenith'. No molecular characterizations of the activity of this gene were published. Finally, Paniego *et al.* (1999) cloned and identified the phlorisovalerophenone synthase (VPS) gene from hop. This gene is a polyketide synthase involved in the primary biosynthetic steps of hop bittering acids. This same gene was later cloned and characterized by Okada and Ito (2001) who observed specific VPS expression in lupulin glands of hop cones.

A few years later, Matoušek *et al.* (2002) cloned and characterized another non-conserved gene sequence coding for chalcone synthase (chs_H1). This gene is also a polyketide synthase with specific activity for condensation of *p*-coumaroyl-CoA with malonyl-CoA producing naringenin chalcone – a precursor to a number of flavonoids. The authors utilized quantitative RT-PCR to assay tissue specificity of expression and observed highly specific expression of this gene in glandular trichomes during hop cone maturation. Lowest levels of expression for this gene were observed in roots. The authors subsequently performed Southern-blot analysis and predicted there were at least six chs_H1-like genes present in the hop genome. An RFLP comparison of chs-like genes from different hop genotypes revealed rearrangements of the gene during evolution with the sequences found in *H. lupulus* ssp. *neomexicanus* being the most distinct from other genotypes.

Sakano *et al.* (2004) cloned and identified the gene for adenylate isopentenyl-transferase (AIPT). This is an important gene in the biosynthesis of cytokinins, which are a well-known group of phytohormones involved in the growth and development of plants (Mok and Mok, 2001). The authors characterized the activity of this enzyme by cloning the putative AIPT cDNA into a bacterial expression vector. Finally, Matoušek *et al.* (2005) recently cloned and performed molecular analyses on a regulatory factor entitled HlMyb1. The myb factors regulate the phenylpropanoid biosynthetic pathway (Stracke *et al.*, 2001) and are involved in a number of processes in plant morphogenesis. The authors were not able to deduce a direct function of this regulatory factor but they did observe elevated levels of activity in mature female and male inflorescences and low expression levels in immature flower structures.

5.3.4 Genetic engineering of hop

The first publication detailing genetic transformation of hop was published by Horlemann *et al.* (2003). They used *Agrobacterium* to transform the 'Tettnanger' with the reporter gene 'GUS' (β-glucuronidase). One of the primary problems encountered by researchers prior to successfully transforming hop is the regeneration of non-meristematic tissue. Horlemann *et al.* (2003) provided an excellent discussion on their solutions to these difficulties, focusing on two critical issues: the condition under which donor tissue was obtained and the hormonal treatments used to induce organogenesis. They found that shoot internodes from plants grown *in vitro* were superior at regeneration. Shoot internodes from plants grown in the greenhouse did not develop beyond the callus stage. They also found that a combination of 1.43 μM IAA and 9.08 μM TDZ (Thidiazuron) in a basal medium of Murashige and Skoog (1962) supplemented with 2% glucose provided the optimal hormonal environment for regeneration. Using this procedure they obtained 40% overall regeneration frequencies of plantlets that formed roots. Of 1440 explants, 83 were selected by kanamycin, and 42 of these 83 tested positive for GUS expression prior to rooting in pots. Twenty-one plants were finally propagated in the greenhouse and were positive for GUS expression and PCR detection of the transgene.

Seigner *et al.* (2005) reported the successful incorporation of the GUS reporter gene along with the transformation of hop with pBin 19 transformation vector (Bevan, 1984) integrated with a chitinase gene (Henning and Moore, 1999), utilizing cultivars 'Saazer' and 'Hallertauer mittelfrueh'. Plants that expressed the integrated chitinase transgene showed decreased growth of powdery mildew but conclusive quantitative data was not yet available. Further studies to verify the stable transformation of these cultivars and to characterize expression of the chitinase gene are currently in progress.

5.3.5 Additional molecular work in hop

Other projects applying molecular tools to hop research have been reported and are in progress. Karlov *et al.* (2003) published research on DAPI-banding and the identification of sex-chromosomes in hop. This process involved chromosome measurements, DAPI-banding and fluorescence *in situ* hybridization (FISH) of tandemly repeated 18S-25S rDNA and 5S rDNA on mitotic chromosomes. These techniques were applied to characterize both male and female hop accessions and differentiate all nine autosomes and the X and Y chromosomes. Prior to the application of these methods, it was difficult to characterize and differentiate all 10 chromosomes (Haunold, 1991). Radisek *et al.* (2003) utilized pathogenicity scores and AFLP analysis to differentiate two strains of *Verticillium* wilt. Field observations demonstrated a non-lethal-type and a lethal-type of *Verticillium* wilt on the same genotype. Susceptible, moderately resistant and resistant hop cultivars were inoculated with each isolate and for the subsequent disease responses were characterized. In addition, AFLP banding patterns of the fungal isolates were examined for genetic differences. These genetic analyses identified two distinct strains correlating with the differences in patho-types. Later research by this group resulted in the identification of AFLP-derived pathotype-specific sequence characterized amplified region (SCAR) markers (Radisek *et al.*, 2004) that have proven highly specific to the pathotypes.

5.4 Conclusion

The evolution of hop breeding could be considered more along the lines of 'revolution' when comparing the past decade to prior research. Less than a half a century ago, some hop breeding programs were focused upon mass selection of clonal replicates. Just a decade ago almost all breeding consisted of simple pedigree methods. Today, scientists are pursuing molecular studies involving transformation of hop lines, isolation and identification of genes involved with important economic and morphological characteristics, genetic fingerprinting to identify potentially heterotic male and female pairs for hybrid work, and the isolation and identification of molecular markers to select for specific traits. Truly, the era of genetics in hops is an exciting and wide-open field.

5.5 Sources of further information

This chapter highlighted a number of past and current studies but could not address all publications, particularly those prior to the early 1990s. Some important sources of information regarding subjects not covered here include the book written by Neve (1991) entitled appropriately *Hops*, a monograph that provides extensive coverage of breeding for plant pathogen and pest resistance. Broad coverage is also provided on the many plant pests including viruses. An additional resource for information on the history, production and breeding of hops is the millennium summary written by Moir (2000). While not exclusively devoted to breeding, this article covers the history and early years of hop production and selection techniques. Older sources of information can be obtained in the book entitled *Hops: Botany, Cultivation and Utilization* by Burgess (1964). Information presented in these three resources provides a fairly complete history of hop breeding and genetics prior to the early 1990s.

5.6 References

ABBOTT, M.S. and M.J. FEDELE. 1994. A DNA-based varietal identification procedure for hop leaf tissue. *J. Inst. Brew.*, 100: 283–285.

BARTH, H.J., C. KINKE and C. SCHMIDT. 1994. *The Hop Atlas*. Joh. Barth & Sohn., Nuremberg.

BEVAN, M. 1984. Binary *Agrobacterium* vectors for plant transformation. *Nucleic Acids Res.*, 12: 8711–8721.

BRADY, J.L., N.S. SCOTT and M.R. THOMAS. 1996. DNA typing of hops (Humulus Lupulus) through application of RAPD and microsatellite marker sequences converted to sequence tagged sites (STS). *Euphytica*, 91: 277–284.

BURGESS, A.H. 1964. *Hops: Botany, Cultivation and Utilization*. Interscience, New York, World Crop Series, p. 300.

CERENAK, A., J. JAKSE and B. JAVORNIK. 2004. Identification and differentiation of hop varieties using simple sequence repeat markers. *J. Am. Soc. Brew. Chem.*, 62(1): 1–7.

DANILOVA, T.V. and J.A. HENNING. 2005. Sex molecular markers reveal differences of Y-chromosome molecular structure among hops of European and American origin. *Acta Hort. (ISHS)*, 668: 85–92.

DELYSER, D.Y. and W.J. KASPER. 1994. Hopped beer: the case for cultivation. *Econ. Bot.*, 48(2): 166–170.

FORE, R.E. and J.D. SATHER. 1947. Breeding techniques with hops, *Humulus lupulus*. *Wallerstein Lab. Commun.*, 10(29): 17–28.

HADONOU, A.M., R. WALDEN and P. DARBY. 2004. Isolation and characterization of polymorphic microsatellites for assessment of genetic variation of hops (*Humulus lupulus* L.). *Mol. Ecol. Notes*, 4(2): 280.

HARTL, L. and S. SEEFELDER. 1998. Diversity of selected hop cultivars detected by fluorescent AFLPs. *Theor. Appl. Genet.*, 96(1): 112–116.

HAUNOLD, A. 1991. Cytology and cytogenetics of hops. In T. Tsuchiya and P.K. Gupta (eds), *Chromosome Engineering in Plants: Genetics, Breeding, Evolution.* Elsevier, New York, pp. 251–263.

HAUNOLD, A., G.B. NICKERSON and S.T. LIKENS. 1983. Yield and quality potential of hop, *Humulus lupulus* L. *J. Am. Soc. Brew. Chem.*, 41(2): 60–63.

HENNING, J.A. and D.L. MOORE. 1999. *Humulus lupulus* endochitinase precursor (HCH1) gene. http://www.ncbi.nlm.nih.gov/entrez/viewer.fcgi?db=nucleotide&val=4960048, GenBank Entry Locus AF147497.

HENNING, J.A. and M.S. TOWNSEND. 2005. Field-based estimates of heritability and genetic correlations in hops. *Crop Sci.*, 45(4): 1469–1475.

HENNING, J.A., A. HAUNOLD and G. NICKERSON. 1997a. Genetic parameter estimates for five traits in male hop accessions: a preliminary study. *J. Am. Soc. Brew. Chem.*, 55(4): 157–160.

HENNING, J.A., A. HAUNOLD and G. NICKERSON. 1997b. Estimates of heritability and genetic correlation for five traits in female hop accessions. *J. Am. Soc. Brew. Chem.*, 55(4): 161–165.

HENNING, J.A., M.S. TOWNSEND and S. KENNY. 2004. Potential heterotic crosses in hops as estimated by AFLP-based genetic diversity and coefficient of coancestry. *J. Am. Soc. Brew. Chem.*, 62(2): 63–70.

HORLEMANN, C., A. SCHWEKENDICK, M. HOHNLE and G. WEBER. 2003. Regeneration and *Agrobacterium*-mediated transformation of hop (*Humulus lupulus* L.). *Plant Cell Rep.*, 22: 210–217.

JAKSE, J., K. KINDLHOFER and B. JAVORNIK. 2001. Assessment of genetic variation and differentiation of hop genotypes by microsatellite and AFLP markers. *Genome*, 44(5): 773–782.

JAKSE, J., D. BANDELJ and B. JAVORNIK. 2002. Eleven new microsatellites for hop (*Humulus lupulus* L.). *Mol. Ecol. Notes*, 2(4): 544.

JAKSE, J., Z. SATOVIC and B. JAVORNIK. 2004. Microsatellite variability among wild and cultivated hops (*Humulus lupulus* L.). *Genome*, 47(5): 889–899.

KARLOV G.I., T.V. DANILOVA, C. HORLEMANN and G. WEBBER. 2003. Molecular cytogenetics in hop (*Humulus Lupulus* L.) and identification of sex chromosomes by DAPI-banding. *Euphytica*, 13(2): 185–190.

KELLER, K.R. and S.T. LIKENS. 1955. Estimates of heritability in hops, *Humulus lupulus* L. *Agron. J.*, 47: 518–521.

KOIE, K., A. OKADA, T. KANEKO and K. ITO. 2005. Construction of the genetic linkage map and QTL analysis on hop (*Humulus lupulus* L.). *Acta Hort. (ISHS)*, 668: 59–66.

LANG, W. 1925. Der falsche Mehltau am Hopfen. *Nachrichtenbl. Deutsch. Pflanzenschutz-dienst*, 5: 63–64.

LINKE, W. and A. REBL. 1950. *Der Hopfenbau.* Hans Carl Verlag, Nuremberg, Germany.

MATOUŠEK, J., V. JUNKER, L. VRBA, J. SCHUBERT, J. PATZÁK and G. STEGER. 1999. Molecular characterization and genome organization of 7SL RNA genes from hop (*Humulus lupulus* L.). *Gene*, 239(1): 173–183.

MATOUŠEK, J., P. NOVÁK, J. BØIZA, J. PATZÁK and H. NIEDERMEIEROVA. 2002. Cloning and characterization of Chs-specific DNA and cDNA sequences from hop. (*Humulus Lupulus* L.). *Plant Sci.*, 162: 1007–1018.

MATOUŠEK, J., L. VRBA, P. NOVÁK, J. PATZÁK, J. DE KEUKELEIVE, J. ŠKOPEK, A. HEYERICK, I. ROLDÁN-RUIZ and D. DE KEUKELEIRE. 2005. Cloning and molecular analysis of the regulatory factor HlMyb1 in hop (*Humulus Lupulus* L.) and the potential of hop to produce bioactive prenylated flavonoids. *J. Agric. Food Chem.*, 53, 4793–4798.

MIYABE, K. and Y. TAKAHASHI. 1906. A new disease of the hop-vine caused by *Peronoplasmopara humuli* n. sp. *Trans. Sapporo Nat. Hist. Soc.*, I, Part II.

MOIR, M. 2000. Hops – a millennium review. *J. Am. Soc. Brew. Chem.*, 58(4): 131–146.

MOK, D.W. and M.C. MOK. 2001. Cytokinin metabolism and action. *Ann. Rev. Plant Physiol. Plant Mol. Biol.*, 52: 89–118.

MURAKAMI, A. 1999. Inheritance of major chemical components in hops. *J. Inst. Brew.*, 105(2): 107–111.

MURAKAMI, A. 2000. Hop variety classification using the genetic distance based on RAPD. *J. Inst. Brew.*, 106(3): 157–161.

MURASHIGE, T. and F. SKOOG. 1962. A revised medium for rapid growth and bioassays with tobacco tissue cultures. *Physiol. Plant*, 15: 473–497.

NESVADBA, V. and K. KROFTA. 2002. New hop variety Agnus as the result of breeding process innovation in the Czech Republic. *Rost. Vroba*, 48(11): 513–517.

NEVE, R.A. 1991. *Hops*. Chapman and Hall, London.

OKADA, Y. and K. ITO. 2001. Cloning and analysis of valerophenone synthase gene expressed specifically in lupulin gland of hop (*Humulus lupulus* L.). *Biosci. Biotechnol. Biochem.*, 65: 150–155.

PANIEGO, N.B., K.W.M. ZUURBIER, S.Y. FUNG, R. VAN DER HEIJDEN, J.J.C. SCHEFFER and R. VERPOORTE. 1999. Phlorisovalerophenone synthase, a novel polyketide synthase from hop (*Humulus lupulus* L.) cones. *J. Biochem.*, 262: 612–616.

PATZAK, J. 2001. Comparison of RAPD, STS, ISSR and AFLP molecular methods used for assessment of genetic diversity in hop (*Humulus lupulus* L.). *Euphytica*, 121: 9–18.

PATZÁK, J. 2003. Assessment of somaclonal variability in hop (*Humulus lupulus* L.) in *in vitro* meristem cultures and clones by molecular methods. *Euphytica*, 131: 343–350.

PATZÁK, J., P. VEJL, S. SKUPINOVA and V. NESVADBA. 2002. Identification of sex in F1 progenies of hop (*Humulus lupulus* L.) by molecular marker. *Rost. Vroba*, 48(7): 318–321.

PEREDO, E.L., R. ARROYO-GARCÍA, J.M. MARTÍNEZ-ZAPATER and M.Á. REVILLA. 2005. Evaluation of microsatellite detection using autoradiography and capillary electrophoresis in hops. *J. Am. Soc. Brew. Chem.*, 63(2): 57–62.

PILLAY, M. and S.T. KENNY. 1996a. Random amplified polymorphic DNA (RAPD) markers in hop, *Humulus lupulus*: level of genetic variability and segregation in F1 progeny. *Theor. Appl. Genet.*, 92(3/4): 334–338.

PILLAY, M. and S.T. KENNY. 1996b. Structure and inheritance of ribosomal DNA variants in cultivated and wild hop, *Humulus lupulus*. *Theor. Appl. Genet.*, 93(3): 333–340.

POLLEY, A., M.W. GANAL and E. SEIGNER. 1997. Identification of sex in hop (*Humulus lupulus*) using molecular markers. *Genome*, 40(3): 357–361.

RADISEK, S., J. JAKSE, A. SIMONCIC and B. JAVORNIK. 2003. Characterization of *Verticillium albo-atrium* field isolates using pathogenicity data and AFLP analysis. *Plant Disease*, 87(6): 633–638.

RADISEK, S., J. JAKSE and B. JAVORNIK. 2004. Development of pathotype-specific SCAR markers for detection of *Verticillium* albo-atrium isolates from hop. *Plant Disease*, 88(10): 1115–1122.

ROBERTS, D.D., W.E. KRONSTAD and A. HAUNOLD. 1980. Genetic variability and association of maturity, yield, and quality characteristics of female hops. *Crop Sci.*, 20: 523–527.

ROYLE, D.J. and H. KREMHELLER. 1981. Downy mildew of the hop. In D. M. Spencer (ed.), *The Downy Mildews*. Academic Press, London, pp. 395–419.

SAKANO, Y., Y. OKADA, A. MATSUNAGA, T. SUWAMA, T. KANEKO, K. ITO, H. NOGUCHI and I. ABE. 2004. Molecular cloning, expression, and characterization of adenylated isopentenyltransferase from hop (*Humulus lupulus* L.). *Phytochemistry*, 65: 2439–2446.

SALMON, E.S. 1949. Two New Hops Resistant to Verticillium Wilt. Wye College, Kent.

SALMON, E.S. and W.M. WARE. 1925. The downy mildew of the hop and its epidemic occurrence in 1924. *Ann. Biol.*, 12(2): 121–151.

SEEFELDER, S., H. EHRMAIER, G. SCHWEIZER and E. SEIGNER. 2000. Genetic diversity and phylogenetic relationships among accessions of hop, *Humulus lupulus*, as determined by amplified fragment length polymorphism fingerprinting compared with pedigree data. *Plant Breeding*, 119: 257–263.

SEIGNER, E., A. LUTZ, H. RADIC-MIEHLE, S. SEEFELDER and F.G. FELSENSTEIN. 2005. Breeding for powdery mildew resistance in hop (*Humulus* L.): strategies at the Hop Research Center, Huell, Germany. *Acta Hort. (ISHS)*, 668: 19–30.

STAJNER, N., J. JAKSE, P. KOZJAK and B. JAVORNIK. 2004. The isolation and characterization of microsatellites in hop (*Humulus lupulus* L.). *Plant Sci.*, 168: 213–221.

STRACKE, R., M. WEBER and B. WEISSHAAR. 2001. The R2R3-MYB gene family in *Arabidopsis thaliana*. *Curr. Opin. Plant Biol.*, 4: 447–456.

SUSTAR-VOZLIC, J. and B. JAVORNIK. 1999. Genetic relationships in cultivars of hop, *Humulus lupulus* L., determined by RAPD analysis. *Plant Breeding*, 118: 175–181.

TOWNSEND, M.S. and J.A. HENNING. 2005. Potential heterotic groups in hop as determined by AFLP analysis. *Crop Sci.*, 45(5): in press.

TOWNSEND, M.S., J.A. HENNING and D.L. MOORE. 2000. AFLP analysis of DNA from dried hop cones. *Crop Sci.*, 40: 1383–1386.

6

The processing of hops

C. Schönberger, Joh. Barth & Sohn GmbH & Co. KG

6.1 Introduction – the processing of hops

Hops have an enormous impact on beer flavour, even though they are only used in comparatively small amounts. Hops have been used in brewing since 1079,[39] and have preservative effects as well as giving beer its characteristic bitterness and aroma. Bitterness in beer is measured in International Bitterness Units (IBU). The typical range used to be between 20 and 50 IBU, with beers having even higher IBUs not considered extraordinary. Nowadays, there is a very clear trend towards mild beers, between 10 and 25 IBU and sometimes even as low as 6 or 7 IBU. Obtaining an intense hop aroma no longer necessitates high bitterness, because of the wide range of hop products now available. Recent research has revealed possible health attributes in hops, which may influence the importance of hops as a raw material, not only for brewing but also for other areas, such as 'nutraceuticals' and functional foods.

The composition of raw hops depends on the variety, the crop year, the growing area, the time of harvest, and the drying and storage conditions.[16,17] Historically, hops for use in brewing were supplied as cone hops. After picking and drying, the cones were compressed into bales and transported to the breweries. Storage conditions (e.g. time and temperature) affected the degradation of aroma and bitterness compounds in the hops, leading to inconsistencies in flavour and quality, and the utilisation rate of cone hops was only about 30%.

Industrial manufacture of hop products began in the 1960s. Many breweries realised that there were major advantages in using hop pellets and hop extracts, such as reduced transportation and storage costs, easier handling and more consistent dosing. Today the vast majority of hops are processed into hop pellets or hop extracts.

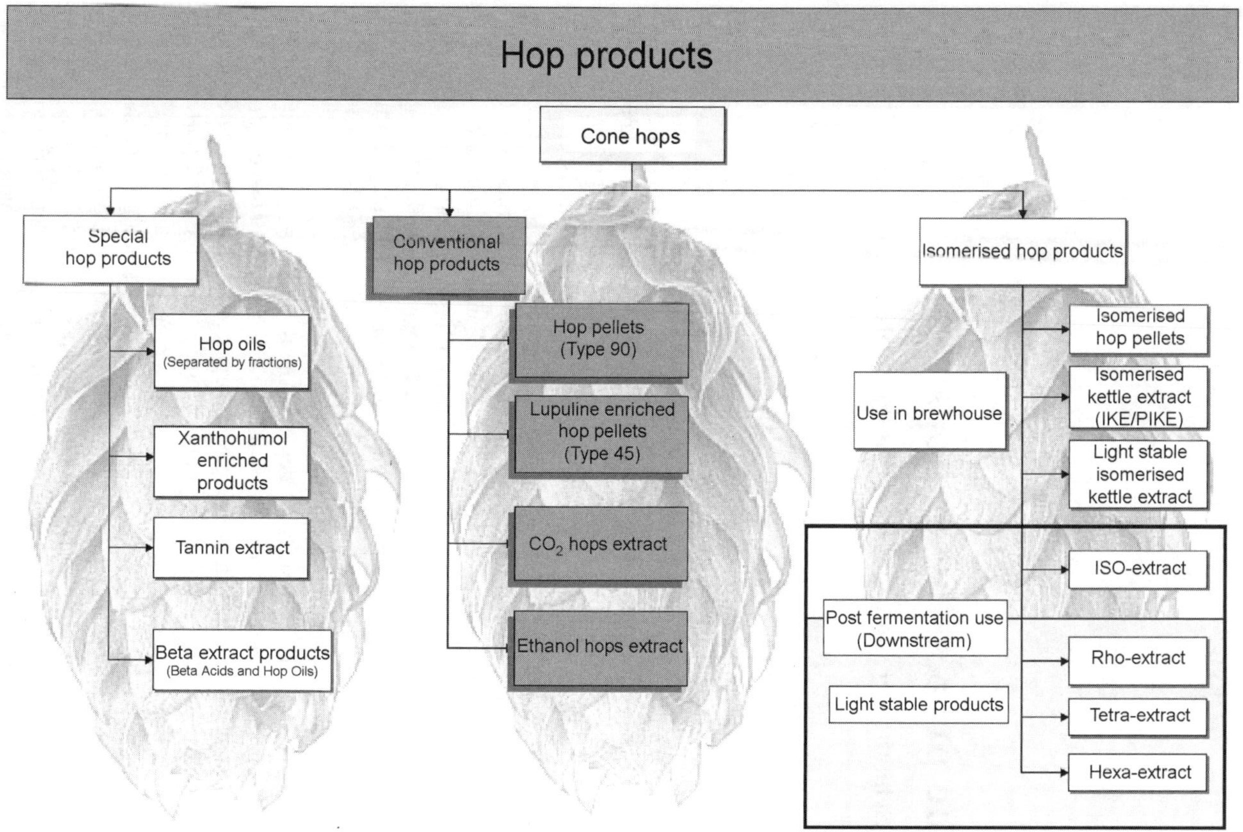

Fig. 6.1 Range of available hop products (*source*: Barth-Haas Group).

The key reaction in the brewing process is the conversion of alpha acid into iso-alpha acid to achieve the desired bitterness. In recent studies, the isomerisation reaction was found to be first order, with reaction rate varying as a function of temperature. Rate constants were determined to be $k_1 = (7.9 \times 10^{11})e^{(-11858/T)}$ for the isomerisation reaction of alpha acids to iso-alpha acids, and $k_2 = (4.1 \times 10^{12})e^{(-12994/T)}$ for the subsequent loss of iso-alpha acids to uncharacterised degradation products,[36] where T = temperature in K. The production of pre-isomerised hop products for use in the brewhouse or after fermentation was a logical consequence of understanding the reactions involved, both economically and in terms of brewing methods.

Apart from hop variety, parameters that influence the utilisation of components with brewing values (e.g. alpha acids, aroma compounds or polyphenols) include the type of hop product selected, dosage, time of dosage, intensity of wort boiling, pH value of the wort, trub separation (hot and cold trub), fermentation temperatures, form of fermentation tanks and the method of filtration and stabilising.[34] Figure 6.1 shows the range of hop products that are available.

More than 95% of the hops grown worldwide are processed into hop products, due to their storage stability, homogeneity, higher efficiency and ease of handling.

6.2 Hop pellets

There are two types of hop pellets, Type 90 and Type 45. Type 90 pellets are simply cone hops compressed into pellets. The name comes from the notion that 100% of the original raw hops will be reduced to approximately 90% due to losses during purification and processing. Nowadays, losses during processing are smaller and the final yield of pellets will be rather more than 90%. Type 45 pellets are lupulin-enriched hop pellets. The number 45 originally indicated a double enrichment of the Type 90 pellets, but since the original alpha acid content in raw hops can limit the degree of enrichment (a high alpha acid content causes problems during processing due to the lupulin-dependent consistence), the number 45 is rather confusing, and these pellets are better referred to as lupulin-enriched hop pellets.

The certification procedure for hop pellets (and hop extracts) includes an identification number for the batch, a code number for the processing plant, and the net weight, ensuring that malpractice is excluded. In contrast to whole hops, the blending of hops of different varieties and growing regions in pellets is permitted, provided that the hops have been grown within the EU and within the same crop year. In this case, the certificate must show clearly the respective percentages of the hops used.

The production of lupulin-enriched hop pellets involves the following steps:[8]

- Preparation: arranging the hop lots ready for processing
- Drying: final drying of selected hops to 8–10% w/w moisture content in a hop kiln

- Cleaning: elimination of extraneous matter, metals, leaves and stems
- Deep freezing: leaf hops are deep-frozen at temperatures between −30°C and −40°C
- Milling: crushing using hammer mills
- Sieving: carried out at −35°C, in several sieving steps, the hop powder is separated into lupulin and bracteole fractions in order to increase the concentration of bitter compounds
- Standardising: setting a specific alpha acid content by means of blending with the respective leaf fraction
- Homogenisation: using suitable orbital screw mixers
- Pelleting: using a pellet press to compress the hop powder into ring dies
- Instant cooling after pelleting: using belt coolers
- Packaging: into foil bags, excluding atmospheric oxygen, and then packing the bags into cartons.

Figure 6.2 shows the production of enriched hop pellets. The chosen level of lupulin enrichment will also determine the polyphenol content in the wort kettle and therefore influence the trub separations, colloidal stability, reductivity and flavour profile of the resulting beer. There are limits to hop dosage, since excessive amounts of spent hops and hot trub can make a complete separation in the whirlpool impossible.

Figure 6.3 shows the production of Type 90 hop pellets. The production of enriched hop pellets includes extra processing steps, including deep freezing, sieving and standardising.

Bitter hops are normally added at the beginning of the wort boiling, which results in a high isomerisation rate during boiling, whereas aroma hops should be added at the end of boiling to avoid the loss of precious compounds due to evaporation.

Pellets should be stored at 0–5°C in order to conserve the aroma compounds and prevent the alpha acids from deteriorating.

6.3 Hop extracts

Hop extracts reduce the mass and volume of natural hops to a higher degree than hop pellets, giving even greater cost savings in transportation and storage. Both factors, together with improved utilisation, offset part of the processing costs.

Many solvents have been used to produce hop extracts (e.g. methanol, dichloromethane and hexane), but only two, carbon dioxide (CO_2) and ethanol, now meet current standards in terms of safety, environmental impact, natural occurrence and cost. The two solvents yield hop extracts with different compositions. Ethanol dissolves a broader range of hop components, while carbon dioxide is more selective for the resins and aroma components. As the solubility behaviour of CO_2 can be fine-tuned by changing temperature and pressure, the composition of the extract can be varied, and fractionating separation or fractionating extraction is used for products with special applications.

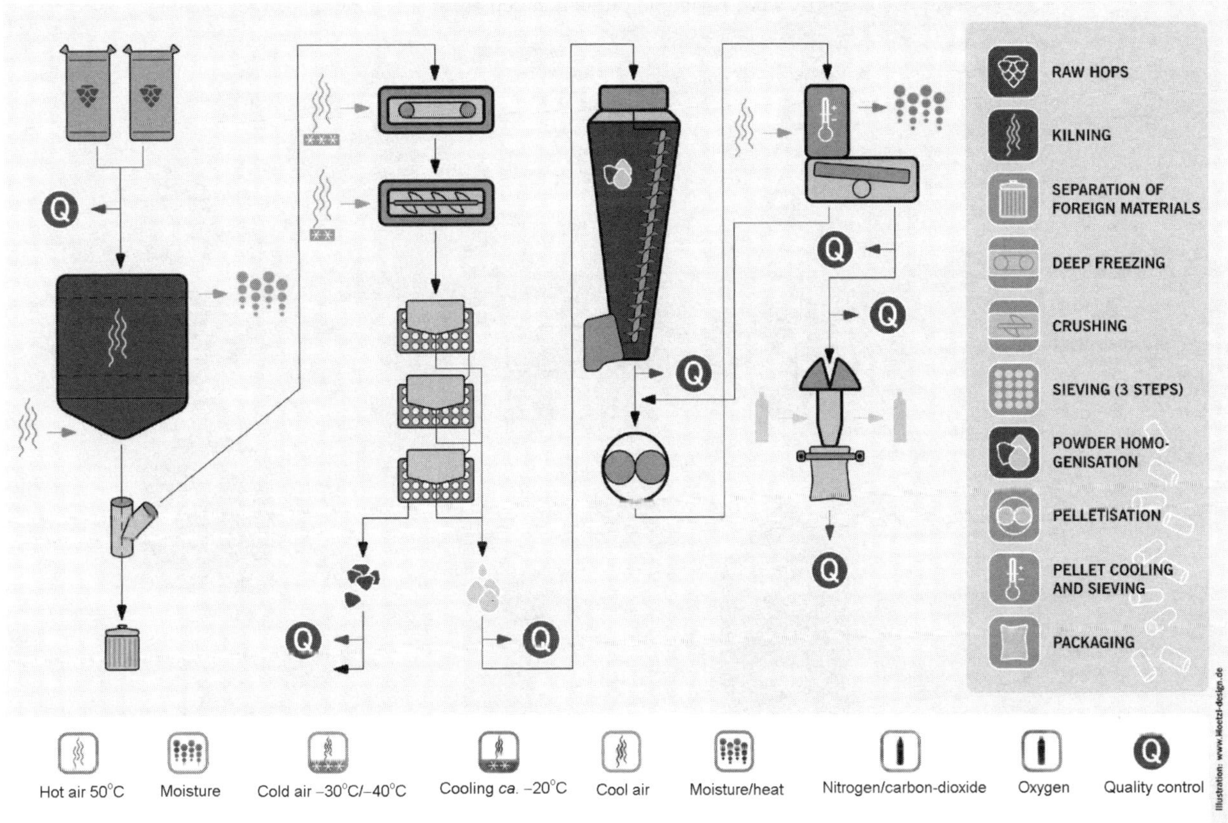

Fig. 6.2 Flow chart of the production of hop pellets (type 45) (*source*: Hopsteiner).

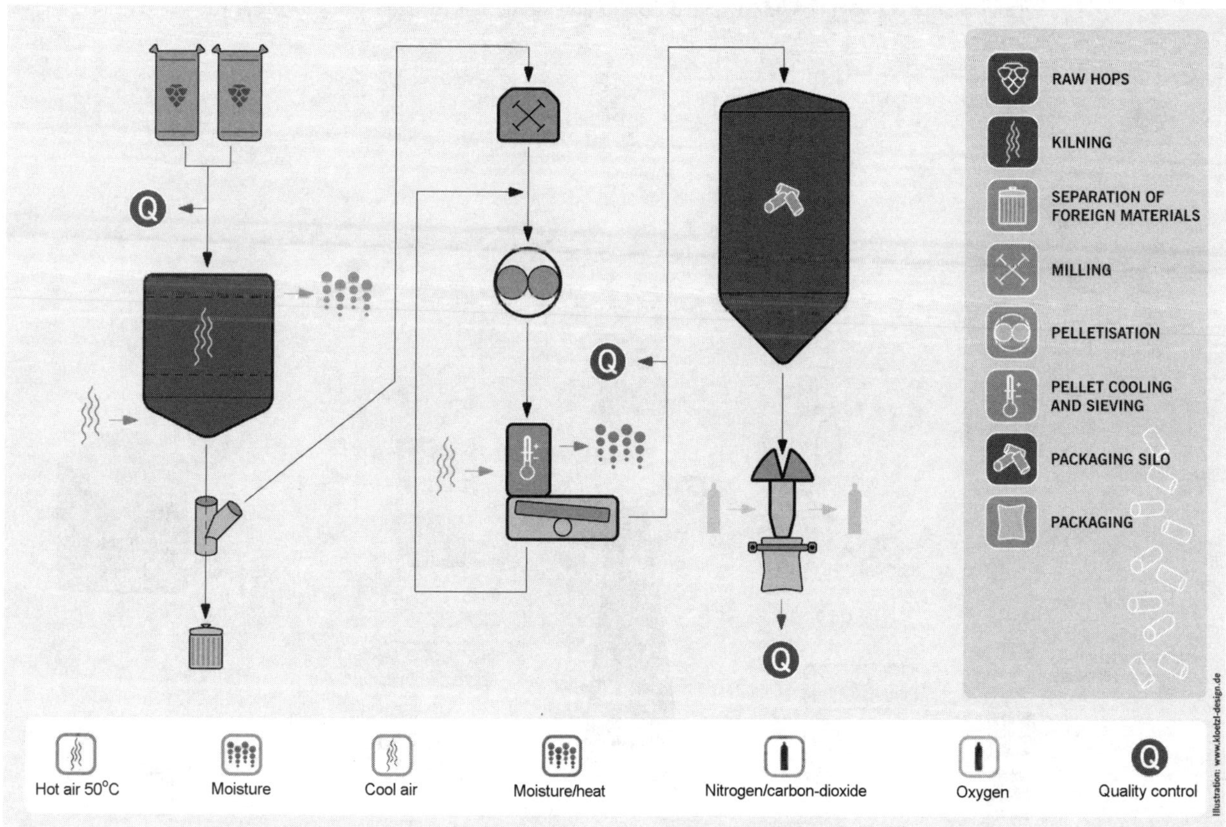

Fig. 6.3 Flow chart of the production of hop pellets (type 90) (*source*: Hopsteiner).

6.3.1 Extraction with carbon dioxide

Extraction with liquid CO_2 began in the 1970s in the UK and Australia. The high selectivity for alpha and beta acids in this type of extraction made the extract suitable for a subsequent isomerisation step. The usual parameters for liquid extraction are 60–65 bar at a temperature range of 5–15°C. The process takes place at constant pressure and the change of density is obtained exclusively by varying the temperature. Liquid CO_2 extracts have a lower content of hard resins and polar bitter substances and less chlorophyll, since a smaller substance spectrum is dissolved.

Hop extraction using liquid carbon dioxide is now far less important than it used to be, with more than 95% of extracts being produced using supercritical CO_2. Application of CO_2 in the supercritical phase (>73 bar; >31°C) gives more flexibility in processing. It is also much more economical because the extraction times are shorter. Figure 6.4 shows a chart of the extraction process. The container is filled with hop pellets and the lid is closed. When the required pressure is achieved (200–300 bar), the solvent passes through the material until extraction is completed. The extraction temperature is typically 40–60°C. The carbon dioxide, containing the extracted components as bitter and aroma substances, reaches the separation tank where the pressure is lowered to 60–80 bar. It is then evaporated in the heat exchanger. The efficiency of extraction can be determined by analysis of either the extract or the spent hop material. The separation is done in batches, but a semi-continuous operation is feasible if several extractors are used, operated either in parallel or in series. Commercial extraction plants generally have three or four extractors. Figure 6.4 shows the production process of CO_2 extraction.

Prior to packaging, the extract is homogenised and analysed. Depending on consumer requirements, the extract is usually filled in cans, although drums or bulk containers are preferred for use in automated dosing systems.[11]

The alpha acid content of CO_2 extract is 35–55%, the beta acid content may vary from 15% to 40%, and the hop oil content from 3% to 12%, depending on the variety.

The major features of extraction using CO_2 are:

- CO_2 is a natural solvent and also a by-product of brewing
- No chemical reactions take place because processing conditions are gentle
- Aroma compounds are obtained quantitatively
- The stability of CO_2 extract is excellent. When stored cold (0–5°C), a shelf-life of up to 8 years is guaranteed. For storing at ambient temperatures (<25°C), the shelf life is 3 years.

6.3.2 Extraction with ethanol

Ethanol extraction involves mixing whole hops with 90% ethanol in a carousel extractor. The ethanol flows through the hop bed counter to the flow of hops, and becomes enriched with hop components. After passing through the extractor, the hops leave as spent material. The solution of ethanol and polar

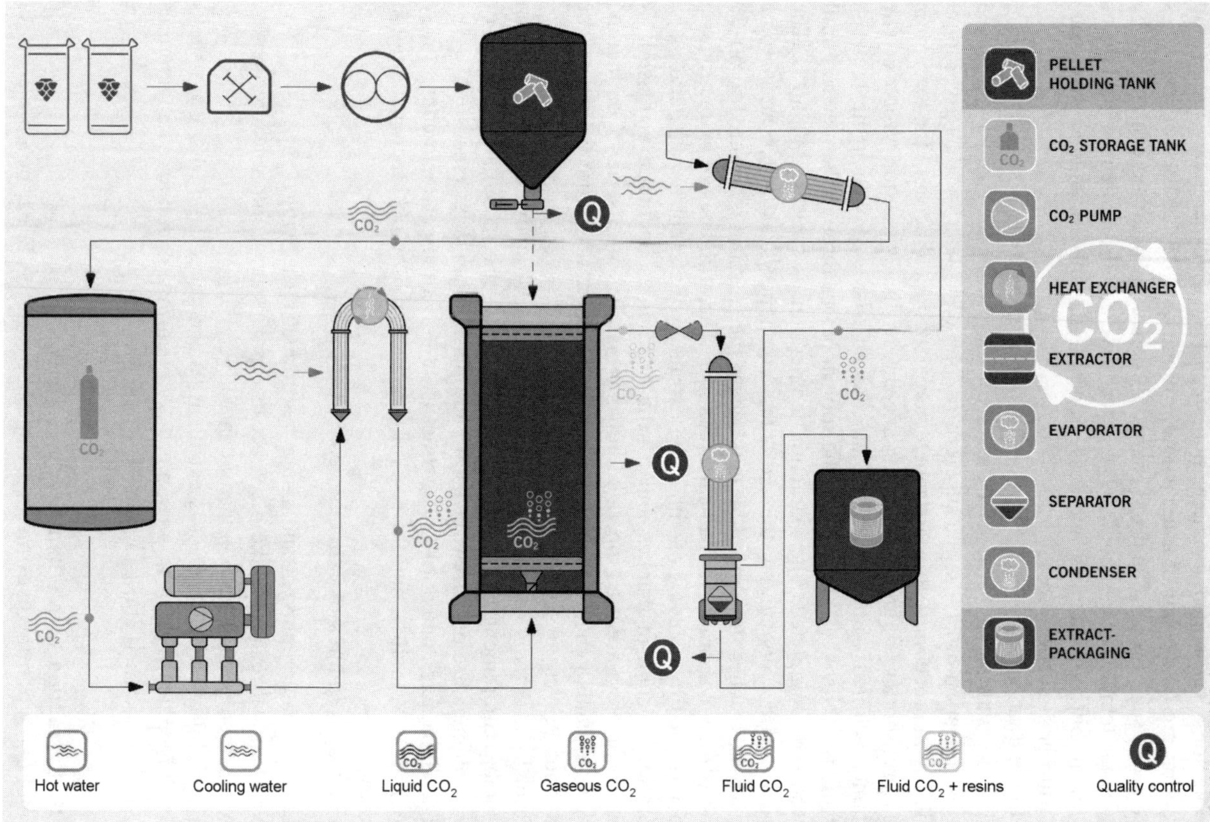

Fig. 6.4 Flow chart of the production of CO_2 extract (*source*: Hopsteiner).

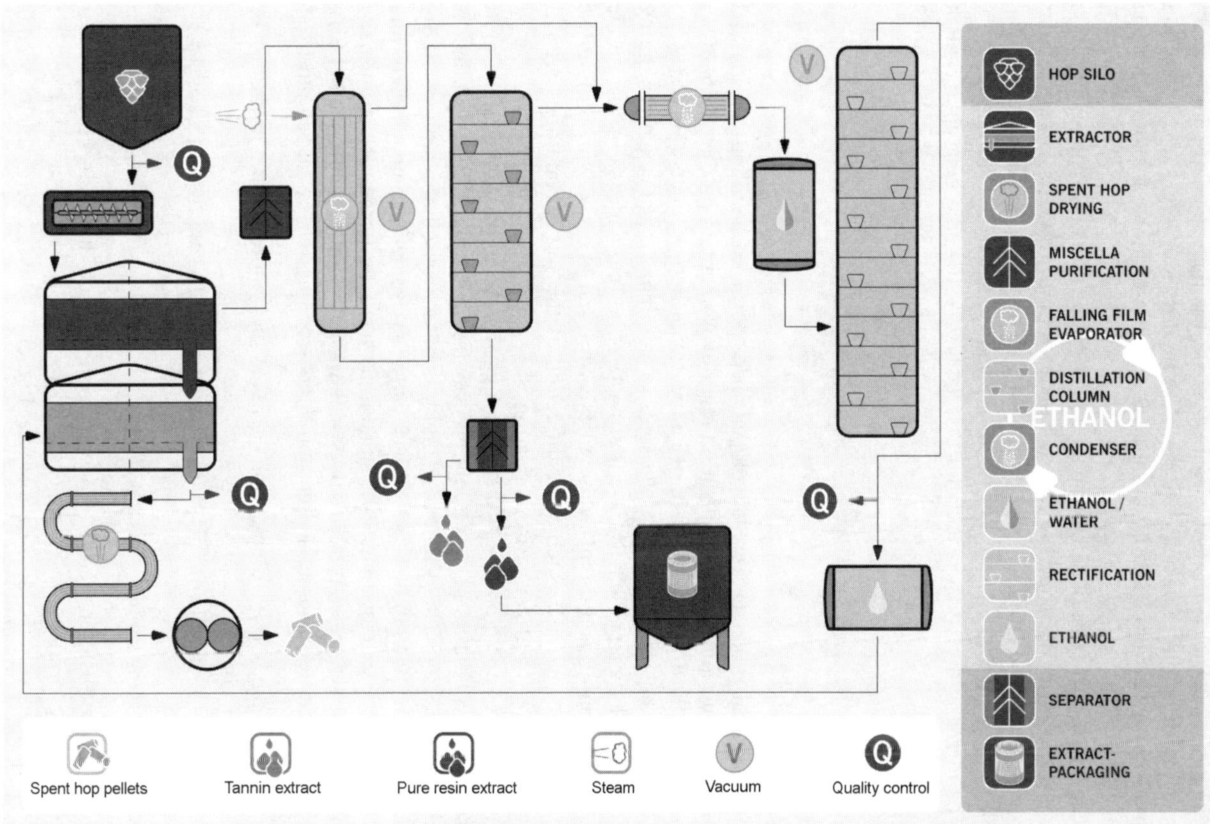

Fig. 6.5 Flow chart of the production of ethanol extract (*source*: Hopsteiner).

hop material is referred to as 'miscella' and is pumped to the evaporation stage, where the ethanol is removed in a vacuum evaporator. The resulting raw extract contains hop acids, hop oils, hard resins and water-soluble components (as polyphenols, salts, proteins and carbohydrates). It is then separated into a resin extract and a tannin extract by means of separators (see Fig. 6.5). Extraction with ethanol is accompanied by a minor conversion of alpha acids to iso-alpha acids (0.5–1.5%).

The benefits of ethanol extraction are lower production costs (no pelletisation step prior to extraction) and the fact that the resulting extract has a very similar composition to the original hops but with reduced polyphenol content. The stability of ethanol extract is very good, especially under cold storage (0–5°C).

The alpha acid content of ethanol extract is 20–50%, the beta acid content may vary from 15% to 40% and the hop oil content from 3% to 12%, depending on the variety.

6.4 Isomerised hop products

Isomerised hop products were developed in response to handling, dosing and economic requirements. The utilisation rates of isomerised products (45–80%) are considerably higher than those of conventional hop products (30–35%).

Normally, the conversion of alpha acids to iso-alpha acids takes place during wort boiling. Efficient isomerisation during wort boiling depends in part on the duration of boiling and the pH value of the wort, but because wort boiling has other objectives as well (such as acidification of the wort), conditions are not always suitable for effective isomerisation.

In isomerised hop products, as the name suggests, the conversion of alpha acids to iso-alpha acids takes place not during brewing, but when the hop product is originally processed. There are two main types of isomerised hop products, those suitable for the use in the brewhouse, such as isomerised hop pellets and isomerised kettle extracts, and those suitable for use after fermentation, normally referred to as downstream products or post-fermentation bittering products (PFB).

The production of isomerised hop products (see Fig. 6.6) is carried out by:[11]

- converting the poorly soluble alpha acids to their more soluble potassium or magnesium salts;
- using alkaline conditions, pH 8–11 (in contrast to the pH in wort of 5.2–5.6); and
- raising the temperature.

Although utilisation rates of isomerised hop products are higher than traditional hop products, it is important to balance the need to use a smaller amount of isomerised product against their higher purchase price, which depends on the varying price of raw materials, processing costs, utilisation rates of the particular brewery, and the expected utilisation of the isomerised product. The impact on flavour must also be taken into account.[34]

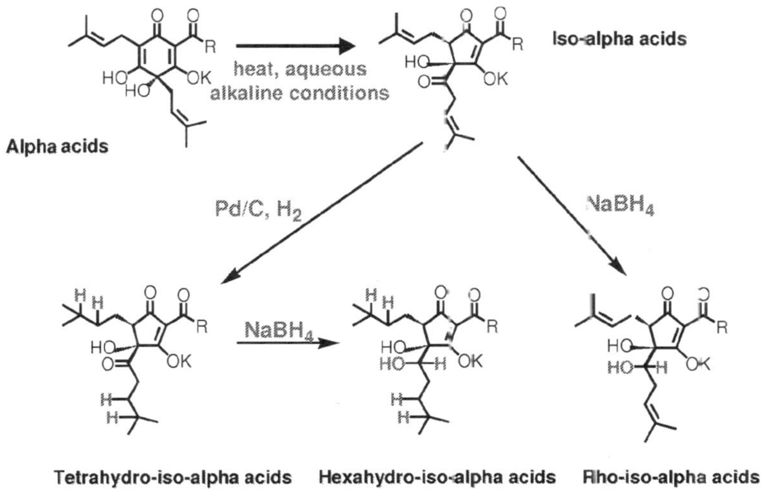

Fig. 6.6 Conversion of alpha acids to their isomerised and reduced forms.

6.4.1 Isomerised hop pellets

The production of isomerised hop pellets is similar to that of hop pellets, except for the addition of food-grade magnesium hydroxide (1–5%) to the hop powder during processing. After processing, the pellets are packed in soft evacuated packs which have been back-flushed with an inert gas such as nitrogen. For 10–14 days, depending on various factors, the pellets are exposed to controlled temperatures between 45 and 55°C. The isomerisation rates are monitored during heat treatment by means of HPLC analysis. The isomerisation rate is normally 90–95%.

The major benefit of isomerised hop products is the increased availability of bittering components (45–60%). Isomerised hop pellets require only a short contact time with the wort (10–15 min) in order to achieve maximum utilisation of bittering constituents.

The stability of iso-alpha acids in isomerised hop pellets is good. At ambient temperatures, losses of iso-alpha acids are likely to be less than 5% over two years. The addition of magnesium hydroxide and subsequent heat treatment change the composition of the aroma substances, with the amounts of low-molecular-weight polyphenols, terpenes, sesquiterpenes and linalool being reduced. Therefore, due to potential flavour effects, isomerised pellets are primarily a bittering replacement. If used as an aroma dosage in the kettle, the flavour impact of the iso-pellets has to be determined by brewing trials.[31]

6.4.2 Isomerised kettle extracts

Isomerised kettle extracts (IKE) increase the utilisation of bitter acids (45–60%) and are produced through CO_2 extraction. There are currently three products:

IKE (Isomerised Kettle Extract), PIKE (Potassium-Form Isomerised Kettle Extract), and LIKE (Light stable Isomerised Kettle Extract), though IKE is the most important. IKE is produced by mixing and heating the pure resin extract with magnesium oxide (3–6%). After isomerisation, the iso-alpha acids are present as magnesium salts; the magnesium is then removed using a strong acid, leaving the iso-alpha acids in their free form. The resulting IKE has handling characteristics very similar to a conventional pure resin extract.

PIKE is produced by heating pure resin extract in contact with aqueous potassium carbonate/hydroxide solution. The result is an isomerised kettle extract in which the iso-alpha acids are present as the potassium salts. The conversion rate for all products is 90–95%. Depending on the processing conditions, the resulting hop oil content may be low. The iso-alpha acid content is typically about 30–55% w/w. The beta acid content is about 15–35% and the oil content is 5–10%, each depending on the variety and the primary extraction conditions.

Isomerised kettle extracts have similar benefits to isomerised hop pellets – increasing the utilisation of bittering constituents. The extraction of the iso-alpha acids into the boiling wort is completed within 10–15 minutes.[34]

LIKE consists of reduced iso-alpha acids (rho-iso-alpha acids) and the other components of the original pure resin extract (beta acids and hop oils). LIKE can be used for beers that are packed in clear bottles and therefore need to be light stable. Although the use of downstream products for this purpose is more common, some brewers prefer to add LIKE in the brewhouse for its additional foaming and anti-microbial properties.

6.4.3 Dosing of hop kettle products

Dosing can be simplified by standardising the alpha (or hop oil content) per foil or container. Foils can be emptied directly into the kettle or via a flushed container; cans have holes punched in them and are then immersed in the wort in a rinsable container or basket. The corresponding amounts of pellets or extract can also be placed in a mixing container, from which the material is automatically dosed by flushing with either water or wort into the kettle. For larger units, suitable emptying devices fill a buffer tank and dosing is conducted via the buffer tank. Mechanical or pneumatic conveyors are used for hop pellets; extract can be pumped.[11]

Automatic preparation and dosing of hop pellets and hop extracts requires suitable bulk containers (up to 200 litres), or foils with a maximum content of 140 kg in a carton. The cartons are cut open automatically and emptied into a buffer tank, and the pellets can then be transported via a scale and a pneumatic feeding unit to the wort kettle. The hop pellets should be kept in a cool area if stored for any length of time. The alpha acid content can be a problem in order to achieve a gentle mechanical feed of pellets, as the pellets tend to stick if the alpha acid content is higher than 14% w/w. Generally, mass flow conveyors are more gentle than screw or pneumatic conveyors.

The tank or barrel of extracts is placed in a heating chamber (approx. 40–45°C) to liquefy the contents, which takes up to 24 hours depending on the type of extract. The barrel is then emptied into a stirred buffer tank and replaced by the next barrel for warming. A pump transfers a defined amount of hop extract from the stirrer tank to the kettle. Alternatively, the extract can be pumped to a mixing tank, in which case it reaches the kettle suspended in water or wort.

6.4.4 Isomerised/reduced downstream products

The brewing industry has to be highly flexible to meet fast changes in consumer habits. Downstream hop products, often referred to as post-fermentation bittering products, are becoming more and more important because they offer, among other properties, the possibility of producing various different kinds of beer from a single brew. The current market share of downstream products is greater than 10%, and further increases in market share are expected.

Since these products are added to the final beer, they must have high purity and be essentially free from insoluble hop resins. For all post-fermentation bittering products, special attention should be paid to the point of addition, which should be close to a region of turbulent flow and well separated from that of any other additives (especially from carbonation). An accurate dosing system is crucial, since the continuous addition is often measured in ml/hl.[21,43] Figure 6.7 gives an example of where the addition of PFB can take place.

Rho-iso-alpha acids are suitable for the production of light-stable beers. Together with tetrahydro-iso-alpha acids or hexahydro-iso-alpha acids, they prevent the formation of the light-induced flavour substance, 3-methyl-2-buten-1-thiol (3MBT, odour threshold 10 ng/l) due to the hydrogenated and reduced

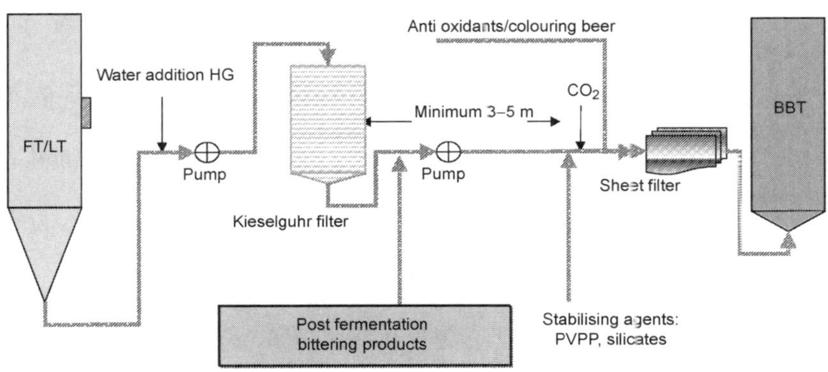

FT: Fermentation tank
LT: Lagering tank
BBT: Bright Beer tank
HG: High gravity

Fig. 6.7 Setup for the addition of post fermentation bittering products.

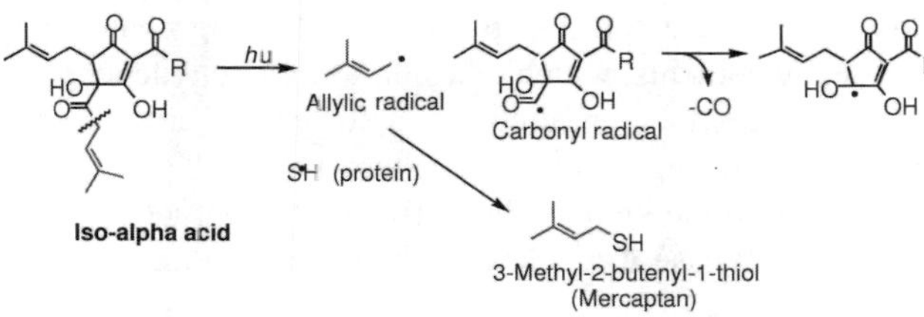

Fig. 6.8 Formation of light struck flavour.

side chains (see Fig. 6.8). These beers must then be bittered with 100% reduced isomerised hop extracts, and no other source of non-reduced iso-alpha acids may be present in the wort/beer. This implies that the yeast has to be absolutely free of iso-alpha acid residues.[46] However, although this prevents the formation of 3MBT, other, similar compounds with possible negative flavour effects may be formed.[24] Furthermore, the use of light-stable hop products does not prevent the formation of other staling flavour compounds in beer that occur naturally due to warm storage temperatures or oxidation reactions.[2] Comparing the flavour stability of conventionally brewed beers with beers produced using reduced hop products revealed that the use of the latter retards the formation of the cardboard flavour but may be associated with a wood-like or a roasted flavour during ageing.[55] It is important to note, however, that the stability of the different homologues varies. The non-reduced *trans*-iso-alpha acids are reported to be significantly less stable than the *cis*-iso-alpha acids, and the fate of *trans*-iso-alpha acids adversely affects the bitterness of beer.[9]

Iso-extract (isomerised hop extract, post-fermentation bittering (PFB), Isohop)
Iso-extract is often used to adjust bitterness prior to final filtration. It is produced using CO_2 extraction. Different production methods include:

- selective removal of alpha acids from beta acids and other resins with alkali prior to isomerisation;
- selective removal of beta acids and other resins from an aqueous solution of iso-alpha acids after isomerisation; or
- selective removal of iso-alpha acids from beta acids and other resins present in an isomerised kettle extract after isomerisation.

The crucial factor is that alpha acids and iso-alpha acids are more acidic than beta acids. Thus, by carefully controlling pH during processing, beta acids can be separated without the use of organic solvents. Recent research (on mice) showed that iso-alpha acids may have health benefits in their ability to significantly suppress fat accumulation and thus body-weight increase, reduce cholesterol and triglyceride content in the liver and reduce blood sugar levels in individuals suffering from mild diabetes.[33,54] Iso extract is commercially supplied in aqueous solutions containing 20–30% pure iso-alpha acids. The utilisation rate in beer is typically about 80%.

Rho extract (reduced iso extract, rho, dihydro-iso-alpha extract, Redihop)
The chemical process for producing Rho extract involves reducing iso-alpha acids to dihydro-iso-alpha acids using borohydride in an aqueous solution at pH 10. After the reduction step, the boron residues are completely removed. The free rho-iso-alpha acids are then solubilised into an aqueous solution by the addition of potassium hydroxide. Rho extract is commercially supplied in aqueous solutions containing 10–30% pure rho-iso-alpha acids. The bitterness properties of rho-iso-alpha acids differ from those of iso-alpha acids, and are reported to be smoother. The bitterness intensity is about 70% that of iso-alpha acids.

Tetra extract (tetrahydro-iso-alpha acids)
There are two basic approaches for the production of tetrahydro-iso-alpha acids, since they can be produced from alpha acids or beta acids. The production of tetrahydro-iso-alpha acids from alpha acids is by far the most important and consists of an isomerisation step followed by hydrogenation. Hydrogenation requires hydrogen gas, pressure and the presence of a catalyst, usually palladium supported on carbon. The hop acids are in aqueous solution during hydrogenation.

The production of tetrahydro-iso-alpha acids from beta acids requires an additional oxidation step prior to isomerisation and hydrogenation. Beta acids are becoming more important in other products due to their anti-microbial properties, and it is no longer considered economical to use them to produce tetrahydro-iso-alpha acids.

The sensory bitterness of tetrahydro-iso-alpha acids is very different from that of the original iso-alpha acids, not only in terms of perceived bitterness (reported to be 100–170% compared to iso-alpha acids, strongly depending on the base beer's attributes) but also bitterness attributes, which are sometimes described as being metallic or harsh.[47,52] The main purpose of using tetra-hydro-iso-alpha acids, apart from their being light-stable, is for their very good foam enhancement.

The typical dosing rate for tetrahydro-iso-alpha acids is 3–5 ppm. Tetrahydro-iso-alpha acids are extremely stable towards oxidative damage,[9,24] and are commercially supplied in an aqueous solution containing approximately 10% pure tetrahydro-iso-alpha acids.

Hexa extract (hexahydro-iso-alpha acids)
Hexahydro-iso-alpha acids are produced from tetrahydro-iso-alpha acids in a reduction reaction with sodium borohydride. Hexa extract is commercially supplied in an aqueous solution containing a 10% concentration of a mixture of hexahydro-iso-alpha acids and tetrahydro-iso-alpha acids in a ratio between 50:50 and 60:40. Since the conversion of tetrahydro-iso-alpha to hexahydro-iso-alpha acids follows a diminishing progression, a higher yield of hexahydro-iso-alpha acids necessitates a longer reaction time and therefore increased costs. The sensory bitterness of hexahydro-iso-alpha acids is reported to be 110% and the mixture is reported to be 130% compared to iso-alpha acids. The bitterness

Table 6.1 Properties of reduced/isomerised extracts

Product	Concentration in commercial products (%)	Light protection	Foam enhancement	Relative bitterness (iso-alpha acids = 1)
Iso extract	20–30	no	no	1
Rho extract	10–35	yes	no	0.6–0.7
Tetra extract	9–10	yes	very good	1.0–1.7
Hexa extract	10	yes	good	1.3

attributes are described as being quite similar to those of iso-alpha acids. Hexa extract also acts as a foam-enhancing agent similar to tetra, but with subtle differences. Table 6.1 summarises the properties of the different extracts.

6.5 Other hop products

6.5.1 Base extracts

After separating the alpha acids from the hop extracts, the remaining beta acids and hop oils are referred to as base extract or beta extracts or by trade names. Extracts containing beta acids and hop oils can be used in the wort kettle to impart a particular hop aroma, and also to prevent excessive foaming in the brewing kettle. They also have anti-microbial properties when the bittering is carried out post-fermentation. For light-stable beers, it is necessary to use base extracts containing no alpha acid or iso-alpha acid residues.

Beta acids have become very important recently due to their anti-microbial properties against Gram-positive and some Gram-negative bacteria (e.g. *Heliobacter pylori*). Beta acids are supplied as anti-microbial solutions to the sugar and ethanol industry in standardised concentrations to prevent bacterial growth during processing.[42]

6.5.2 Hop oils/hop oil products

Normally the dosage of hops is based on the amount of alpha acids. In order to achieve the right hop aroma, however, a dosage based on the hop oil content, or the content of compounds that affect character, such as linalool, is preferable.[28] Methods to influence the hop aroma in beer include late dosage of hops and lowering the wort temperature to <90°C prior to the whirlpool rest.[25,34]

Hop oils are prepared using fractionating CO_2 extraction, setting the pressure at different levels in order to separate the hop oil from the resins and the CO_2. The separation is not complete and is therefore followed by distillation (vacuum and/or molecular distillation) to obtain the required composition of hop oil compounds according to the specific variety. Though hop oils are highly concentrated, they are still extremely costly. Blends and standardised hop oils of various compositions are commercially available.

Table 6.2 Identified compounds in hop oil[40]

	Number	Monoterpenoids	Sesquiterpenoids
Hydrocarbons	85	18	49
Alcohols	78	12	34
Aldehydes	20	2	1
Ketones	52	1	3
Carbon acids	34	1	0
Esters	62	5	0
Bases	9	0	0
Sulphuric compounds	41	3	5
Acetals	1	0	0
Ether	3	2	0
Halogenic compounds	1	0	0
Furans	7	3	0
Epoxides, pyrans, coumarins	16	2	11
Total	409	49	103
%		12	25

Adding hop oil in the kettle will result in low recoveries; addition before fermentation will result in a different hop oil character due to the chemical reactions of volatile compounds during fermentation and the impact of yeast metabolism. For addition prior to filtration, hop oils have to be dispersed in a suitable carrier (ethanol or food-grade emulsifiers). Adding hop oil to the kettle will require 1–5 g/hl, a pre-fermentation addition will require 0.5–2 g/hl, and adding prior to filtration requires 0.05–0.3 g/hl, depending on the intensity of aroma and taste required and the characteristics of the base beer.

Hop oils can be divided into a non-polar (hydrocarbon) fraction (40–80%) and a polar (oxygenated and sulphur-containing) fraction.[35,38] Screening by Nijssen and others has identified over 400 compounds[40] (see Table 6.2).

The hydrocarbon fraction mainly consists of terpenes as monoterpenes (e.g. linalool, myrcene, geraniol), mono- and bicyclic monoterpenes (e.g. limonen and beta pinen) and sesquiterpenes (e.g. humulene and charyophyllene). A terpenoid alcohol, 2-methyl-3-buten-2-ol, has sedative hypnotic effects and is often used together with valerian.[1] Moir[38] reported that the following compounds affected beer character: linalool, linalooloxid, citronellol, geraniol, geranylacetat, alpha-terpineol, humulen-8,9-epoxid, alpha-eudesmol, t-cadinol and humulenol, though only linalool, geraniol and humulen-8,9-epoxid had an impact on aroma. Using flavour dilution analysis, Schieberle and others have identified 20 odorants in the variety 'Spalter Select'. Table 6.3 shows the relative potency of the more important compounds.[44]

Further research on the flavour impact of linalool (described as 'floral') revealed that the enantiomeric R form has a flavour threshold in beer of 2–5 μg/l.[27] The threshold of R- compared to S-linalool in air is 80 times lower. The chiral distribution of linalool in raw hops and in hop oil is 94:6 (R:S-linalool).[48]

Table 6.3 Potent aroma compounds in hop oil[44]

Most potent	*Trans*-4,5-epoxy-(E)-2-decenal[1]
↓	R-linalool, myrcene
	Ethyl 2-methylpropanoate[1]
	Methyl 2-methylbutanoate[1]
	(Z)-1,5-octadien-3-one[1], nonanal
	(E,Z)-1,3,5-undecatriene[1]
	1,3-(E),5(Z),9-undecatetraene[1]
	Propyl 2-methylbutanoate
Least potent	4-ethenyl-2-methoxyphenol, 1-octene-3-one

[1] Previously unknown hop components.

During the brewing process, this ratio can change to a final ratio of linalool in beer of 52:48 (R:S-linalool).[27] Ongoing research projects are aimed at clarifying whether these aroma compounds are present in hops in their free form, or linked to other compounds.

Hop aroma products (hop essences)
A range of beer-soluble hop aroma products have been developed based on advances in production of hop essences after late or dry hopping. These products are supplied in food-grade solvents (ethanol or propylene glycol) and are used as post-fermentation bittering products before or after final filtration. They are produced from hop oils using chromatographic fractionation techniques, and the resulting standardised aroma products are named after hop varieties or specific flavour attributes (floral, herbal, citrus, etc.).[22,32,47] The ability to add a specific hop aroma without bitterness makes these products very suitable for low-alcohol or alcohol-free beverages.

6.5.3 Xanthohumol and others

Polyphenols comprise a vast spectrum of compounds that have one or more phenolic functional groups. They are 'secondary plant components', meaning that they are used not for plant growth but for other qualities, such as pigmentation or protection against pests. Polyphenols are reported to influence haze formation in beer. To increase the physical stability of beer, polyphenol-free hop extracts can be used, or the specific polyphenols that influence haze can be removed using PVPP during filtration.[13,14] Figure 6.9 shows the classification of the different groups of polyphenols. They may exhibit the following effects on human health:[5]

- Anti-oxidative
- Anti-carcinogenic (reducing the risk of cancer)
- Anti-inflammatory (inhibiting inflammation)
- Anti-microbial
- Modulating the immune reaction

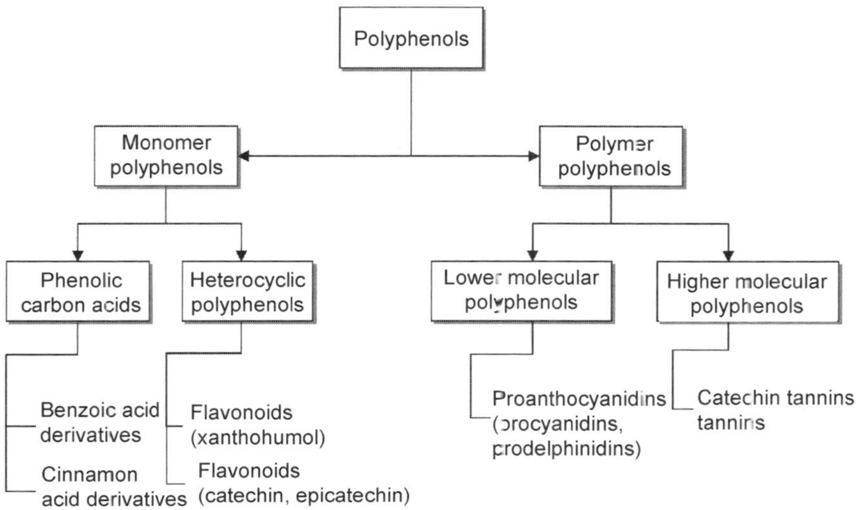

Fig. 6.9 Classification of polyphenols.[29]

- Controlling blood pressure
- Regulating blood sugar.

Recent research findings show that aroma varieties with a low alpha acid content generally have a higher polyphenol content than high alpha acid varieties.[29] Considerable research has been carried out over the last five years on xanthohumol, a prenylated chalcone, and an astonishing diversity of advantageous effects have been demonstrated, particularly an anti-carcinogenic potential.[4,5,15,19,49,51] Xanthohumol has also been shown to have the following properties:

- Antiviral
- Anti-fungal
- Anti-plasmodial
- Against osteoporosis
- Anti-atherogen.

As a logical consequence of these findings, all major hop processors are offering xanthohumol-enriched extracts. The taste thresholds of xanthohumol and iso-xanthohumol were found to be 0.9 mg/l for xanthohumol (described as being slightly astringent and raspy in aftertaste) and 0.5 mg/l for isoxanthohumol (described as being slightly bitter).[18] Depending on the extraction process, the xanthohumol content in these products varies from 2% to 40%.[15] Xanthohumol is offered by various manufacturers in its natural form.

The xanthohumol content in hops varies between 0.2% and 1.0%, depending on the variety.[8] The xanthohumol content in conventional hop products is lost during the brewing process, with a few exemptions,[51] so commercially produced beer contains only very small quantities of xanthohumol (less than 0.2 mg/l). Losses are due to conversion to iso-xanthohumol, precipitation and absorption by yeast cells or haze particles, as well as the poor solubility of xanthohumol. With isomerisation, xanthohumol loses most of its positive characteristics. Using xanthohumol-enriched hop products combined with a late hop dosage during wort boiling is quite effective at increasing the xanthohumol content in the final beer. However, it has been observed that the xanthohumol content may influence fermentation by slowing down the growth of yeast cells during the first 20 hours of fermentation.[41] A low pitching rate and omitting stabilisation also help to increase the xanthohumol content in beer. Newer findings showed that high molecular substances in roasted malts are able to form complexes with xanthohumol. Adding xanthohumol dissolved in roast malt extracts resulted in beers with more than 10 mg/l xanthohumol. On an industrial scale, a wheat beer with more than 1 mg/l xanthohumol was produced.[53]

Another prenylated flavonoid with remarkable properties appears to be 8-prenylnaringenin (8-PGN). 8-PGN is a highly active phyto-oestrogen, having similar effects to the female sex hormone oestrogen, and therefore could become important for hormone replacement therapy.[6,22] Recently, the well-known polyphenol resveratrol (which accounts for the major benefits associated with measured red wine consumption) was also found in hops, though in only very small quantities.[7] It was also found that a mixture of all hop proanthocyanidins showed higher anti-oxidative activity than single flavonoids or proanthocyanidins.[5] Certain fractions of higher molecular hop polyphenols in tea, which can also be found in hops, are reported to have anti-bacterial and anti-caries effects.[37] Figure 6.9 shows how polyphenols are classified. All these findings have led to an increase in the number of patents dealing with the extraction of polyphenols from hops and other applications.

Polyphenol-enriched hop products can be produced from spent hops, so the raw material for these products is at present remarkably inexpensive, and any possible products therefore highly profitable. Apart from all the health-related benefits, trials with specific polyphenol fractions (proanthocyanidins, flavonol glycosides and prenylated flavonoids) have revealed significant effects on the mouthfeel of beer, increasing the 'fullness' of beer, improving flavour stability and reducing lautering time.[20] The use of products enriched in xanthohumol, 8-PGN, or polyphenols in general, seems at present to be due to their associations with various health effects rather than for conventional brewing reasons. Table 6.4 shows the contents of different polyphenols in dried hops.[5]

6.6 Hop, hop products and relevant beer analyses

The analysis of hop components has been a major challenge for over 100 years. Current methods typically measure single components, and are made possible by our improved chemical understanding of the nature and complexity of hop components and the sophistication of available instruments. Analyses are

Table 6.4 Contents of different polyphenols in dried hop cones[5]

Substance	Concentration (%)
Total polyphenols	8
Proanthocyanidins	5
Flavonoids	2
Catechin	0.3
Quercetin	0.2
Phenolic carbon acids	0.04
Ferula acid	0.01
Resveratrol	<0.01
8-prenylnaringenin	0.01
Xanthohumol	1

primarily conducted to control the bitterness of the beer. However, since the relationship between bitterness and the content of bitter acids is non-linear, and the taste perception of bitterness and its interaction with other taste qualities is not yet completely understood, such analysis still presents a challenge.[23] Hop analyses have to cover four categories:

- Hop orders (alpha acid contents)
- Quality Assurance (QA), alpha acids, iso-alpha acids, HSI, hop aroma compounds, etc.
- Food safety, pesticides, heavy metals, mycotoxins, nitrate, etc.[45]
- Research.

An extensive compilation of the chemical structures of all hop components and their properties relating to different methods of analysis is given by Verzele and de Keukeleire.[50] Accurate determination of the alpha-acid content is a crucial part of every contract, but different methods will always give different results. There are basically two approaches for analysing the alpha acid or iso-alpha acid content: non-specific methods using conductivity or spectro-photometric methods, and HPLC analysis. Non-specific methods have been used for more than 100 years, HPLC methods only for about 25 years. HPLC methods are the most accurate, but many breweries may not be able to afford the equipment. Reference methods are found in EBC Analytica or the ASBC methods of analyses.[3,10] Table 6.5 shows an overview of the most important methods. The results of the various methods are not interchangeable; for instance, EBC 7.4 tends to produce relatively low values, and 7.5 relatively high values. ASBC – Hops 6 also tends to produce relatively high values. The HSI (hop storage index) indicates whether the hops are fresh or aged (0–1) (<0.31 regarded as fresh), and in this context, variations in varieties depending on the crop year have to be considered. EBC 7.9 is used to analyse or compare post-fermentation bittering products.

Analyses for food safety are gaining more importance due to the increasingly strict regulations regarding consumer protection and traceability of products.

Table 6.5 Important analytical methods to evaluate hops and hop products

Method	EBC	ASBC
Conductometric value	7.4	Hops-6B
Conductometric value (Woellmer modification) for hop and hop products	7.5/7.6	Hops-8B(II)
Alpha acids (spectrophotometric)	–	Hops-6A, -8B(i)
Alpha acids (HPLC)	7.7	Hops-14
Alpha, beta and iso-alpha acids (HPLC)	7.8	Hops-15, -9C,-9D
Conductometric bitter value (for ethanol extract)	7.6 plus 50% of iso-alpha acids (7.8)	–
Iso-alpha acids and reduced iso-alpha acids in hop products by HPLC	7.9	–
Hop oil	7.10	Hops-13
Hop storage index (HSI)	–	Hops-12

Hops are a natural product and consequently exposed to environmental influences during the vegetative phase, with substances such as metals, radionuclides and mycotoxins being among possible environmental contaminants.[12,45] Other substances, such as nitrates and zinc, enter the plant when the soil is fertilised. Pesticide analyses are carried out on a regular basis to ensure that all residues in the hops and hop products are within legal limits. Additional analyses, not yet fully regulated by law, are also carried out to determine the contents of nitrates, heavy metals or mycotoxins as an additional benefit for customers and consumers. Traceability, as required under EC 178/2002, is straightforward because every hop product can be traced back to the respective lots and the respective farmers due to the extensive certification procedures.[11]

Aspects of bitterness analysis in beer

The major reason for using hop bitter acids is to control the bitterness of beer, therefore analytical methods are needed to measure bitterness. But sensory perception is not linear, and bitter is not well developed as a taste attribute in human beings (for instance compared to herbivores). Hence there will always be discrepancies between measured and perceived bitterness, even if the most accurate methods, e.g. HPLC, are used.[23,30]

There are only a few reference methods for quality control of hop bitterness and aroma in wort and beer. Bitterness is usually measured using EBC method 9.8, spectrophotometric measurement of the absorbency at 275 nm working with an empirical factor of 50. This method is not specific since less polar substances, such as unchanged alpha and beta acids and their degradation and rearrangement products, are co-extracted. Also, substances unrelated to bitterness might absorb light at 275 nm, so that even beers with no hops will have IBU values between 2 and 4. The factor 50 is based on the assumption that about 70% of the absorbency is caused by iso-alpha acids, though this only applies in the case of beers brewed with fresh hops. The IBU counts of beers brewed with deteriorated hops will therefore be higher than the perceived bitterness, and using the same factor

Table 6.6 Conversion factors for spectrophotometric methods to determine IBU in beer[11]

Bittering compound	Conversion factor
Kettle hopping with fresh hops (pellets or extracts)	50
Kettle hopping pre-isomerised hop products	55–58
PFB	65–70
PFB (reduced products)	70–73

while brewing with isomerised hop products will lead to lower IBU values than that perceived.[52] Table 6.6 indicates the appropriate conversion factors if using spectrophotometric methods. HPLC analyses are always recommended to determine the exact contents of the various acids.[11]

6.7 Future trends

Future trends in hop processing will clearly focus on the potential of hop polyphenols, and within that group the flavonoids in particular. Another research sector will focus on additional properties and applications of beta acids. The potential of other hop constituents such as anti-oxidants and anti-microbials will also be explored.[26] Findings will lead to advanced hop products for many other markets, apart from the brewing industry, and will focus on applications within 'nutraceuticals', functional food products or pharmaceutical products. Xanthohumol-enriched products will also be a challenge for the brewing industry. Research efforts will be aimed at overcoming the difficulties presented by xanthohumol in the brewing process (poor solubility and rapid isomerisation to iso-xanthohumol).

For global players within the brewing industry, the importance of flexibility and the ability to create new brands will continue, and it is likely that the trend towards very mild beers will also continue. Therefore it will be extremely important to have advanced hop products available for very specific and independent effects in beer (bitterness, aroma, foam enhancement, flavour stability, preservative effects, and increasing the content of specific hop compounds). Beers in clear bottles have been a great success in various countries, therefore research into flavour stability will continue to be important.

6.8 Sources of further information

www.BarthHaasGroup.com
www.Hopsteiner.com
www.hvg-germany.de
www.yakimachiefs.com
www.deutscher-hopfen.de

6.9 References

1. ABOURASHED E (2004), 'In vitro binding experiments with a valerian, hops and their fixed combination', *Phytomedicine*, 11, 633–638.
2. ARAKI S (2001), 'A new parameter for determination of the extent of staling in beer', *J Am Soc Brew Chem*, 60(1), 26–30.
3. ASBC (1992), Methods of Analysis (8th edn), ASBC, St Paul, MN.
4. BIENDL M (2002), 'Xanthohumol content in hops', *Hopfenrundschau International*, 02/03, 72–75.
5. BIENDL M (2005), 'Polyphenols – plant components with marketing potential', *Hopfenrundschau International*, 05/06, 22–27.
6. BIENDL M (2004), 'Estrogenic activity of hop components', *Brauwelt International*, I, 72–75.
7. CALLEMIEN D, JERKOVIC V, ROZENBERG R (2005), 'Hops as an interesting source of resveratrol for brewers', *J Agric Food Chem*, 53(2), 424–429.
8. CMA BROCHURE (2005), *The Spirit of Beer – Hops in Germany*, CMA, Bonn, www.deutscher-hopfen.de
9. DE COOMAN L (2000), 'Alterations of the profiles of iso-alpha acids during beer ageing marked instability of *trans*-iso-alpha acids and implications for beer bitterness consistency in relation to tetrahydro-iso-alpha acids', *J Inst Brew*, 106(3), 169–178.
10. EBC (2005), *Analytica-EBC*, Hans Carl Verlag, Nuremberg, Germany.
11. EBC (1997), *Manual of Good Practice, Hops and Hop Products*, Hans Carl Verlag, Nuremberg, Germany.
12. FORSTER A (1988), 'Zur Nitratdosage durch Hopfen und Hopfenprodukte', *Brauwelt*, 128(6), 188–190.
13. FORSTER A (1995), 'Untersuchungen zu Hopfenpolyphenolen', in EBC, *Proceedings of the 25th EBC Congress, Brussels*, Oxford University Press, Oxford, 143–150.
14. FORSTER A (1999), 'Hopfenpolyphenole – mehr als nur Trübungsbildner in Bier', *Hopfenrundschau International*, 99/00, 68–74.
15. FORSTER A (2003), 'A new xanthohumol-enriched hop extract and its use for unfiltered beers', *Hopfenrundschau International*, 03/04, 65–71.
16. FORSTER A (2000), ' Die Bedeutung des Jahrganges bei der qualitativen Bewertung von Hopfenprodukten', *Brauwelt*, 140(18), 698–703.
17. FORSTER A (2004), 'Wie bestimmt sich der durchschnittliche alpha Säuregehalt einer Hopfensorte', *Brauwelt*, 22, 654–660.
18. FRITSCH H (2005), 'Sensory characterisation of xanthohumol and isoxanthohumol', in *EBC, Proceedings of the 30th EBC Congress, Prague*, Hans Carl Verlag, Nuremberg, Germany.
19. GERHÄUSER C (2002), 'Cancer chemopreventive activity of xanthohumol, a natural product derived from hop', *Molecular Cancer Therapeutics*, 959–969.
20. GOIRIS K (2005), 'Hop polyphenols: potential for beer flavour and flavour stability', in *EBC, Proceedings of the 30th EBC Congress, Prague*, Hans Carl Verlag, Nuremberg, Germany.
21. GUZINSKI J A (1994), 'Practical considerations of reduced hop extracts' in EBC, *Monograph XXII, EBC Symposium on hops, Zoeterwoude*, Hans Carl Verlag, Nuremberg, Germany, 105–113.
22. HEYERICK A, DE KEUKELEIRE D (2002), 'Modulation of the phytoestrogenicity of beer by monoterpene alcohols present in various hop oil fractions', *J Inst Brew*, 108(1),

64–101.

23. HUGHES P (2005), 'Hop bitter acid analyses: past present and future', in EBC, *Proceedings of the 30th EBC Congress, Prague*, Hans Carl Verlag, Nuremberg, Germany.
24. HUVAERE K (2004), 'Photooxidative degradation of beer bittering principles: product analysis with respect to lightstruck flavour formation', *Photochem Photobiol Sci*, 854–858.
25. KALTNER D (2000), 'Untersuchungen zur Ausbildung des Hopfenaromas und technologische Maßnahmen zur Erzeugung hopfenaromatischer Biere', Freising-Weihenstephan, Dissertation TU München.
26. KALTNER D (2001), 'Investigation of the influence of hop products on the microbiological stability of beer', in EBC, *Proceedings of the 28th EBC Congress, Budapest*, Hans Carl Verlag, Nuremberg, Germany.
27. KALTNER D (2003), 'R-linalool als Schlüsselaromastoff für das Hopfenaroma in Bier und sein Verhalten während der Bieralterung', *Monatsschrift für Brauwissenschaft*, 11/12, 192–196.
28. KALTNER D (2005), 'Moderne Aspekte zur Hopfung des Bieres', *Brauindustrie*, 2, 22–25.
29. KAMMHUBER K (2005), 'Differentiating between the world range of hop varieties according to bitter compounds and polyphenols', *Hopfenrundschau International*, 05/06, 42–46.
30. KANEDA H, WATARI J, TAKASHIO M (2003), 'Adsorption and desorption of beer and coffee on a lipid membrane as related to sensory bitterness', *J Inst Brew*, 109, 27–33.
31. KETTERER M (2003), 'The influence of isomerised hop pellets on beer quality', in *Proceedings of the 29th EBC Congress, Dublin*, Hans Carl Verlag, Nuremberg, Germany.
32. KOEN G (2003), 'The oxygenated sesquiterpenoid fraction of hops in relation to the spicy hop character of beer', *J Inst Brew*, 108(1), 86–93.
33. KONDO K (2003), 'Proventive effects of dietary beer on lifestyle', in EBC, *Proceedings of the 29th EBC Congress, Dublin*, Hans Carl Verlag, Nuremberg, Germany.
34. KROTTENTHALER M (2005), 'Hopfen', in Heyse K A, *Praxishandbuch der Brauerei*, Behr's Verlag/Hans Carl Verlag, Nuremberg, Germany.
35. LERMUSIEAU G (2001), 'Use of GC-olfactometry to identify the hop aromatic compounds in beer', *J Agric Food Chem*, 49(8), 3867–3874.
36. MALOWICKI M, SHELLHAMMER T H (2005), 'Isomerisation and degradation kinetics of hop acids in a model wort-boiling system', *J Agric Food Chem*, 53, 4434–4439.
37. MATSUMOTO M (1999), 'Inhibitory effects of oolong tea extract on caries-inducing properties of mutants streptococci', *Caries Res*, 33(6), 441–445.
38. MOIR M (1994), 'Aromatic compounds', in EBC, *Monograph XXII, EBC Symposium on Hops, Zoeterwoude*, Hans Carl Verlag, Nuremberg, Germany.
39. MOIR M (2000), 'Hops – a millenium review', *J Am Soc Brew Chem*, 58(4), 131–146.
40. NIJSSEN L M (1996), 'Volatile compounds in food. Qualitative and quantitative data: hop oil', Central Institute for Nutrition and Food Research, Zeist, The Netherlands, No. 62.
41. PIETERCELIE A (2005), 'Investigations into the effects of hop compounds on yeast behaviour with a new micro-scale fermentation method', in EBC, *Proceedings of the 30th EBC Congress, Prague*, Hans Carl Verlag, Nuremberg, Germany.
42. POLLACH G, HEIN W, BEDDIE D (2004), 'The concept of different natural antibacterials

for the sugar industry – Ein Konzept mit verschiedenen natürlichen antimikrobiellen Hilfsstoffen für die Zuckerindustrie', *Zuckerindustrie*, 129(8), 555–564.

43. SELDESLACHTS D (1999), 'The use of high tech hopping in practice', in EBC, *Proceedings of the 27th EBC Congress, Cannes*, Oxford University Press, Oxford.
44. SCHIEBERLE P (2000), 'Comparison of the most odour active compounds in fresh and dried hop cones, variety Spalter Select, based on GC-Olfactometry and odour dilution techniques', *J Agric Food Chem*, 48, 1776–1783.
45. SCHMIDT (2003), 'Umweltkontaminanten in Hopfen', *Brauwelt*, 143(45), 1493–1497.
46. SCHÖNBERGER C (2003), 'Bedeutung nicht-flüchtiger Geschmacksstoffe in Bier', Dissertation TU München, Norderstedt, Verlag BOD.
47. SCHÖNBERGER C (2005), 'Evaluations of pure hop aromas in alcohol free beer', *Brauwelt International*, 23, 181–184.
48. STEINHAUS M (2003), 'Quantitation of R- and S-linalool in beer using solid phase microextraction (SPME) in combination with a stable isotope dilution assay (SIDA)', *J Agric Food Chem*, 51(24), 7100–7105.
49. STEVENS J F, PAGE J E (2004), 'Xanthohumol and related prenylflavonoids from hops and beer', *Phytochemistry*, 65, 1317–1330.
50. VERZELE M, DE KEUKELEIRE D (1991), *Chemistry and Analysis of Hop and Beer Bitter Acids* (1st edn), Elsevier Science, Amsterdam.
51. WALKER C (2004), 'Investigation into the high levels of xanthohumol found in stout and porter style beers', *Brauwelt International*, II, 100–104.
52. WEISS A , SCHÖNBERGER C (2002), 'Sensory and analytical characterisation of reduced, isomerised hop extracts and their influence and use in beer', *J Inst Brew*, 108(2), 236–242.
53. WUNDERLICH S (2005), 'Enrichment of xanthohumol in the brewing process', *Mol Nutr Food Res*, 49(9), 874–881.
54. YAJIMA H (2004), 'Isohumulones, bitter acids derived from hops, activate both peroxisome proliferator-activated receptor alpha and gamma and reduce insulin resistance', *J Biol Chem*, 279(32), 33456–33462.
55. ZUFALL C (2005), 'Flavour stability and ageing characteristics of light-stable beers', in EBC, *Proceedings of the 30th EBC Congress, Prague*, Hans Carl Verlag, Nuremberg, Germany.

7

Yeast genetics in brewing: new insights and opportunities

D. E. Quain, red.ts Ltd, UK

7.1 Introduction

Today any mention in the brewing world of 'yeast genetics' often results in a rolling of the eyes and all too obvious disinterest! It was not always like this: the last 25 years have seen an extravagant rise in interest in genetics followed by an equally dramatic decline. The drivers for this reflect on similarly abrupt changes in the worldwide brewing industry. The collective impact of legislation, consumer trends, customer purchasing power, globalisation and consolidation has changed the landscape beyond all recognition. Regrettably but inevitably, these changes have hit brewing science hard, with a substantial contraction in resources, both people and financial, working in what was fertile ground. To make this real from a 'UKcentric' viewpoint, in 1980 the big brewers – Bass, Scottish & Newcastle, Courage, Whitbread, Guinness and Allied Breweries – all had thriving 'research laboratories' mixing the 'commercially sensitive' with open, publishable work. At the same time fundamental research was being conducted at the Brewing Industry Research Foundation (BIRF) and at the 'brewing schools' of Heriot Watt University in Edinburgh and at the University of Birmingham. These universities also played a key role in annually delivering graduates and postgraduates knowledgeable in the ways of brewing science and technology. Today, without exception, the above brewing companies and their technical resources have been lost through 'consolidation' and the Birmingham brewing school has closed. BIRF has become BRi (Brewing Research International), a smaller organisation focused on food safety and contract research. Graduate-level education in brewing remains in the hands of Heriot Watt/ICBD (International Centre for Brewing and Distilling) together with a new provider in 2006, the University of Nottingham.

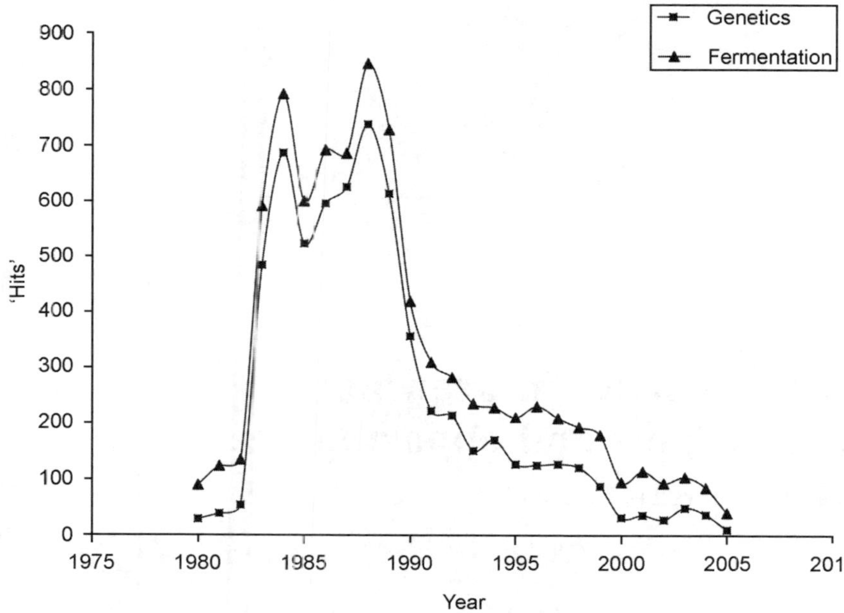

Fig. 7.1 Publications by year cited in BREW database (source: live link 'expert search' of BRi BREW database. Searches of 'genetics + *Saccharomyces cerevisiae*' and 'fermentation + *Saccharomyces cerevisiae*' against 'publication date').

The above journey provides a partial explanation for the rise and fall of enthusiasm in fundamental and applied brewing research. Whilst undeniably UK focused, the same pressures and consequences have been experienced elsewhere in the brewing world. Drilling down to this chapter, it is perhaps no surprise that the recent history of brewing yeast genetics mirrors that of research. Indeed, as can be seen in Fig. 7.1, the number of publications on 'genetics + *Saccharomyces cerevisiae*' cited year by year in BRi's excellent BREW database peaked during the halcyon days of brewing research and has been in decline thereafter. Similar searches for 'fermentation + *Saccharomyces cerevisiae*' paint a similar picture, although it can be argued that the subject of 'genetics' outpunches its weight compared to the process of 'fermentation'.

From the personal perspective of involvement, it is with hindsight surprising that the genetic technologies that were touted to offer so much have had so little impact on the brewing industry. In particular the early dream of customised and 'to order' genetically modified yeast has become an easy technical reality but has been well and truly 'parked' in today's competitive world of big beer brands and global ambition, so much so that genetically modified brewing yeast has yet – and is unlikely in the foreseeable future – to find application in commercial fermentations.

With this troubled preamble in mind, the ambition of this chapter is to focus on developments in the 'big science' genetics of *Saccharomyces cerevisiae* that have immediate or future relevance to brewing yeast. Without exception these diverse and frequently astonishing developments are due to a major milestone in biology – the sequencing of the genome of *S. cerevisiae* published in 1996. However, it is both revealing and disappointing that such a major event should have passed by almost unnoticed in the worldwide brewing industry. Accordingly, this chapter will not only update the reader on the yeast genome story but touch on other areas – genetic modification, genetic instability and taxonomics which are hopefully interesting, stimulating and relevant to today's brewing technologists.

7.2 Fundamentals

Popularly known as *Saccharomyces cerevisiae*, yeast is a single-celled organism and a member of the fungi kingdom. Unlike bacteria, yeast contains a nucleus and is described as a lower eukaryote, with man (*Homo sapiens*) being a higher (highest!) eukaryote. Indeed, one of the reasons for yeast's pre-eminent role in biology is that it is an excellent starting place and model for what *may* happen in man. In terms of numbers, the nuclear genome consists of 16 chromosomes of just over 12 million base pairs (bp) (equivalent to about 6000 genes) and a mitochondrial genome of almost 86 000 bp (see Section 7.3). The number of 'sets' of chromosomes is described as 'ploidy' with brewing strains being polyploid, having up to three (triploid) or four (tetraploid) sets. In reality strains are 'aneuploid' – where the 'ploidy' is not a perfect multiple of the haploid state – which allows for extra or a reduced number of specific chromosomes (see Section 7.5). Taxonomically, the current view is that ale yeasts are *S. cerevisiae* and lager yeasts are *S. pastorianus*, which is a hybrid of *S. cerevisiae* and the closely related species *S. bayanus* (see Section 7.6).

7.3 The yeast genome

The genome incorporates all the hereditary information encoded in DNA located in the cell nucleus or in the mitochondrion. This includes sequences that code for genes (potential proteins) and for non-coding regions. Simply, unravelling the DNA sequence of an organism is the first (but big!) step to building a holistic understanding of how it works and equally what happens when things go wrong. Since 1995, when the bacterium *Haemophilus influenzae* became the first genome to be sequenced, 332 genome sequences have been completed including *S. cerevisiae* (1996), *Escherichia coli* (1997), *Drosophila melanogaster* (2000), *Homo sapiens* (2001), *Gallus gallus* or chicken (2004) and *Canis familiaris* or dog (2005). At the time of writing more than 1300 other genomes across nature are in the process of being sequenced – see the genomes online database (www.genomesonline.org).

The yeast genome sequence was released in April 1996 (Goffeau *et al*., 1996) and, although the third genome to be sequenced, it was the first eukaryote. In many respects the techniques, strategies and communications developed for yeast paved the way for the many subsequent sequencing projects that – in terms of scale – culminated in the human genome project. However, the yeast genome was delivered by an international team of over 600 scientists and took in excess of four years. Today, not surprisingly, genome sequencing has scaled up and become automated, so much so that to sequence the yeast genome today would require just a few days of full-scale production (Salzberg, 2003).

The use of the Internet and 'modern infomatics technology' (Goffeau *et al.*, 1996) was lauded as being critical to the success of the yeast project. Like many aspects of life today, the use of the World Wide Web is now taken for granted in collaborative projects, be they big or small. A splendid example of this is the '*Saccharomyces* Genome Database' or SGD found at www.yeastgenome.org. This awe-inspiring site is the 'home' of the yeast genome project and a 'one stop shop' for information about genes and proteins together with utilities to draw comparisons and insights. A new example is the 'fungal alignment viewer' (Christie *et al.*, 2004) that enables sequences from *S. cerevisiae* to be compared to other closely related *Saccharomyces* species (see Section 7.3) – described as 'comparable genomics'. Although in a state of continuous change and refinement, the genome of 16 chromosomes consists of 6604 potential proteins or ORFs ('open reading frames') of which 67% are currently characterised, with the remainder being 'uncharacterised' (21%) or 'dubious' (12%). This 'genome shapshot' report (Hirschman *et al.*, 2006) can be further cut into categories of (i) molecular function, (ii) biological processes and (iii) cellular components, and then further quantified by the number of genes responsible for 'organelle organisation and biogenesis', 'transport' and so on. Although of huge ongoing value to the yeast 'professional', the reader is warmly encouraged to visit this site and 'surf' its diverse and fascinating content. It is no surprise that the SGD has received over 57.5 million hits since August 1994!

Although a remarkable and stimulating resource, the provenance of the yeast whose genome was sequenced (S288C) was predominately derived from strain EM93 that was isolated from Californian rotting figs in 1938. As noted by Mortimer and Johnston (1986), 'whether these strains were originally part of the figs' flora or were spoilage organisms originating as commercial baking and/or brewing yeasts is unknown'. Indeed, as noted in a brief report by Meaden (1996), there are clues that the sequenced yeast had little or no involvement in brewing. Accordingly, although by no means undermining the value of the genome project, its application to brewing yeast is 'directional'. As with closely related species, some comparative genomics is required of other 'industrial strains' of *S. cerevisiae*. Help is at hand (Table 7.1) in that five different isolates of *S. cerevisiae* are currently being sequenced, of which two are yeasts used in brewing. Indeed, preliminary headlines for the sequence of Weihenstephan lager yeast have been published (see Section 7.5, Nakao *et al.*, 2003). Hopefully, a fuller description, together with detailed insights of the points of difference between this lager yeast and its likely parental species (*S. cerevisiae* and *S. bayanus*), will be published (see Section 7.5.1).

Table 7.1 Live yeast sequencing projects, January 2006

Strain	Provenance	Status	Organisation	Reference
S. cerevisiae S288C	Rotting fig	Complete	International collaboration	
S. cerevisiae Kyokai no. 7	Sake	Ongoing	National Institute of Technology and Evaluation	www.bio.nite.go.jp
S. cerevisiae RM11	Californian vineyard	Ongoing	Broad Institute	www.broad.mit.edu
S. cerevisiae YJM789	Lung of AIDS patient	Ongoing	Stanford	http://med.stanford.edu
S. pastorianus Weihenstephan	German lager?	Ongoing	Kitasato University	http://genome.ls.kitasato-u.ac.jp
S. pastorianus	Danish lager?	Ongoing	Carlsberg Research Centre	

Source: www.genomesonline.org

7.3.1 Microarrays

The sequencing of yeast and other genomes has changed the mindset of biologists from a local (gene) to a global (genome) perspective. This thinking has logically extended to the linkage (see Fig. 7.2) between the genotype and phenotype with the creation of the new disciplines of 'transcriptomics', 'proteomics' and 'metabolomics' that focus on gene expression, its regulation and the output of metabolites. Of these, analysis of transcription at the genomic scale has found application in beer fermentations.

The motivation and thrust behind 'transcriptomics' is 'knowing when and where a gene is expressed often provides a strong clue as to its biological role. Conversely, the pattern of genes expressed in a cell can provide detailed information about its state' (DeRisi *et al.*, 1997). The development of DNA microarrays ('gene chip', 'DNA chip') has enabled the analysis of transcription at the global, genomic scale to provide a 'gene expression profile' or 'signature' (for a review see Lucchini *et al.*, 2001). The microarray is a glass slide 'printed' with 6400 genes/ORFs representing the entire genome of *S. cerevisiae* S288C. Gene expression is tested by extraction of RNA, conversion back to DNA, amplification and labelling prior to application to the microarray. Genome activity is assessed by hybridisation to complementary DNA, which is typically visualised and quantified via fluorescent dyes. As would be anticipated from experiments involving the simultaneous probing of 6000 genes, data analysis and mining require sophisticated software treatments and controls.

Microarrays have been enthusiastically embraced by the wider yeast community (Lucchini *et al.*, 2001) and have been applied to events in brewery

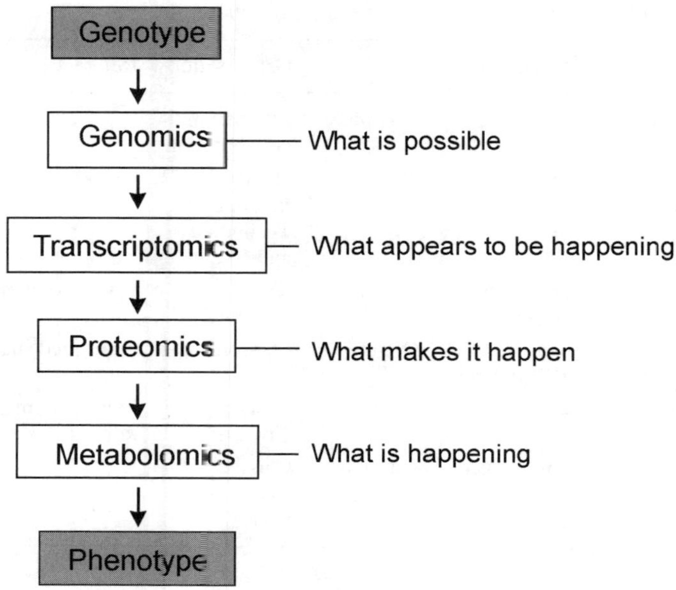

Fig. 7.2 The journey from genotype to phenotype.

fermentation. Of note, Higgins and co-workers (2003) monitored yeast transcription in pilot plant and commercial scale fermenters. Comparison of yeast after 1 and 23 hours of fermentation showed that RNA levels of 100 genes were at least threefold higher after 1 hour compared to later in fermentation. Although the induction of genes responsible for the formation of sterol and unsaturated fatty acids would be anticipated, a number of genes involved in protection against oxidative stress are of notable interest. Other studies using microarray technology in brewing have probed the impact of nitrogen dilution in a comparison between adjunct and malt worts (Kondo *et al.*, 2003), whereas Peng *et al.* (2003) quantified the induction and repression of genes in early, mid- and late fermentation.

7.4 The rise and fall of genetic modification

The promise of 'modifying' brewing yeast strains to perform better or differently was the focus of many research groups worldwide in the 1980s. Indeed, the peak in publications (Fig. 7.1) on 'yeast genetics' captured by the BREW database was fuelled chiefly by the optimism surrounding the evolving technologies underpinning genetic modification. Looking back, one cannot help a wry smile to read that Tubb (1981) concluded that 'in 1983 it will be 100 years since Hansen introduced a pure culture pitching yeast for the first time. It would be fitting indeed if we could celebrate that anniversary with successful introduction of a genetically improved strain of yeast into commercial practice'.

Twenty-five years later, the following quote is perhaps more appropriate: 'it is clear that today the odds are stacked against the commercial exploitation of genetically modified yeast in the brewing industry' (Boulton and Quain, 2001). With this gloomy prognosis in mind, the aim of this section is to review the ambitions and current status of genetic modification of yeast used in the production of alcoholic drinks. However, it is not my intention to detail the well-established genetic methods that are used to introduce 'foreign' DNA into recipient yeasts, for which the reader is referred elsewhere (Meaden, 1986; Walker, 1998; Hammond, 1998, 2003). In passing it is worth noting that 'genetic engineering' is now an element of 'metabolic engineering', which adopts a broader holistic approach to modification and considers the metabolic implications and opportunities of the directed change (Penttila, 2001). In many respects metabolic engineering recognises that the 'isolationist' approach of the early genetic modification tended to ignore the implications of gene manipulation for the wider cell physiology and metabolism.

The literature on the genetic modification of brewing yeast is full of lists of targets and opportunities that were perceived – pretty much from the outset – to be achievable. By way of example, Table 7.2 is a distillation of the 'art of the possible' published over the years (Tubb, 1981, 1984; Lancashire, 1986; Walker, 1998; Boulton and Quain, 2001; Briggs *et al.*, 2004). Whilst undoubtedly bullish, a number of these technical targets were achieved, albeit without the 'stamp of approval' of commercial use in production. For example, Hammond (2003) reviews the success of introducing into brewing yeast a variety of novel characteristics including glucoamylases, β-glucanases, flocculation, 'killer' anti-contaminant properties and modified beer flavour (diacetyl, esters, sulphur compounds). In terms of potential impact, the vast body of work on the 'diacetyl problem' remains an engaging opportunity. As Hammond (2003) makes clear, many teams using many approaches addressed this challenge with varying degrees of success. Then as now, removing or reducing the need for diacetyl reduction at the end of fermentation through genetic manipulation (Hammond 1998) is an extremely attractive goal for two reasons, firstly because of the pressure on process intensification or cycle time reduction, and secondly because the more efficient use of fermenters can minimise the need for vessel replacement, resulting in significant capital savings.

However, despite the apparent upsides in targeted strain improvement, no brewery worldwide has taken the step of introducing genetically modified yeast. Although regulatory approval is doubtless onerous and labelling a little explicit, the fear of a highly negative public reaction has dissuaded any one company from being first to market. In an attempt to combat this view, there was a concerted effort in the mid-1990s to 'talk up' the technology via BRi's 'demonstrator' project with a dextrin degrading amylolytic brewing strain and consequent beer, 'Nutfield Lyte' (for details of the genetics and approval process see Hammond, 1998). Whilst gaining approval from the UK regulatory authorities together with widespread media coverage, the yeast has never been used commercially.

Table 7.2 Targets for the genetic modification of brewing yeast

Driver	Objective	Target	Comments
Beer quality	Modify beer flavour Reduce maturation/ conditioning time	Manipulate specific target genes	Capability to control the concentration of desired flavour substances, e.g. esters, H_2S, SO_2
Beer quality	Reduce/block ethanol production	No/low alcohol products	Better quality?
Cost	Avoid the need for addition of exogenous enzymes (processing aids) Avoid the need for adding mixed enzyme 'soups'	E.g. β-glucanases, proteases and amylases	Value not high!
Cost	Add value to spent yeast	Introduce 'value adding' gene that can be switched on after use in brewery	Opportunity with the market for spent yeast in decline
Efficiency	Ferment non-fermentable wort dextrins	Introduce amylolytic enzymes	Hydrolyse linear/ branched oligo-saccharides (>G3) to fermentable sugars
Efficiency	Utilisation of other grain derived carbohydrates	E.g. pentoses, cellobiose	Improve alcohol yield
Hygiene	'Self cleaning' yeast	Introduce 'killer factor'	Selectively kill contaminating wild yeasts
QA	Labels to easily identify and differentiate commercial strains	Introduce unique markers	Useful in troubleshooting, strain QA
Vessel utilisation	Ferment very high gravity worts at the normal rate without compromising beer quality, yeast viability or serial repitching	Ethanol tolerance Osmotic pressure General yeast physiology	Collection gravity >1100 Final ABV >10%v/v
Vessel utilisation	Ferment at elevated temperatures without compromising beer quality, yeast viability or serial repitching	Thermotolerance General yeast physiology	Lager fermentations at 20–30°C
Vessel utilisation	Avoidance of diacetyl 'rest' or 'stand' in fermenter or downstream in maturation/conditioning tank	Introduce foreign enzyme that circumvents the formation of diacetyl or increases flux through pathway	Depending on the fermentation process, diacetyl stand range from a few days to weeks
Yeast handling	Modify non-flocculent 'powdery' strain to flocculent strain	Flocculence gene(s)	Move from yeast cropping ex-centrifuge to conventional cone cropping

Source: expanded from Boulton and Quain (2001).

So why should the worldwide public take against the concept of genetic modification? Certainly there has been much in the way of negative media coverage, which has stimulated public concern. However, communication of the concept has been poor. For example, the language – genetic engineering, recombinant DNA technology, 'GM', transformation – is certainly not user-friendly and feels that it is not the sort of thing that should be associated with beer. Indeed, as noted by Bamforth (2005), 'scientists, generally, are not terribly smart about conveying their ideas and discoveries. As a result they attract suspicion, fear and antipathy'. A further argument is that the technology was not 'sold' to consumers, who are much more amenable to 'change' if they understand and value the benefit. For example, consumers seem to have no problems with recombinant human serum albumin or insulin being sourced from modified yeast. However, looking at the list of targets for genetic modification (Table 7.2), it is hard to identify any compelling direct consumer benefit. The reality is that the ambitions for this technology focused on what it could do for the brewer! With hindsight, if the 'demonstrator' project had focused on reducing cycle time, the concept could have been sold to consumers as improving 'freshness', or more importantly reducing the 'cost of goods' which could translate as a benefit in reducing the price of a pint!

Fast-forward to today and the lesson of selling-in to the consumer appears to have been learnt by the North American wine industry. Two genetically modified yeasts have been developed at the Wine Research Centre at the University of British Columbia, which the US Food and Drug Administration consider as 'Generally Regarded As Safe' (GRAS). The first yeast (*S. cerevisiae* ML01) has been modified to perform a timely malolactic fermentation and thereby minimise the possibility of the presence of biogenic amines that cause headaches and other allergenic reactions in consumers. The second yeast, which achieved GRAS in January 2006, has been manipulated to virtually eliminate the formation of ethyl carbomate which is a potential carcinogen found in wine. For further details of these two yeasts, see 'Suppliers of modified wine yeasts' in the References, Section 7.10.

Whether or not these two fresh examples of genetically modified yeasts being badged as GRAS convinces North American wine producers and consumers remains to be seen. However, the principle of majoring on the consumer 'benefit' has clearly been recognised and is the primary message in all communications, with 'genetic modification' featuring lower down the agenda. It will be fascinating to watch developments here, which clearly could provide a template for changing public perception and acceptance of the technology in brewing.

7.5 Genetic instability – problem or opportunity?

Reports of genetic changes in brewing yeasts – although infrequent – suggest 'instability' to be fact rather than fiction. Indeed, from the global perspective of 'big science' such changes are a manifestation of natural evolutionary events in response to some selective pressure that results in improved 'fitness'. In yeast, 'changes' include 'gene-local amplifications, changes in chromosome copy number, and intrachromosomal and interchromosomal translocations' (Dunham *et al.*, 2002). Of these, engineered yeast with (interchromosomal) translocations between chromosomes 'out-competed' the reference yeast without any translocation, reinforcing the improved fitness theme (Colson *et al.*, 2004). Indeed, chromosomal changes appear to be common in yeast, such that in a detailed review, Boulton and Quain (2001) tabulated changes reported in the literature to each of the 16 chromosomes. Without wishing to dwell on the genetic detail, in a brewing context the occurrence of such genomic rearrangements is unlikely to be observed unless they have an 'obvious' impact on yeast performance. This leads to the contrary position that 'silent' genetic changes can and will exist in brewing yeast without being observed or recognised.

The brewing literature on genetic instability splits neatly into observations of phenotypic and/or genetic change. In the former camp are a series of publications from workers at Guinness, who between 1963 and 1996 reported 'spontaneous' changes in maltotriose utilisation and, in particular, switches in flocculence (for details see Boulton and Quain, 2001). From a genetic perspective, chromosomal length polymorphisms have been reported in bottom fermenting lager yeasts in Japanese breweries at Sapporo (Sato *et al.*, 1994) and Kirin (Tanaka and Kobayashi, 2003). Casey (1996) in a fascinating paper reported on changes to production yeast from the Stroh Brewery stored between 1958 and 1985. In all, seven distinct chromosome patterns (karyotype) were found with changes restricted to four chromosomes, which in terms of fitness for purpose carry genes for flocculence, glycolysis, maltose utilisation and diacetyl production. Finally, some observations link both the phenotype with genetic change in a production lager yeast from Coors (previously Bass) Brewers in the UK. Here (Table 7.3) initial observations of atypically heavy flocculence were eventually linked with genetic changes by studies at ICBD (Heriot Watt University) and the University of Manchester. It is particularly noteworthy that seemingly the same variant was found some seven years later in a brewery 300 miles away! Hopefully in the future there will be further instalments that will unravel new insights into this fascinating story.

Given the 'difficulties' of exploiting genetic modification (see Section 7.4) it is surprising that naturally derived genetic changes in brewing yeast have not been intentionally encouraged by application of selective forces. Such strain improvement would, although quite legitimate, be limited to the exaggeration or minimisation of existing characteristics typical of brewing yeast. Such a wish list might include parameters such as flocculation, rates of sugar uptake, flavour production, diacetyl reduction, ethanol tolerance and so on. Just how these needs might be met would require careful selection of appropriate selective forces. However, taking a leaf out the geneticist's book, genome rearrangements in yeast have been selected for using continuous culture regimes (Adams *et al.*, 1992; Dunham *et al.*, 2002).

Table 7.3 Genetic instability – a case history

Observation	Reference
In 1991, during process trials on yeast oxygenation, a heavily flocculent line was isolated. Pumped solids were 60–75% (normal = 40%). Laboratory studies showed the variant (BB11 '56') to be much more sensitive to calcium promoted flocculence than the 'parent' yeast (BB11).	Boulton & Quain (2001)
DNA (RFLP) fingerprinting demonstrated small but definite genetic differences between BB1 and BB11 '56'.	Wightman *et al.* (1996)
In 1998, a similarly heavily flocculent yeast (BB11 'W') of the same strain recovered from a different brewery.	–
Genomic analysis shows both BB11 '56' and 'W' to have an additional copy of chromosome VII compared to BB11.	Lockhart (2003)

7.6 Genetic differences between yeast species

The science of taxonomy seeks to 'develop a "natural" classification that reflects the evolutionary history and phylogenetic relationships of contemporary organisms and to develop procedures whereby specimens of individual species may be unambiguously identified' (Edwards-Ingram *et al.*, 2004). Inevitably this is a complex and specialised field that is subject to frequent changes of name which 'has long made taxonomists objects of ridicule amongst those doing biological research' (Barnett *et al.*, 2000). The churn, contraction and expansion of names reflect two key factors. Perhaps the most significant and unpredictable is the growing insight into the complexity of the natural world. For example the two major texts (Kurtzman and Fell, 1998; Barnett *et al.*, 2000) on yeast taxonomy describe about 700 species, which is now thought to extend to at least 1200 which, alarmingly, may represent only 1% of the species on the planet (Boekhout, 2005). The other drive for change in taxonomy has been the evolution of testing methods used for differentiation. Traditionally, taxonomists focused on the phenotype with testing for growth on diverse substrates together with any clues from colony morphology and microscopic appearance (for an example of this approach, see Barnett *et al.*, 2000). However, with the advent in the 1990s of 'molecular taxonomy', classification has moved to the genome and DNA-based methodologies. A particularly profitable approach has been to base the analysis of evolutionary development ('phylogenetics') of the DNA sequence encoding all or part of the ribosomal RNA. Not surprisingly as sequencing technologies evolve, more robust phylogenetic trees are being generated from the comparison of complete (or near complete) genome sequences (Edwards-Ingram *et al.*, 2004).

Although close to the hearts of the brewing and baking communities, *S. cerevisiae* is – as noted above – only one of currently 1200 yeast species! To explain, 'species' can be defined as yeasts which can (potentially) interbreed, whereas a genus (*Saccharomyces*) is a group of closely related species. The number of species within the *Saccharomyces* genus has waxed and waned over recent decades from a high of 41 in 1970 to a low of 10 in 1990 (see Boulton and Quain, 2001 for details). The current picture (Table 7.4) suggests 14–19 species, which is further divided into two groups, those closely related species (*sensu stricto*) and those more diverse (*sensu lato*). The 'domesticated' species – *S. cerevisiae, S. bayanus, S. pastorianus* – are found in the *sensu stricto* group which is characterised by having the same number (16) of chromosomes and similar genome organisation. However, in an important paper (Vaughan-Martini and Kurtzman, 1985), analysis of the DNA homology between these species showed *S. pastorianus* to have a very good 'fit' with *S. bayanus* (72%) and to be reasonably close to *S. cerevisiae* (53%), whereas *S. cerevisiae* and *S. bayanus* have little or no commonality. Further, this and other studies showed the genome of *S. pastorianus* to be 50% bigger than that of *S. cerevisiae.* These insights lead to the conclusion that *S. pastorianus* is not a single species within the *Saccharomyces* genus but a natural hybrid of *S. cerevisiae* and *S. bayanus*.

Table 7.4 The *Saccharomyces* genus

Saccharomyces	Source	Barnett *et al.* (2000)	Kurtzman and Fell (1998)	Naumov *et al.* (2000)	*Sensu stricto* complex
S. barnettii	Sauerkraut, soft drink	✓	✓		
S. bayanus	Fruit juice, beer, perry, grape must	✓	✓		✓
S. cariocanus	*Drosophila* sp.			✓	✓
S. castellii	Soil, baboon caecum, buttermilk	✓	✓		
S. cerevisiae	Wine, beer, fruit, soil, soft drinks, man	✓	✓		✓
S. dairenensis	Fermenting grapes, dry fruit	✓	✓		
S. exiguous	Grape must, sewage, soil	✓	✓		
S. kluyveri	Soil, *Drosophila* sp., tree exudate	✓	✓		
S. kudriavzevii	Decayed leaf			✓	✓
S. kunashirensis	Soil near hot spring	✓			
S. martiniae	Fermenting mushroom	✓			
S. mikatae	Decayed leaf, soil			✓	✓
S. paradoxus	Oak tree exudates, soil	✓	✓		✓
S. pastorianus	Beer	✓	✓		✓
S. rosinii	Soil	✓	✓		
S. servazzii	Soil, man with HIV	✓	✓		
S. spencerorum	Soil, larval gut	✓	✓		
S. transvaalensis	Soil	✓	✓		
S. unisporus	Kefyr, cheese	✓	✓		

Although subsequent studies (Casaregola *et al.*, 2001) have reinforced this proposal – e.g. coexistence of chromosomes from each 'parent' in *S. pastorianus* (Tamai *et al.*, 1998) – the sequencing of this species (Nakao *et al.*, 2003, see Section 7.3) will, over time, open a Pandora's box of insight, opportunity and experiment.

7.6.1 Taxonomy of lager yeast

Although an interesting example of the joys of taxonomy, the above focus on *S. pastorianus* reflects this yeast's previous label of *S. carlsbergensis*, which of course is more familiar as lager yeast! In passing it is noteworthy that the 'type' or definitive strain of *S. pastorianus* syn. *carlsbergensis* held in culture collections worldwide (see Chapter 8) is that of 'Carlsberg bottom yeast no. 1' which dates back to the end of the nineteenth century. Although lager yeast originated in Bavaria, in 1845 Jacob Christian Jacobsen 'managed to secure two pots of yeast from Brewer Sedlmayr' and transport it 600 miles north, to what became the Carlsberg brewery in Copenhagen. Some 50 years later, Emil Christian Hansen – a contemporary of Pasteur and the 'Father' of the pure yeast culture – in a 'shrewd career move' (Campbell, 2000) named this yeast *S. carlsbergensis*.

Although long accepted as a matter of fact, the distinctions between ale (*S. cerevisiae*) and lager yeast (*S. pastorianus* syn. *carlsbergensis*) have always been headlined as differences in process performance (e.g. bottom vs top fermenting), physiology (maximum growth temperature) or laboratory tests (e.g. utilisation or not of melibiose) (see Table 7.5). Fortuitously the focus on comparative yeast genomics, particularly between *S. cerevisiae* and other species within the *sensu stricto* complex, is providing fresh insights into the differences between ale (*S. cerevisiae*) and lager (*S. pastorianus*) yeasts. Although as befits their grouping into the *Saccharomyces sensu stricto* complex, both species have much in common (the *S. cerevisiae* genome), *S. pastorianus* contains some different genetic material from *S. bayanus*. Quite what *S. bayanus* contributes to the genetic mix is unclear, but it is notable that this wine yeast is 'cryophilic' and is found in low-temperature wine fermentations (Naumov, 1996). Accordingly, it is tempting to conclude that the adaptation of lager strains to lower temperatures is a reflection of the *S. bayanus* contribution to the lager

Table 7.5 Headline differences between ale and lager yeasts

	Ale yeast	Lager yeast
Species	*S. cerevisiae*	*S. pastorianus*
Genome size	1	1.5
Maximum growth temperature (°C)	$\geq$37	$\leq$34
Melibiose hydrolysis	No	Yes
Fructose transport	Facilitated	Active
Maltotriose utilisation	Generally poor	Generally good
Growth between 6–12°C	Poor	Good

yeast genome. This has been confirmed experimentally by hybridisation experiments with *S. cerevisiae* and *S. bayanus* (Sato *et al*., 2002). More recently, the unravelling of this saga has continued with the early headlines (Nakao *et al*., 2003) for the whole genome sequence for a polyploid German lager yeast (Weihenstephan 34/70). Whilst having either two or three copies of each of the 16 chromosomes, 15 chromosomes were of the *S. cerevisiae* type, 12 of the 'non-*S. cerevisiae* type' (presumably *S. bayanus*) and eight chromosomes were a hybrid of both species. Further, it has been shown (Piskur *et al*., 1998) that in *S. pastorianus* the mitochondrial genome is derived from the non-*S. cerevisiae* parent.

In terms of evolution the two species are differentiated by time. *S. cerevisiae* is estimated (Wolfe and Shields, 1997) to have arisen about 100 million years ago, with *S. bayanus* (the presumptive other parent of *S. pastorianus*) diverging 5–20 million years later (Kellis *et al*., 2003). In passing and to complete the story, the use of yeast by man – initially in grape fermentations – is relatively very recent, dating back to Neolithic times about 7000 years ago (Mortimer, 2000). This has been further refined in a fascinating paper by Cavalieri *et al*. (2003) who, from residue present inside one of the earliest known wine jars from Egypt, have demonstrated the presence of ribosomal DNA from *S. cerevisiae*. Whilst demonstrating the role of yeast in wine making in 3150 BC, the authors also speculate that ancient literary evidence implies that grapes generally provided the yeast for cereal fermentations.

The work of Kellis and coworkers (2003) provides an important test of whether (or not) the hybrid genome of *S. pastorianus* is of any real genetic significance. Here the genome sequences of three *sensu stricto* species (*S. bayanus, S. paradoxus* and *S. mikatae*) were compared to that of *S. cerevisiae*. Whilst on the one hand, 'all 16 chromosomes from each of the three newly sequenced genomes map beautifully onto *S. cerevisiae*' (Salzberg, 2003), *S. bayanus* exhibited genomic rearrangements (five reciprocal translocations and three inversions) plus 19 novel genes. To make this real, the nuances of the full comparative analysis indicate that the sequence divergence between *S. cerevisiae* and *S. bayanus* 'is similar to that between human and mouse' (Kellis *et al*., 2003)!

Finally, to close this section, the exploration of genomic differences between *S. cerevisiae* and *S. bayanus* will continue at a pace and will doubtless start to close the gap in understanding between the yeast genotype and its expression as a phenotype (see Fig. 7.2). Surely, coming out of all this endeavour will be snippets of insight that relate to both ale (*S. cerevisiae*) yeast and the hybrid (*S. pastorianus*) lager yeasts. Indeed, it will be fascinating to use this knowledge to drill down to the more subtle differences between strains of brewing yeast which through genomic heterogeneity (see Casaregola *et al*., 2001) are sufficient to be differentiated through DNA fingerprinting techniques. Hopefully, some 10 years after the publication of the 'yeast genome', the brewing industry will actively seek to benefit from this insight and apply it to a better understanding of the process behaviour of ale and lager yeasts.

7.7 Future trends

With *S. cerevisiae* reinforcing its status as biology's favourite model eukaryote, the future is clearly full of great promise and excitement. Certainly, the focus on the relationship between the genotype and the phenotype will continue at a pace along with the expansion in comparative genomics of (hopefully) brewing strains and closely related and more distant species. Clearly, the humble yeast cell is a great candidate for the ultimate insight of understanding how a cell works. However, it is a moot point just how much of the current and upcoming new knowledge is translated into value for the brewing industry. Admittedly the track record since the genome project in 1996 is poor, with little significant industry traction other than in Japan and the USA. While those companies that have been engaged will doubtless continue, it is of concern that so many choose to ignore the developments in genetics as being of little relevance or interest. Perhaps it is timely for industry bodies and organisations to – if nothing else – pull together interested individuals to coordinate a watching brief on behalf of the industry, be it country or worldwide. Ironically, should the wine industry and its consumers take to genetically modified yeast, metabolic engineering could be reborn as a disruptive technology!

7.8 Sources of further information

For a fuller understanding of the advances and fundamentals of yeast genetics, the interested reader is directed to more expansive (reasonably) recent reviews (Walker, 1998; Boulton and Quain, 2001; Hammond, 2003; Briggs *et al.*, 2004). For more 'real time' knowledge, popular science journals (*Nature*, *New Scientist*) provide downloads of this increasingly complex science and, of course, the Internet remains a fantastic but diverting resource!

7.9 Acknowledgements

I would like to thank the Directors of Coors Brewers for permission to publish this review and Sue, Ben, Rosie and Sophie for their support and understanding.

7.10 References

ADAMS, J., PUSKAS-ROZSA, S. and WILKE, C.M. (1992), 'Adaptation and major chromosomal changes in populations of *Saccharomyces cerevisiae*', *Curr. Genet.*, 22, 13–19.

BAMFORTH, C. (2005), 'The gene dream – a nightmare'?, *Brew. Guard.*, July/August, 24–26.

BARNETT, J.A., PAYNE, R.W. and YARROW, D. (2000), *Yeasts: Characteristics and Identification* (3rd edn), Cambridge, Cambridge University Press.

BOEKHOUT, T. (2005), 'Gut feeling for yeasts', *Nature*, 434, 449–451.

BOULTON, C. and QUAIN, D. (2001), *Brewing Yeast and Fermentation*, Oxford, Blackwell Science.

BRIGGS, D.E, BOULTON, C.A., BROOKES, P.A. and STEVENS, R. (2004), *Brewing Science and Practice*, Cambridge, Woodhead.

CAMPBELL, I. (2000), 'Brewing yeast selection', *Microbiol. Today*, 27, 122–124.

CASAREGOLA, S., NGUYEN, H-V., LAPATHITIS, G., KOTYK, A. and GAILLARDIN, C. (2001), 'Analysis of the constitution of the beer yeast genome by PCR, sequencing and subtelomeric sequence hybridization', *Int. J. Systematic and Evolutionary Micro.*, 51, 1607–1618.

CASEY, G.P. (1996), 'Practical applications of pulsed field electrophoresis and yeast chromosome fingerprinting in brewing QA and R&D', *Tech. Quart. Master Brewers Assoc. Americas*, 33, 1–10.

CAVALIERI, D., McGOVERN, P.E., HARTL, D.L., MORTIMER, R. and POLSINELLI, M. (2003), 'Evidence for *S. cerevisiae* fermentation in ancient wine', *J. Mol. Evol.*, 57, S226–S232.

CHRISTIE, K.R., WENG, S., BALAKRISHNAN, R., COSTANZO, M.C., DOLINSKI, K., DWIGHT, S.S., ENGEL, S.R., FEIERBACH, B., FISK, D.G., HIRSCHMAN, J.E., HONG, E.L., ISSEL-TARVER, L., NASH, R., SETHURAMAN, A., STARR, B., THEESFELD, C.L., ANDRADA, R., BINKLEY, G., DONG, O., LANE, C., SCHROEDER, M., BOTSTEIN, D. and CHERRY, J.M. (2004), '*Saccharomyces* Genome Database (SGD) provides tools to identify and analyze sequences from *Saccharomyces cerevisiae* and related sequences from other organisms', *Nucleic Acids Res.*, 32, D311–D314.

COLSON, I., DELNERI, D. and OLIVER, S.G. (2004), 'Effects of reciprocal chromosomal translocations on the fitness of *Saccharomyces cerevisiae*', *EMBO Reports*, 5, 392–398.

DERISI, J.L., IYER, V.R. and BROWN, P.O. (1997), 'Exploring the metabolic and genetic control of gene expression on a genomic scale', *Science*, 278, 680–686.

DUNHAM, M.J., BADRANE, H., FEREA, T., ADAMS, J., BROWN, P.O., ROSENZWEIG, F. and BOTSTEIN, D. (2002), 'Characteristic genome rearrangements in experimental evolution of *Saccharomyces cerevisiae*', *Proc. Nat. Acad. Sci.*, 99, 16144–16149.

EDWARDS-INGRAM, L.C., GENT, M.E., HOYLE, D.C., HAYES, A., STATEVA, L.I. and OLIVER, S.G. (2004), 'Comparative genomic hybridization provides new insights into the molecular taxonomy of the *Saccharomyces* Sensu Stricto complex', *Genome Res.*, 14, 1043–1051.

GOFFEAU, A., BARRELL, B.G., BUSSEY, H., DAVIS, R.W., DUJON, B., FELDMAN, H., GALIBERT, F., HOHEISEL, J.D., JACQ, C., JOHNSTON, M., LOUIS, E.J., MEWES, H.W., MURAKAMI, Y., PHILIPPSE, P., TETTELIN, H. and OLIVER, S.G. (1996), 'Life with 6000 genes', *Science*, 274, 546–567.

HAMMOND, J. (1998), 'Brewing with genetically modified amylolytic yeast', in Roller, S. and Harlander, S. (eds), *Genetic Modification in the Food Industry*, London, Blackie, 129–157.

HAMMOND, J.R.M. (2003), 'Yeast genetics', in Priest, F.G. and Campbell, I. (eds), *Brewing Microbiology* (3rd edn), New York, Kluwer Academic/Plenum Publishers, 67–112.

HIGGINS, V.J., BECKHOUSE, A.G., OLIVER, A.D., ROGERS, P.J. and DAWES, I.W. (2003), 'Yeast genome-wide expression analysis identifies a strong ergosterol and oxidative stress response during the initial stages of an industrial lager fermentation', *Appl. Environ. Microbiol.*, 69, 4777–4787.

HIRSCHMAN, J.E., BALAKRISHNAN, R., CHRISTIE, K.R., COSTANZO, M.C., DWIGHT, S.S., ENGEL, S.R., FISK, D.G., HONG, E.L., LIVSTONE, M.S., NASH, R., PARK, J., OUGHTRED, R., SKRZYPEK,

M., STARR, B., THEESFELD, C.L., WILLIAMS, J., ANDRADA, R., BINKLEY, G., DONG, O., LANE, C., MIYASATO, S., SETHURAMAN, A., SCHROEDER, M., THANAWALA, M.K., WENG, S., DOLINSKI, K., BOTSTEIN, D. and CHERRY, J.M. (2006), 'Genome Snapshot: a new resource at the Saccharomyces Genome Database (SGD) presenting an overview of the *Saccharomyces cerevisiae* genome', *Nucleic Acids Res.*, 34, D442–D445.

KELLIS, M., PATTERSON, M., ENDRIZZI, M., BIRREN, B and LANDER, E.S. (2003), 'Sequencing and comparison of yeast species to identify genes and regulatory elements', *Nature*, 423, 241–254.

KONDO, H., YOMO, H., TAKATANI, T., IZUMI, T., HATANAKA, H., JUJITA, A. and FUKUI, N. (2003), 'Gene expression analysis of brewing yeasts during beer fermentations using DNA microarray', *Proc. 29th Congr. Eur. Conv.*, Dublin, 531–543.

KURTZMAN, C.P. and FELL, J.W. (1998), *The Yeasts, a Taxonomic Study* (4th edn), Amsterdam, Elsevier.

LANCASHIRE, W.E. (1986), 'Modern genetics and brewing technology', *Brewer*, September, 345–348.

LOCKHART, L. (2003), *Genomic analysis of variability in brewers' yeast*, PhD thesis, University of Manchester, Faculty of Science and Engineering.

LUCCHINI, S., THOMPSON, A. and HINTON, J.C.D. (2001), 'Microarrays for microbiologists', *Microbiol.*, 147, 1403–1414.

MEADEN, P. (1986),'Genetic engineering of yeast', *Brew. Guard.*, July, 23–27.

MEADEN, P.G. (1996), 'Yeast genome now completely sequenced', *Ferment*, 9, 213–214.

MORTIMER, R.K. (2000), 'Evolution and variation of the yeast (*Saccharomyces*) genome, *Genome Res.*, 10, 403–409.

MORTIMER, R.K. and JOHNSTON, J.R. (1986), 'Genealogy of principal strains of the yeast genetic stock center', *Genetics*, 113, 35–43.

NAKAO, Y., KODAMA, Y., NAKAMURA, N., ITO, T., HATTORI, M., SHIBA, T. and ASHIKARA, T. (2003), 'Whole genome sequence of a lager brewing yeast', *Proc. 27th Congr. Eur. Conv., Dublin*, 524–530.

NAUMOV, G.I. (1996), 'Genetic identification of biological species in the *Saccharomyces sensu stricto* complex', *J. Indust. Micro.*, 17, 295–302.

NAUMOV, G.I., JAMES, S.A., NAUMOVA, E.S., LOUIS, E.J. and ROBERTS, I.N. (2000), 'Three new species in the *Saccharomyces sensu stricto* complex: *Saccharomyces cariocanus*, *Saccharomyces kudriavzevii* and *Saccharomyces mikatae*', *Int. J. Systematic and Evolutionary Micro.*, 50, 1931–1942.

PENG, X., BOWER, P., BARNEY, M., NAVARRO, A., SEABROOKS, J.R. and RYDER, D.S. (2003), 'Patterns of yeast gene expression during fermentation and in response to stress', *Proc. 29th Congr. Eur. Conv.*, Dublin, 490–502.

PENTTILA, M. (2001), 'Metabolic engineering approaches – opportunities for brewing', *Proc. 28th Congr. Eur. Conv., Budapest*, 505–513.

PISKUR, J., SMOLE, S., GROTH, C., PETERSEN, R.F. and PEDERSEN, M.B. (1998), 'Structure and genetic stability of mitochondrial genomes vary among the genus *Saccharomyces*', *Int. J. Systematic and Evolutionary Micro.*, 43, 1015–1024.

SALZBERG, S.L. (2003), 'Yeast rises again', *Nature*, 423, 233–234.

SATO, M., WATARI, J., SAHARA, H. and KOSHINO, S. (1994), 'Instability in electrophoretic karyotype of brewing yeasts', *J. Am. Soc. Brew. Chemists*, 52, 148–151.

SATO, M., KISHIMOTO, M., WATARI, J. and TAKASHIO, M. (2002), 'Breeding of brewer's yeast by hybridization between a top fermenting yeast *Saccharomyces cerevisiae* and a cryophilic yeast *Saccharomyces bayanus*', *J. BioSci. Bioeng.*, 93, 509–511.

SUPPLIERS OF MODIFIED WINE YEASTS – available from Lesaffre (www.lesaffreyeast

corp.com) and from First Venture Technologies (www.firstventuretech.com).

TAMAI, Y., MOMMA, T., YOSHIMOTO, H. and KANEKO, Y. (1998), 'Co-existence of two types of chromosomes in the bottom fermenting yeast, *Saccharomyces pastorianus*', *Yeast*, 14, 923–933.

TANAKA, K. and KOBAYASHI, O. (2003), 'Analysis of karyotypic polymorphisms in a bottom-fermenting yeast strain by polymerase chain reaction', in Smart, K. (ed.) *Brewing Yeast Fermentation Performance* (2nd edn), Oxford, Blackwell Science, 3–12.

TUBB, R. (1981), 'Developing control over yeast', *Brew. Guard.*, July, 21–24.

TUBB, R.S. (1984), 'Genetic development of yeast strains', *Brew. Guard.*, September, 34–37.

VAUGHAN-MARTINI, A. and KURTZMAN, C.P. (1985), 'Deoxyribonucleic acid relatedness amongst species of the genus *Saccharomyces stricto sensu*', *Int. J. Systematic Bacteriology*, 35, 508–511.

WALKER, G.M. (1998), *Yeast Physiology and Biotechnology*, Chichester, John Wiley & Sons.

WIGHTMAN, P., QUAIN, D.E. and MEADEN, P.G. (1996), 'Analysis of production brewing strains of yeast by DNA fingerprinting', *Lett. Appl. Micro.*, 22, 90–94.

WOLFE, K.H. and SHIELDS, D.C. (1997), 'Molecular evidence for an ancient duplication of the entire yeast genome', *Nature*, 387, 708–713.

8

Yeast supply and propagation in brewing

D. E. Quain, red.ts Ltd, UK

8.1 Introduction

In the theatre of brewing, yeast is viewed, quite inappropriately, as a 'supporting actor'. But unlike the more commercial leading actors water, malt and hops-yeast appears not once but in many successive productions! Further, with fermentation being the longest chunk of the process, yeast is on stage longer. Finally and critically, as an agent of change, yeast facilitates the most dramatic event in the whole production – transformation of wort to beer. Given this impressive history, it is a surprise that yeast is not lauded and celebrated as the real star of the production of beer.

This chapter builds on this thinking, and is focused on the principles and best practices behind in-brewery propagation of yeast together with emerging 'work-arounds' such as the advent of active dried yeast.

8.1.1 The need for propagation?

There is now general acceptance for the periodic introduction of freshly propagated yeast in most if not all breweries. This has been driven by breweries managing a number of discrete strains and, worldwide, the growth of franchise brewing. However, there is no hard and fast rule as to the frequency of propagation. Experience suggests two camps: those that replace stocks after 15–20 generations and those who operate in a band between five and 10 generations. Although best practice, the current drivers for this process are loosely defined around (i) hygiene and (ii) consistency of yeast performance. How compelling these arguments are is a matter of debate. Certainly with closed vessels, effective yeast handling and optimal cleaning regimes, 'hygiene' is now perhaps less of an issue for breweries in the early twenty-first century. Similarly, the case for deterioration in yeast performance with advancing generations has received little focused study. Anecdotally, some brewers observe 'weakness' or 'sickness' (Jones, 1997) whereas others will claim that over time a yeast strain becomes 'better' and more 'manageable'. Possibly these contrasting views reflect the inevitable selection that cone cropping brings. The growing argument around cell ageing (see Powell *et al.*, 2000 for a review) suggests that older, larger cells are disadvantaged and that cone cropping selects for these cells. However, the 'jury is still out' with regard to 'ageing' with contrasting reports as to its significance (Deans *et al.*, 1997; Quain *et al.*, 2001). However, there is no doubt that brewers – typically ale – whose yeast is a mixture of two or occasionally three strains do have a clear case for inconsistency developing over time. Here, the mixture of strains will change as, depending on flocculation characteristics and cropping regimes, one strain predominates at the expense of another. A further contributor to this theme is that of genetic change or drift. The argument that 'mutation' (Maule, 1980) is a reason for replenishment of yeast is now substantiated by a number of observations of phenotypic (typically flocculation) change, which increasingly are associated with genetic changes (Boulton and Quain, 2001) (see Chapter 7). Just how widespread and frequent these events are is not clear as these insights are hard-won technically and reflect gross 'obvious' changes such a shift in flocculence. Despite this caveat, genetic changes are, leastways directionally, an increasingly good argument for the principle of periodic yeast replacement.

8.2 Storage and supply of yeast cultures

Key elements of the propagation 'offer' are that the yeast is supplied consistently as microbiologically pure and of the correct strain. As ever, these twin demands are best achieved by the use of a defined process. Good examples used by UK brewers have been described (Quain, 1995; Kennedy *et al.*, 2003) that apply both best-practice microbiological testing and DNA fingerprinting to assure yeast quality. For both organisations, the importance of this documented and traceable process is highlighted by its inclusion in an ISO accredited Quality System. The hierarchy of an approach (Quain, 1995) to yeast supply is illustrated in Fig. 8.1. The outcome of both approaches is the generation of yeast slopes of assured quality that are supplied to breweries for propagation. The acceptable shelf-life of such slopes ranges from 16 weeks (Kennedy *et al.*, 2003) to 26 weeks (Quain, 1995).

Underpinning yeast supply is the need to store yeast stocks securely and in a way that minimises (or preferably removes) any threat to viability or genetic integrity (for a review see Boulton and Quain, 2001). Unequivocally, the 'gold standard' for the storage of microorganisms is immersion in liquid nitrogen at −196°C (Kirsop and Doyle, 1991), an approach used with brewing yeasts by

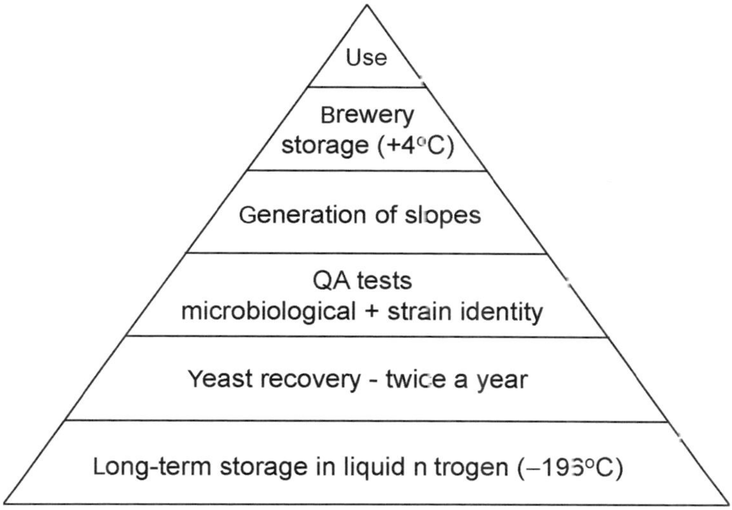

Fig. 8.1 Yeast supply process.

Labatts (Wellman and Stewart, 1973), South African Breweries (Hulse *et al.*, 2000), Bass/Coors since 1990 (Quain, 1995), Scottish Courage since 1983 (Jones, 1997, Kennedy *et al.*, 2003) and the National Collection of Yeast Cultures in the UK.

Storage in liquid nitrogen requires adherence to strict protocols to ensure no impact on cell viability or genetic stability. As ever, the physiology of yeast is important and, in this case, yeast is best grown oxidatively prior to freezing. After resuspension (*ca.* 100 × 10^6/ml) in fresh media containing the cryoprotectant glycerol, small aliquots (typically <0.5 ml) of yeast are carefully cooled from room temperature to −30°C followed by immersion in liquid nitrogen at −196°C. The cooling process, which takes 2 hours, cannot be fast-tracked without severely damaging the viability of the yeast population. Conversely, recovery from storage in liquid nitrogen requires the ampoule or straw to be immersed in water at between 20°C and 37°C.

Although invariably more convenient than cryopreservation, other techniques for the long-term (more than six months) storage of yeast cannot be recommended (Quain, 1995). Storage on agar slopes or in broth, with periodic sub-culture, while simple, suffers from poor viability and damningly from genetic and phenotypic changes. By way of example, Hulse *et al.*, (2000) clearly showed cryopreservation to outperform slope storage in terms of consistency and attenuation. Similarly, the more complicated freeze-drying or lyophilisation has had a particularly bad press, with reports of catastrophically low viability and enrichment in genetic variants. In contrast, the UK's National Collection of Yeast Cultures (www.ncyc.co.uk) stores and supplies freeze-dried strains.

Although the method of choice, the application of cryopreservation suffers from technical complexity and cost (both equipment and regular deliveries of liquid nitrogen). Given this, it is not surprising that this approach to the storage of yeast has typically been taken up, as noted above, by 'central' laboratories of brewing groups or by commercial yeast collections. A more convenient approach that is increasingly finding favour is long-term storage in freezers at −70°C, an approach used by South African Breweries (Hulse *et al.*, 2000), White Labs (C. White, personal communication) and many academic laboratories.

As an alternative to in-house yeast supply and storage, a growing option is to outsource the responsibility to a third party. Although perhaps lacking flexibility, third party storage and supply removes the need for the associated laboratory facilities and specific expertise and labour. Current examples of this approach include Danbrew (www.danbrew.dk) (formally Alfred Jorgensen Laboratories) and the collaboration between Cara Technology (www.cara-online.com) and the National Collection of Yeast Cultures. Building on this, rather than opting for storage and supply of 'home' strains, production yeast strains can be sourced from a third party. Indeed, a growing diversity of brewing strains typically with a known pedigree are available from commercial yeast collections, some of which, together with the number of yeasts offered, are detailed in Table 8.1.

In some cases breweries without propagation facilities or capacity receive bulk slurries from donor breweries either from propagator or cropped from fermenter. These stainless steel tanks, typically 8 hl, need to be hygienically designed and capable of steam sterilisation and CiP. Cooling is an issue as these tanks are typically not lagged and consequently the yeast slurry warms up during

Table 8.1 Commercial collections of brewing yeasts

Collection	Location and Internet address	Strains[1]
Belgian Co-ordinated Collections of Micro-organisms (BCCM)	Belgium bccm.belspo.be	25
Danbrew	Denmark www.danbrew.dk	350
National Collection of Yeast Cultures	UK www.ncyc.co.uk	344
Research Institute of Brewing and Malting (RIBM)	Czech Republic www.beerresearch.cz	115
Siebel Pure Yeast Library	USA www.siebelinstitute.com	8
VTT	Finland www.vtt.fi	190
White Labs	USA www.whitelabs.com	64

[1] Number of brewing strains in each collection as of November 2005.

transit. As with storage in general, the physiology and viability of yeast can only deteriorate with time. Accordingly, and specifically with bulk transport, every effort should be made to minimise time in transport and awaiting transfer. For a fuller description of interbrewery transport see Boulton and Quain (2001).

8.3 Propagation

Emil Christian Hansen is recognised as being the 'Father' of the pure yeast culture (*aka* 'propagation') in the brewing industry. Since Hansen's introduction of this early 'disruptive technology' in Copenhagen's Carlsberg brewery in 1884, brewery propagation processes have evolved slowly. Whilst quite rightly focusing on the importance of hygiene and strain purity, the propagation process has, until relatively recently, been more aligned to a small-scale fermentation than a process for the growth of yeast. Equally, 'many propagation vessels are little more than hygienically designed fermenters!' (Boulton and Quain, 2001). The simple realisation that yeast propagation is fundamentally about the growth of yeast rather than the production of beer has resulted in a new 'aerobic' philosophy where yeast is intentionally grown, both in the laboratory and in the brewery, in the presence of oxygen. The physiological implications of aerobic growth are reviewed in Section 8.3.2 and at length elsewhere (Boulton and Quain, 2001), but suffice to say such conditions support the extensive growth of yeast that, importantly for pitching on, is 'lipid replete'. This move to 'aerobiosis' has resulted in significantly greater yields of yeast and a simplification of a hitherto complicated and protracted propagation process. Similarly, the long-held rule that lager yeast is propagated at lager fermentation temperatures has been successfully challenged. Today's high yield aerobic propagators are typically at 25°C for both ale and lager strains. Both an increase in temperature and the move to aerobiosis have significantly accelerated process turnaround time for yeast propagation.

As a process and, irrespective of whether the process is fundamentally aerobic or anaerobic, yeast propagation involves a series of biomass scale-ups culminating in pitching a production scale fermenter. It would without doubt be accepted by all practitioners that the key performance indicators for this process are strain purity, viability, microbiological cleanliness and, most challengingly, meeting process and product specifications in first-generation fermentation.

8.3.1 Laboratory propagation

Whether pure yeast strains on slope or liquid culture are sourced internally or externally, the objective of the laboratory phase of propagation is to grow sufficient biomass to enter the brewery propagation phase without issue. This is achieved by serial transfers of increasing culture volume that meet the twin needs of scale-up and hygiene. Although the timeline of these steps is clearly important, the potential threat to hygiene overrides all considerations and,

Table 8.2 Laboratory (L) and plant (P) propagation processes

Coors[1]	Scottish Courage[2]
L – yeast slope	L – yeast slope
↓	↓
L – 2 × 10 ml – static – 24 h at 25°C	L – 2 × 250 ml – shaken – 48 h at 27°C
↓	↓
L – 2 × 100 ml – shaken – 72 h at 25°C	L – 20 litres – constant oxygenation – 48/72 h at 20°C
↓	↓
L – 3 litres – constant aeration – 72 h at 25°C	P – 15hl – constant oxygenation – 48/72 h at 20°C vessel 1
↓	↓
L – 20 litres – constant aeration – 72 h at 25°C	P – 75 hl – constant oxygenation – 72/120 h at 15°C vessel 2
↓	↓
P – 8 hl – constant aeration – 24/30 h at 25°C vessel 1	P – 1000 hl fermentation
↓	↓
P – 140 hl – constant aeration – 24/30 h at 25°C vessel 2	P – 2000 hl fermentation
↓	
P – 1600 hl fermentation	

[1] Boulton and Quain (2001), Boulton and Mielenenski (2006).
[2] Jones (1997).

accordingly, this determines a 'safety-first' mix of culture volume, yeast concentration and incubation time.

The two examples of laboratory (and plant) propagation described in Table 8.2 neatly capture today's approach to the 'safety-first' principles outlined above. Both approaches, whilst different, are fit for the purpose of yeast propagation as described by two UK brewing companies (Jones, 1997; Boulton and Quain 2001; Boulton and Mielenenski, 2006). Although a direct comparison reveals subtle differences of emphasis, both approaches focus on ensuring the aerobic growth of yeast. Typically this is achieved initially in shake flasks ($\leq$2 litres) on a rotary shaker (200 rpm) with headspace of at least 60%. From this point, the next and final stage of the laboratory process is performed in a flat-bottomed, cylindrical stainless steel vessel with an operating volume of 20–25 litres. These 'Carlsberg flasks' (Kunze, 1999) are hygienically designed culture vessels, which can be autoclaved or subject to direct heat treatment. Best practice is to continuously aerate or oxygenate the culture with filtered air or oxygen so as to achieve high cell densities (*ca.* 200 $\times$ 10^6/ml). These gases, nitrogen or carbon dioxide, are also used as a motor gas in the transfer of the culture to the propagator seed tank. Commercially, 'Carlsberg flasks' of a broadly common design are available from a number of suppliers (see 'suppliers of Carlsberg flasks' in the References, Section 8.8). A modified Carlsberg flask has been described (Boulton and Quain, 1999) (Fig. 8.2), which achieves higher rates of oxygen transfer through a

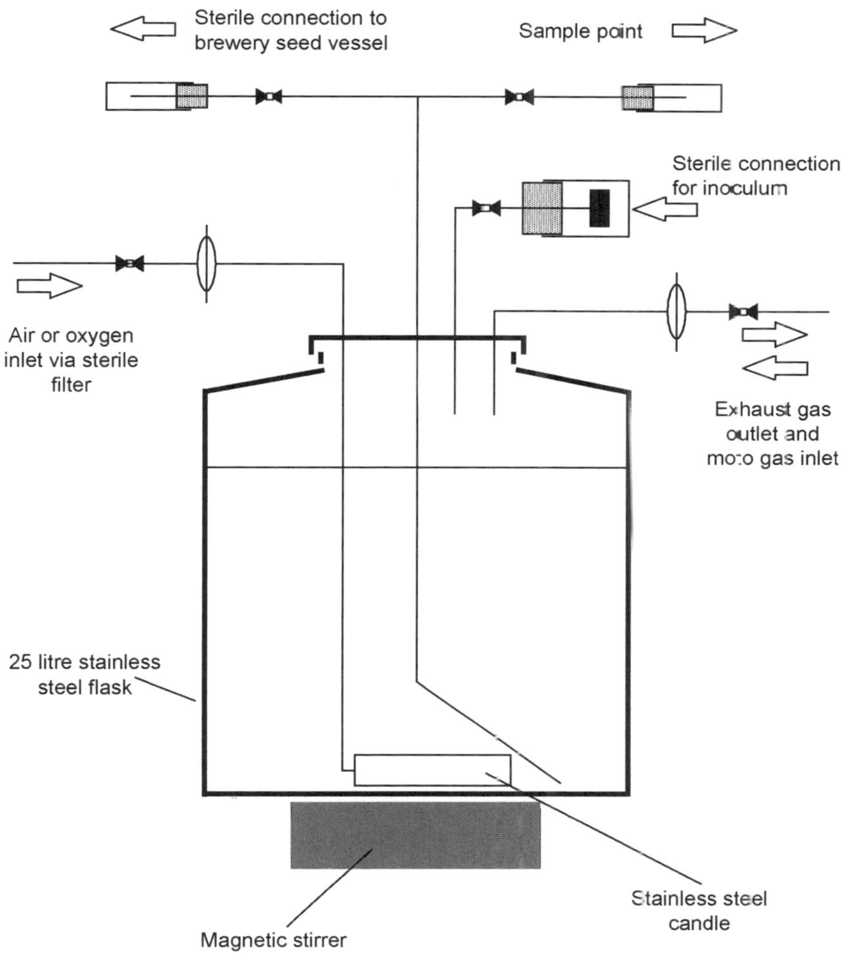

Fig. 8.2 Final stage propagation vessel.

combination of flow rate (through a sinter) together with mixing via a stirrer. For practical reasons of convenience these vessels are usually filled with sterile wort, whereas earlier steps in the laboratory are normally conducted with a semi-defined medium such as YEPG (yeast extract, peptone, glucose) or MYGP (malt extract, yeast extract, glucose, peptone) or commercial media such as YM (yeast mould broth similar to MYGP).

As ever, attention to detail is critical to the success of the laboratory phases of yeast propagation. Indeed, the maxim 'get the small things right and the big things look after themselves' is particularly appropriate to this process. For example, it is important that the entire slope is sacrificed at the first step. It is a false and dangerous economy to 'sample' a slope and then inoculate the starter culture. Such bad practice can result in slope contamination, which can then compromise subsequent propagations. This step, like those in the yeast supply process, is best performed in the presence of an 'observer' who validates that the correct yeast strain has been selected for propagation. In general, microbiological best practice is a given, and with a robust process underpinning the process of slope generation (see Section 8.2), there should be no need for exhaustive sampling and testing to check for potential contaminants. In addition, the risk of microbiological contamination should be minimised by every culture transfer being performed in a laminar or Class 2 safety cabinet. To ensure maximum yield, every effort should be made to ensure effective oxygen transfer into the culture at every stage of the process. This approach ensures rapidity and removes any potential microbiological vulnerability of low cell counts at the start of a fresh culture.

8.3.2 Brewery propagation

The major driver for the development of aerobic propagation systems is dissatisfaction with the performance of the 'first generation' fermentation. Typically with 'traditional' anaerobic propagators the cell yield is too low to achieve a satisfactory pitching rate, which results in a sluggish, protracted fermentation producing non-standard beer. 'Patches' such as rousing and aeration can be applied but usually to no great effect! Overlaid on this sorry scenario is a further contributor to the problem, the increasing size of fermenters. To make this real, in the past, yeast was pitched on into relatively small fermenters such as 640 hl (Maule, 1980) or 656 hl (Molzahn, 1977) with worts at comparatively lower specific gravity (*ca.* 1050/12.5° Plato). Today, with few exceptions, the demands of production have resulted in standard cylindroconical vessels in the range 1600–4000 hl with higher-gravity worts (1.060/15° Plato or higher). As a consequence of this increased scale, the output from an anaerobic propagator (say 80 hl at 70×10^6 cells/ml) is unsatisfactorily low. In 1600 hl of wort this would equate to a pitching rate of 3.5×10^6 cells/ml against a specification of 15×10^6 cells/ml for a high gravity (1.060/15° Plato) wort. Given this it is no surprise that such a process regime would cause frustration and complexity and ultimately require 'reengineering' to become more reliable and consistent.

The transition from anaerobic to aerobic propagation was stimulated by the 'related world' of bakers' yeast production. This process is at the extremity of 'aerobic' with the yeast being in an oxidative rather than a fermentative physiological state. This is achieved via a fed-batch process with the substrate molasses being added throughout the process at a rate that ensures that the sugar concentration is maintained at a low level. As described at length elsewhere (Boulton and Quain, 2001), this, together with high rates of oxygen transfer, results in catabolite derepressed or oxidative physiology. Compared to fermentative metabolism, catabolite derepressed yeast is highly efficient at converting sugar to new yeast biomass, theoretically achieving 54% conversion compared to a maximum of 7.5% under anaerobic conditions. Indeed, in bakers' yeast production the formation of ethanol is a marker for the overfeeding of molasses and a loss of yield of yeast (Reed, 1982).

Taking the bakers' yeast process as stimulus, the aerobic propagation of brewers' yeast is focused on providing sufficient oxygen to achieve maximum cell division. In the case of the more traditional anaerobic propagators, oxygen was provided at the beginning of the process, which reinforced the principle that propagation was based on 'fermentation' principles rather than growing yeast. In some instances propagators would receive supplementary aeration or oxygenation via, at best, sintered lances sporadically throughout the process. However, gas transfer under such conditions is extremely poor with the majority of added oxygen breaking out in the vessel headspace. This reflects on the fundamental rules of gas transfer in cultures which is influenced by agitation, velocity or increasing the partial pressure of the gas. Of these three parameters, agitation is the most effective variable in facilitating gas transfer (for a still relevant review see Finn, 1954). The gas-absorbing capacity of a fermentor is measured as 'K_La', which is described by Finn as 'the only satisfactory way to characterise the performance of laboratory or industrial devices'.

Accordingly, the ambition of an aerobic brewers' yeast propagator is to maintain a minimum but detectable dissolved oxygen level throughout the process. It is this idea that is at the core of this process, which is delivered by a combination of gas flow rate and, critically, effective agitation. Such conditions ensure that the propagating yeast is not oxygen limited. The outcome of this approach is cell yields of *ca.* 200 $\times$ 10^6 cells/ml compared to 50–70 $\times$ 10^6 cells/ml from a conventional anaerobic propagator.

A well-documented example of an aerobic propagator is that now in use in two UK breweries (Table 8.2) (Boulton and Quain, 1995, 1999, 2001; Boulton and Mielenenski, 2006). The stimulus for this approach is to meet a performance brief of going from a 140 hl propagation directly into 1500hl of wort at the target pitching rate of 15 $\times$ 10^6 cells/ml. This was achieved using a two-stage aerobic system (see Fig. 8.3) with a 13 hl seed vessel (working volume 8 hl) feeding the 220 hl propagator (working volume 140 hl). Both vessels are distinguished by the presence of perforated stainless steel sparge rings together with a heavy-duty top-mounted variable speed agitator positioned above the oxygenation ring. To ensure turbulent flow, internal baffles are positioned within each vessel. During commissioning, precise operating conditions were established to meet the specified gas transfer rate (K_La) so as to ensure guarantee aerobic conditions were maintained throughout the process (Boulton and Mielenenski, 2006). In this case the maximum agitation speed was 58 rpm with oxygen flow rates between 10 and 100 l/min. In routine use, oxygen is not directly measured but strain-specific ramped flow rate is applied based on direct measurement during plant commissioning. This configuration, together with wort at 12.5°P and a temperature of 25°C, achieves the desired cell yields of 180–200 $\times$ 10^6 cells/ml (equivalent to 15 $\times$ 10^6 cells/ml in 1500 hl of wort) at viabilities of 95% or greater within 24 hours. Figure 8.4 shows the key parameters.

The hygienic robustness of this design is borne out by no microbiological failures in respectively five and 10 years of operation with both breweries handling four yeast strains. Best practice measures include dedicated single-use

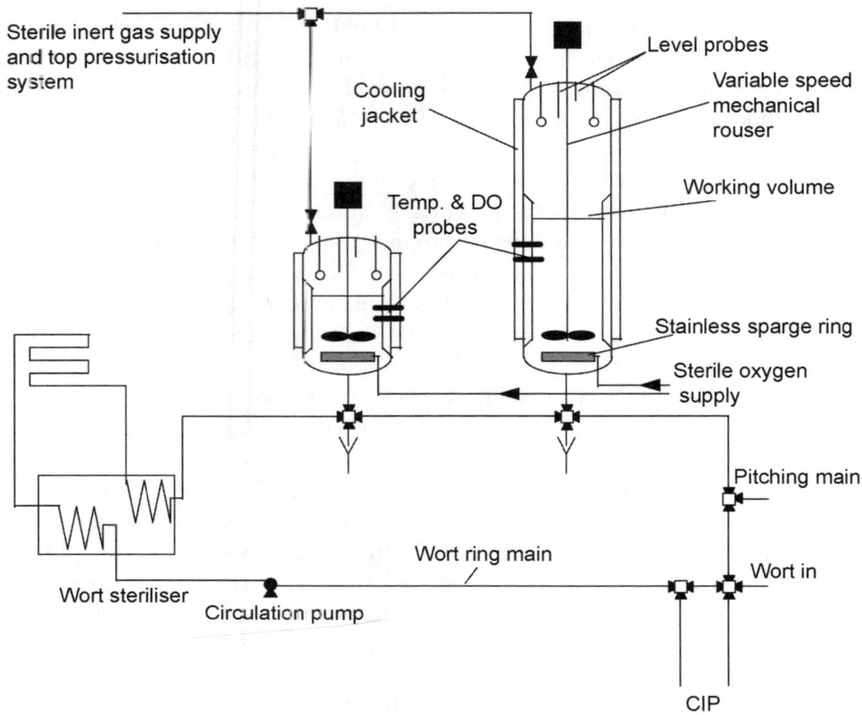

Fig. 8.3 Schematic design of aerobic propagator.

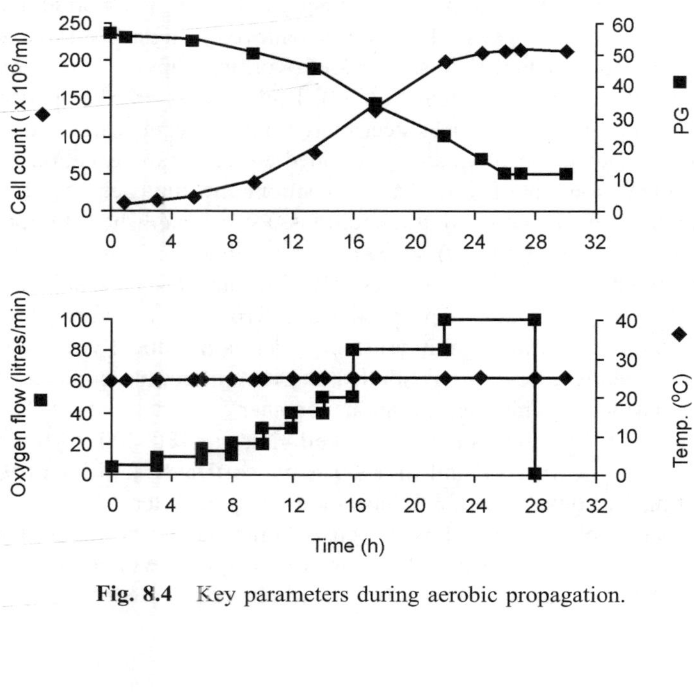

Fig. 8.4 Key parameters during aerobic propagation.

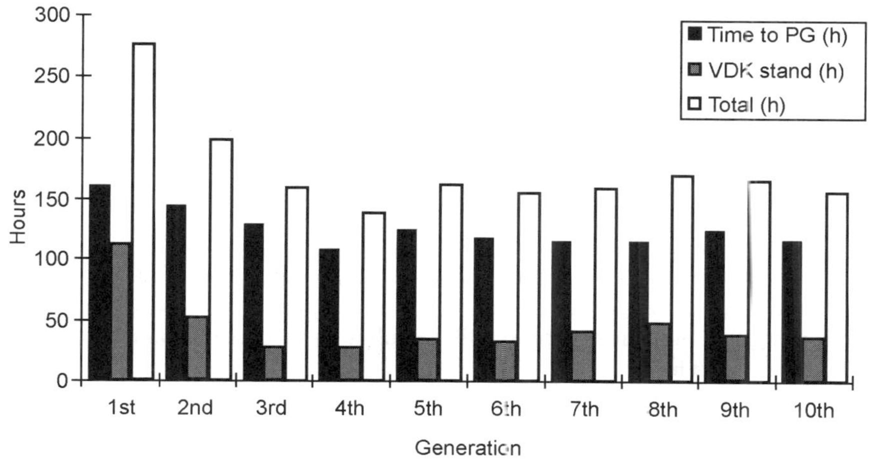

Fig. 8.5 Fermentation performance ex aerobic propagator.

CiP, sterile wort, steam sterilisation of vessels and associated pipework, gas microbiological filters, and steam-treated sample points. Foam is managed through initially the vessel freeboard, followed by application of top pressure and, where required, reduction in agitation and gas flow rate.

Although broadly successful, this approach to aerobic propagation has not fully met expectations. Although the target pitching rate has been achieved in fermenter, the first-generation – and to lesser extent the second-generation – cycle times were longer than standard (see Fig. 8.5). However, yeast crops, viability and beer quality from first-generation fermentations were indistinguishable from subsequent generations (Boulton and Quain, 1999; Boulton and Mielenenski, 2006). The reasons for this disparity are not clear, but Boulton and Mielenenski (2006) have proposed a persuasive argument built around aerobically grown cells being smaller than anaerobic cells. Accordingly, although comparable cell numbers are pitched, the cell volume of anaerobic cells is almost three times that of aerobic cells and their surface area nearly twice that of aerobic cells. Although the direct significance of these observations is not clear, it is tempting to conclude that these differences may well explain the performance of aerobically propagated yeast in fermenter.

Similarly detailed accounts of production-scale aerobic propagators are few and far between. One report (Table 8.2, Jones, 1997) describes the need for 'effective aeration or oxygenation throughout fermentation' without any process description. Commercial solutions have been described (Wainwright, 1999; Nielson, 2005) from the perspective of aeration and process optimisation. Without quantifying the $K_{L}a$, Nielson (2005) noted the significance of agitation rate over gas flow and concluded that a dissolved oxygen tension of 8% was optimal for aerobic yeast production. Interestingly, yields of 120 $\times$ 10^6 cells/ml were considered to be physiologically superior to cell yields of 170 $\times$ 10^6/ml, suggesting that the 'yeast becomes less vital' with extended propagation. This is in contrast to the supplier's website (www.alfalaval.com), which describes one of the benefits of its propagation plant as 'cell counts of up to 150–200 million cells per millilitre'. The other report (Wainwright, 1999) takes a different perspective, focusing on the opportunities for continuous propagation via their 'Conti-Prop' process. Here in conjunction with a 'Turbo air' aeration jet, up to 90% of the propagated yeast volume is transferred to fermenter at knowingly low pitching rates as this 'yeast is very active'. The remaining 10% of the original propagation is topped up with fresh wort and the process restarts. This topping-up approach ('Drauflassen') naturally lends itself to breweries with only one production strain and, according to the supplier's website (www.esau-hueber.de), the company is successfully installing this technology in a number of European breweries.

8.4 Active dried yeasts

Active dried yeast (ADY) has long been used in the baking industry as an alternative to compressed, fresh yeast. The advantages of ADY are its availability and shelf-life. The production processes of ADY are well established around an incremental fed-batch fermentation of molasses followed by drying via continuous tunnel belt driers or air lift/fluidised bed driers. Although loss of viability is not a major concern for bakers' yeast, additives are frequently included to improve rehydration and shelf-life and to enhance viability.

Although active dried bakers' yeast has been lauded as a 'technical triumph' (Reed, 1982), dried brewing yeasts have fared poorly because of poor and inconsistent viability (O'Connor-Cox and Ingledew, 1990). However, a series of publications in the late 1990s has forced a reconsideration of the opportunities that active dried brewers' yeast (ADBY) might bring to commercial brewing. Although initially focused on primary fermentation, the marketing of ADBY has shifted towards playing a role in supplementing or replacing yeast propagation. Certainly it is attractive to consider the benefits such an approach might bring. Clearly with simple handling and storage at 10ºC for up to two years, the major advantage is of availability and production flexibility. Depending on scale and demand, ADBY can easily be envisaged as replacing the laboratory steps of yeast propagation. The opportunities for ADBY in plant propagation are perhaps more complex. Where plant is already available it is unlikely to be replaced through this route, though, in the absence of propagation facilities ADBY can clearly play a role but obviously with a revenue cost. Less obvious but related opportunities are in the provision of yeast for sporadically brewed 'speciality' beers or for franchise brewing operations.

Without exception in the last decade, the numerous publications (see Boulton and Quain, 2001) on the application of ADY in brewing have used ale and lager strains from a leading supplier of dried yeast. Although a portfolio of strains of known, respectable provenance, most of the reported work has focused on three lager strains ('Saflager') with one exception (van den Berg and Van Landschoot,

2003) that extended the theme to top-fermenting strains ('Safale' and 'Safbrew').

Increasingly, ADBY is positioned as being 'multi-functional' with its application and opportunities for its use changing with brewery size, from 'micro' through 'regional' to 'national' (Reckelbus *et al.*, 2000). As discussed above, the scale of the opportunity relates to the extent to which dried yeast can replace existing activity. What is clear is that when pitching rates are corrected for viability, ADBY can perform acceptably (beer quality and fermentation metrics) in the first and subsequent generations. However, inevitably all breweries are different, and clarity on the cost-effective entry point for ADBY requires appropriate analysis to allow economic justification. Conversely, van den Berg and Van Landschoot (2003) see the opportunity differently, with ADBY playing a potential role in Belgium with secondary fermentation in bottle ('refermentation').

Despite some obvious opportunities, the downsides of ADBY require comment. Of primary concern is viability, which post rehydration is anywhere between 50% and 80% (reviewed in Boulton and Quain, 2001: Finn and Stewart, 2002: van den Berg and Van Landschoot, 2003), and which is potentially exacerbated by viability measurements being optimistic at lower viabilities. Although the pitching rate can simply be corrected for viability, a significant amount of the pitched biomass is dead, which is not good practice in terms of consistent beer quality (Finn and Stewart, 2002). Secondly, ADBY, given its scaled production, is not surprisingly less clean microbiologically than yeast propagated through laboratory and plant procedures. However, there is a view (van den Berg and Van Landschoot, 2003) that the bacterial contaminants are not beer spoilers and therefore of less concern. Although arguably the case, the situation regarding possible contamination with wild yeasts is less clear. A further consideration in the argument is the process required to rehydrate dried yeast which, importantly, should adhere to specific guidelines of the supplier.

All in all, ADBY does present some real opportunities to reinvent the propagation process in the brewing industry. The key determinants of success are, not surprisingly, yeast viability and freedom from contaminants be they other brewing yeasts, wild yeasts or bacteria. Glibly and simply, the former requires to be raised and the latter reduced! A further consideration, given the concerns on yeast drying on a small scale, is to understand the implications of ADBY on the genetic stability of the yeast being dried. Finally, the rehydration process should be made robustly 'foolproof' and, tellingly, many major brewers will require their own yeasts to be confidentially dried to fully accept the opportunities of ADBY.

8.5 Future trends

Although it would be stimulating to speculate on the application of fed-batch propagators or genuinely continuous aerobic systems, the reality is that yeast propagation is likely to evolve in incremental steps rather than through step change. For example, the efficiency (but not the complexity!) of propagation could be significantly improved by changing the feedstock from wort to mannitol (Quain and Boulton, 1987) or corn steep liquor (Taidi *et al.*, 2001). The global future of the potentially disruptive technology of active dried brewers' yeast awaits significant improvements in viability and hygiene.

8.6 Sources of further information

A fuller, more detailed appreciation of the diverse aspects of the supply of yeast cultures and subsequent propagation can be found in recent texts on fermentation (Boulton and Quain, 2001) and brewing (Kunze, 1999; Briggs *et al.*, 2004; Ockert, 2006). Information on the current commercial solutions for aerobic propagators can be found on the Internet (see 'Suppliers of yeast propagators' in the References, Section 8.8). Similarly, information on the supply of active dried brewers yeast can be found on suppliers' web pages (see 'Suppliers of active dried yeast' in the References).

8.7 Acknowledgements

I would like to thank the Directors of Coors Brewers for permission to publish this review. Additionally, a big thank you to Dr Chris Boulton of Coors Brewers for our many years of rewarding collaboration, to Professor Charlie Bamforth for forbearance and to Dr Chris White of White Labs for sharing valuable insight.

8.8 References

BOULTON, C. and MIELENENSKI, A. (2006), 'Aerobic yeast propagation', in Smart, K. (ed.), *Brewing Yeast Fermentation Performance*, 3rd edn, Oxford, Blackwell Science, in press.

BOULTON, C.A. and QUAIN, D.E. (1995), 'Propagation of brewing yeast under aerobic conditions', UK patent GB2 298 651.

BOULTON, C.A. and QUAIN, D.E. (1999), 'A novel system for propagation of brewing yeast', *Proc. 27th Congr. Eur. Conv., Cannes*, 647–654.

BOULTON, C. and QUAIN, D. (2001), *Brewing Yeast and Fermentation*, Oxford, Blackwell Science.

BRIGGS, D.E., BOULTON, C.A., BROOKES, P.A. and STEVENS, R. (2004), *Brewing Science and Practice*, Cambridge, Woodhead.

DEANS, K., PINDER, A., CATLEY, B.J. and HODGSON, J.A. (1997), 'Effects of cone cropping and serial repitch on the distribution of cell ages in brewery yeast', *Proc. 26th Congr. Eur. Conv., Maastricht*, 469–476.

FINN, D.A. and STEWART, G.G (2002), 'Fermentation characteristics of dried brewers yeast:

effect of drying on flocculation and fermentation', *J. Am. Soc. Brew. Chem.*, 60, 135–139.

FINN, R.K. (1954), 'Agitation-aeration in the laboratory and in industry', *Bact. Rev.*, 18, 254–274 (full article available through PubMed, www.pubmed.gov).

HULSE, G., BIHL, G., MORAKILE, G. and AXCELL, B. (2000), 'Optimisation of storage and propagation for consistent lager fermentations', in Smart, K. (ed.), *Brewing Yeast Fermentation Performance*, Oxford, Blackwell Science, 161–169.

JONES, H. (1997), 'Yeast propagation – past present and future', *Brewers' Guardian*, October, 24–27.

KENNEDY, A.J., TAIDI, B., AITCHISON, A. and GREEN, X. (2003), 'Management of multi-strain, multi-site yeast storage and supply', in Smart, K. (ed.) *Brewing Yeast Fermentation Performance*, 2nd edn, Oxford, Blackwell Science, 131–137.

KIRSOP, B.E. and DOYLE, A. (eds) (1991), *Maintenance of Microorganisms and Cultured Cells: A Manual of Laboratory Methods*, Academic Press, London.

KUNZE, W. (1999), *Technology Brewing and Malting*, 2nd edn, Berlin, VLB.

MAULE, D.R. (1980), 'Propagation and handling of pitching yeast', *Brewers Digest*, February, 36–40.

MOLZAHN, S. (1977), 'The propagation of brewing yeast', *Brewers Guardian*, July, 22–24.

NIELSON, O. (2006), 'Control of the Yeast Propagation Process – how to optimize oxygen supply and minimize stress', *Tech. Quart. Master Brewers Assoc. Americas*, 42, 128–132.

O'CONNOR-COX, E. and INGLEDEW, M. (1990), 'Commercial active dried yeasts demonstrate wide variability in viability and in two common indicators of yeast physiological quality', *Brewers Digest*, June, 34–36.

OCKERT, K. (ed.) (2006), *MBAA Practical Handbook for the Speciality Brewer, Vol. 2, Fermentation, Cellaring, and Packaging Operations*, St Paul MN, Master Brewers Association of the Americas.

POWELL, C.D., VAN ZANDYCKE, S.M., QUAIN, D.E. and SMART, K.A. (2000), 'Replicative ageing and senescence in *Saccharomyces cerevisiae* and the impact on brewing fermentations', *Microbiol.*, 146, 1023–1034.

QUAIN, D.E. (1995), 'Yeast supply the challenge of zero defects', *Proc. 25th Congr. Eur. Conv., Brussels*, 309–318.

QUAIN, D.E. and BOULTON, C.A. (1987), 'The propagation of yeast on oxidative carbon sources – advantages and implications', *Proc. 21st Congr. Eur. Conv., Madrid*, 417–424.

QUAIN, D., POWELL, C., HAMILTON, A., RUDDLESDEN, D. and BOX, W. (2001), 'Why warm cropping is best', *Proc. 28th Congr. Eur. Conv., Budapest*, 388–396.

RECKELBUS, B., FELS, S., GOSSELIN, Y. and DEBOURG, A. (2000), 'Optimising the use of dried yeasts in breweries', *Tech. Quart. Master Brew. Assoc. of the Americas*, 37, 21–25.

REED, G. (1982), 'Production of bakers' yeast', in Reed, G. (ed.) *Prescott & Dunn's Industrial Microbiology*, 4th edn, Byfleet, Macmillan, 593–633.

SUPPLIERS OF CARLSBERG FLASKS – SCANDI BREW (www.alfalaval.com), Tuchenhagen (www.tuchenhagen.de), Keofitt (www.keofitt.de) and DME (www.dme international.com).

SUPPLIERS OF YEAST PROPAGATORS – SCANDI BREW (www.alfalaval.com), Tuchenhagen (www.tuchenhagen.de) and Esau & Hueber (www.esau-hueber.de).

SUPPLIERS OF ACTIVE DRIED YEAST – Fermentis (www.fermentis.com) and Lallemand (www.lallemand.com).

TAIDI, B., MAZIKE, H.G. and HODGSON, J.A. (2001), 'Use of corn steep liquor to increase the yield of brewing yeast obtained from propagation', *Proc. 28th Congr. Eur. Conv., Budapest*, 297–305.

VAN DEN BERG, S. and VAN LANDSCHOOT, A. (2003), 'Practical use of dried yeasts in the brewing industry', *Cerevisiae*, 28, 25–30.

WAINWRIGHT, T. (1999), 'Yeast Propagation and Yeast Aeration in Particular the Esau & Hueber Systems', *Ferment*, April/May, 40–42.

WELLMAN, A.M. and STEWART, G.G. (1973), 'Storage of brewing yeasts by liquid nitrogen refrigeration', *Appl. Microbiol.*, 26, 577–583.

9

Water in brewing

M. Eumann, EUWA Water Treatment Plants, Germany

9.1 Introduction

9.1.1 Water as a raw material

Very often water is considered a utility in a brewery. As water is an important constituent of beer, though, making up more than 90% of the product, it is worth looking at water from a raw material perspective.

Historically the characteristics of different beer styles were influenced by the composition of the water used in their manufacture (see Table 9.1). To produce today's most prominent beer style, Lager or Pilsner beer (a bottom-fermented beer, light in colour and with a more or less pronounced hop flavour), it is necessary to treat the water used to make it, if the naturally available water source does not meet the composition accepted today as the standard.

Table 9.1 Different raw water compositions

		Munich	Dortmund	Vienna	Burton on-Trent	Pilsen
Total hardness	ppm $CaCO_3$	264	737	689	980	28
m-alkalinity	ppm $CaCO_3$	253	300	551	262	23
Non-carbonate hardness	ppm $CaCO_3$	11	437	138	718	5
Ca	ppm $CaCO_3$	189	655	407	880	18
Mg	ppm $CaCO_3$	75	82	282	100	10
Residual alkalinity	°G	10.6	5.7	22.1	−0.2	0.9
Dry solids	ppm	284	1110	948		51

9.1.2 Raw water sources

One can distinguish between surface water, which is water derived from lakes, rivers, man-made reservoirs or dams, and well water from underground. Surface water is more susceptible to seasonal changes due to rainfall or drought periods. It usually contains small quantities of minerals and high organic loads and is microbiologically contaminated. Well water from a suitable depth is usually protected from the surface and therefore its supply fluctuates far less. Depending on the hydrogeological situation, though, it can contain a lot of minerals. Its microbiological quality is usually good. Care has to be taken to use a well properly to prevent overuse which eventually can cause it to dry up.

9.2 Water for use in breweries

9.2.1 Water analysis and evaluation, definitions

Very often it is not acceptable to rely on a fluctuating water supply. Apart from obtaining hydrogeological data to determine whether water is sufficiently available, analysis should ideally also be carried out over at least a one-year period to determine seasonal fluctuations. But what should be analysed to evaluate the water composition? Table 9.2 lists the relevant components. Of course, the analysis is not confined to these parameters and specific site-related problems may require in-depth analysis. In most cases, though, analysing the parameters shown in Table 9.2 would be sufficient.

As an aside, total hardness of water is defined as the sum of all earth alkali ions, in this context Ca^{2+} + Mg^{2+}. Total hardness can be divided into carbonate and non-carbonate hardness, the former being $Ca(HCO_3)_2$ and $Mg(HCO_3)_2$ and the latter $CaCl_2$, $CaSO_4$, $Ca(NO_3)_2$, $MgCl_2$, $MgSO_4$ and $Mg(NO_3)_2$. Alternative terms used are temporary and permanent hardness, because a strong acid like HCl can make the $Ca(HCO_3)_2$ disappear and form CO_2, whereas the HCl has no effect on the non-carbonate hardness and is thus permanent.

Depending on the system used, hardness and alkalinity can be defined as ppm $CaCO_3$, °F (degree French hardness), °G (degree German hardness) or

Table 9.2 Important water analysis parameters

Smell	Residual alkalinity
Taste	Na
Colour	Ca
Turbidity	Mg
Temperature	HCO_3
Conductivity	Cl
$KMnO_4$ consumption	SO_4
Fe	NO_3
Mn	SiO_2
As	NO_2
NH_4	THMs

Table 9.3 Definition of hardness units

1°F = 10 mg $CaCO_3$/l
1°G = 10 mg CaO/l
1 meq/l = 50 ppm as $CaCO_3$

meq/l. The SI measurement is mol/m^3. See Table 9.3 for the definition of hardness units.

9.2.2 Filtered water

The minimum requirement for water used in a brewery should be that it conforms to potable water standards such as the European drinking water regulations, WHO (World Health Organization) requirements, FDA (Federal Drug Administration) requirements or equivalent official standards. There are always two areas which need attention when deciding which type of water to use: firstly the process requirements, and secondly the integrity of materials in contact with the water, which are affected mainly by scaling and corrosion.

Filtered water of a suitable standard should be used for cleaning. If water is not heated, there is no risk of scaling. There are limitations, though, in the levels of chloride acceptable in filtered water. The chloride level should not exceed 100 ppm to avoid corrosion of stainless steel. Complete guidelines for the composition of filtered water are in Table 9.4.

Table 9.4 Guideline for filtered water quality

		Range	
		Min.	Max.
Total hardness	ppm $CaCO_3$		
Ca	mg/l		
Mg	mg/l	0	10
Na	mg/l	0	200
HCO_3	ppm $CaCO_3$		
Cl	mg/l	0	100
SO_4	mg/l	0	250
NO_3	mg/l	0	50
SiO_2	mg/l		
THM	μg/l	0	10
Fe	mg/l	0	0.1
Mn	mg/l	0	0.05
NH_4	mg/l	0	0.5
NO_2	mg/l	0	0.1
BrO_3	mg/l	0	0.01
H_2S	μg/l	0	5

Other parameters according to WHO/EU drinking water guideline.

Table 9.5 Guideline for service water quality

		Range	
		Min.	Max.
Total hardness	ppm $CaCO_3$	50	90
Ca	mg/l		
Mg	mg/l		
Na	mg/l	0	200
HCO_3	ppm $CaCO_3$		
Cl	mg/l	0	50
SO_4	mg/l	0	250
NO_3	mg/l	0	50
SiO_2	mg/l		
THM	μg/l	0	10
Fe	mg/l	0	0.1
Mn	mg/l	0	0.05
NH_4	mg/l	0	0.5
NO_2	mg/l	0	0.1
BrO_3	mg/l	0	0.01
H_2S	μg/l	0	5

Other parameters according to WHO/EU drinking water guideline.

9.2.3 Service water

This quality of water should be used whenever the water is heated but not used in the brewing process. Primary examples would be hot CIP (Clean In Place) and final rinse water for packaging applications. Table 9.5 outlines the composition of service water. It is important that hardness in this water is limited to prevent scaling, which can, for example, lead to the blockage of spray nozzles in the bottle washer. It is also important that the chloride content is limited to about 50 ppm. This is to avoid corrosion of stainless steel. If hot service water starts stainless steel corrosion in one place, the same water can also have an impact on other stainless steel parts of the brewery which come into contact with that water, even when it is no longer hot, because of the FE dissolved in the water. Furthermore, service water has to be free of microbiological contamination.

9.2.4 Brewing water

As mentioned earlier, more than 90% of beer is water. Therefore it is of utmost importance that the quality of the brew water is high. A very important ion in brew water is calcium. During mashing it reacts with the phosphate buffer from the malt, influencing the pH level. In order for the enzymes to work properly, a pH of 5.2–5.4 is optimal. More than 100 years ago Kolbach came up with an empirical formula to determine whether the pH will rise or fall. This is called residual alkalinity and is defined according to the formula in equation 9.1:

$$RA = m - \frac{Ca + 0.5Mg}{3.5} \tag{9.1}$$

where RA = residual alkalinity, ppm $CaCO_3$
m = m-alkalinity, ppm $CaCO_3$
Ca = calcium concentration, ppm $CaCO_3$
Mg = magnesium concentration, ppm $MgCO_3$

A residual alkalinity of zero means that there is no pH influence. A positive residual alkalinity will lead to an increased pH, and a negative residual alkalinity will decrease the pH. Today in many cases artificial acidification by addition of lactic acid or mineral acid (mostly phosphoric acid) is used to adjust the pH. The calcium content during mashing and sparging can be influenced by dosing with calcium chloride and/or calcium sulphate.

Another important reason why calcium is required in the brew water is for oxalate precipitation. It is necessary to have sufficient calcium in the water to facilitate oxalate precipitation, which of course is also aided by low temperatures in storage vessels. If not removed, oxalate crystals can cause gushing in the finished product, but brewing with water of the correct composition will prevent this being an issue. Therefore it is worth aiming at a negative residual alkalinity in the brew water and sufficient calcium content.

Magnesium ions, although important for human health, are not necessarily a sufficiently positive influence on pH in the concentrations usually present in brew water. Because magnesium phosphates are more soluble than calcium phosphates, about twice the amount of magnesium is required to achieve the same effect as is achieved with calcium. Further, for reasons of taste the use of high levels of magnesium salts is not encouraged.

Because sodium salts are very soluble they do not cause the phosphates to precipitate. As there is no interaction with the phosphate buffer, subsequently there is no pH change.

On the anion side, chloride is a factor that influences the corrosivity of the water, though this does not happen when the water is already mixed with the wort. Exactly why this is the case cannot be clearly determined, but this is of advantage to breweries. Beer containing predominantly chloride anions has a milder flavour.

Having sulphate ions in the brew water is of no disadvantage. It is even reported that elevated levels of sulphate can facilitate sulphite formation. Sulphite acts as an oxygen scavenger and thus increases the ageing potential of a beer. A pronounced bitterness is associated with higher sulphate contents in the beer.

Nitrate can be reduced in anaerobic conditions to NO_2^-, which poisons cells and therefore destroys the brewing and fermentation process. Therefore levels of NO_2^- in brew water have to be very low.

Bicarbonate increases the pH of water as it absorbs acidity. Nearly all water contains a certain natural level of bound CO_2 and bicarbonate. Bicarbonate buffers any acidity, for example that caused by CO_2 or other acids like lactic or phosphoric acid. Before the artificial acid can bring down the pH, the bicarbonate has to be destroyed first. Bicarbonate, though, can also be responsible for

Table 9.6 Guideline for brewing water quality

		Range		Target
		Min.	Max.	
pH		5	9.5	
Ca	mg/l	70	90	80
Mg	mg/l	0	10	
Na	mg/l	0	20	
HCO_3	ppm $CaCO_3$	10	50	25
Cl	mg/l	30	80	50
SO_4	mg/l	30	150	100
NO_3	mg/l	0	25	
SiO_2	mg/l	0	25	
Residual alkalinity	ppm $CaCO_3$		20	<0
THM	μg/l	0	10	
Fe	mg/l	0	0.1	
Mn	mg/l	0	0.05	
NH_4	mg/l	0	0.5	
NO_2	mg/l	0	0.1	
BrO_3	mg/l	0	0.01	
H_2S	μg/l	0	5	
Turbidity	NTU	0	0.5	

Other parameters according to WHO/EU drinking water guideline.

formation of a layer of lime on mild steel piping which protects the piping from corrosion. This is of less importance nowadays, though, as stainless steel is the preferred material in today's brewery environment. It also should be noted that when water comes into contact with mild steel in the form of pipelines, reservoirs or vessels, free aggressive CO_2 has to be removed beforehand by a CO_2-trickler and the so-called +p (phenolphthalein) alkalinity (pH > 8.2) has to be adjusted.

Another component worth looking at in brew water is silica. Silica also has limited solubility in water and therefore should be removed to levels below 25 ppm. It goes without saying that oxidizing agents like chlorine in its various forms or ozone should not be present in the brewing liquor. THMs (Tri Halo Methanes) are also substances of concern in the brew water stream and should be removed to as great an extent as possible, preferably to below 10 ppb. Table 9.6 describes brew water.

9.2.5 Dilution water

It is important that the calcium content of dilution water is not higher than the calcium content of the concentrated beer to be diluted, in order to avoid oxalate precipitation. There are also strict requirements regarding oxygen content, which should mainly be less than 20 ppb, and THM levels are also of great importance. Some breweries distinguish between brewing liquor and dilution liquor in terms

Table 9.7 Guideline for dilution water quality

		Range		Target
		Min.	Max.	
pH		5	9.5	
Ca	mg/l	20	40	30
Mg	mg/l	0	10	
Na	mg/l	0	20	
HCO_3	ppm $CaCO_3$	10	50	25
Cl	mg/l	0	50	
SO_4	mg/l	0	50	
NO_3	mg/l	0	25	
SiO_2	mg/l	0	25	
Residual alkalinity	ppm $CaCO_3$		20	<0
THM	μg/l	0	10	
Fe	mg/l	0	0.02	
Mn	mg/l	0	0.02	
NH_4	mg/l	0	0.5	
NO_2	mg/l	0	0.1	
BrO_3	mg/l	0	0.01	
H_2S	μg/l	0	5	
Turbidity	NTU	0	0.5	

Other parameters according to WHO/EU drinking water guideline.

of levels of THMs permitted, though others do not. The microbiological composition of the dilution water is very important as this water is not necessarily boiled. Treatment with a proper disinfection system and a UV (ultraviolet) system prior to use is indispensable (see Table 9.7).

9.2.6 Boiler feed water

For the boiler house it is necessary that the feed water is of adequate quality and free of hardness. The cheapest method to obtain good boiler feed water is to recover as much condensate as possible, as the condensate usually contains very low levels of minerals and thus virtually no hardness. It is important that the feed water contains HCO_3 at a level of less than 50 ppm (expressed as $CaCO_3$), otherwise the intense heat in the boiler leads to the formation of NaOH and CO_2 from $NaHCO_3$. The CO_2 will be corrosive and eventually destroy the boiler. The mineralization of the boiler feed water also impacts on the frequency of blowdown and thus the energy efficiency of the boiler. Hot brew water can be used as a source of boiler feed water, if appropriately treated for hardness, as this can help to provide water of the correct alkalinity and save energy.

Boiler feed water must also be thoroughly deaerated to achieve an oxygen content ideally of less than 20 ppb. It should also be conditioned properly with caustic to adjust the pH, with phosphate for hardness scaling prevention and with sodium bisulphite for oxygen scavenging.

Table 9.8 Guideline for boiler and boiler make-up water quality

		Boiler feed water		Boiler water for $\leq$10 bar boilers	
		Min.	Max.	Min.	Max.
Total hardness	ppm $CaCO_3$	0	2		
Conductivity	μS/cm			0	5000
pH				10.5	12
HCO_3	ppm $CaCO_3$	0	50		
SiO_2	mg/l	0	20		
Fe	mg/l	0	0.1		
Mn	mg/l	0	0.5		
PO_4	mg/l PO_4			10	20
$KMnO_4$ index	mg/l $KMnO_4$			0	10
SO_3	mg/l			4	10
p-alkalinity	meq/l			1	12

For energy efficiency it is also worth exploring the possibilities of a closed condensate system which is operated under pressure, as considerable energy savings can be achieved. A frequent boiler water analysis must be executed as well. Table 9.8 provides guidelines for boiler feed and boiler water. It should be noted that the correct composition of the boiler and boiler feed water also depends on the pressure and the size of the boiler.

9.3 Treatment technologies

Water treatment in the brewery differs from treatment of other components used in the brewing process, as it is completely site related. The treatment process cannot easily be standardized unless breweries have a common water source. There are always several options to achieve the desired water quality. It is very important to have a good analytical data pool prior to designing a water treatment plant.

Several of the technologies described in the following sections can give similar results and therefore the choice of which technology to use should be driven by economics and constraints of space. There is no such thing as the ideal water treatment system and every brewery is unique. This is also reflected in the fact that different water treatment companies are likely to offer different solutions to problems and it is not easy for the one-off purchaser of a treatment system to see the pros and cons of each solution.

9.3.1 Disinfection

Depending on the source, water is likely to be microbiologically contaminated to differing degrees and surface water is far more likely to be affected. However, to

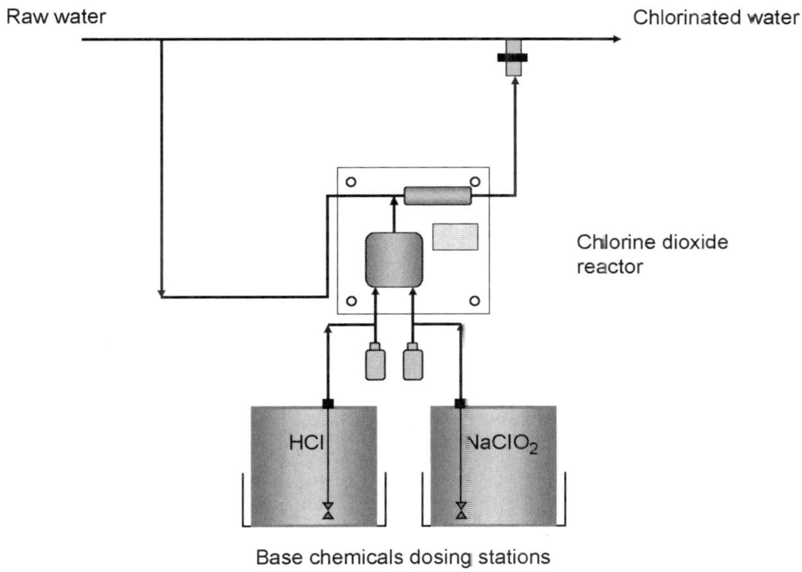

Fig. 9.1 ClO_2 plant based on the acid-chlorite process.

avoid any risk it should be the norm to disinfect all water coming on site. Even if the water is already treated by the municipality and thus most likely to have been exposed to some sort of disinfection process, the piping systems are often in such a poor state that a reliable and constant level of disinfectant cannot be detected all the time. The preferred disinfection method is chlorine dioxide (ClO_2). It works according to the following principle:

$$5NaClO_2 + 4HCl \rightarrow 4ClO_2 + 5NaCl + 2H_2O \quad (9.2)$$

The layout of a ClO_2 plant is as shown in Fig. 9.1. The advantages of ClO_2 plants are that they are easy and safe to use and virtually no by-products (THMs) are formed. Disinfection with sodium hypochlorite or chlorine gas is far more hazardous and unwanted by-products are created. It is usually too expensive only to use ozonization as a disinfection method. This method should be used only in cases where oxidation is required as well. It should also be noted that UV and ozone do not deposit anything in the water but do not possess any residual disinfection ability, so recontamination is possible.

9.3.2 Oxidation/aeration

For the removal of certain components, especially iron and manganese, oxidation is necessary. Furthermore, aeration might be necessary to strip out odours like H_2S.

In most cases, a simple injection of air with a static mixer is enough to supply sufficient oxygen to facilitate the reaction. To remove H_2S, though, or correct a low pH due to excess CO_2, open aeration is necessary. It has to be considered that an intensive aerator without filling material is much easier to clean than a trickler column, as the oxidized components might readily precipitate. It is beneficial to design the aeration system so it can be cleaned in place. It is also possible to use ozone as an oxidizing agent as mentioned in the previous section. However, it is quite expensive and therefore its application in this way is usually confined to special cases.

Only low levels of chlorine are allowed legally, therefore oxidation by chlorination is in most cases not possible because the amount of chlorine required exceeds the legal limit.

Should the water contain too much ammonia, it is necessary to provide enough oxygen for nitrate to form. Oxidation to form nitrate requires the presence of the bacteria *Nitrosomonas* and *Nitrobacter*, and therefore it can only be performed in conjunction with a carrier material like a sand filter. This is a rather slow reaction and pilot tests have to be carried out in many cases for this route to be followed. Alternatives to this process are strongly acidic cation exchange and reverse osmosis.

9.3.3 Particle filtration

Particle filters are used to remove suspended solids. These solids either originate from pipelines or have been created by the previous oxidation process or by a flocculation or precipitation process (e.g. lime softening). The most common filter is the sand filter in which different grades of sand and gravel are layered within a mild steel-coated vessel.

It might also be necessary to install a precipitation reactor prior to filtration, if for instance iron levels are very high. A flocculant, in most cases PAC (Poly-Aluminium-Chloride), is beneficial for the flocculation process. It might also be possible to use lime, as use of this compound both raises pH and serves as a crystallization point for flocs. It has been noted that chlorine has an inhibiting influence on floc formation. Figure 9.2 shows a sand filter plant.

Apart from sand filters, multilayer filters consisting of different grades of sand, gravel and anthracite are also used. Anthracite is a coal-based black material which allows in-depth filtration, so that particles are removed not only from the surface but also from within the bed. When anthracite is used, the removal capacity is higher and the backwash is less frequent. However, it must be ensured that all filtration devices can be backwashed with air. Sand filters are backwashed with air and water at the same time. Multilayer filters work separately, first with air, then water. They also facilitate the removal of H_2S.

If arsenic is present, it can be removed in GEH (Granulated Ferric Hydroxide) filters. GEH is a filter material based on synthetic granular ferric hydroxide. It is an adsorbent with a high porosity and a large inner surface. The material is rather expensive and has to be replaced or regenerated when exhausted. It should not be used for particle filtration, therefore, as this would reduce its adsorption capacity.

Fig. 9.2 Sand filter plant (Source: EUWA).

The alternative to classic sand and multilayer filters is ultrafiltration technology. Hollow fibre membranes made out of poly(ether)sulphone or cellulose triacetate are used in dead-end or crossflow filtration. The filtered water quality is quite good, but issues with wastewater and investment costs have so far prevented widespread use of this technology in breweries.

In many cases it is worth exploring the opportunity to recycle the backwash water from filtration. This will be described in more detail in Section 9.4.

9.3.4 Activated carbon filters

Activated carbon filters are used for dechlorination, adsorption and H_2S removal. Dechlorination is a very rapid process in which activated carbon causes free chlorine to be converted into chloride. The reaction happens within seconds so probably takes place in the top 10 cm of the bed. It has been reported that chlorine dioxide cannot be removed by activated carbon filters, but this is not the case. Using activated carbon filters is a safe and reliable method to dechlorinate water containing chlorine dioxide as well as water containing free chlorine.

To adsorb (and therefore remove) substances such as THM or odours and colours from water, the activated carbon filter acts as a chromatographic bed. The time taken to adsorb contaminants depends on the concentration and polarity of the components to be removed and the way they interact with each other. The contact time, though, is usually 10–20 minutes.

Recently, activated carbon plants equipped with steam-stripping devices have been constructed. The steam-stripping devices use high temperatures of 140°C to strip out volatile components from the carbon filters and therefore the lifespan of the carbon is extended up to five-fold. Whether or not a steam-stripping device is in use, it is important in any case to construct the activated carbon filter plant in such a way that it can be sterilized properly with either hot water (min. 95°C) or steam (up to 140°C). The only material from which the plant can be made is stainless steel, the grade depending on the chloride content of the water. A strainer plate for a proper distribution of water and steam also needs to be in place and the strainer nozzles must be covered with gravel. Steam sterilization should happen at least once a week, or when microbiological conditions deteriorate. The corresponding flow diagram is shown in Fig. 9.3. If there are several carbon filters in use, they should be run in parallel and taken offline only for sterilization and/or backwashing, as continuous operation is the best method to prevent premature microbiological growth on the carbon.

In some cases also, organic scavengers in the form of polystyrene resins are used for adsorption. These can be regenerated, for example with caustic soda among other substances. Using organic scavengers, however, is a rather expensive method and is used only in special cases. Figure 9.4 shows an activated carbon filter plant installation.

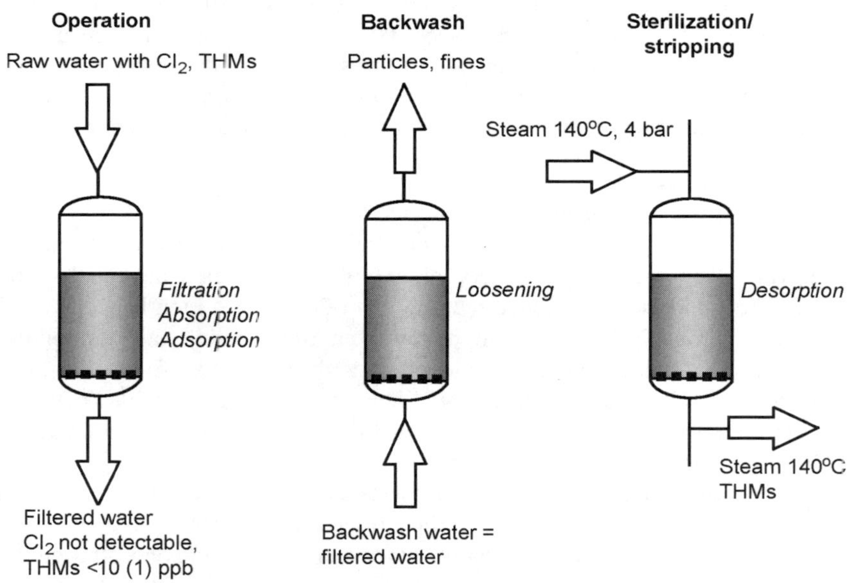

Fig. 9.3 Activated carbon filter – process steps (Source: EUWA).

Fig. 9.4 Activated carbon filter plant (Source: EUWA).

9.3.5 Lime precipitation

Lime precipitation is the oldest method commonly used for water treatment in breweries. Although some consider this method old-fashioned, it has not lost its appeal. It remains an attractive method even today if it is carried out properly. It is used for waters containing mainly carbonate hardness. By the addition of $Ca(OH)_2$, either as saturated lime water in smaller plants or as lime milk in larger plants, most of the carbonate hardness is precipitated. The corresponding equation is as follows:

$$Ca(HCO_3)_2 + Ca(OH)_2 \rightarrow 2CaCO_3\downarrow + 2H_2O \quad (9.3)$$

Open lime softening (also called one-stage lime softening) and lime precipitation under pressure are two different lime precipitation processes used in breweries. In a one-stage lime softening plant, the reaction takes place in a cylindroconical tank, whereas in lime precipitation under pressure, $CaCO_3$ is precipitated at so-called reactor sand, through which water and lime $Ca(OH)_2$ are run. It is worth mentioning that iron and manganese are removed as well when carrying out one-stage lime softening, whereas when using the lime precipitation under pressure method, these components need to be treated beforehand.

It is important to take into account the ratio of magnesium to non-carbonate hardness for successful decarbonization. $FeCl_3$ will react with carbonate hardness and increase the non-carbonate hardness, thus influencing the ratio between

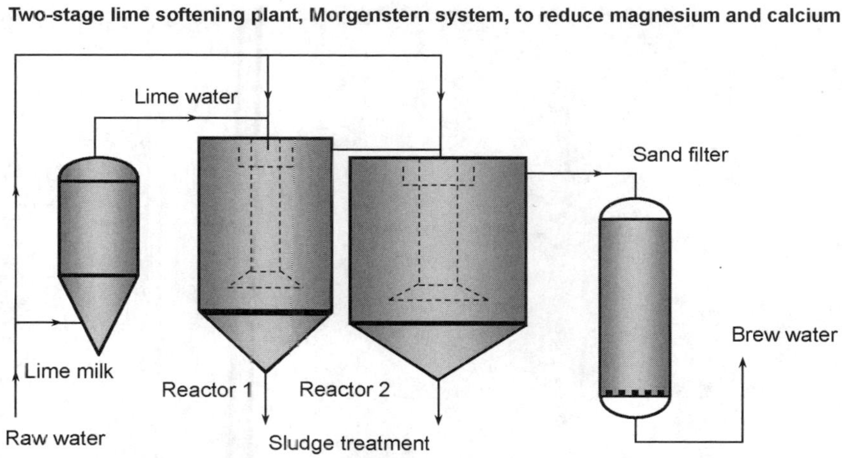

Fig. 9.5 Lime softening process steps.

Mg and non-carbonate hardness. If $Mg(HCO_3)_2$ needs to be removed, it is necessary to use a two-stage lime softening plant with reactor and refiner following the Morgenstern system (Fig. 9.5). As can be seen from the following equation, the amount of $Ca(OH)_2$ doubles, as magnesium has to be precipitated as hydroxide and not as bicarbonate. Figure 9.5 shows a flow diagram of this process.

$$Mg(HCO_3)_2 + 2Ca(OH)_2 \rightarrow 2CaCO_3\downarrow + Mg(OH)_2\downarrow + 2H_2O \qquad (9.4)$$

Open lime softening can also be used for the treatment of water containing humic acids, to increase the pH and to flocculate organic water components. The pH of the water is relatively high, but this is not a matter of great concern, since the water's buffer capacity is considerably reduced by lime softening, in some cases to a m-alkalinity of <25 ppm as $CaCO_3$. This is usually <50 ppm as $CaCO_3$.

After lime softening, the water always passes through a sand or multilayer filter for the removal of insoluble matter and further dealkalization at the sand surface. By adding a rinse water recycling system to recycle filtration water and by dewatering sludge or treating it in chamber filter presses, the percentage of water wasted can be reduced to <1%.

The operating costs of lime softening plants are very low and the systems are very robust. A disadvantage of this method is the large floor space required for a lime softening plant and the fact that automation is relatively costly. The limitations of the quality of the raw water to be treated also need to be considered.

There are several exciting new applications of lime softening plants. One of these combines a lime softening plant with a reverse osmosis plant, for example to remove nitrate. The quantity of wastewater is reduced using this technology to <5%. If lime softening plants are connected to Ca-addition systems, the water quality achieved improves considerably. Figure 9.6 shows a lime softening plant with 150 m^3/h throughput.

Fig. 9.6 Lime softening plant (Source: EUWA).

9.3.6 Ion exchange

Ion exchange is a very suitable method of changing the composition of the water. It can be applied specifically and can be used to convert a wide range of substances.

Weakly acidic cation exchanger

Following this method, Ca^{2+} and Mg^{2+} ions bound to HCO_3^- are removed (dealkalization). The advantages of this method are its high capacity and the fact that it uses regeneration chemicals very efficiently. Regeneration is carried out with HCl or H_2SO_4 in counter-current, and backwash in the vessel is possible. A disadvantage is that it is not possible to monitor this process satisfactorily. There is also a danger of over- or under-regeneration.

Strongly acidic cation exchanger

Following this method, all cations are exchanged with H^+ ions (using HCl or H_2SO_4 for regeneration) or with Na^+ ions (using NaCl for regeneration). When an acid is used for regeneration, one advantage is the possibility of monitoring by differential conductivity. Conductivity is measured before and after the safety zone and transmitted to an instrument where the differential conductivity is displayed. Hence the exact exhaustion point can be determined. The regeneration should be effected in counter-current, in order to achieve a constant high water quality combined with a low chemical requirement and low wastewater production. This can be achieved with a special design which integrates quality monitoring with differential conductivity measuring. This design is also the only counter-current system which enables backwash within the vessel, as it does not require an upper nozzle plate.

If the water contains carbonate- and non-carbonate hardness, using a so-called layer bed exchanger is the optimal solution. When a layer bed exchanger is used, strongly and weakly acid exchange material is layered in a vessel. This system therefore combines the advantages of weakly and strongly acidic exchangers. Figure 9.7 shows, from left to right, a weak exchanger, a layer bed exchanger and a strong exchanger

By regenerating strongly acidic cation exchangers with NaCl in counter-current, savings of >40% compared to conventional systems are achievable.

Weakly basic anion exchanger

Following this method, anions of strong acids are removed and replaced with OH^- ions. A strongly acidic cation exchanger always has to be installed in front of a weakly basic anion exchanger, and in that combination demineralization can be achieved. The advantages of this method are its high capacity and efficient use of chemicals. Monitoring is by conductivity measurement, and regeneration uses NaOH.

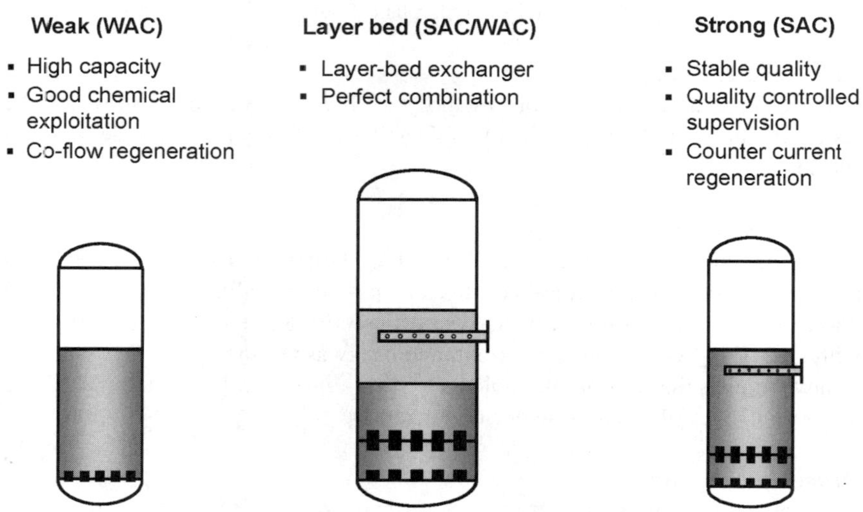

Fig. 9.7 Ion exchange principles – weak, layer bed and strong exchangers.

Strongly basic anion exchanger
Anions from strong as well as weak acids (especially silica and carbonic acid) are removed by this exchanger, which is regenerated with NaOH. The regeneration is effected in counter-current, and is similar to the regeneration of strongly acidic cation exchangers. For the NaCl regeneration the ion exchange is conducted in a Cl^- cycle. In addition, specific nitrate exchangers are available which remove nitrate. These can be combined with strongly acidic cation exchangers, using the anion of the acid for regeneration. When using HCl and H_2SO_4 for regeneration, the Cl/SO_4 ratio can be adjusted in the treated water.

In all ion exchange systems, calcium is usually added to neutralize the acidity. The calcium levels are always adjusted back to the required level. CO_2 removal is really only necessary if the corrosivity of the water is a problem for mild steel piping reservoirs and vessels. Only food-grade resin should be used in all brewery applications, except, of course, in the boiler house where non-food-grade resin can be used.

An ion exchanger installation can be seen in Fig. 9.8.

Fig. 9.8 Ion exchange plant (Source: EUWA).

9.3.7 Membrane technology

Apart from the ultrafiltration process described earlier, nanofiltration and especially RO (reverse osmosis) have gained a significant market share in water treatment. RO is the best available filtration method. In reverse osmosis, in order to overcome the osmotic pressure of an aqueous solution, a pressure is applied to a semi-permeable membrane. Therefore, most of the molecules and ions will be rejected, whereas pure water can pass through the membrane. The saline solution is called the concentrate, whereas the demineralized portion is called the permeate. The ratio between permeate and feed water is called the recovery rate. The higher this value is, the less wastewater is produced.

The efficiency of the membranes can be affected by fouling or scaling. Fouling is the term used to describe the presence of organic deposits on the membrane, which can be at least partly removed from the membrane surface by a suitable cleaning regime. In this context the Silt Density Index (SDI) serves as an indicator. Scaling is the precipitation of inorganic salts at the membrane surface. A suitable selection of cleaning agents can be used to reverse this effect if they are applied before too much scale has formed.

For the operation of a reverse osmosis plant to be trouble-free, it is essential to pretreat the water sufficiently. The water should be free of particles, which can be guaranteed by installing polishing filters, and the amount of organic substances should also not be too high. Furthermore – due to the organic polymer structure of membranes – oxidants (e.g. Cl_2 or O_3) must not be present, as their presence can lead to irreversible damage of the membrane. Therefore, in most cases at least sand filters and often also activated carbon filters are used to pretreat the water.

In order to minimize the wastewater, the recovery rate should be as high as possible. To achieve this, anti-scalants or scale inhibitors are used. They prevent crystallization although the solubility limit of salts has been exceeded so that a remarkable reduction in operating costs can be achieved. Furthermore, adding mineral acid or carbon dioxide can lead to a better recovery rate. However, the recovery rate always depends on the quality of the raw water available. Usually the recovery rate is 80–90%, but in some cases it can be increased up to 95%. Figure 9.9 is a flow diagram of the reverse osmosis process.

Technological developments have also reduced the necessary pressure to less than 13 bar, or in some cases even to less than 10 bar. Nanofiltration requires even less pressure. Since more porous membranes are used in nanofiltration, however, ions, such as smaller monovalent ions, e.g. sodium and chloride, can pass through the membrane and it is therefore not ideally suited for brewing applications. Nanofiltration can theoretically be used to soften water, but the technology is still expensive compared to an ion exchange system with a strongly acidic cation exchanger regenerated with sodium chloride. For special cases such as ion exchange wastewater recycling, nanofiltration can be used successfully.

Reverse osmosis plants should also be built in conjunction with calcium addition systems, as the mineral content of the water produced is minimal. The

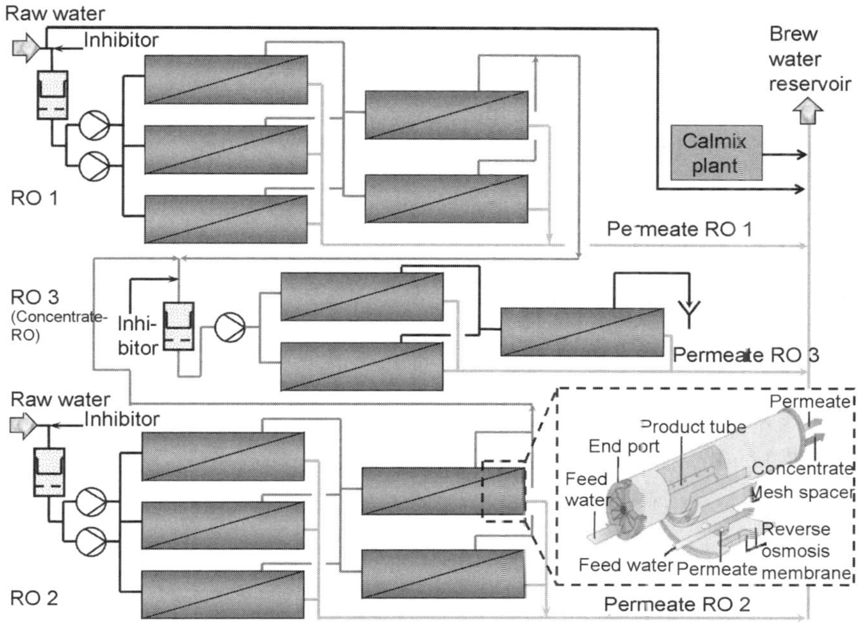

Fig. 9.9 Reverse osmosis flow diagram.

advantage of RO is that from very fluctuating raw water conditions a very stable so-called permeate is produced, therefore variability in water quality from the beer production process is no longer such an issue.

Figure 9.10 shows a reverse osmosis plant with 80 m^3/h permeate capacity.

Fig. 9.10 Reverse osmosis plant (Source: EUWA).

9.3.8 Calcium blending

Various forms of calcium blending are possible. The presence of calcium ions is a decisive factor in the brewing process. They are required both for mashing and sparging to influence the pH via the phosphate buffer of the mash and also for precipitation of oxalates to avoid gushing. There are different ways in which calcium can be added.

If the water comes from an ion exchanger, for example a strongly acidic cation exchanger, then clear lime water produced from lime powder or lime milk in a lime saturator is used to neutralize mineral acids. $CaCl_2$ and $CaSO_4$ are formed according to the following equations:

$$2HCl + Ca(OH)_2 \rightarrow CaCl_2 + 2H_2O \quad (9.5)$$

$$H_2SO_4 + Ca(OH)_2 \rightarrow CaSO_4 + 2H_2O \quad (9.6)$$

In a second step, the carbonate hardness can be adjusted to the desired value, though a surplus of CO_2 remains in the water:

$$3CO_2 + Ca(OH)_2 \rightarrow Ca(HCO_3)_2 + CO_2 \quad (9.7)$$

If free CO_2 cannot be tolerated due to its corrosive nature, the system can be combined with a CO_2 trickler. For this purpose the pH has to be adjusted to >8.2 (−p value).

If the mineral acidic salinity of the water is not sufficient, if the water is very soft or if only demineralized water (from reverse osmosis or ion exchange) is available, the addition of hydrochloric acid and/or sulphuric acid is recommended to solve the problem and ensure that enough calcium is produced. This can be done both before and after the lime saturation process.

It is also possible to mix lime milk, hydrochloric acid and/or sulphuric acid with the main brew water flow. Lime milk is added first, then HCl and/or H_2SO_4, before the mixture goes through the reactor and is then mixed with the main brew water flow. When this method is followed, $CaCl_2$ and $CaSO_4$ can be added separately. This water can then be used directly for mashing and sparging or can be added into the wort. Figure 9.11 is a flow diagram demonstrating this process. The advantage of this process compared to conventional $CaCl_2$ and $CaSO_4$ dosing is its complete automation. No manual interference is required, which helps to ensure that the water quality is stable and reproducible.

Because of the positive influence of this process on the pH of the water, less lactic or mineral acid is then required to be added in the brewhouse. Furthermore, it is worth noting that because of its low solubility $CaSO_4$ cannot be dosed in a liquid and completely dissolved form except when following the system described above. This, of course, also influences the effectiveness and economics of the application of $CaSO_4$. This novel process opens up new possibilities of tailoring the water quality according to the requirement of specific beer brands and beer types. Since nowadays it is the norm rather than the exception to produce a wide range of products with different characteristics, variable and flexible treatment processes that meet these demands are indispensable. Figure 9.12 shows a picture of such a system for a 500,000 hl/year brewery.

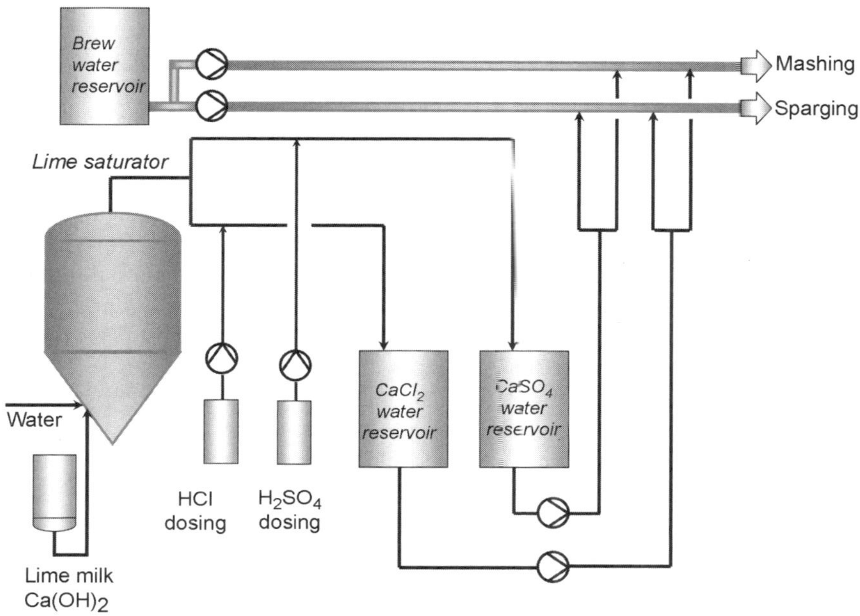

Fig. 9.11 Calmix® flow diagram.

Fig. 9.12 Calmix® plant (Source: EUWA).

9.3.9 Deaeration

Deaeration of water is necessary for various applications. Especially when using a high-gravity brewing process, a very low oxygen level is required. Originally deaeration was achieved by applying a vacuum and using a stripping gas (CO_2). The residual oxygen levels achievable with this method are no longer sufficient for today's requirements (<10 ppb). The hot deaeration process is a different method which uses a raised operating temperature. The process design features are the same as described above, but the O_2 levels achieved are lower. This process, though, has a much higher energy requirement. A further possibility is a packed bed column, which can be used either at ambient temperatures with a higher residual O_2 content or hot with a lower residual O_2 level. The risk of precipitation increases if the hot process is used.

A membrane deaeration system is another approach used. At its core are hollow fibre membrane modules, which have a large surface area. The membranes are semi-permeable, i.e. permeable to gas but not to water. By using a stripping gas (CO_2 or N_2) and applying a vacuum, nearly complete oxygen removal at ambient temperatures can be achieved. The water flows upstream, whereas the stripping gas is led downstream through the membrane by the vacuum pump. Due to the modular setup, the capacity of a membrane deaeration system can easily be expanded. The membranes can be used in parallel and/or in series. It is essential to integrate a disinfection as well as a filtration unit into the membrane treatment system. The deaerated water will also need to be cooled down to <1°C. Compared to other systems, the membrane system offers lower operating costs, precipitation is avoided and the plant size is compact. Figure

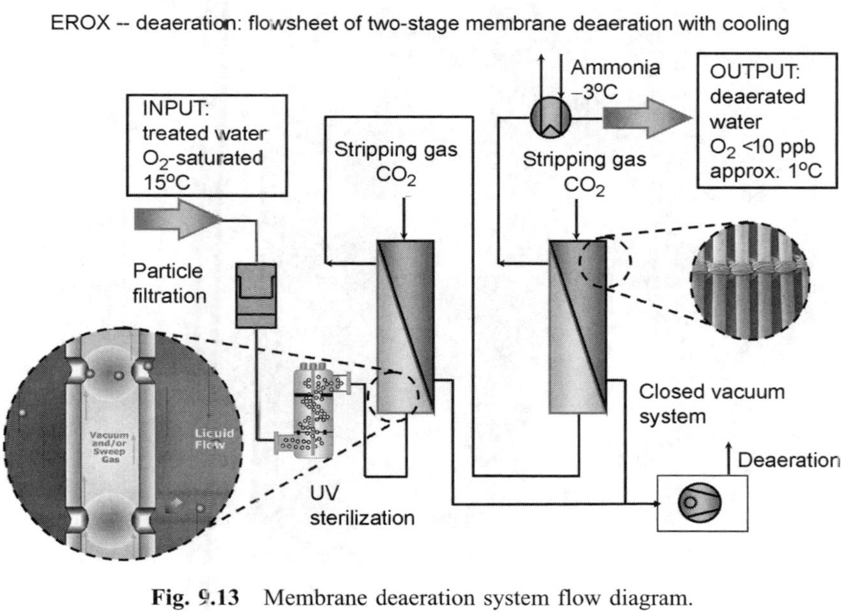

Fig. 9.13 Membrane deaeration system flow diagram.

Fig. 9.14 Erox deaeration plant (Source: EUWA).

9.13 shows a flow diagram of this process and a picture of the plant can be found in Fig. 9.14.

There has also been debate on whether to use deaerated water for mashing and sparging as well, processes in which deaerated water is not usually used. This possibility has not yet been explored fully and therefore the benefits compared to the costs involved have not been evaluated on a full-scale basis.

9.4 Recycling

As water can be a scarce commodity, recycling is becoming more and more important. To treat the whole effluent stream of a brewery so that it can be recycled as potable water requires major efforts and the efficiency of the process will inevitably be low. It is more sensible to focus on specific areas where the contents of the water are clearly defined and can be tackled more specifically. However, further developments in membrane technology will certainly widen recycling possibilities.

One area that produces relatively large amounts of wastewater is the water treatment plant itself. Therefore it is worth considering recycling backwash water from particle filtration systems. Some recycling systems are shown in Fig. 9.15. It is also worth considering dewatering the sludge from lime softening systems. In ion exchange systems nanofiltration of the regenerant effluent from

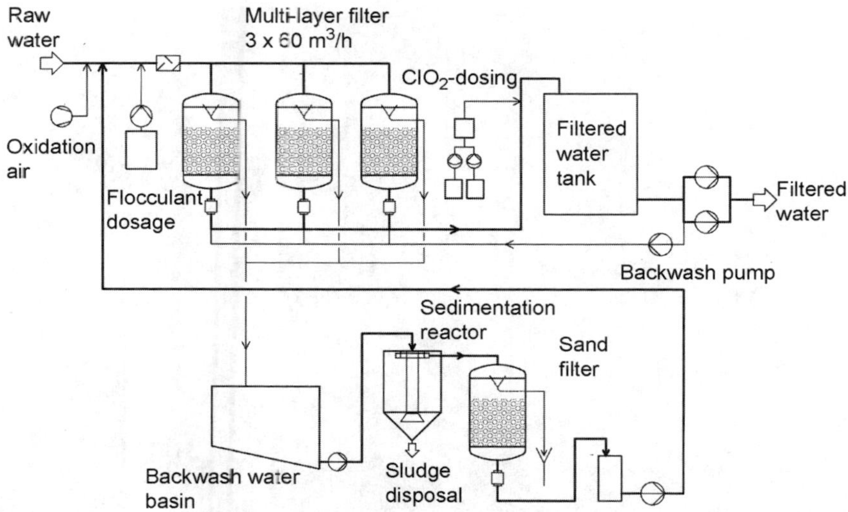

Fig. 9.15 Backwash water recycling.

weakly acidic cation exchangers can be worthwhile. Recycling systems in bottle washers are already widely used, but there is not always a balance between the cost of the extra equipment required and the savings achieved.

There have been attempts at recycling some of the wastewater created during reverse osmosis. To do this, the efficiency of RO is kept artificially low, and then the concentrate stream can be used in other applications in the brewery. This, however, requires very balanced, integrated water reticulation within the brewery and large reservoirs to buffer water fluctuations. Therefore this system has not been adopted in many breweries.

9.5 Future trends

Brewery water treatment systems of the future will be very flexible, allowing breweries to tailor-make their water for different products. At the same time, these future water treatment systems will aim to achieve optimum efficiency in terms of operating cost and especially wastewater produced. The advances in analysis techniques will inevitably lead to further challenges, as it will be possible to detect certain components that are not an issue today but will then need to be removed. It will also continue to be vital for brewers to pay attention to their water supply to avoid surprising and unexpected quality defects in the finished product.

9.6 Sources of further information

WZW Forschungszentrum Weihenstephan für Brau- und Lebensmittelqualität; Alte Akademie 3, D-85354 Weihenstephan, contact: Dr Karl Glas (k.glas@wzw.tum.de).

IBD – The Institute of Brewing & Distilling, 33 Clarges Street, London W1J 7EE (www.ibd.org.uk).

VLB Berlin Research and Teaching Institute for Brewing in Berlin, Seestrasse 13, D-13353 Berlin, contact: Dr Alfons Ahrens (ahrens@vlb-berlin.org).

MBAA – Master Brewers' Association of the Americas, 3340 Pilot Knob Road, St Paul, MN 55121-2097 (www.mbaa.com).

EUWA H.H. Eumann GmbH Water Treatment Plants, Daimlerstrasse 2-10, D-71116 Gaertringen (web: www.euwa.com, e-mail: info@euwa.com).

9.7 Bibliography

EBC: *Manual of Good Practice, Vol. 8, Water in Brewing*, Hans-Carl Verlag, Nürnberg, 2001.

A. GROHMANN: *Die Trinkwasserverordnung* [German Drinking Water Regulation]: *Einführung und Erläuterung für Wasserversorgungsunternehmen und Überwachungsunternehmen*, Erich Schmidt Verlag, Berlin, 2003.

WILLIAM A. HARDWICK: *Handbook of Brewing*, Marcel Dekker, New York, 1995.

WOLFGANG KUNZE: *Technology Brewing and Malting* (3rd edn), VLB, Berlin, 2004.

PROF DR L. NARZIß: *Die Technologie der Würzebereitung* (7th edn), Ferdinand Enke Verlag, Stuttgart, 1992.

10

The brewhouse

J. M. H. Andrews, Briggs of Burton Plc, UK

10.1 Introduction

The purpose of the brewhouse is to receive the main ingredients of beer, being malted barley, cereal or sugar adjuncts, hops, and water, and to process these materials to produce a hopped cold wort ready for fermenting. There have been various attempts to design a continuous process, none of which have found acceptance, so the current and foreseeable technologies are a batch process.

The performance of the brewhouse is normally measured by the quantity of cold wort per brew and the number of brews per week including downtime for cleaning and maintenance. In addition the overall extract from the raw materials is measured as a percentage of the extract which results from a standard laboratory procedure such as the European Brewery Convention (EBC) method. Lastly some measure of quality of cold wort is usually specified in terms of suspended solids and/or turbidity, where low figures are considered beneficial.

All of these measures are affected by the type and quality of the malted barley and cereal adjuncts, and as a consequence engineers have sought to design equipment which can accommodate the natural variation in incoming materials. There are two main technical approaches to the brewhouse.

10.1.1 The lauter tun brewhouse

Milling of malted barley is carried out in a six-roller dry mill or a two-roller wet or steep-conditioned mill and the resulting grist is mixed with mashing liquor in a pre-mashing device to make the mash. This is processed in a mash conversion vessel to convert starch to sugar and the whole mash is transferred by pump or gravity to the lauter tun. The function of this equipment is to extract the liquid component from the grain component so as to maximise extract, and to dispose of the spent grain to silos for collection and removal from site. Thereafter the liquid is boiled with hops addition in the wort kettle during which water is evaporated to some pre-determined percentage of the brew length. Some protein denaturing takes place and further solids removal is required usually in a whirlpool to allow hot wort at low solids content to be cooled through a plate heat exchanger to fermenting temperature.

There are a number of variations to this basic model, the most important being the use of decoction and double mashing. In this method, mashing is carried out at a lower temperature, and periodically a proportion of the mash is transferred to a separate vessel, raised to boiling point, and transferred back to the mash. The purpose is to raise the temperature of the mash by defined steps which are determined by the enzymic reactions required to maximise extract.

10.1.2 The mash filter brewhouse

The process steps are identical to the lauter tun version but commence with a size reduction hammer mill which has the effect of reducing the malted barley and cereal adjunct to a powder. The mash conversion process is substantially the same, but the separation of the liquid sugar from the grain solids is carried out in a thin bed mash filter, the technology of which has been refined in the last 20 years. The spent grain is retained on filter cloths, and the filtration path is very small (e.g. 50 mm) (Hermia, 1992) when compared to a lauter tun with a grain bed depth up to 500 mm (Andrews, 1996). In addition a pressure difference is created across the filter bed to accelerate filtration, and often the filter plates are equipped with a membrane which can be inflated to squeeze the spent grain. All these techniques are used to overcome the resistance to filtration inherent in the powdered grist with its very small particle size.

The perceived benefits of mash filters compared with lauter tuns are clearer wort, measured extract of 1–2% improvement, and lower water usage resulting in cold wort of a higher specific gravity. The negative issues are high capital cost, the extraction of undesirable compounds (e.g. polyphenols) from the husk fraction of the malt, and some element of the measured extract at the last runnings not being fermentable and having an adverse effect on flavour stability (Schild, 1936).

The remainder of the brewhouse steps of wort boiling and wort clarification are common to both technologies.

10.2 Milling

The purpose of milling is to reduce the size of the malt particles and to expose the endosperm to attack by enzymes during the mashing process, so that the greatest conversion of starch to fermentable sugar is achieved in the shortest possible time. Particle size is important for the performance of the mash separation system, and for lauter tun performance it is considered beneficial to preserve the husk portion of the malt grain. For mash filters, this is not critical as the filter bed is comparatively thin and the pressure differential across the filter back is comparatively high.

For a lauter tun, poorly modified malt should be milled more finely to give enzymes access to the starch which can be shrouded behind cell wall material which has not been broken down adequately in the malting process. However, the smaller particle size impacts negatively on the rate of run-off of wort. Conversely a well-modified malt can be milled more coarsely to permit faster separation without sacrificing extract performance.

10.2.1 Dry milling

There is a wide variety of dry mills incorporating two, three, four, five, or six rollers, but in larger breweries today they are almost exclusively of the six-roll variety arranged in three pairs with the gap between each successive pair being less than the preceding pair. The first pair is intended to crack open the malt grain and remove the husk which is then separated on a vibrating sieve and arranged to bypass any subsequent milling operation. The top rolls may be fluted and rotate at typically 300–400 rpm with a diameter of 250–300 mm, and, the gap between them for dry malt should be 1.6–1.8 mm. The successive pairs are similar but the gaps are 0.8–1.2 mm and lastly 0.5–0.8 mm. By adjusting the gaps and analysing the results through a set of sieves, whilst monitoring brewhouse performance for yield and throughput, it is practical to optimise mill settings. Furthermore, different settings can be utilised for varying recipes of raw materials, and the modern versions of six-roller mills are capable of the gaps being set remotely from the control system as a recipe-derived parameter.

10.2.2 Conditioned dry milling

It is desirable to preserve the malt husk for fast separation in the lauter tun, but the husk is a very friable material which shatters easily, and conditioned milling was developed to overcome this by adding a small amount of water or steam to the malt through a screw conveyor just before milling. Typically, 1% of the malt by weight is added and this is taken up entirely by the husk without penetrating the kernel; the moisture content of the husk is raised to perhaps 10% by weight making it flexible. The kernel must remain substantially dry and brittle so that the necessary size reduction can be carried out effectively. Conditioned dry milling has been reported to reduce cycle time with poorly modified malt by up to 10 minutes in a 3-hour lauter tun cycle, or alternatively can run with a deeper grain bed and therefore longer brewlength for the same cycle time (Andrews, 1996).

10.2.3 Steep conditioned milling

In steep conditioned milling the malt is fed into a conditioning chamber prior to milling where it is sprayed with warm mashing liquor at 50–70°C for a controlled time, typically 60 s, such that the uptake of moisture by the husk is considerably enhanced to perhaps 25% by weight to make it very pliable and to preserve it through subsequent milling operations. The mill normally consists of a pair of rollers with the gap between them set at 0.3–0.5 mm. Immediately after milling all the balance of the mash liquor is added and the mash is pumped to the mash conversion vessel (Herrmann, 1998). The main benefit is that the husk is almost entirely preserved and the lauter tun bed therefore has an enhanced porosity. This allows a much higher specific bed loading, typically 20% more than with dry milling, and therefore the lauter tun can have a smaller diameter and lower capital cost.

The main reported disadvantages are fourfold. Firstly that the power consumption is typically four to six times higher than for a dry mill for the same brewlength. This is because it is essentially a combined mill and mashing device and mashing of necessity should be carried out in an absolute maximum of 30 minutes, whereas the grist from a dry mill can be milled over 2–3 hours and held in a grist case for mashing later.

Secondly, air is entrained in the milling and mashing process which can lead to lipoxygenase activity with pale lager malt, resulting in possible undesirable flavours in the beer. This can be overcome by flooding the milling and mashing chambers with carbon dioxide or nitrogen. Thirdly, the turbidity of the wort is higher, which is a function of the fine mill gap setting. Fourthly, the cleaning of the mill and mash chambers must be rigorous as any residue of mash can lead to infection (Wilkinson, 2001).

10.2.4 Hammer milling

Hammer milling is a dry process used exclusively for the preparation of grist for use in a mash filter. The machine consists of a perforated cylinder or sieve, and inside this is a rotor mounted with freely swinging 'hammers'. The malt is introduced to the centre of the cylinder and is pulverised between the rotor and the sieve so as to reduce it to powder. There is no requirement to adjust the particle size of the grist. The power requirement is very high, 100 kW being typical in a one million hectolitre per annum brewery. Consequently the vibration and noise outputs are high and require special housings to absorb them. Maintenance must be regular and replacement of the sieve is not uncommon after 300–400 batches, with hammers being replaced after 1000 batches.

10.2.5 Submerged disc milling

Recently a fine milling process for mash filters analogous to wet roller milling has been pioneered (De Brackeliere, 2000). Thus milling and mashing are carried out at the same time between two counter-rotating stainless steel ribbed discs with a gap of 0.3–0.5 mm between them. The ensuing mash is pumped to the mash conversion vessel. Very little data is yet available on the benefits or otherwise of this system, but it has been developed by Meura, the leading producer of mash filters.

10.3 Mashing devices

Wet milling techniques embody the mashing in process within them, but dry milling is normally used to create a buffer stock of dry grist in a grist case, which is then mashed within 10–15 minutes directly into the mash conversion vessel. The key attribute of an effective device is to achieve complete wetting of the dry grist by mashing liquor at a precisely controlled temperature to optimise conversion. The mashing liquor is usually derived from hot and cold liquor in an in-line static mixer using a modulating valve on the cold liquor and a temperature feedback signal downstream. Modern designs also ensure that all dust is contained, and oxygen pick-up minimised, although taken to a logical conclusion this requires brewing liquor to be deaerated and grist cases to be purged with inert gas, and it is a fact that hardly any brewing companies go to this extreme (Fig. 10.1).

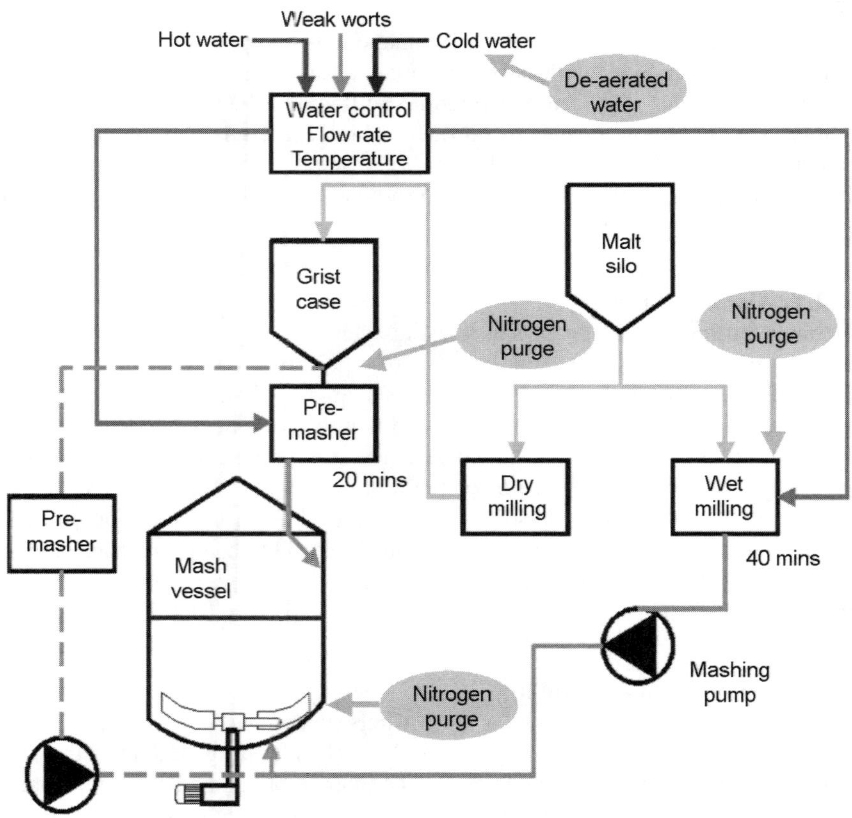

Fig. 10.1 Sources of oxygen at mashing are from grist 600 g/te (grams per metric tonne) of malt and from mashing water 10 g/te of malt. They can be eliminated by purging with inert gas such as nitrogen and by the use of deaerated water.

A typical mashing device has no moving parts and is arranged as two concentric tubes, the central one of which introduces the grist at a controlled rate, the annular space receiving the attemperated mashing liquor. The inner tube can either be perforated to allow the liquor to diffuse into the grist, or be shorter than the outer one which has a converging nozzle at the delivery end that allows the liquor to penetrate the grist. The mashing device is normally located on the top of the mash conversion vessel and the mash is introduced down the side wall of the vessel to minimise splashing and oxygen pick-up.

Due to interest over the last few years in the impact of oxygen in the brewhouse, some breweries introduce a small mash mixing vessel, from where the mash can be gently pumped into the conversion vessel from the bottom to minimise oxidative effects. However, it is clear that to have a significant effect on oxygen pick-up, steps must be taken to eliminate oxygen from brewing liquor and grist.

10.4 Mash conversion vessels

Mash cooking and mash conversion are carried out in similar separate vessels as a batch operation, mash cookers being used for decoction mashing processes and also for cereal (e.g. rice, maize) cooking to be added to the malt mash. The objectives of mash conversion are to solubilise substances, through enzymic action, that are not soluble in their natural state, to dissolve substances that are immediately soluble, and to facilitate chemical changes through enzymic action. Most importantly, all the starch must be converted to sugars which are largely fermentable. Thus the process is a mixing–heating reaction, starting with the receipt of the mash which must be intimately mixed to a homogeneous state at a carefully controlled temperature in order to optimise enzymic activity. The temperature is then increased, either by transferring boiling mash from the mash cooker or by using external heating. At each temperature rise the brewer seeks to optimise the process depending on recipe beer type and raw materials, and ultimately the temperature is raised to 76–78°C before transfer to the mash separation phase. The purpose of the final temperature rise is to halt further enzymic action, and also to reduce the viscosity of the liquid element to aid filtration.

Over the last decade there has been recognition by all suppliers of the key features of a successful mash conversion vessel. It is required to handle mash gently so as not to degrade particle size, nor to aid oxygen pick-up. It is also now known that the impermeable substance beta glucan can be released from cell wall material by shear forces into a gel-like substance with negative effects on filtration. In conflict with this is the requirement for intensive mixing and blending, and the ability to raise the temperature rapidly and homogeneously whilst avoiding burning onto heating surfaces. Thus the interaction of heating jackets and agitators is critical to good performance.

A typical mash conversion vessel (Fig. 10.2) is constructed of stainless steel and is cylindrical with a conical or dished bottom and with a large diameter

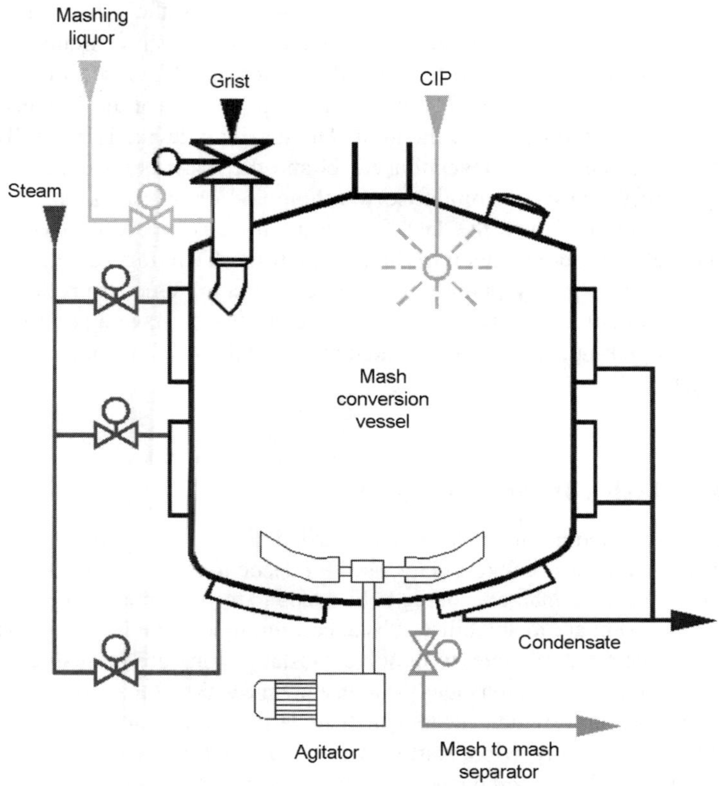

Fig. 10.2 Typical modern mash conversion vessel. Off-centre large-diameter agitator with tilted bottom dish gives good mixing without baffles. Agitator tip speed is less than 3.8 m/s for roller-milled grist and less than 3.0 m/s for hammer-milled grist. Agitator drive is variable.

agitator. This is bottom mounted and designed to run at slow speed, typically 10–40 rpm, and to mix effectively without the need for side-wall baffles which would be a source of unacceptable shear forces.

Heating is normally provided through steam jackets which have two or three zones to enable processing of different batch sizes. They are constructed as either dimple or limpet coil and are invariably of stainless steel. The steam pressure is limited to a maximum of 3 bar to minimise the risk of burning the mash.

Recently there has been the revival of an old idea to provide heating by direct tangential steam injection, which can be used to aid mixing as well as heat. This eliminates the risk of burning on a jacket surface but introduces the risk of locally subjecting the mash to a much higher temperature than the enzymes can withstand. In addition, the steam boiler and pipework must be clean and free from contaminants and the boiler feed water must be pure. Whereas these criteria can be met in a small packaged boiler, a large brewery would have some difficulty in achieving them.

10.4.1 Mash transfer

Often modern brewhouses are constructed on one level due to capital cost constraints. The consequence of this is that mash must be pumped from the mash conversion vessel to the mash separation device. Minimising shear forces during this transfer is very important to the performance of the mash separation stage, as the combination of oxygen pick-up, beta glucan release, and particle size reduction reduces filtration rates. Consequently pumps are selected to run at slow speed with open impellers and mash transfer pipework is sized for a transfer velocity not exceeding 1.5 m/s. In addition the length of pipework is kept to a minimum and bends are of very large radius with no abrupt changes of direction.

10.5 Mash separation

A modern state-of-the-art brewhouse typically produces batches 10–12 times per 24-hour day for 5–6 days and carries out intensive cleaning at weekends. It is normal that the rate-limiting process in the brewhouse is mash separation, and two main technologies are employed at this stage, namely the lauter tun and the mash filter. The principles of filtration were established in the nineteenth century, and liquid flow was found to be proportional to filter bed permeability, pressure differential, and filter bed area, and inversely proportional to viscosity of filtrate and filter bed depth. The two leading technologies for mash separation seek to exploit these principles in very different ways. The lauter tun maintains permeability by coarse milling and the use of rakes to loosen the bed, but otherwise depends on gravity for the run-off. It is normally operated at a bed loading of 150–200 kg/m^2 for 10–12 brews per day. On the other hand, the mash filter makes no attempt to maintain permeability but uses a thin mash bed, typically at one tenth the loading of a lauter, supported on polypropylene cloths. The pressure differential is provided by pumping at up to 1 bar pressure, and also by squeezing the grain bed using flexible membranes incorporated into the filter. As a consequence the mash filter tends to produce better measured extracts, shorter cycle times, and higher wort gravity than the lauter tun, but at higher capital cost and less flexibility. Tables 10.1 and 10.2 compare the two techniques.

Both lauter tuns and mash filters have been used for many decades, but have been refined and developed such that they continue to be the technologies of choice, with lauter tuns being dominant, producing possibly 75% of the world's beer. The choice between the leading technologies is influenced by many factors

Table 10.1 Typical grist analysis by EBC sieve % in each fraction, using 100% malt comparing lauter tun with membrane mash filter

Sieve no.	Mesh size (mm)	Lauter tun (%)	Mash filter
1	1.270	18	1
2	1.010	8	4
3	0.547	35	9
4	0.253	21	26
5	0.152	7	19
6	–	11	14

including geography, franchise agreements, raw material factors, cost of capital, brewery capacity and diversity of recipes, to name but a few.

The mash filter can produce up to 14 brews in 24 hours against the more usual 10–12 brews for a lauter tun. It should be noted that the latter produces quicker cycles at lower bed loadings, but the more correct comparison should be based on total wort production per week when a 15-tonne lauter tun at eight brews per day produces the same output as a 10-tonne mash filter at 12 brews per day (see, for example, Table 10.3).

It is common for lauter tuns to sacrifice a small amount of extract in the last part of the run-off in order to improve cycle times, but it should be noted that the universal method of measuring extract is by specific gravity, which assumes that any measured gravity above that of water has value to the brewer. This is not the case and many of the world's largest brewers deliberately cut the run-off early on the basis that below 2°–3° Plato, much of the measured extract is not fermentable, and it also contains a higher proportion of unwanted materials such as polyphenols and astringent substances. The very high extract performance of

Table 10.2 Comparison of mash separation for lauter tun and modern mash filter. The flow rate is governed by the D'Arcy equation:

$$Q = KA\,\Delta P/\mu L$$

where Q is the flow rate, K is the bed permeability, A is the filter bed area, ΔP is the pressure difference across the bed, μ is the liquid viscosity, and L is the bed depth

	Lauter tun	Mash filter
K (Coarse fraction – plansifter)	25	5
P (typical, barG)	0.02	0.5
A (M_2 filter surface area)	50	333
U (inverse of mash, l/kg ratio)	0.3	0.4
L (bed depth, metres)	0.36	0.04
Q (relative flow rate)	231	53031
Relative flux rate (flow/M_2)	4.6	159

Note: barG = bar gauge

Table 10.3 Comparison of cycle time for lauter tun and modern mash filter

	Lauter tun	Mash filter
Typical cycle time	180	120
Plate flood	5	–
Mash transfer	10	5
Recirculation	10	–
	25	5
Strong wort collection	50	25
Dilute wort collection	55	65
Two deep-bed rakes	20	–
	125	90
Drain/grainout	20	15
Underplate wash	5	–
	25	15
Totals	**175**	**110**

the mash filter is due to the hammer-milled grist giving more than 100% of the extract obtainable by standard methods in the laboratory. This performance cannot be achieved unless the last runnings are collected to less than 1° Plato.

The flexibility of the mash charge in a lauter tun is between −50% and +30% of its nominal output, but the mash filter is more restricted at −20% to +10%. Consequently the mash filter is most effective with a relatively small recipe portfolio and fixed brewlength. Technically it is possible to insert a dividing plate into a mash filter to accommodate smaller brews, but in practice this is a time-consuming task and rarely undertaken.

The wort quality produced by the mash filter compares favourably with the lauter tun, particularly in respect of colloidal haze and settleable solids, but it tends to higher levels of polyphenols as a direct result of hammer-milling the husk and acrospire (Table 10.4).

The spent grain from the mash filter is appreciably drier at 70% moisture than that from the lauter tun at 78–80%. The automatic removal of spent grain from a mash filter can be a problem when it sticks to the plates and requires manual intervention. By contrast, spent grain removal from modern lauter tuns is extremely efficient.

The issue of costs is finely balanced, with the mash filter's advantages on cycle time, liquor usage, and extract performance counterbalanced by its significantly higher capital cost. The regular maintenance costs are also substantial, with filter cloths requiring replacement every 1000 brews, and frequent inspections to check the integrity of the membranes. At 10 tonnes of

Table 10.4 Comparison of wort quality for lauter tun and modern mash filter

		Average value during wort collection	
		Lauter tun	Mash filter
Settleable solids	Mls/l	<1.0	<0.2
Haze	°EBC	<10	<2
Lipids	Mgs/l	18	17
Polyphenol	Mgs/l	180	195

Note: Mls/l = millilitres per litre; Mgs/l = milligrams per litre; °EBC = degrees of haze using the European Brewery Congress method.

dry goods, the choice is marginal either way on a pure financial basis, but at larger brewlengths, two mash filters are required against only one lauter tun, and the financial case favours the lauter tun.

The future will continue to be driven by competitive pressures. The pursuit of absolute cost efficiency and the value of flexibility will be a matter for individual company strategies. The bigger companies are likely to hold more and more of the market, and production facilities will continue to be rationalised into larger breweries. The smaller brands will not disappear but will need flexible strategies to succeed. As a consequence there will be a substantial market for both mash filters and lauter tuns.

10.5.1 Lauter tuns

The lauter tun is a circular vessel made of stainless steel with a very high diameter to height ratio. The true bottom may be flat or contain a series of valleys. The wort outlets pierce the real bottom and are collected together into a manifold for transfer to the wort boiling vessel or pre-run vessel. Although the flow rate in individual run-off pipes, and therefore the head loss, is small, considerable care has been devoted to ensuring equal flow rates, including bell mouth entry and equalising pipe runs. This is to ensure that extraction is uniform across the whole grain bed (see Fig. 10.3 on page 224).

The false bottom is made of stainless steel and is designed to support the mash and hold back the granular material, allowing clear wort to filter through. Typically it is made of segments for ease of lifting to inspect the space underneath, and slotted either by machining solid plate to give an open area of approximately 8% with a slot width of 0.7 mm and a length of up to 50 mm, or by using wedge wire which is resistance-welded to crossbars with a slot width of 0.5–0.7 mm, giving an open area up to 18%. Machined or wedge wire plates provide no significant resistance to flow relative to the mash bed and there is no detectable difference in run-off rates between the two designs. However, wedge wire is more prone to mechanical damage, being of a much less robust construction.

The underplate space historically has been prone to clogging with fine material which finds its way through the false bottom and solidifies on the real bottom. However, modern cleaning systems incorporate high-pressure spray jets located in the underplate space, and at the end of every brew complete cleaning of debris is achieved in 2–3 minutes by pulsing these jets three to four times with a drain-down period between pulses. As a consequence, false bottom plates are lifted for inspection purposes only once a year.

It is normal to add brewing liquor to the lauter tun to completely fill the space between the false and real bottom so that mash can be introduced gently either through the bottom or low down on the side wall, to minimise splashing, oxidation, and shear forces. It also pre-heats the real bottom of the lauter tun, which is advantageous to the first part of the run-off to maintain lower wort viscosity. It is desirable to minimise the underplate space, as liquor used for this purpose reduces the amount available for sparging or washing the grain. Current best practice indicates that the depth of this space has been reduced successively to approximately 20 mm which equates to a volume equal to perhaps 1–2% of the total run-off.

The rake gear consists of lauter blades mounted vertically on cross-beams which are mounted on a central shaft that can be both rotated and raised and lowered. Lauter blades are designed to improve filter bed permeability and to counteract the compression of the bed due to gravity and wort flow. They must not cut slots through the bed as this would simply allow sparge water to run through without washing or leaching the grain. Typical operating philosophy requires minimal use of the rakes for low wort turbidity, and often the strong first wort is removed at low flow rate without raking at all. The rake gear also carries some means of discharging the spent grain at the end of the cycle, and this is generally a plough bar which can be slowly lowered so that it sweeps the grain towards the outlet ports which are mounted in the bottom of the tun.

In operation lauter tuns are fully automated. Typically wort flow is defined by a profile within the recipe control, and after strong wort is removed at constant flow rate, sparge is introduced through spray nozzles from above at a controlled flow rate. The dilution effect allows the wort flow to be ramped up, to a factor of three by the end of run-off. Feedback signals from differential pressure transmitters and also from a turbidity meter in the run-off pipe provide control to the rakes which are introduced at variable rotational speed and depth, with the purpose of maintaining differential pressure and minimising turbidity. Deep bed raking is only required if the differential pressure reaches a preset maximum, and it would be normal to stop forward wort flow in this situation to avoid a burst of very turbid wort going forward.

10.5.2 Mash filters

The mash filter in its current form utilises very finely ground grist which maximises the extract from the conversion process. The filter is made from rectangular polypropylene plates, which are pressed together by a hydraulic ram to create water-tight pockets between them. Alternate plates are equipped with elastic membranes which can be inflated with compressed air to squeeze the mash against a permeable polypropylene cloth. The largest machine can accept up to 11 tonnes of grist and contains 60 plates. The space between the uninflated membrane and the filter cloth is only 40 mm, so the filtration path is initially 40 mm and is reduced to possibly 25–30 mm when compressed by the membrane.

Each of the plates contains ports in the corners which, when pressed together, form the channels by which mash is introduced into the bottom of each chamber and by which sparge liquor can be introduced. Other channels are created between the plain plates and the filter cloths by which wort can be removed. It is important that all the chambers are completely filled with mash, as underfilling can lead to 'channelling' with the filter cake and loss of extract. In practice this means that for efficiency they need to operate constantly at one brewlength, and it is common to produce 12–14 brews per day, so they are better suited to breweries with a capacity of more than one million hectolitres per year.

The method of operation of a mash filter is as follows. The filter is filled with mash from the bottom and the chambers are vented until the filter is filled, at which point the vents are closed and the wort outlets opened. With strong wort being withdrawn, the mash continues to be pumped in, depositing the mash solids onto the filter cloths which builds the filter bed quickly enough for bright wort to be produced even in the first runnings. When the mash transfer is complete, the inlets are closed and the membranes inflated to compress the grain bed against the filter cloths, squeezing out the strong wort. Once the membranes are deflated, sparge liquor is introduced through the mash ports into the space between the membranes and the filter cakes. After sparging is complete the membranes are inflated again to squeeze the spent grain cake to a moisture content of around 70%. Lastly the plates are separated automatically so that the spent grain can drop under gravity into a hopper with a screw conveyor.

10.6 Wort boiling

Wort boiling is perhaps the process stage of brewing which is least understood in terms of the reactions taking place, testing the ability of the plant designer to select the desirable aspects and suppress the undesirable ones. There is general acceptance that the criteria expected to be met must include volatile removal, hop isomerisation, protein denaturing and flocculation, sterilisation, enzyme inactivation, flavour formation, and concentration by evaporation. However, the process specifications for the plant designer are always reduced to the desired evaporation rate and perhaps the number of brews required between cleaning cycles. Because the desired criteria are based on complex reactions of which our knowledge is insufficient, plant design is often reduced to a simple heat transfer approach for minimum capital cost. However, empirical evidence shows that some solutions to the design requirements meet the brewers' criteria better than others (Andrews, 1992).

Typically wort is boiled for 60–90 minutes with the addition of hops, and the hot break is subsequently removed. Experience over the last 20 years has shown that many changes in boiling conditions can frequently be reflected in a negative impact on beer quality (Schwill-Miedaner, 2002). Some of the technologies introduced can now be seen to have failed with the benefit of hindsight. Consequently, high temperature wort boiling, microwave systems, and high pressure pumped external systems will not be discussed here.

The main functions of wort boiling together with the parameters which are believed to affect them are as follows:

- Volatile removal
  - evaporation
  - two-phase flow
- Isomerisation
  - temperature
  - time
- Flocculation
  - two-phase flow
    low shear
- Sterilisation
  - temperature
  - time
- Enzyme inactivation
  - temperature
  - time
- Concentration
  - evaporation

The evaporation of water is not often a substantial issue in many breweries and not of primary interest, and the only other function requiring evaporation is the removal of unwanted volatiles. There is no good reason to assume that the evaporation of water in itself correlates to effective evaporation of volatiles.

It is clear that vigorous mixing is essential to cause coagulation of proteins and form flocks, but shear forces must be minimised to prevent disruption of flocks. Over-heating should be avoided as it will adversely affect foam retention, will 'burn-on' and caramelise sugars, and will lead to the formation of usually undesirable flavours (Andrews, 2003).

10.6.1 The principles of boiling wort

Nearly all modern systems use condensing steam to give up latent heat to the wort. The water vapour evaporated from the wort is at a lower pressure and temperature than the original steam, but it contains all the heat in the original steam less the sensible heat left behind in the condensate. It is therefore simple to derive the exact mass of steam required to generate a given evaporation rate irrespective of the type of boiling system, as a function of wort volume at the start of boil, desired evaporation rate, desired boiling time, steam supply pressure and wort pressure. It is practical to meter the exact quantity of steam required, and the plant design is reduced to the task of providing sufficient surface area to ensure that the required quantity of steam will condense in the required time.

All heat transfer is based on Fourier's law, which states that $q = ua\Delta T$, where q is the rate of heat flow, u is the overall heat transfer coefficient, a is the heating surface area, and ΔT is the temperature difference between the steam and the wort. It is advisable to maintain a low ΔT and certainly the use of saturated steam at a pressure above 3 bar, equating to 144°C, is not advisable because of the negative effect of Maillard reactions. Therefore the determination of the overall heat transfer coefficient is the key to establishing the correct surface area and this is dominated by the wort side fouling and boundary layer, which in turn is dictated by the velocity of wort flow. Pumping the wort improves the heat transfer but has a negative influence on hot break formation. Each successive brew between cleans introduces some fouling of heating surfaces and changes the conditions.

10.6.2 Internal boiling systems

The most common type of wort kettle is equipped with an internal tube bundle as heater. The wort passes upwards through the tubes into a constriction which emerges above the wort level and is terminated with a spreader plate which the wort strikes and thereafter returns to the wort surface, while the volatiles escape out of the vent stack at the top of the kettle. The main variations on this basic model are those systems which attempt to improve the heat transfer by circulating wort through a pump into the bottom of the kettle immediately under the heater to increase the velocity of flow. This is claimed to have the effect of increasing the number of brews between cleans from 4–6 in the classic design to about 14 (after Steinecker).

The other main variation is dynamic low-pressure boiling, which is designed to achieve a more rapid evaporation of volatiles, with the corollary that the overall water evaporation can be reduced, saving primary energy. In this system the pressure is raised and lowered six times per hour, between pressures of 1.0 bar and 1.2 bar corresponding to temperatures of 100–102°C and 104–105°C. Each time the pressure is released, the total contents of the kettle instantaneously boil, producing a large quantity of flash steam which carries away the volatiles. When the pressure is raised once again, volatiles are converted from their precursors more quickly than at atmospheric pressure. With this system it is claimed that good volatile stripping can be achieved with total evaporation of 4–5% (Kantelberg, 2000).

A completely different type of internal heater is the thin-film evaporator (Stippler, 2000). The wort is boiled and evaporated in a thin film passing over a steam heated cone located in the upper chamber of a two-part vessel, the lower part being the whirlpool. Typically wort is passed over the cone four times in 40 minutes and returned continuously through a tangential inlet to the whirlpool, and hot break is removed continuously. After the final whirlpool rest, during which there is a further creation of unwanted dimethylsulphide volatile, the wort is passed again over the heated cone before being cooled to fermentation temperature. To achieve the heat transfer with steam at 2.8 bar (130°C) the cone is very large but it is claimed that good beer quality can be achieved at 4–5% overall evaporation.

10.6.3 External boiling

The engineering problem of producing a high rate of heat flow can be solved by increasing the heating surface area, rather than concentrating on the heat transfer coefficient. There is a physical limitation to surface area with internal heaters which does not apply if the heater tube bundle is taken external to the kettle, and this configuration has been developed to an advanced state.

The currently available system utilises a vertical tube bundle contained within a steam chest with a wort inlet pipe from the base of the kettle to the bottom of the tubes, and a return pipe from the top to the kettle above the wort level. Very large surface areas are now used, up to five times more per unit of wort than can be achieved internally (Andrews, 2003). Consequently very low steam pressures can be used, as low as 0.7 bar (113°C), which is very positive for foam retention in finished beer, and also results in virtually no fouling so that 40 or mores brews can be achieved between cleans. It has also been demonstrated that such a low ΔT has an advantageous impact on flavour stability and good results are achieved at 4% overall evaporation.

There are many advantages with external systems. They can be used over a wide range of brewlengths, and heat-up can commence almost immediately while the kettle is filling, whereas an internal heater cannot be used until it is submerged below wort level. It can be cleaned with only a small amount of caustic solution. It lends itself very readily to vapour recompression energy-saving technology (see Section 10.7), and can be used as a combined kettle/whirlpool (Fig. 10.3).

The kettle/whirlpool configuration is a very attractive proposition in certain circumstances, combining the functions of a pre-run vessel, kettle, and whirlpool in one single vessel, which can produce large time savings with optimum results. It is also practical to employ one external wort heater shared between two kettle/whirlpools and it is then possible to produce eight to nine brews per day.

10.6.4 Volatile stripping

An alternative approach to volatile stripping (Braekeleirs, 2001) is to use a steam-stripping column positioned after the whirlpool stand during which it is possible that more volatiles, such as DMS, can be formed if the precursor has not been completely converted in the wort boiling stage. In the stripper, wort is sprayed from above onto a packed column filled with ceramic rings to provide a very large surface area. Steam is supplied to the bottom of the column and moves upwards, carrying away volatiles at a total evaporation of 1–2%.

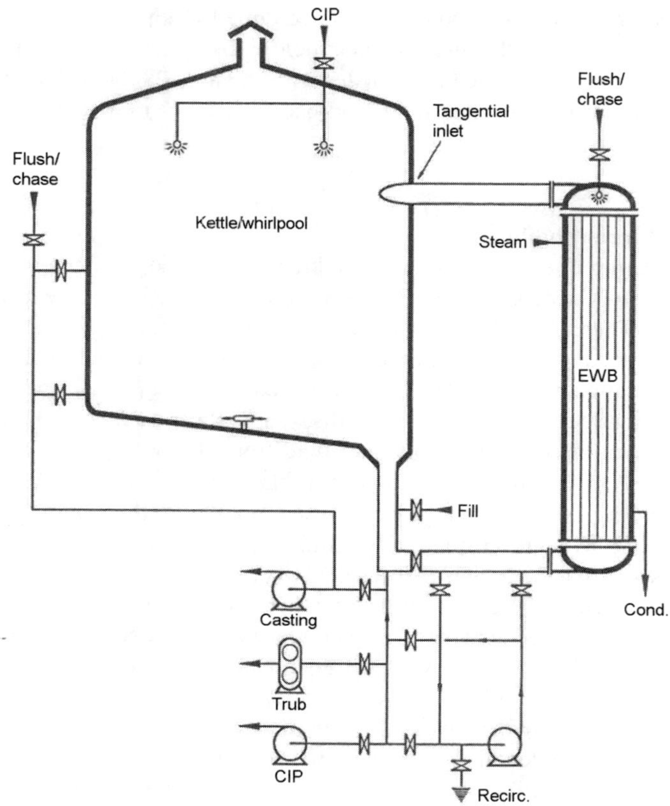

Fig. 10.3 Kettle/whirlpool with flat bottom and trub recovery.

10.7 Energy recovery systems

Wort boiling uses typically 20% of the total energy consumption in a brewery and therefore is a significant target for conservation or recovery techniques. Wort boiling technology has provided a number of approaches to reducing energy consumption by reducing the evaporation rate from 6–8% down to 4–5%. There are a number of ways of recovering the energy from the remaining vapours, the most simple and widespread being to condense the vapours in a heat exchanger and make hot water. This can be stored in a vertical tank in which the hotter, less dense water stratifies to the top and can be used in an in-line heat exchanger to raise the temperature of the wort between mash separation and the wort kettle. The cooler water returning from the heat exchanger is returned to the bottom of the vertical storage tank where, being denser, it stratifies to the bottom. In this way nearly all the energy in the vapour at 4% evaporation is recovered and reused within the brewhouse.

Alternative methods of energy recovery require the vapour at atmospheric pressure to be compressed to a higher pressure and temperature and then used ideally in an external wort boiler directly. There are two main approaches to compressing the vapour, the lower cost method being to inject high pressure steam, through an ejector nozzle, thereby resulting in a mix of new steam and evaporated vapour at an intermediate pressure of between 1.2 and 1.8 bar. The drawbacks of this system are that high pressure steam is required, ideally more than 10 bar, and this is rarely available, and also it is impossible to recycle all the kettle vapours, 50–70% recovery being typical. This method is known as 'thermal vapour recompression'. The other approach is to compress the vapour with a mechanical compressor driven by an electric motor and to reuse the vapour compressed to typically 1.5–2.0 bar. The main drawback is that the size of the motor drive is very large and the saving in steam energy costs is not usually more than the energy cost of running the motor, with electrical energy unit cost normally being much higher than steam energy unit cost.

10.8 Hot wort clarification

During wort boiling, hot break forms largely from the denaturing of proteins which should form into large particles if the wort boiling system is working properly. These large particles are held together by weak attractive forces (Findlay, 1971) which can be aided by the use of copper finings, but it is important that shear forces are kept to a minimum both in wort boiling and in the transfer to the clarification stage. There will also be significant quantities of hop debris and this, together with hot break, should be capable of settling by gravity with 30 minutes. Although hot wort centrifuges are used in a few breweries in conjunction with a whirlpool, by far the most common technology is the standalone whirlpool. This consists of a vertical cylindrical tank with a height to diameter ratio of 0.5–1.0 into which the wort is transferred through a tangential port. Transfer should take no more than 10 minutes in order to impart the angular momentum necessary to achieve a satisfactory rotational speed. As the rotation slows due to wall friction, the solid particles are swept to a cone in the centre, a process that takes approximately 30 minutes. Hot bright wort is removed normally through branches at successively lower levels, with the final one in the base of the tank. It is usual to use a speed-controlled pump so that when the trub cone is uncovered the rate of withdrawal of wort can be reduced to minimise the risk of the trub cone breaking up.

The problems which arise from poor whirlpool performance are nearly always associated with factors upstream in the brewhouse, the most common being poor wort boiling or over-aggressive pumping, but a consequence of these problems is that a very large number of different shaped bases have been trialled. The most commonly used has a flat bottom with a 2–3° slope to the outlet, which allows the trub cone to be drained quite dry, but nevertheless there is a loss of around 1% of extract when the trub is discarded. The alternative approach is to use a shallow cone bottom and recover the wort from the trub slurry through a centrifuge, which reduces wort loss to less than 0.5%.

The trub cone itself is normally broken up using water jets so that it can be withdrawn through the bottom outlet, and it is most easily disposed of by pumping onto the spent grain from mash separation. It is important that it is not allowed to settle for too long after the hot wort has been removed, as in the presence of air it can set very hard and be difficult to dislodge.

A few very large breweries have introduced hot wort clarification by centrifuge, but this approach is very capital and maintenance intensive, and as a consequence has not been adopted widely.

10.9 Wort cooling

Hot wort must be cooled rapidly under aseptic conditions to the temperature at which fermentation will start when yeast is introduced. Very typically the temperature is in the range 8–12°C for lagers and 16–20°C for ales, although other temperatures are used. Cooling is universally achieved with a plate heat exchanger which can be arranged with one or two stages. The main coolant is brewing water at ambient temperature, which by cooling the hot wort, is raised to typically 80°C and used for subsequent brews. The typical flow ratio of liquor to wort is 1.1 to 1 and this ensures that all the required brewing liquor is heated by the outgoing wort.

To achieve cold wort temperatures below 20°C it is common practice to arrange a second stage of cooling in the plate heat exchanger to use either chilled water at 2–3°C or a secondary refrigerant such as ethylene glycol at −3°C to −4°C.

Cleaning of the plate heat exchanger is critical, both to maintain its cooling efficiency because fouling at the hot end is substantial, and to minimise the possibility of microbial infection at the cold end. A dedicated cleaning system, or at least a dedicated caustic tank, is best practice, and although some breweries clean after every brew, it is more common after 3–8 brews.

10.10 Cold wort clarification

Cold break is a haze that forms in wort during cooling and consists of particles, generally less than one micron in diameter, made up of proteins, polyphenols, and carbohydrates which precipitate out of the solution. Some proportion may be hot break which has not been previously removed.

The removal of cold break before yeast pitching is by no means universal. It is reported by some that removal has no effect on downstream operations or on final beer quality, while others report that cold break leads to higher fermentation rates (Narziss, 1971; Crompton, 1991). Many breweries no longer remove any cold break, but for those that do it may be by kieselguhr filtration, centrifugation, or settling tank, or by flotation, which is the most widely used approach. Cold wort is held in a flotation tank, often after pitching with yeast, and sterile air is introduced through the base in the form of a stream of fine bubbles. As they rise they collect the cold break particles and a layer of foam forms on the surface over a period of 6–12 hours. The wort is then removed from the bottom of the vessel, leaving the foam and about two-thirds of the cold break behind. The whole process includes the oxygenation and first fermentation stages.

10.11 Cold wort oxygenation

Cold wort ready for yeast pitching must be saturated with dissolved oxygen in order for the yeast to perform in the first stage of fermentation. The injection of sterile filtered air is still widely used, but it is difficult to achieve more than 8–10 ppm by this method, whereas a good guide figure is for the oxygen parts per million to be equal to the degrees plato (°P) of the wort. In modern breweries brewing at high gravity, this often means 15–16°P and the necessary level of oxygen can only be achieved by the use of pure bottled oxygen.

Typically this is introduced in a fine stream of bubbles into the throat of a venturi tube carrying the wort flow, with an in-line static mixer downstream to provide a high level of turbulence to aid the solution of the gas. Controlling the amount of oxygen is often by flowmeter and control valve giving a fixed ratio to the wort flow, and normally an oxygen meter is mounted downstream to monitor the resulting oxygen level.

10.12 References

ANDREWS J M H (1992) *Proc. 22nd Conv. Inst. of Brewing (Australia and New Zealand Section)* Melbourne, p. 65.

ANDREWS J M H and WILKINSON N R (1996) *Ferment* **9** (5), 257.

ANDREWS J M H and AXCELL B C (2003) *MBAA Technical Quarterly* **40** (4), 249.

BRAEKELEIRS R and BAUDUIN C L (2001) *Proc. 8th Brew. Conv., Institute of Brewing (Central and Southern Africa Section)*, Sun City, p. 36.

CROMPTON I E and HEGARTY P K (1991) *Proc. 23rd Congr. Eur. Brew. Conv.*, Lisbon, p. 625.

DE BRACKELIERE C, HARMEGNIES F, TIGEL R and MENDES J P (2000) *Brauwelt International* **18** (5), 372.

FINDLAY W P K (1971) *Modern Brewing Technology*. Macmillan Publishers, London, pp. 60–82.

HERMIA J and RAHIER G (1992) *Ferment* **5** (4), 280.

HERRMANN H, KANTELBERG B and WIESNER R (1998) *Brauwelt International* **16** (1), 44.

KANTELBERG B, WIESNER R, JOHN L and BREITSCHOPF J (2000) *Brewing and Distilling International* **31** (3), 16.

NARZISS L, KEININGER H and REICHENDER E (1971) *Proc. 13th Congr. Eur. Brew. Conv.*, Estoril, p. 197.

SCHILD D (1936) *Wochenschr. Brau* **53**, 345–353.

SCHWILL-MIEDANER A and MIEDANER H (2002) *Brauwelt International* **2002** (1), 19.

STIPPLER K (2000) *Ferment* **13** (1), 34.

WILKINSON N R (2001) *Brewers' Guardian* **130**, (4), 29.

11

Fermentation of beer

C. Boulton, Coors Brewers Ltd, UK

11.1 Introduction

The modern brewing industry is dominated by a small number of large companies. These operate on an international stage and produce national or global brands, typically pale pilsner-type lager beers. This market is highly competitive. The need to sustain or grow these global brands within this highly competitive marketplace has provided much of the impetus for the introduction of new fermentation technologies and changes in fermentation practice.

Undoubtedly the major driver for changes in fermentation practice and the underlying technology is cost. Since fermentation is frequently the rate-determining step in the brewing process, much effort has been devoted to identifying ways of increasing productivity. Batch sizes may be increased by using ever bigger fermenters; however, approaches that increase the productivity of existing vessels are particularly attractive since they avoid capital expenditure. Such strategies include the use of ultra-high gravity brewing, ensuring high efficiencies of conversion of sugar to alcohol and developing methods for reducing cycle times.

Good fermentation practice requires precise control of the variables that are influential on the progress and outcome of the process. This is especially important in the case of global brands produced simultaneously at several sites. Precise control of fermentation is an essential prerequisite to ensuring that different product streams generate consistent beer. In recent years advances have been made in improving the ability to monitor and control fermentation progress. These developments have been underpinned by a better understanding of the processes that occur during fermentation. In particular, much progress has been made in unravelling the complex interrelationships that exist between wort composition, yeast physiology, fermenter design, fermentation management and beer quality. A review and discussion of these developments is provided in this chapter.

11.2 Current developments

11.2.1 Fermenter design

The cylindroconical vessel remains the fermenter of choice for the majority of large-scale commercial brewers. Alternative designs such as spheroconicals or Asahi vessels, where the cone is replaced by a sloping base, have been used by some but have not seen widespread adoption. The main features of a typical modern cylindroconical fermenter are shown in Fig. 11.1. The ubiquitous nature of cylindroconical fermenters is due to the many advantages that they possess. Thus, they have a small footprint relative to their capacity and they are easily built into tank farms serviced by a common system of mains for filling, emptying and cleaning. They can be used to produce any beer quality and, providing appropriate cooling is available, they can be used in uni-tank operations. They are relatively inexpensive to construct and run compared with

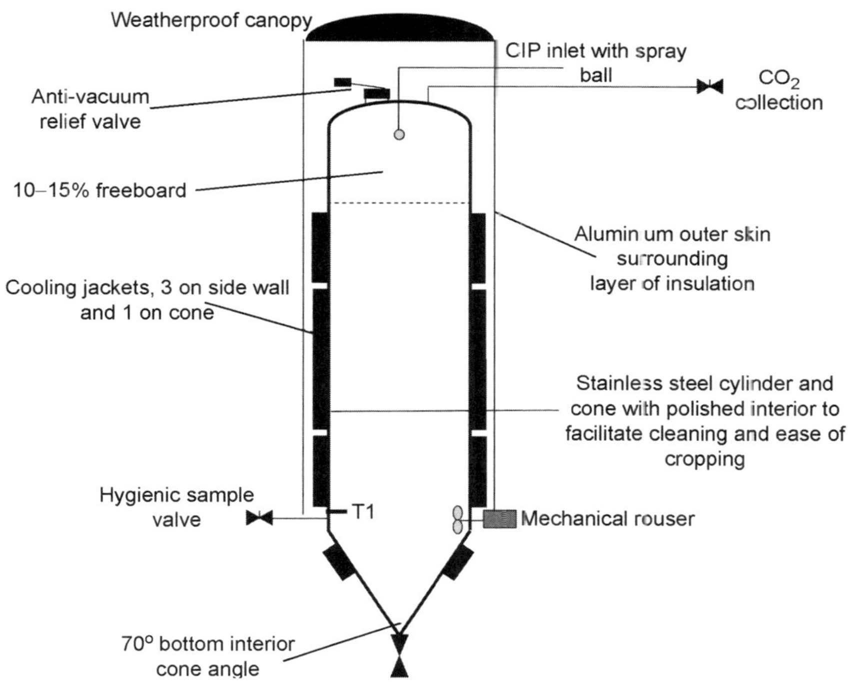

Fig. 11.1 Major features of a typical modern cylindroconical fermenter. A single thermometer probe (T1) is shown. Usually others are fitted at different heights in the vessel. Level and pressure probes are usually also fitted but are not shown.

their volume output. The enclosed nature of the vessels and their stainless steel construction make them easy to clean and favour good hygiene. Cooling is reasonably efficient and, if required, the vessels may be top-pressured and collection of CO_2 is possible. The separation of the yeast crop is efficient and favours low beer losses. Although sufficient freeboard is required to accommodate foaming during primary fermentation, the vessels usually show good hop utilisation efficiencies compared to other designs.

Apart from the introduction of new methods for monitoring and controlling fermentation, as will be described subsequently, the fundamental design of cylindroconical fermenters has remained unchanged. This is surprising in that there is scant information in the literature regarding the effects of vessel geometry on fermentation performance. The capacity of individual cylindroconical fermenters is typically within the range 1500–3000 hl, although smaller and much larger ones are used by some brewers. The capacity of vessels chosen by individual brewers is usually based on grounds which bear little or no relation to the process of fermentation, since the ability of the brewhouse to supply wort is usually the deciding factor. In this regard it is common that there is a mismatch such that more than one batch of wort is required to fill large vessels. This has consequences for fermentation management and performance, as discussed in Section 11.3.

Vessel aspect ratios are typically between 1:3 and 1:5 (diameter to height). It used to be the case that squatter vessels were favoured for lager production since this more accurately mirrored more traditional horizontal types. This is now less so. Very tall vessels are not generally favoured, since the combination of high pressure due to the hydrostatic head and gas stripping may cause adverse flavour effects such as reduced ester formation at the expense of elevated higher alcohols. However, whether or not these observations can be attributed in such a simple cause-and-effect manner is perhaps questionable (see Section 11.3).

It is perhaps true to say that current fermenter design and geometries owe more to good engineering practice than a fundamental consideration of the nature of the biological process that occurs within them. In addition, more prosaic but nevertheless essential considerations such as cost also exert a controlling influence. Thus, it would not be uncommon that the aspect ratio of a cylindroconical vessel might be dictated by factors such as the size of cylinder that is readily available and easily transportable to the brewery. The consequences of this are that the detailed management of such vessels which provides the desired combination of fermentation performance and beer quality tends to have evolved via empirical observation.

11.3 Fermentation management

There are four aspects to fermentation management, excluding cleaning:

1. Establishment of desired conditions at the completion of fermenter fill
2. Monitoring and control of the progress of fermentation

3. Identification of the endpoint
4. Removal of the yeast crop and emptying the fermenter.

11.3.1 Fermenter fill

At the beginning of fermentation it is necessary to fill the vessel with a measured volume of wort of a chosen concentration and temperature, containing dissolved oxygen at a desired concentration and an evenly distributed suspension of yeast cells at a desired viable concentration. Control of temperature and wort concentration in a modern brewery should not be problematic. Wort paraflows are capable of controlling wort cooling to achieve a temperature in the fermenter within ±0.5°C of the set-point. Similarly, modern brewhouses and dilution systems allow control of wort concentration to better than ±0.25°P. The variation in total extract delivered to the fermenter is an important criterion, since many pitching systems assume that this is not a variable. Clearly the control of total extract delivered to the fermenter cannot be better than the accuracy of all the controlling and monitoring component parts of the process. A brewhouse run according to current best practice should be capable of delivering better than within ±5% of the target total extract.

The advent of high gravity brewing has resulted in the use of initial dissolved oxygen tensions greater than that achievable by dosing with air (*ca.* 8 mg/l). Initial dissolved oxygen tensions of 15–25 mg/l are now usual and this requires the use of pure oxygen. An automatic oxygen dosing system is shown in Fig. 11.2. Systems such as the one illustrated based on thermal mass flow meters are very accurate and capable of achieving dissolved oxygen tensions better than within ±5% of target. Such systems undoubtedly provide a precise means of controlling dissolved oxygen tension at the point of addition in the wort main. However, it is a moot point as to whether or not the precision of oxygen control persists through to the filled fermenter. Wort mains usually operate under relatively high pressures (*ca.* 3–5 bar) and this aids oxygen solution. When such wort is pumped into an empty fermenter at atmospheric pressure it is likely that there is some gas breakout and loss of oxygen to the atmosphere.

Current state-of-the-art automatic yeast pitching systems rely on the use of biomass sensors for the instantaneous quantification of viable yeast concentration. These sensors utilise the passive radiofrequency dielectric properties of viable cells such that yeast biovolume is quantified as measured capacitance (Harris and Kell, 1986; Harris *et al.*, 1987). An automatic pitching system is shown in Fig. 11.3. Systems such as these are suitable for use with any yeast strain and in conjunction with accurate flow meters are capable of controlling pitching rates to better than within ±2% of the target viable count. The benefits of accurate pitching rate control on fermentation cycle time consistency are illustrated in Fig. 11.4.

The manner in which operations such as yeast pitching and wort oxygenation are performed relative to the filling of fermenters has an important and sometimes unrecognised influence on fermentation performance and beer flavour.

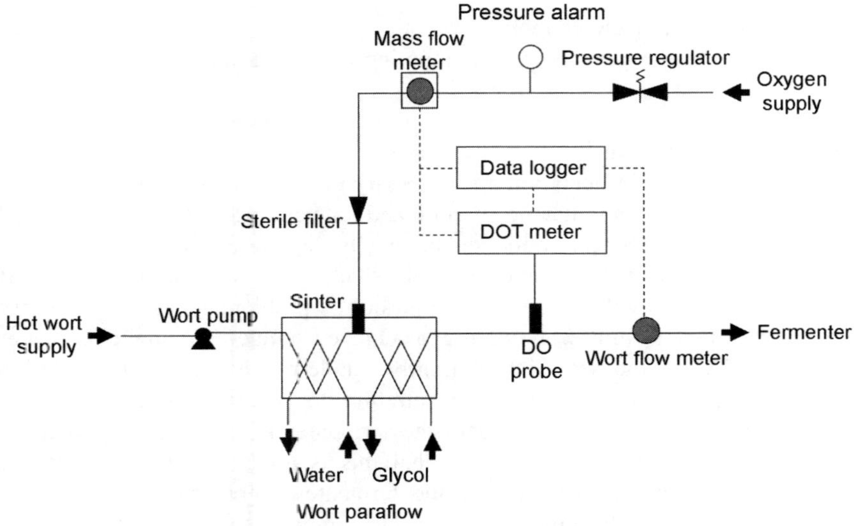

Fig. 11.2 System for the automatic control of oxygen addition to wort during fermenter fill. Before addition of oxygen the system performs an automatic pressure check. If satisfactory, sterile oxygen is added via a mass flow meter, usually between the first and second stages of the paraflow. An in-line dissolved oxygen probe located in the wort main downstream of the paraflow checks the achieved oxygen concentration and compares this with a predetermined set-point.

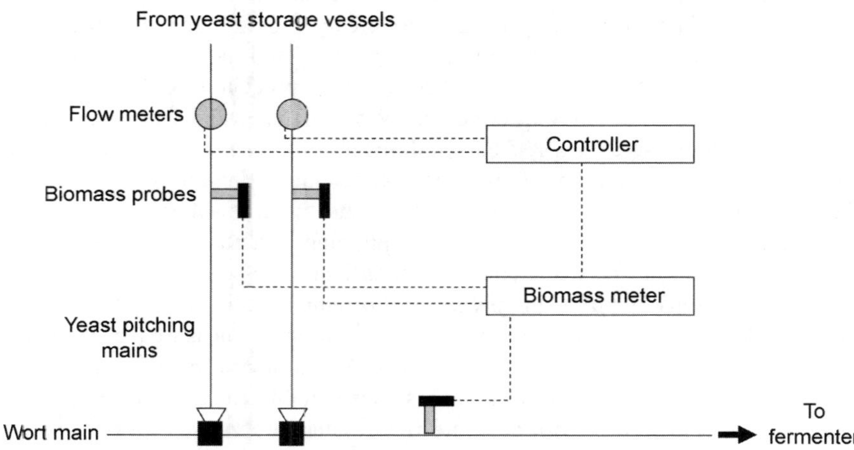

Fig. 11.3 System for in-line automatic control of yeast pitching rate. Biomass meters may be multiplexed to several probes to cater for multiple pitching/wort mains. Probes are usually located in pitching mains but may be placed in the wort main. A controller selects calibration for the appropriate yeast strain and logs total yeast pitched using outputs of instantaneous viable yeast concentration from the biomass meter and total slurry added from in-line flow meters.

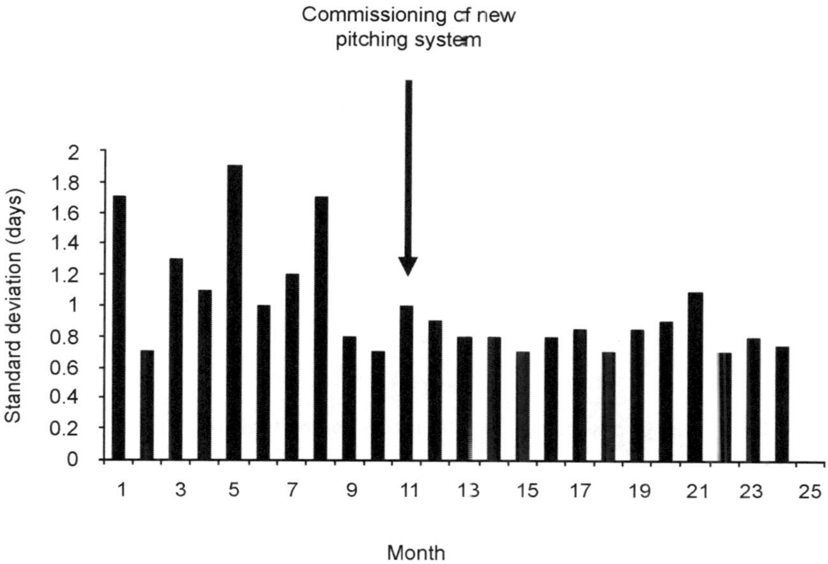

Fig. 11.4 Effect of the installation of an automatic pitching rate control system on the consistency of fermenter cycle time. The data show the standard deviation on average monthly cycle times for 28 × 1500 hl conical fermenters. Prior to month 11, pitching rate was controlled based on sampling and laboratory analysis of yeast removed from storage vessels. At month 12 an automatic in-line system was commissioned.

This is particularly the case with very large vessels requiring multiple batches of wort delivered over several hours. The ideal of wort collection is to ensure that all the yeast cells are exposed simultaneously to all the oxygenated wort. In practice this is not possible. In order to ensure that wort and yeast are intimately mixed, some brewers choose to dose yeast into the wort throughout the entire period of vessel filling. This is not advisable, particularly where the time to fill is lengthy. In this situation at the completion of fill the wort may contain a heterogeneous population of yeast exhibiting a continuum of physiology from active growth to early lag phase. In effect the result is under-pitching, since the yeast added latterly is unable to compete with that pitched initially.

Preferably all the yeast should be pitched over as short a time as possible and very early in wort collection. This minimises the risk of microbiological contamination and ensures a rapid start to fermentation. This approach also has perhaps unforeseen consequences. In this instance aerobic conditions can persist for a considerable period. It has been demonstrated that in yeast ATF1, the gene encoding for an alcohol acetyltransferase and an important source of beer esters, is repressed by molecular oxygen (Fujii *et al.*, 1997). Control of beer ester levels is commonly accomplished via selection of the initial oxygen concentration, thereby regulating yeast growth and by implication the availability of ester-forming substrates. However, since oxygen is capable of modulating the expression of genes coding for ester producing enzymes, it is clear that the time

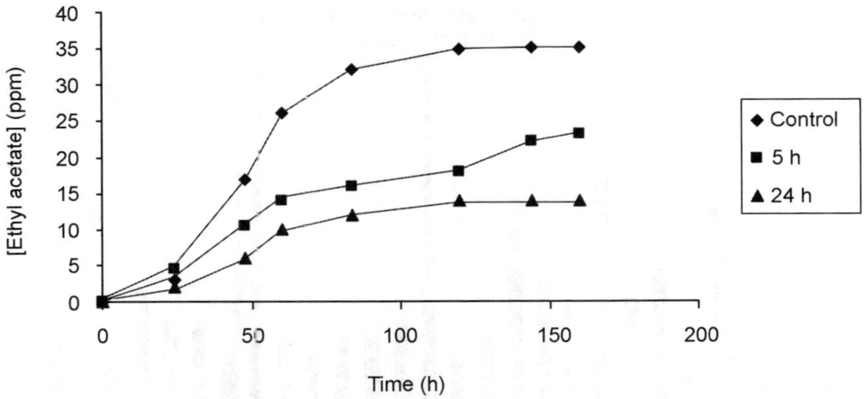

Fig. 11.5 Effect of post-collection oxygenation on the formation of ethyl acetate. Data were obtained from triplicate 8 hl pilot-scale fermentations of a 15° Plato lager wort pitched at 15 × 10^6 viable cells per ml. All fermentations were collected with an initial wort dissolved oxygen concentration of 15 ppm. Trial fermentations were sparged with pure oxygen via a sinter located in the cone for the times indicated.

for which yeast is exposed to aerobic conditions is also an important factor. Post-collection oxygenation can be used to reduce beer ester levels, as illustrated in Fig. 11.5. It is assumed that this is a consequence of molecular oxygen repressing the ATF1 gene, as described above.

11.3.2 Monitoring and controlling fermentation progress

In the majority of breweries control of fermentation remains an essentially passive process. Thus, care is taken to establish desired conditions at fermenter fill, as described in the previous section, after which fermentation rate is controlled via the application of cooling. Attemperation of whole tank farms is usually managed by a single microprocessor-controlled supervisory system such as ACCOS (Wogan, 1992). This will monitor the temperature of all the vessels under its jurisdiction and compare actual values with those expected for the particular stage in the process that each vessel has reached. Out-of-specification values cause an alarm to be raised to prompt remedial action. An important consideration of fermentation/conditioning tank farms is to ensure that sufficient refrigeration is available to match peak demand. Commonly this is not achieved.

In cylindroconicals attemperation is via external cooling jackets. Temperature is measured using platinum resistance-type probes conforming to BS 1904 class A or B and capable of an accuracy better than ±0.2°C. Multiple jackets are provided (Fig. 11.1), the use of which is dependent on the phase of fermentation or conditioning in unitank operations. During primary fermentation it is common to apply cooling to the top jacket only in an attempt to set up thermal gradients and thereby drive mixing of the vessel contents via convective currents in addition to the mechanical mixing provided by CO_2 evolution (Maule, 1976). For primary fermentation the temperature probe is usually located near the cone of the vessel and therefore distant from the cooling jacket. The efficiency or lack thereof of this approach is discussed in Section 11.4.

Measurement of wort specific gravity performed on samples removed periodically from the fermenter remains the most common method of monitoring the progress of primary fermentation. The plot of decrease of specific gravity with time can be compared with a pre-established standard profile, and deviation from non-ideal behaviour can be identified. Several commercial systems are available which allow automatic on-line measurement of wort specific gravity (Moller, 1975; Cumberland *et al.*, 1984; Dutton, 1990; Sugden, 1993, Hees and Amlung, 1997). The most successful of these devices rely on the measurement of differences in pressure at fixed locations within the vessel. These differences can be related to the specific gravity of the fermenting wort. For example, the system described by Hees and Amlung (1997) utilises two submerged pressure sensors, one placed in the cone and one near the surface of the liquid. The differential pressure between these allows computation of wort specific gravity and volume. A third sensor is located in the head space. Output from this provides compensation for changes in top pressure. The authors claimed a precision of $\pm 0.2\%$ compared with off-line laboratory analysis. The advantages of these in-line systems are that they provide continuous measurements and by implication rapid warning of non-ideal behaviour, and in addition they provide an output that is suitable for use in an interactive control system, typically one in which the rate of decline in specific gravity is linked to the vessel attemperation system. This allows primary fermentation rate to be modulated by changes in temperature such that a desired profile is followed.

The practice of monitoring fermentation progress via measurement of wort specific gravity (SG) owes much to the fact that brewers became used to determining sugar concentrations because of the requirements of some taxation systems. Changes in other related parameters may be used and indeed have been proposed, for example, rates of CO_2 evolution (Stassi *et al.*, 1991), exothermy (Ruocco *et al.*, 1980), fall in pH (Leedham, 1983), suspended viable yeast count (Carvell *et al.*, 2001) and ethanol formation in fermenter head space (Anderson, 1990). With the exception of the use of pH and yeast concentration, all these approaches have the advantage that they are non-invasive. This overcomes the potential problem of vessel heterogeneity such that measurement taken from a discrete point within the fermenter may not be representative of the whole. All these approaches provide outputs suitable for use in interactive control systems. However, including on-line SG determination, all provide little or no data during the early stages of the process. This is disadvantageous, since any interactive control system requires information as soon as possible in order that non-ideal behaviour can be quickly identified and remedial actions taken.

During the lag phase of fermentation the yeast consumes the initial charge of oxygen. The rate of oxygen uptake is a function of temperature, yeast concentration and yeast physiological state. In particular, there is a positive correlation between oxygen consumption rate and yeast sterol content It has been suggested that oxygen uptake rate could be measured in-line after the pitching yeast has been dosed into wort during fermenter fill (Boulton and Quain, 1987). Providing the yeast concentration and temperature are known, the oxygen consumption rate can be used to assess yeast physiological condition and thereby adjust wort oxygenation so as to select the optimum concentration for consistent fermentation performance. This approach has been tested at pilot scale but has yet to be applied in commercial fermentations.

Cylindroconical fermenters are usually filled only to 80–85% of their total capacity in order to leave sufficient freeboard to accommodate the foam head that is formed during active primary fermentation. This represents a reduction in potential productivity for each fermenter. Apart from loss of product, uncontrolled foaming also causes a loss of bittering components and therefore a lowering in the efficiency of hop utilisation. Antifoams may be used in sparing amounts where permitted, although care must be taken not to compromise the head potential of finished beer by overdosing. Use of a camera mounted in the top-plate of fermenters has been recommended (Wasmuht and Weinzart, (1999). This allowed direct observation of fermenting wort such that progress could be monitored. Such a system would be useful for monitoring and if necessary controlling foaming. A true interactive system for controlling foaming has been described by Ogane *et al.* (1999). This consists of a laser beam generator and sensor mounted in the top-plate of the vessel which when properly set up detects both the liquid surface and top of the foam head. By difference the foam height can be calculated and antifoam dosed in automatically, if required.

11.3.3 Identification of the end-point of fermentation

In order to maximise vessel utilisation efficiencies it is necessary to identify the end-point of fermentation as rapidly as possible. In the case of many modern rapid lager fermentations this is marked by the achievement of a minimum concentration of total vicinal diketones (VDK). After this the yeast can be cropped and the beer chilled and either retained for a further period of cold conditioning, if a unitank procedure is followed, or moved to a separate conditioning tank. Measurement of total VDK remains problematic, since in the majority of breweries analysis is performed in the laboratory. In large breweries with many fermenters these analyses may take several hours to perform. Delays of 8–12 hours between sampling and delivery of results are not uncommon.

This loss of fermenter productivity could be remedied by more timely VDK analysis. Little progress has been made in this area. Ideally, VDK would be measured automatically in the fermenter such that attainment of the desired threshold concentration would cause a step onto the next stage with the minimum of delay. No such sensors currently exist, undoubtedly because a number of significant technical problems would need to be overcome. Thus, it is likely that the sensor would analyse fermenter headspace gas since in this way a single system with a suitable manifold of piping could service a whole tank farm, providing that it was demonstrated that a defined relationship existed between VDK concentration in the beer and headspace gas. This would be a complex and difficult-to-clean arrangement. There is the additional problem that the sample would need to be treated in such a way that the pool of α-acetolactate precursor is converted into free diacetyl prior to analysis. A simpler alternative would be to develop a simple and rapid method of diacetyl measurement. A possible candidate is based on a change in fluorescence due to the reduction of NADH by a specific diacetyl reductase. It is suggested that this could be incorporated into an automatic fluorimeter which would not require the use of skilled operators and would be capable of producing a result in less than 1 hour (J. Carvell, personal communication).

11.3.4 Removal of the yeast crop and emptying the fermenter

Prior to emptying cylindroconical fermenters it is necessary to remove the yeast crop which has settled in the cone. This may be a manual operation based on visual inspection of the green beer as it passes a sight glass located in the cropping main. Automatic cropping systems using the capacitance yeast biomass monitor have been described (Boulton and Clutterbuck, 1993). Here viable yeast concentration is measured in-line during fermenter run-down. Depending on the measured yeast concentration, the process flow may be directed automatically to waste, to yeast storage tank or to the green beer centrifuge. Automatic cropping

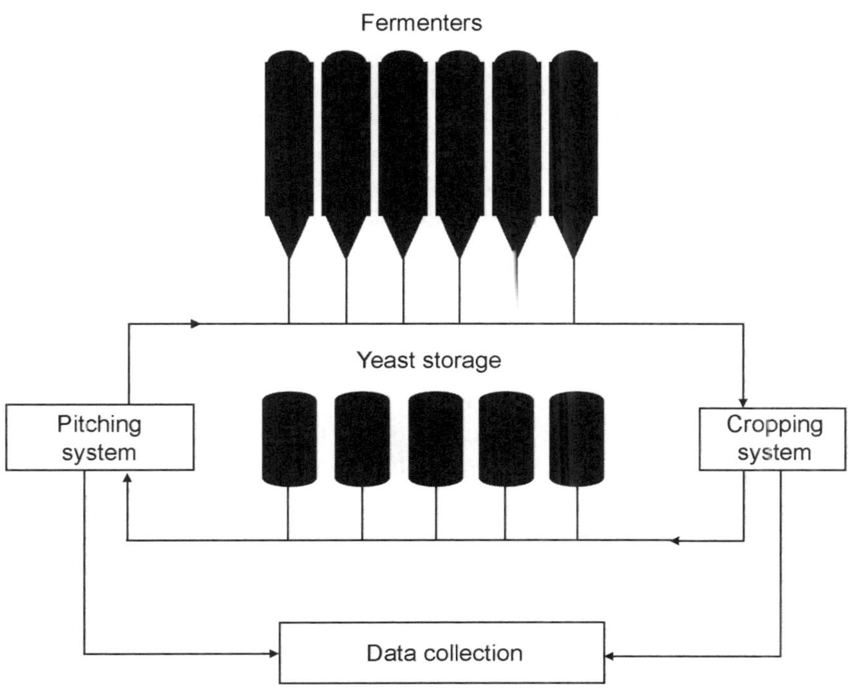

Fig. 11.6 Total in-line yeast management system based on automatic in-line radiofrequency capacitance biomass probes for control of yeast pitching and cropping.

Table 11.1 Outputs from integrated total yeast management system based on automatic in-line measurement of pitching and cropping

Parameter	Information available
Yeast identity	Which strain
Age	Generation number (number of fermentation cycles)
Yeast pitching	From which storage vessel To which fermenter
Yeast cropping	From which fermenter To which storage vessel
Fermentation efficiency	Calculated yeast growth (from pitch:crop ratio)
Yeast stock	Total yeast available for pitching Prompt for disposal at maximum generation number Prompt for propagation

and pitching systems may be used in combination (Fig. 11.6). With a suitable data logger the output from both yeast biomass probes provides a convenient method of tracking all yeast within the brewery at any given time. The data obtainable from a yeast management system such as this are shown in Table 11.1.

In a single-tank fermentation process after removal of the yeast crop the green beer is transferred to a conditioning tank. During transfer most of the remaining suspended yeast is removed by centrifugation. The efficiency of the latter has improved compared to earlier models such that beer losses have been reduced. Some of the cropped yeast slurry not needed for repitching purposes may be dosed back into the pre-centrifuge product stream. More efficient centrifuges which generate a relatively dry yeast cake are used by some to satisfy the supply of pitching yeast without the need for prior cropping. This simplifies fermenter run-down but presumably exposes the yeast to considerable mechanical stress. Shear damage to yeast by centrifugation has been reported and is associated with filterability problems and the release of haze materials and possibly foam-negative lipids into beer (Siebert *et al.*, 1987). In addition, possible changes in flocculence characteristics after centrifugation have been claimed (Harrison and Robinson, 2001).

An undoubted drawback of cylindroconical fermenters is the difficulty of selecting the proportion of the yeast crop most suitable for repitching. In addition, once settled into the cone, efficient attemperation of yeast is difficult. Cahill *et al.* (1999) observed thermal gradients in yeast crops such that there was a differential of 11°C between yeast near to cooling jackets and that located in the centre of the cone. To counteract this threat to yeast and beer quality, crops may be removed early and before cooling has been applied (Loveridge *et al.*, 1999). Yeast distribution during the time course of fermentation is discussed in more detail in Section 11.4.

During cropping it is usual to discard the first runnings, since these tend to be contaminated with high levels of trub. Predictably there is evidence of further heterogeneity in sedimented yeast crops due to the influence of gravity. Hodgson *et al.* (1999) observed a gradient of increasing cell age from the surface to the bottom of yeast cone crops, based on bud scar staining with the fluorochrome, calcofluor. This is explained as mean yeast cell age and diameter are positively correlated. Fractionation of the crop and repitching revealed that cells recovered from the middle to the top of the crop (middle-aged to young) performed better than larger/older cells. This suggests that more of the first cropped yeast should be discarded before retaining that destined for repitching. Powell *et al.* (2004) confirmed considerable heterogeneity in yeast cone crops in terms of both yeast cells and environmental variables such as pH, ethanol concentration and temperature. These authors observed that cells at the base of the crop were less healthy than those located elsewhere. Interestingly, highly flocculent and aged cells were located in the central region of the cone. This led the authors to suggest that caution should be exercised in selecting which portion of a crop should be retained for repitching purposes.

11.4 Yeast distribution in the fermenter

The distribution of yeast cells in cylindroconical vessels throughout fermentation is assumed as opposed to actually being measured. The majority of cylindroconicals rely on natural mixing by CO_2 evolution and thermal currents. Mechanical agitators are not usually fitted unless vessels are used in a unitanking process or where the yeast is exceptionally flocculent. The assumption is therefore that mixing is sufficiently good throughout primary fermentation to ensure homogeneity. Yeast sedimentation at the end of the process is a result of the disappearance of fermentable sugars and this is aided by crash cooling.

In this respect brewing fermenters are distinct from other biotechnology industries where provision of mechanical agitation is the norm. Previous work, at pilot or smaller scale, has shown that mechanical agitation results in acceleration of fermentation. For example, Boswell *et al.* (2002) reported that in the case of 500 ml fermentations in vessels fitted with Rushton-type turbines there was a threshold power input value of *ca.* 0.25 kW/m^2 below which there was no effect on fermentation performance. At values above the threshold, fermentation rate increased and attenuation times decreased. This power rating was claimed to be roughly equal to that generated by CO_2 evolution in a 400 m^2 commercial fermenter. In addition, ester formation was reduced and higher alcohols elevated compared to non-stirred controls. These effects were attributed to higher yeast metabolic rates as a consequence of enhanced mass transfer driven by the improved turbulence.

Similar heterogeneity due to lack of mixing has been observed in commercial-scale cylindroconical fermenters during fermentation of a high gravity lager wort (Boulton *et al.*, 2005). Both temperature and viable yeast concentration were monitored in a 1500 hl fermenter using eight radiofrequency capacitance probes immersed at different locations within the vessel. Throughout the first 12 h of the fermentation there was evidence of lack of mixing of yeast and wort. The vessel contents were thoroughly mixed between 12 h after pitching and up to the point at which the wort was approximately half attenuated. After this time the yeast began to settle into the cone. The temperature of the yeast in the cone increased approximately 3°C above the set point of 20°C. Crash cooling reduced the beer temperature to 3°C but the yeast in the cone to between 12 and 15°C, only.

The relative proportions of yeast in the cone and suspended in the body of the beer throughout the fermentation are shown in Fig. 11.7. Up until 50 h after the completion of fermenter fill, the yeast was evenly distributed throughout the vessel. After this time (*ca.* 60% wort attenuation) yeast began to settle into the cone until at 65 h (*ca.* 84% attenuation) half the total yeast was in the cone. This increased to 70% at 75 h (*ca.* 90% attenuation). After this time no further changes were observed.

Pitching of this vessel occurred over a short period just after the start of wort pumping. It may be assumed that wort flow into the vessel was relatively non-turbulent and this accounted for the lack of mixing during this early phase. The lack of mixing persisted until yeast growth commenced with concomitant CO_2 evolution. Yeast sedimentation was largely independent of the decrease in gravity and was not influenced by the application of cooling. Furthermore, the decrease in VDK concentration in the latter part of the fermentation was apparently catalysed

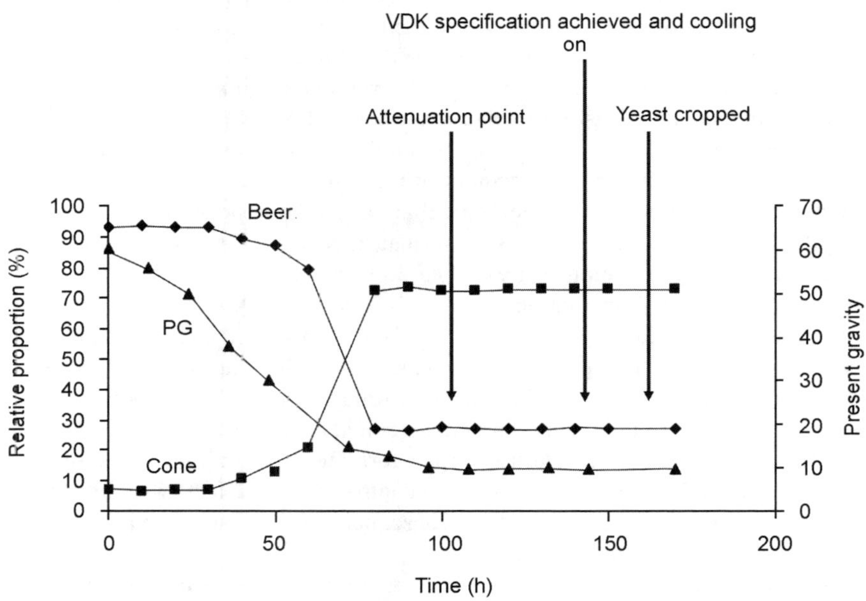

Fig. 11.7 Graphical representation of the relative proportions of the total yeast population in the cone and the body during the course of a 1500 hl cylindroconical lager fermentation. The times of achievement of attenuation gravity, VDK specification, application of chilling and yeast crop removal are indicated by arrows.

by the small proportion of yeast that remained suspended in the beer and not apparently by the majority which was settled in the cone. Some 80 h elapsed between the achievement of attenuation gravity and removal of the yeast crop. In view of the poor attemperation of the settled yeast and its apparent non-participation in diacetyl removal, this provides confirmation that in the interests of yeast quality early cropping would be recommended. Cooling of the large volume of beer in a jacketed cylindroconical is relatively inefficient. The data shown here would suggest that since cooling did not precipitate yeast settlement there would be no reason why the fermenter should not be emptied while still at fermentation temperature and cooled using a more efficient in-line refrigeration system. Typically this would reduce fermenter residence times by 12–24 h.

These results were surprising in that the yeast used was relatively non-flocculent. In the case of the combination of this yeast and wort fermented in this vessel, the underlying assumptions made regarding yeast behaviour and fermentation management were clearly misplaced. The fermenter used in these trials was fitted with a mechanical agitator mounted axially at the junction of the cone and the base of the cylinder. When this was switched from the start of wort collection up until the wort became half attenuated, there was an acceleration of both primary fermentation and reduction of VDK compared to unstirred controls (Fig. 11.8). The total time saving was approximately 12 h, apparently confirming that lack of homogeneity due to poor mixing of the fermenter was a rate-limiting factor.

Improved mixing of fermenters might be accomplished by retro-fitting of mechanical agitators or by gas rousing using evolved CO_2. Both of these may have undesirable side-effects. Mechanical agitators may cause hygiene issues and gas rousing could lead to flavour perturbations due to stripping of volatiles. In the case of new vessels, alternative and perhaps more elegant designs might be attempted. Andrews (1997) proposed the use of vertical cooling jackets where refrigerant could be applied to only part of the total circumference of the vessel. The author suggested that this approach would drive mixing via the generation of powerful thermal gradients in a manner that avoided intrusive mechanical agitation. The approach has yet to see commercial exploitation.

11.5 CO_2 collection

It seems inevitable that those brewers who do not yet collect CO_2 for reasons of economy will be forced to do so because of environmental considerations and the likelihood that uncontrolled emissions will attract a tax levy. In the recent past there have been quality issues with purchased CO_2, notably contamination with impurities such as benzene. CO_2 derived from brewery fermentation using properly managed and designed plant has the advantage of purity and a 'green' premium.

Recovery systems from the fermenter are initiated 8–12 h after pitching to allow time for air to be purged from the head space of the fermenter and the wort

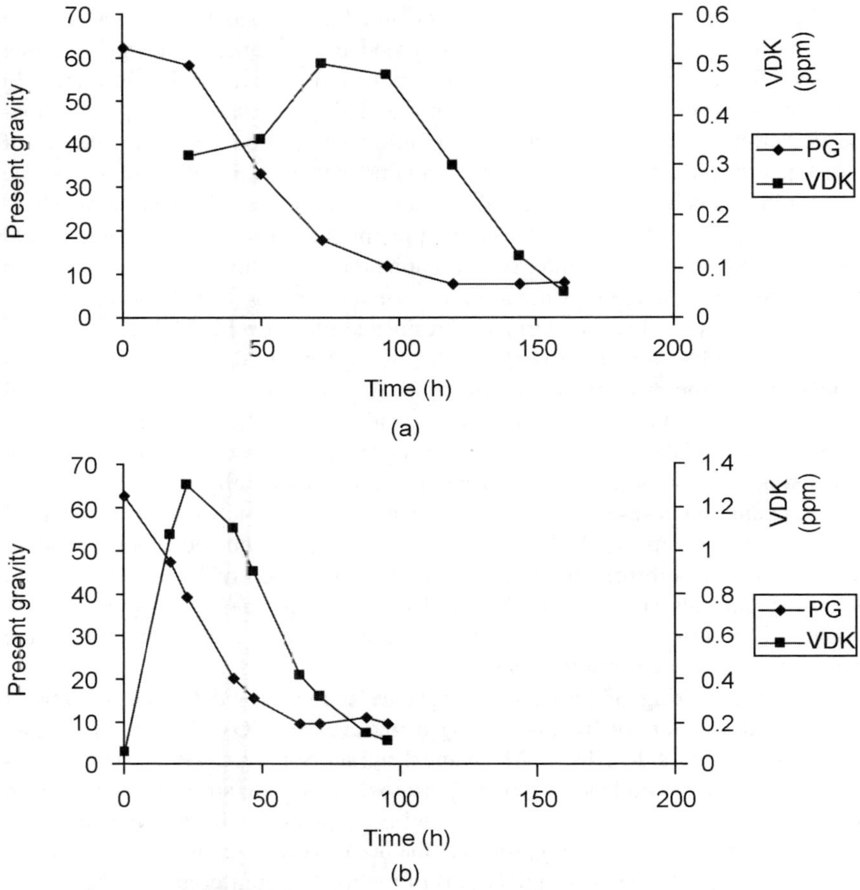

Fig. 11.8 Comparison of fermentation profiles of 1500 hl cylindroconical lager fermentations (a) with and (b) without mechanical agitation.

to become saturated with CO_2. This reduces actual yields to approximately 70% of theoretical. Purified collected CO_2 may be stored as a gas for use in the brewery. More usually it is liquefied. Plants such as this are capable of generating CO_2 with a purity of greater than 99.99%.

11.6 Modelling fermentation

The progress and outcome of fermentation are reliant on, amongst other things, two very complex variables, wort composition and yeast physiological state. The current inability to assess the significance of these two variables at the onset of fermentation explains why the majority of brewers still practise passive control strategies dependent upon precise regulation of those parameters amenable to manipulation. The level of inconsistency in both fermenter cycle times and the resultant product, common in many breweries, suggest that these passive control strategies fall a long way short of the ideal.

Modelling of brewery fermentation may overcome these problems (Marin, 1999). The intention of such models is to simulate the sum of events which together comprise fermentation and from this identify a few preferably easily measured parameters, the values of which are predictive of performance. Providing the prediction is accurate, setting and controlling the values of the relevant parameters should produce consistency of process and product. True interactive control of fermentation is possible by on-line measurement of chosen parameters and adjusting variables such as temperature and pressure in response to deviation from the pre-identified norm.

Two general types of model have been proposed. Kinetic types rely on a detailed knowledge of the sum of reactions that underpin yeast growth on wort and how this is influenced by environmental conditions. Providing the model is accurate, it can be used to identify otherwise hidden relationships between measurements made at the start of fermentation and the ultimate outcome. Self-teaching types, typified by the neural network, do not contain any actual models but instead rely on comparison of values of various inputs with actual outputs such that the modelling system learns by experience. As more data sets are acquired the network gains greater cumulative experience and the model gains in accuracy. Usually back-propagation neural networks are used in which actual and desired outcomes are compared and any discrepancy is used to fine-tune the individual weighting elements and increase overall accuracy.

Both approaches have advantages and disadvantages. Kinetic-based models are reliant on the accuracy of the underlying biochemical equations from which they are constructed. In many cases these have not been fully elucidated, in particular relationships between fermentation conditions and the development of important groups of yeast-derived beer flavour components. Neural networks avoid much of the initial complexity of kinetic-based models since they merely compare inputs and outputs. However, actual inputs which have been chosen for proposed models have often been parameters which are not usually available at the start of commercial fermentations, such as wort FAN.

Many published modelling systems have yet to be applied in commercial brewing and are based on laboratory-scale studies. For example, one group has derived models for the prediction of the formation of diacetyl, isoamyl alcohol, phenyl ethanol, isoamyl acetate, ethyl acetate and ethyl hexanoate. Models were based on measured formation of CO_2 at different ranges of temperature, pressure, wort concentration and initial yeast pitching rate (Titica *et al.*, 2000; Trelea *et al.*, 2001). The authors suggest that models could be used to predict when diacetyl specification would be achieved and the final concentrations of some important beer flavour compounds. In addition, by modulating pressure and temperature in response to CO_2 evolution rate, fermentation cycle time and some aspects of beer flavour could be controlled. In another report (Yuen and Austin, 2000) a combination of kinetic modelling for primary fermentation and a neural network model for assessment of diacetyl formation and dissimilation was used to predict the effect of temperature on fermentation performance at commercial scale.

It is difficult to quantify the value of such models. They are of academic interest and there is always the possibility that through their use new relationships might be identified which lead to practical control systems. At this stage it seems likely that their greatest strength lies in prediction as opposed to control. This is useful for production scheduling but possibly disguises the real requirement which is to identify and eliminate sources of inconsistency that fuel the need for prediction.

11.7 Continuous fermentation

The burgeoning of interest in continuous fermentation which occurred during the 1960s and 1970s did not result in widespread adoption. This situation remains largely unchanged, although the 1990s saw some resurgence of interest based on reactors containing immobilised yeast – sufficiently so, in fact, to justify an entire EBC symposium ('Immobilised yeast applications in the brewing industry', *EBC Monograph XXIV*, Espoo, Finland, 1995). A small number of research groups continue to develop new types of reactor and processes but these have not seen enthusiastic adoption by commercial brewers. At commercial scale there are small islands of zealous proponents of continuous fermentation set in a broad sea of indifference. The Dominion Group in New Zealand have operated with much success continuous fermenting systems at several of their breweries (Stratton *et al.*, 1994).

The advantages of continuous fermentation, namely rapidity and consistency, should be attractive to brewers producing large volumes of a single beer quality, as is the case with many of the international brands which currently have a large share of the total world market. However, batch fermentation continues to predominate, largely due to inertia. Thus, the case for change is not sufficiently compelling to justify scrapping existing fermentation plant and spending capital on a replacement continuous system.

A single-tank continuous fermentation system cannot be easily used for both primary and secondary fermentation. In order to produce a beer that is a match for the batch produced counterpart, it is necessary to separate primary and secondary fermentation. Thus, wort attenuation and VDK removal are essentially linear time-based processes which are not easily engineered to occur simultaneously and continuously.

Removal of diacetyl using a reactor containing immobilised yeast is a well-established process. In view of the need to improve the productivity of primary batch fermenters these represent an attractive proposition and are likely to see wider acceptance. Several such systems are described in the EBC symposium alluded to already. Prior to delivering the beer to the fermenter, it is necessary to ensure that the pool of the diacetyl precursor, α-acetolactate, is converted to

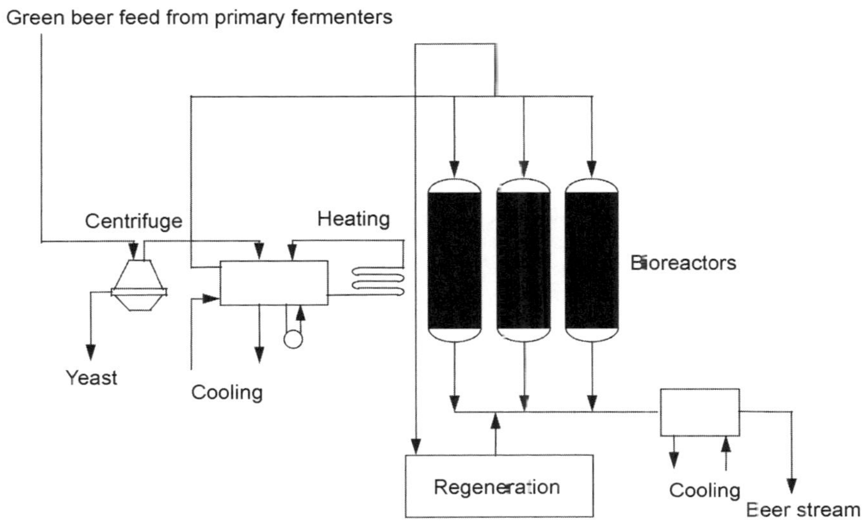

Fig. 11.9 Continuous immobilised yeast reactor for removal of VDK (after Pajunen and Gronqvist, 1994).

diacetyl. This represents a significant drawback, since this conversion is usually accomplished by a heat treatment. This is costly and needs careful control to avoid deterioration of beer. A continuous immobilised yeast maturation reactor used at commercial scale is shown in Fig. 11.9. In the system shown the beer is removed from primary fermentation and passed through a continuous centrifuge to remove most of the yeast. After this it is subjected to a heat treatment (90°C for 7 min), then cooled to 15°C and delivered into the immobilised yeast reactors. During the heat treatment dissolved oxygen levels must be kept as low as possible in order to avoid thermal damage to beer. The combination of anaerobiosis and heat favours the direct conversion of some α-acetolactate to acetoin and also serves to reduce microbial loadings. Residence time in the bioreactor is approximately 2 h. During this time free diacetyl is converted to acetoin and 2,3-butanediol. No changes in the concentrations of important flavour compounds such as esters and higher alcohols were observed. After passage through the reactor the beer is treated to achieve colloidal stability and filtered, as normal.

Continuous primary fermentation remains beset by technical problems. Attemperation is made difficult by the presence of the carrier and relatively high yeast concentration, CO_2 evolution disrupts some immobilisation carriers. Since primary fermentation is an ordered process, uptake of sugars and formation of flavour compounds may be altered compared to the batch process. In many cases these problems have been addressed by the use of multiple tanks, for example Kronlof and Virkajarvi (1999). These allow physical separation of different stages of the batch process and afford opportunities during transfer from one to the other to regulate temperature and remove CO_2.

Descriptions of refinements to immobilised reactors designed for both primary fermentation and diacetyl removal continue to appear in the brewing literature. These apparently resolve most of the problems associated with this approach, although usually at pilot or laboratory scale. It seems certain that eventually commercial systems capable of producing beers that are an organoleptic match for those made by a conventional batch process will be available. In the current competitive market these are unlikely to see adoption other than at greenfield sites.

11.8 High gravity brewing

High gravity brewing has become the most used strategy for maximising fermenter productivity for large-scale commercial brewing. Initial wort concentrations of 15° to 17° Plato are commonly used. The upper limits for high gravity brewing are set by the maximum wort concentration that can be delivered by the brewhouse without compromising efficiency and by the effects of highly concentrated worts on yeast and fermentation performance. At present the perceived negative effects on fermentation performance outweigh brewhouse considerations.

High gravity brewing is associated with altered levels of ester formation such that with very concentrated worts unbalanced flavours may arise, post-dilution, due to high levels of mainly ethyl acetate and isoamyl acetate. In addition, highly concentrated worts reportedly exert high levels of stress on yeast, resulting in low viability crops and excretion of proteases which may reduce concentrations of foam positive beer proteins (Stewart, 2001). These undesirable effects result in the upper limit of wort concentration being set at round about 16–18° Plato. In the interests of fermenter productivity it would be beneficial if this limit could be raised.

The adverse effects on yeast metabolism are usually ascribed as being due to a combination of two yeast stresses, namely high osmotic pressure (low water activity) and ethanol toxicity. A consensus is now arising that this explanation is only partially correct. Undoubtedly these stresses are real and have the potential to exert deleterious effects on yeast; however, these can be ameliorated by appropriate fermentation management and manipulation of wort composition.

Hammond *et al.* (2001), using a synthetic growth medium, demonstrated that maltose concentrations of up to 250 g/l, equivalent to nearly 28° Plato, did not inhibit fermentation rates in the initial stages. Inhibition was observed later and this was ascribed to ethanol toxicity. Furthermore, increasing osmotic pressure by the addition of sorbitol, a non-metabolised sugar alcohol, was not inhibitory and levels of glycerol, a marker of yeast osmotic stress, were not elevated. These authors concluded that the total nutritional status of the medium was crucial. In the work described, the medium had a free α-amino nitrogen content five-fold higher than that observed in a typical 16° Plato wort. The requirement to optimise wort composition for ultrahigh gravity fermentation has been noted by others. In particular, the need to use increased levels of divalent metal ions, especially Mg^{2+}, has received particular attention. Thus, Walker *et al.* (1996) concluded that supplementing high gravity worts with Mg^{2+} increased fermentation rates and reduced the inhibitory effects of ethanol. The ratio of magnesium to calcium was considered crucial.

These results suggest that further work is needed to optimise the composition of very high gravity worts. As is ever the case, caution should be exercised when manipulating wort concentrations via the use of sugar syrups. This is illustrated by the observations of Stewart (2001) that high esters associated with concentrated worts can be reduced by ensuring that the ratio of glucose to maltose is low. Esters levels may also be modulated by controlling oxygen exposure post-pitching as shown in Fig. 11.5. In addition, It seems likely that increasing fermenter productivity via high gravity brewing will require the use of nutritional supplements in order to safeguard yeast. This can be aided by the benefits of early cropping, as described in Section 11.4, thus removing yeast from the hostile environment of the fermenter as soon as possible.

11.9 Yeast physiology and fermentation performance

Yeast physiological condition and fermentation performance are intimately related. Two particular aspects of this subject have received considerable attention in recent years: firstly, an assessment of yeast stress responses and how these have been elicited by efforts to intensify the productivity of the fermentation process, secondly, identification of how inconsistent yeast handling can lead to variable-pitching yeast physiology and methods that have been developed to assess this variability.

11.9.1 Fermentation and yeast stress

As yeast progresses through the cycle of pitching, fermentation, cropping and storage it is subject to numerous distinct stresses. For example, during the storage phase the yeast is starved in the presence of high ethanol concentration and possibly subjected to mechanical shear forces. If acid-washed before repitching it is exposed to a pH shock. On pitching into the fermenter there is an abrupt shift from anaerobic to aerobic conditions, possibly resulting in oxidative shock. Simultaneously, the yeast shifts to conditions of high osmotic pressure, low water activity, high sugar concentration and possibly an increase in temperature of 10–15°C. As fermentation progresses there is a reverse transition from aerobiosis to anaerobiosis, high ethanol and dissolved CO_2 concentrations and high hydrostatic pressures in tall vessels. In late fermentation these stresses are supplemented by the onset of starvation, possibly at elevated temperature if much of the yeast has settled into the cone during warm conditioning.

Reports of the physiological responses of yeast to applied stresses are legion in the brewing literature. For example, in the proceedings of the 29th EBC

Congress which took place in Dublin in 2003 more than six papers are devoted to this subject. Similarly, the proceedings of the second conference on brewing yeast and fermentation performance published in 2003 (ed. K. Smart, Blackwell Science, Oxford, UK) devotes four consecutive contributions to yeast physiology, stress responses and brewery fermentation.

The practical significance of these studies is not clear. The global changes that occur in response to applied stresses are undeniable. The most revealing studies have used DNA micro-array technology to examine the response of the whole yeast genome to environmental transitions. Peng *et al.* (2003) reported that of 6200 genes examined during fermentation 1700 were induced more than four-fold and 2100 repressed more than four-fold. When yeast was subjected to individual stresses 900 genes were either up- or down-regulated. Interpretation of the genomic response to applied stress depends upon the standpoint that is taken. On the one hand the terminology of stress perhaps implies that yeast cells suffer damage when they are exposed to the conditions that pertain during fermentation and storage. The converse view might be that brewing yeast cells have a genome that is highly adaptable and able to express or repress genes quickly in response to environmental changes such that they are able to survive or even prosper in a variety of niches.

Beer is the spent medium that remains after the growth of yeast on wort. The conditions that yeast experiences during this growth phase, whether they are stressful or not, are presumably essential if fermentation is to proceed as desired and beer is to have the required composition. Conditions only become truly stressful in the real sense of the word when the viability of the crop is compromised. This is of course easily detectable by simple staining techniques and microscopic examination. As already discussed, process intensification via acceleration of fermentation rate and high gravity brewing coupled with inappropriate yeast handling are all individually capable of causing yeast damage. It is essential to identify these real causes of yeast stress and eliminate them from the process.

11.9.2 Assessing yeast physiological condition

Enthusiasm for the development of new methods for assessing yeast viability and physiological condition (vitality tests) continues unabated; for reviews see Lentini (1993), Bendiak (2000) and Smart (2001). The tests for physiological status probe many aspects of yeast metabolism and structure that would be expected to be relevant to fermentation performance. These include markers of stress, yeast age, cell wall structure (marker for flocculence), membrane function, reserve carbohydrates and fermentative ability.

The practical utility of these tests has yet to be proven. In the majority of commercial breweries assessment of yeast quality uses the procedure of determination of viability via staining with methylene blue. The result of this test produces two useful outputs. Firstly, it allows correction for viability in manually derived pitching rate calculations. Secondly, it is itself a vitality test since yeast with a low viability is likely to be of compromised quality and rejected for use in fermentation. Several alternatives to the methylene blue method which claim improved accuracy have appeared in the literature; for example see White *et al.* (2003). The best of these new methods are those based on fluorescent dyes suitable for use in apparatus capable of automatic cell enumeration. This eliminates operator error and probably increases the productivity of individual laboratory workers.

Do vitality tests add value to simple assessments of viability? In a commercial brewing laboratory using current passive fermentation control regimes, the answer is probably not a great deal. Tests which produce as output a measure of fitness to pitch are probably of no greater use than viability tests, as discussed already. Tests which offer the most promise are those that probe those aspects of yeast physiology which influence the optimum pitching rate and requirement for oxygen which will produce consistent fermentation performance and product. This approach has not been generally taken up; instead, more effort has been placed in controlling yeast handling such that inconsistencies are eliminated and by inference the need for testing.

At present vitality tests probably have most value as research tools as opposed to routine production aids. Thus, they can be used to underpin developments which, for example, seek to increase process productivity. If it is suspected that use of an altered process or wort composition is compromising yeast quality, use of the appropriate cellular probe might identify the cause and thereby suggest a remedy. From this point of view, it is now well recognised that brewing yeast populations are heterogeneous. Bulk vitality tests are therefore of limited value since they only provide an average assessment of the population being tested. Flow cytometers overcome this problem by separating and counting populations and with appropriate dyes assess the physiological state of individuals. Flow cytometry has been applied to brewing yeast and appears to offer great promise as a powerful tool for assessing populations of cells (Hutter and Lange, 2003).

11.10 Future trends

Polarisation within the brewing industry between many small traditional producers and a few very large multinational operators seems likely to continue to become more entrenched. The larger brewers are likely to drive innovation in fermentation and other areas because of the need to compete on cost. This is likely to fuel a greater need to improve productivity via the use of even more concentrated worts fermented in as a short time as possible, as discussed already. Since these very large batch sizes will represent a considerable investment, there will be a need for the greater use of reliable on-line sensors linked to control systems. A further driver for automation will be the promise of reducing manning levels. In order to accomplish productivities greater than those achievable at present, it will probably be necessary to redesign current fermenters in order to eliminate some of their undesirable characteristics, for example improved attemperation and mass transfer. It is already the case that the mainstream brewing industry has adopted an iconoclastic position regarding some of the cherished beliefs of traditional lager producers. It seems likely that this will continue, possibly with borrowing from other industries such as fuel alcohol manufacture.

Undoubtedly, the technical problems associated with continuous fermentation will be resolved. These developments will be augmented by the introduction of brewhouses with increased productivity and perhaps capable of continuous wort production. In the event of a new brewery build, a continuous process would merit serious consideration. However, since the beer market is flat or even reducing in many countries, the likelihood of new brewery builds is perhaps small.

Developments in yeast technology are likely to be modest. The use of dried yeast for contract or occasional brewing is becoming increasingly popular. This trend may continue and possibly some brewers may choose to purchase dried yeast for mainstream brands or even generate their own, in-house, as an alternative to propagation. Consumer resistance means that at present there is no likelihood of using genetically modified strains. This seems an obvious way of removing undesirable traits or introducing desirable ones. Indeed many strains which possess useful characteristics have been produced and are already available for use. In the present climate they will not be. Further developments, therefore, are likely to be restricted to a greater understanding of yeast biochemistry. In this regard techniques which allow assessment of the response of whole genomes and proteomes to changes in environment are likely to prove to be very powerful tools.

11.11 Sources of further information

It is beyond the remit of this chapter to provide sources for further information of mainstream yeast research. For this the mainstream academic literature should be searched using appropriate key words. Nevertheless, the series of books *The Yeasts* (eds A.H. Rose, A.E. Wheals and J.S. Harrison, Academic Press, London, UK), the second edition of which now runs to six volumes, provides an excellent resource for all matters concerning yeast. For cutting-edge research the Internet continues to grow in stature, and to facilitate this the various Internet search engines are invaluable. The same method can provide rapid access to a vast archive of information regarding developments in brewing science and engineering.

A comprehensive description of brewing fermentation and yeast biochemistry may be found in Boulton and Quain (2001). The proceedings of the biennial conferences on the same subject held at Oxford, UK (published under the banner *Brewing Yeast and Fermentation Performance*, ed. K.A. Smart, Blackwell Science, Oxford, UK) provide detailed up-to-date coverage of many recent developments. In addition, they give a platform where students can present preliminary results of studies which might or might not reach the point of commercial exploitation. Similarly, many new developments are presented at the major brewing conferences such as the EBC Congresses; the proceedings of the latest, held at Prague in 2005, are currently in press. Other mainstream brewing journals such as the *Journal of the Institute of Brewing*, the *Journal of the American Society of Brewing Chemists* and the *Master Brewer of the Americas Technical Quarterly*, as well as the rest of the more populist trade literature, provide coverage of modern developments.

11.12 References

ANDERSON, R.G. (1990), 'Aspects of fermentation control', *Ferment*, 3, 242–249.

ANDREWS, J.M.H. (1997), 'Temperature profiles in dual purpose tanks', *Proc. 6th Conv. Inst. Brew. (Central and South African Section), Durban*, 47–54.

BENDIAK, D. (2000), 'Review of metabolic activity tests and their ability to predict fermentation performance', in *Brewing Yeast and Fermentation Performance*, ed. K.A. Smart, pp. 34–45, Blackwell Science, Oxford, UK.

BOSWELL, C.D., NIEWNOW, A.W. and HEWITT, C.J. (2002), 'Studies on the effect of mechanical agitation on the performance of brewing fermentations: fermentation rate, yeast physiology and development of flavour compounds', *J. Am. Soc. Brew. Chem.*, 60, 101–106.

BOULTON, C.A. and CLUTTERBUCK, V.J. (1993), 'Application of a radiofrequency permittivity biomass probe to the control of yeast cone cropping', *Proc. 24th Cong. EBC, Oslo*, 509–516.

BOULTON, C.A. and QUAIN, D.E. (1987), 'Yeast, oxygen and the control of brewery fermentation', *Proc. 21st EBC Cong., Madrid*, 401–408.

BOULTON, C.A. and QUAIN, D.E. (2001), *Brewing Yeast and Fermentation*. Blackwell Science, Oxford, UK.

BOULTON, C.A., WILSON, C., PETERS, A. and WRIGHT, J. (2005), 'An investigation into the distribution of viable yeast mass and temperature variation in cylindroconical vessels during fermentation', *Proc. 30th EBC Cong., Prague*, in press.

CAHILL, G., WALSH, P.K. and DONNELLY, D. (1999), 'A study of thermal gradient development in yeast crops', *Proc. 27th EBC Cong., Cannes*, 695–702.

CARVELL, J., TODD, R., CUNNINGHAM, S. and YONEZAWA, T. (2001), 'Real time monitoring of viable yeast concentration in production fermenters using novel RF impedometric probes', *Proc. 28th Cong. EBC, Budapest*, 79.

CUMBERLAND, W.G., MACDONALD, D.M. and SKINNER, E.D. (1984), 'Automated fermenter control at Moosehead Breweries Ltd', *MBAA Tech. Quart.*, 21, 39–45.

DUTTON, J. (1990), 'FV control via real time SG monitoring', *Brew. Dist. Int.*, May, 20–21.

FUJII, T., KOBAYASHI, O., YOSHIMOTO, H., FURUKAWA, S. and TAMAI, Y. (1997), 'Effect of aeration and unsaturated fatty acids on expression of the *Saccharomyces cerevisiae* alcohol acetyltransferase gene', *Appl. Environ. Microbiol.*, 63, 910–915.

HAMMOND, J.R.M., DAVIS, D., LEE, M. and STOREY, K. (2001), 'Does osmotic pressure affect yeast performance in high gravity fermentations', *Proc. 28th EBC Cong., Budapest*, 33.

HARRIS, C.M. and KELL, D.B. (1986), 'The estimation of microbial biomass', *Biosensor*, 1, 17–84.

HARRIS, C.M., TODD, R.W., BUNGARD, S.J., LOVITT, S.J., MORRIS, J.G. and KELL, D.B. (1987), 'Dielectrical permittivity of microbial suspensions at radiofrequencies, a novel method for the real time estimation of microbial biomass', *Enz. Microb. Technol*, 9, 181–186.

HARRISON, S.T.L. and ROBINSON, A. (2001), 'Disk stack centrifugation and fermentation performance', *Proc. 8th Conv. Inst Brew. (Africa Section), Sun City, South Africa*,157–164.

HEES, M. and AMLUNG, A. (1997), 'Automation of fermentation using continuous extract content measurement in the cylindroconical fermenter', *Brauwelt*, 37, 1549–1551.

HODGSON, J., PINDER, A., CARLEY, B.J. and DEANS, K. (1999), 'Effect of cone cropping and serial re-pitch on the distribution of cell ages in brewery yeast', *MBAA Tech. Quart.*, 36, 175–177.

HUTTER, K.J. and LANGE, C. (2003), 'Flow cytometry: a new tool in brewing technology', in *Brewing Yeast and Fermentation Performance*, ed. K.A. Smart, pp. 169–173, Blackwell Science, Oxford, UK.

KRONLOF, J. and VIRKAJARVI, I. (1999), 'Primary fermentation with immobilised yeast', *Proc. 27th Cong. EBC, Cannes*, 761–770.

LEEDHAM, P.A. (1983), 'Control of brewery fermentation via yeast growth', *Proc. 19th Cong. EBC, London*, 153–160.

LENTINI, A. (1993), 'A review of the various methods for available for monitoring the physiological status of yeast, yeast viability and vitality', *Ferment*, 6, 321–327.

LOVERIDGE, D., RUDDLESDEN, J.D., NOBLE, C.S. and QUAIN, D.E. (1999), 'Improvement in brewery fermentation performance by 'early' yeast cropping and reduced yeast storage time', *Proc. 7th Inst. Brew. Conv. (Africa Section), Nairobi*, 95–99.

MARIN, M.R. (1999), 'Alcoholic fermentation modelling: current state and perspectives', *Am. J. Enol. Vitic.*, 50, 166–178.

MAULE, D.R. (1976), 'Cylindroconical tanks – a study of cooling jacket performance', *The Brewer*, May, 140–144.

MOLLER, N.C. (1975), 'Continuous measurement of wort/beer extract in a fermenter', *MBAA Tech. Quart.*, 12, 41–45.

OGANE, O., YOKOYAMA, F., KATSUMATA, T., TATEISI, Y. and NAKAJAMA, I. (1999), 'Measurement of foam height in cylindroconical fermenters using a laser-based distance sensor', *MBAA Tech. Quart.*, 36, 155–157.

PAJUNEN, E. and GRONQVIST, A. (1994), 'Immobilised yeast fermenters for continuous lager beer fermentation', *Proc. 23rd IOB Conv. (Australia and New Zealand Section), Sydney*, 101–113.

PENG, X., BOWER, P., BARNEY, M., NAVARRO, A., SEABROOK, J.R. and RYDER, D.S. (2003), 'Patterns of gene expression during fermentation and in response to stress', *Proc. 29th Cong. EBC, Dublin*, 45.

POWELL, C.D., QUAIN, D.E. and SMART, K.A. (2004), 'The impact of sedimentation on cone heterogeneity', *J. Am. Soc. Brew. Sci.*, 62, 8–17.

RUOCCO, J.L., COE, R.W. and HAHN, H.W. (1980), 'Computer assisted exotherm measurement in full-scale brewery fermentations', *MBAA Tech. Quart.*, 17, 69–76.

SIEBERT, K.J., STENROOS, L.E., REID, D.S. and GRABOWSKI, D. (1987), 'Filtration difficulties resulting from damage to yeast during centrifugation', *MBAA Tech. Quart.*, 24, 1–8.

SMART, K.A. (2001), 'Biomarkers of yeast condition predicting fitness to ferment', *Proc. 6th Brew. Conv., Inst. Guild Brew. (Africa Section), Sun City, South Africa*, 138–149.

STASSI, P., GOETZKE, G.P. and FEHRING, J.F. (1991), 'Evaluation of an insertion thermal mass flowmeter to monitor carbon dioxide evolution rate in large scale fermentations', *MBAA Tech. Quart.*, 28, 84–88.

STEWART, G.G. (2001), 'Fermentation of high gravity worts its influence on yeast metabolism and morphology', *Proc. 28th EBC Cong., Budapest*, 36.

STRATTON, M.K., CAMPBELL, S.J. and BANKS, D.J. (1994), 'Process developments in the continuous fermentation of beer', *Proc. 23rd IOB Conv., Asia Pacific Section, Sydney*, 96–100.

SUGDEN, S. (1993), 'In-line monitoring and automated control of the fermentation process', *Brewers' Guardian*, June, 21–32.

TITICA, M., LANDAUD, S., TRELEA, I.C., LATRILLE, E., CORRIEU, G. and CHERUY, A. (2000), 'Modelling of the kinetics of higher alcohol and ester production based on CO_2 emission with a view to control beer flavour by temperature and top pressure', *J. Am. Soc. Brew. Chem.*, 58, 167–174.

TRELEA, I.C., LANDAUD, S., LATRILLE, E. and CORRIEU, G. (2001), 'Prediction of confidence limits for diacetyl concentration during beer fermentation', *J. Am. Soc. Brew. Chem.*, 59, 77–87.

WALKER, G.M., BIRCH, R.M., CHANDRASENA, G. and MAYNARD, A.I. (1996), 'Magnesium, calcium, and fermentative metabolism in industrial yeast', *J. Am. Soc. Brew. Chem.*, 54, 13–18.

WASMUHT, K. and WEINZART, M. (1999), 'Observing fermentation with the help of a new control system referred to as *TopScan*', *Brauwelt Intenational*, 17, 512–513.

WHITE, I.R., RICHARDSON, K.E., SCHIEWE, A.J. and WHITE C.E. (2003), 'Comparison of yeast viability/vitality methods and their relationship to fermentation performance', in *Brewing Yeast and Fermentation Performance*, (2nd edn), ed. K.A. Smart, pp. 138–148, Blackwell Science, Oxford, UK.

WOGAN, R. (1992), 'Process control and MIS the ACCOS family', *Brew. Dist. Internat.*, August, 20–21.

YUEN, D. and AUSTIN, P. (2000), 'Fermentation modelling in commercial beer manufacture, including temperature effect', in *Brewing Yeast and Fermentation Performance*, ed. K.A. Smart, pp. 177–188, Blackwell Science, Oxford, UK.

12

Accelerated processing of beer

I. Virkajärvi, VTT Technology, Finland

12.1 Introduction

The brewing industry has evolved during recent years and now consists mainly of large companies with large production capacities producing large batches. Current batch sizes range between 6000 and 10 000 hl. As lagering is the most time-consuming part of the brewing process, tank volume requirements are very large. Therefore any acceleration of lagering and maturation is of great economical and logistical importance.

Due to historical development the terms used for the process of diacetyl reduction are many: lagering, maturation, storage and conditioning. In this chapter the definitions from the *EBC Manual of Good Practice Fermentation and Maturation* (1999) are used, though the definitions may overlap depending on country and technology. *Lagering* consists of secondary fermentation, maturation and stabilization. *Secondary fermentation* refers to the process following the main fermentation in which most of the yeast is removed from the beer. At this stage there are some fermentable carbohydrates present and the aim is to reduce undesirable flavour compounds to acceptable levels. *Maturation* is a process in which no utilization of carbohydrates occurs. During *stabilization*, haze-forming materials are removed from beer so that it has the desired shelf-life. In a modern cylindroconical fermentation vessel brewery, the distinction between these processes is somewhat difficult to determine. The properties of beer after the main fermentation and after the secondary fermentation are given in Table 12.1.

The aims of traditional lagering are (Delvaux, 1996) natural carbonation of beer, precipitation of haze-forming complexes, sedimentation of yeast, stripping of unwanted volatiles (H_2S, DMS), reduction of aldehydes and ketones, release

Table 12.1 The properties of beer at the end of the main fermentation and at the end of the secondary fermentation

	End of main fermentation	End of secondary fermentation
Fermentable extract (°P)	1	0.5
Temperature (°C)	5–7	–2
Yeast concentration (10^6 cells/ml)	15–25	<5
Carbon dioxide (g/dm^3)	>2.5	>5
Diacetyl (mg/dm^3)	<0.15	<0.05
Oxygen (mg/dm^3)	<0.1	<0.1

Source: *EBC Manual* (1999).

of amino acids, phosphates, fatty acids and sulphur compounds from yeast, and 'marriage' of all remaining components to create a good mouthfeel.

During lagering many undesirable flavour and aroma compounds are reduced to acceptable levels. The most critical of these compounds is diacetyl There are many reactions known to take place during lagering other than those related to diacetyl reduction (see Section 12.5).

12.2 Diacetyl reduction

Diacetyl is considered an off-flavour in lager beers, creating a buttery taste. The removal of diacetyl is the critical element when considering 'maturity' of lager beer. Diacetyl may be part of the flavour of some beer brands, typically in some ales, but usually in lager beer it is an off-flavour.

12.2.1 Current practice

In current brewing of lager beer the temperatures – and times – are brewery dependent and a rather large variation is found. In general terms, pitching is performed at 7–8°C and then the temperature is allowed to rise to 10–11°C (some breweries go to even 14–15°C). At the end of primary fermentation the beer is cooled down to 4–5°C for secondary fermentation (and maturation). At the end of secondary fermentation the temperature is further reduced to near or just below 0°C.

Secondary fermentation in single-vessel production starts when fermentable extract has reached a desired level. Then the temperature of the cone of the cylindroconical vessel is lowered from the fermentation temperature (12–14°C) to around 5°C and yeast starts to settle as the rate of carbon dioxide evolution decreases. Quite often the head pressure is allowed to increase (0.3–0.5 bar) in order that there is the correct level of dissolved carbon dioxide in the green beer. The low temperature of the cone allows the yeast to stay in a good metabolic condition while the cylindrical part of the vessel is higher in temperature, causing more rapid diacetyl reduction. When the yeast has settled it is removed through the bottom valve. After removal of the yeast, which should occur about 24 hours after cooling of the cone, beer is allowed to stay in the vessel until the desired level of diacetyl is reached. Then the whole content of the vessel is slowly brought to 5–7°C. The maturation period in this temperature is usually 2–8 days (*EBC Manual*, 1999). Yeast is removed for a second time before cooling the beer to the stabilization temperature (–1°C). This cold stabilization temperature is held for 2–3 days before final yeast removal and subsequent filtration.

12.3 New techniques

12.3.1 Diacetyl reactions

The increased understanding that has accumulated during the last decades has facilitated the use of new technologies to speed up lagering. The following is a short summary of reactions that are related to formation and reduction of diacetyl and other vicinal diketones.

Diacetyl (butane-2,3-dione) and its chemical homologue pentane-2,3-dione with their precursors are called vicinal diketones. They are not the only vicinal diketones found in beer, but the most important ones. Diacetyl is more significant than its pentane homologue in beer. The origin of these two compounds in beer is yeast amino acid metabolism. Diacetyl is formed as a side product of valine biosynthesis, and pentane-2,3-dione as a side product of isoleucine biosynthesis (Villanueba *et al.*, 1990).

The routes are schematically presented in Fig. 12.1. Yeast utilizes amino acids sequentially (Pierce, 1987). Valine, which is a Group II amino acid, is synthesized, while yeast uses the Group I amino acids. The need for synthesis applies when the Group II amino acids are exhausted later in fermentation: once again valine must be synthesized. This leads to the normally observed two-peak

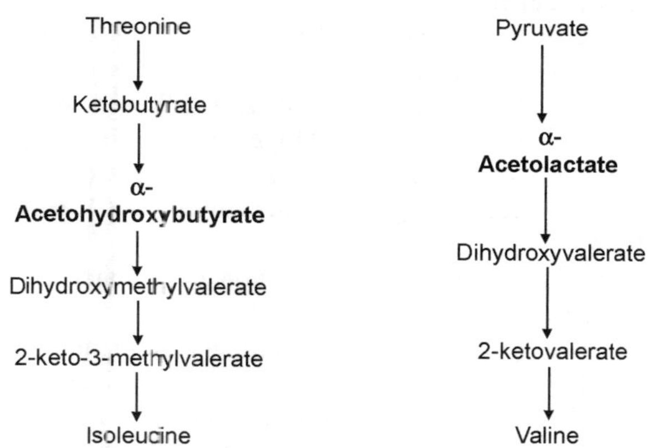

Fig. 12.1 The formation of vicinal diketone precursors (in bold) during beer fermentation.

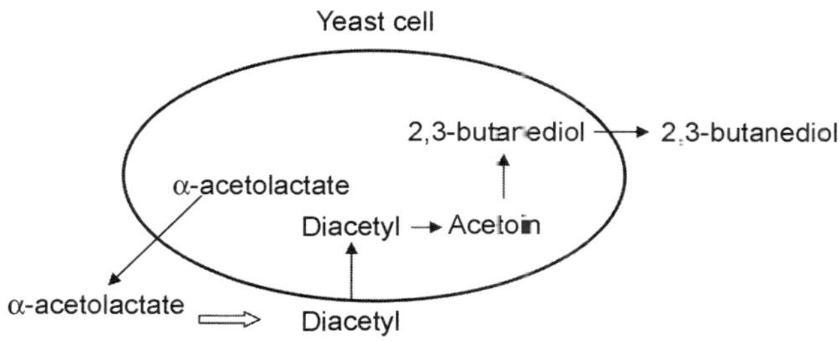

Fig. 12.2 The reduction of vicinal diketones during beer fermentation.

appearance of total diacetyl in beer fermentation. The upper part of the biosynthetic pathway functions faster than the lower part: alpha-acetolactate synthetase, the synthesis of which is controlled by ILV2 genes, is more efficient than the reducto isomerase enzyme which is controlled by ILV5 genes (Dillemans *et al.*, 1987; Villanueba *et al.*, 1990). The discrepancy in rates leads to leakage of alpha-acetolactate and alpha-acetohydroxybutyrate from mitochondria and yeast cells (Penttilä and Enari 1991).

Once outside the cell, alpha-acetolactate is chemically decarboxylated to diacetyl. This is the slowest reaction, mostly because the fermentation temperature is low. After the decarboxylation step, yeast takes up diacetyl and reduces it to acetoin (Fig. 12.2).

The new techniques can be divided into two categories: those that do not use genetic engineering of the yeast (or barley) and those that do. Of course other divisions can be made, too. Consumers may not be comfortable with use of genetic engineering techniques. For some insights see Caporale and Monteleone (2004).

12.3.2 Techniques not employing genetic engineering

Warmer temperatures

The idea of using warmer temperatures in lagering rather than traditional temperatures, very near 0°C is by no means a new one (Linko and Enari, 1967; see Fig. 12.3). This accelerated method produces beer with the same analytical properties as traditionally produced beer (Pajunen and Enari, 1978).

There exists, of course, an upper limit for lagering temperature above which beer quality will suffer due to, among other things, yeast autolysis. Masschelein (1986) showed that the amounts of caproic and caprylic acids in beer increase faster at higher temperatures. The fatty acids have low taste threshold values and give the beer goaty, soapy or harsh taste (Vinh *et al.*, 1981). Therefore yeast removal from the tank must be timed properly.

CO_2 purge/mixing

The level of diacetyl was found to be lower during the main fermentation stage when the fermenting wort was circulated by a peristaltic pump. The 12-litre

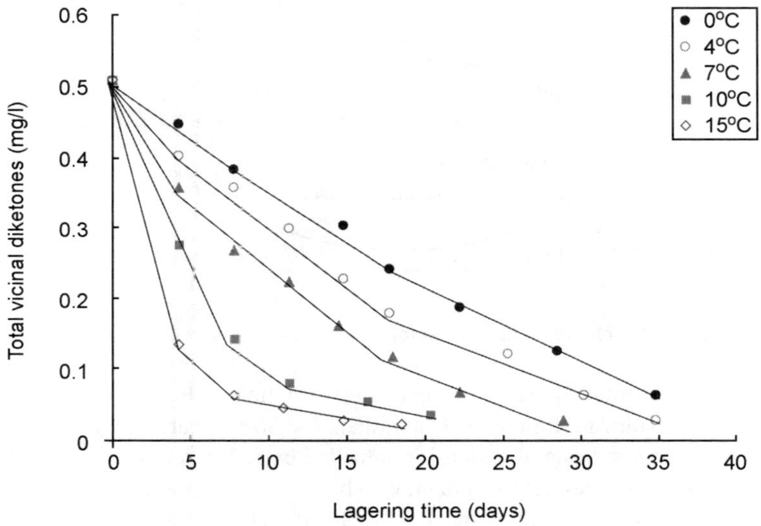

Fig. 12.3 The reduction of total diacetyl during lagering at different temperatures (redrawn from Linko and Enari, 1967).

fermentation vessel was agitated by circulating wort at the rate of 40 ml/min (Pajunen and Mäkinen, 1975). The maximum vicinal diketone concentration in agitated fermentation was only 60% of that in non-agitated fermentation. By the end of main fermentation vicinal diketones were at a level of only 50% of that in non-agitated fermentation. The authors found increased yeast growth in the agitated fermentations, which is very probably due to increased oxygenation due to the silicone tubing used.

Lourenço *et al.* (1971) used carbon dioxide washing to reduce storage time (lagering). The best results were achieved when CO_2 washing was performed a few days after the main fermentation. It was found that the lagering time could be reduced using the same yeast strains without decreasing the flavour stability.

Continuous diacetyl reduction

Process biotechnology has been advanced enormously by science-oriented and innovative brewmasters. Continuous fermentation is an innovation; application of immobilization is another. Immobilization means limiting free movement; enzymes or cells, for example, are bound to a defined space. Immobilization has been used for a long time in producing vinegar (Mitchell, 1926), but the first technical papers using immobilized enzymes on a laboratory scale appeared in the 1950s (Levin *et al.*, 1964). The first immobilized enzyme process on an industrial scale was the production of L-amino acids in 1969 (Sato and Tosa, 1993).

Christensen (1997) defined disruptive technological change as an event that market-leading, profitable and well-managed companies fail to see and react to and as a consequence lose market dominance. Typically the technology associated with a disruptive change originally has a lower performance than existing technologies, but its faster rate of improvement rapidly changes the situation. Products manufactured using new technologies have features that customers value: they are cheaper, simpler, smaller and more convenient to use. It remains to be seen whether continuous diacetyl reduction is a disruptive technological change.

Baker and Kirsop (1973) described a continuous immobilized diacetyl reducing process as early as 1973. In this process yeast-free beer was heated to speed up alpha-acetolactate decarboxylation to diacetyl, cooled down and then fed into a reactor where yeast was immobilized on kieselguhr. Earlier Narziss and Hellich (1971, 1972) had reported a speeded-up process for beer fermentation and maturation, but their process lacked the heat treatment stage.

In Finland the basic concepts of the process were applied in the 1980s (Pajunen *et al.*, 1987) and after laboratory and pilot scale development phases it was applied on an industrial scale in 1990 (Pajunen, 1995). The present capacity at this particular Finnish brewery is 1 million hectolitres per year (Pajunen and Gronquist, 1994). There is another Finnish brewing company also using this technique, see below (Hyttinen *et al.*, 1995).

In continuous diacetyl removal, the main fermentation stage is a normal batch fermentation process. After the main fermentation yeast is removed from beer by centrifugation. Then the beer is heat treated at 64–90°C for 7–20 minutes during which alpha-acetolactate is decarboxylated partly to diacetyl and partly directly to acetoin (Kamiya *et al.*, 1993, Pajunen *et al.*, 1989). The beer is then cooled to 15°C and pumped through a reactor where yeast is immobilized on a carrier. The residence time in the reactor is 2 hours. Thus the diacetyl removal is achieved in just over 2 hours. The process is presented schematically in Fig. 12.4.

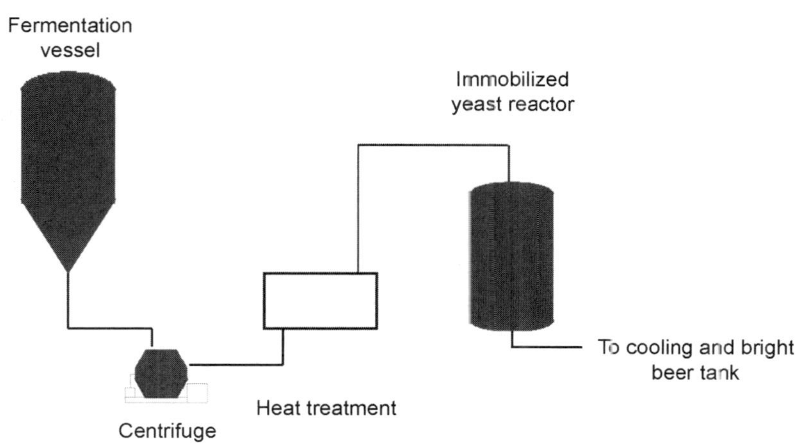

Fig. 12.4 Continuous diacetyl reduction process with immobilized yeast (redrawn from Pajunen, 1995)

The aim of centrifugation is to remove the yeast from beer as completely as possible (under 10^4 cells/ml) to avoid yeast autolysis in the subsequent heat treatment. This also reduces the beer losses in the surplus yeast. Another requirement for successful heat treatment is avoidance of oxygen pick-up (Inoue *et al.*, 1991). Heat treatment is for 7 minutes at 90°C (Pajunen, 1995) after which the beer is cooled to 15°C and fed into the upper part of the immobilized yeast reactor.

The yeast bound on the carrier rapidly reduces the remaining diacetyl into acetoin, but also ferments the remaining sugars and reduces other carbonyl compounds. The evolved carbon dioxide is kept in the beer by a pressure of 3 bar. The total residence time in the reactor is 2 hours. In this brewery there are four parallel reactors of $7\,m^3$ giving flexibility to the maturation process. The yeast concentration of the outflowing beer is low: less than 10^5 cells/ml (Pajunen, 1995). Then the beer is cooled to the stabilization temperature of −1.5°C.

The yeast immobilized in the reactors is the brewery's normal yeast, either grown by normal propagation or taken from a previous fermentation. The inoculation of the reactors is achieved by pumping the yeast slurry into the reactor, and even distribution is achieved by circulating beer in the reactor. After inoculation no separate yeast growth period is needed. During the maturation process a very small growth of yeast will occur as the amount of fermentable sugars is low (and controlled). This low growth is advantageous as it ensures long operational times of the reactors. On the other hand, a reactor or all the reactors can be left in stand-by state just by shutting off the beer feed (Pajunen, 1995).

Another Finnish brewery uses the same principle for its continuous maturation system, but the engineering details are different (Hyttinen *et al.*, 1995). The first installed system uses downward flow and a non-porous carrier material. The second uses upward flow and porous glass beads as the carrier material. Some practical aspects of the porous material are reported by Gronieck *et al.* (1997). The non-porous carrier is modified cellulose (diaminodiethyl DEAE) (US Patent 4915959). The glass beads are available commercially.

Although continuous diacetyl removal is reported to have economic advantages (Pajunen *et al.*, 1991) and has been tested in a few breweries in Europe (Mensour *et al.*, 1997), in Japan (Nakanishi *et al.*, 1985; Yamauchi *et al.*, 1994) and Brazil (Nothaft, 1995), its use is not widespread. In non-alcoholic beer fermentation the immobilized process has gained a better foothold in Europe (Mensour *et al.*, 1997). An economic comparison between immobilized and traditional maturation processes is given by Pajunen (1995) in Table 12.2. The cost of carrier material is excluded in the comparison.

The carrier material plays an important role in this process and its properties should be considered carefully. Factors such as price, convenience for immobilization, type of immobilization possible, cell load, mass transfer properties, whether it channels or blocks reactors, stability, rigidity, regeneration properties, sterilization properties, whether it binds contaminating micro-organisms, possibility of use in various reactor designs, possibility of fluidization and approval for food use are important. Some factors are interdependent: if the price is very low, there is not necessarily a need for any regeneration.

Table 12.2 An economic comparison between the immobilized process and the traditional maturation process, carrier cost excluded

	Immobilized process	Traditional process
Maturation time (weeks)	0.006[1]	2
Maturation tank volume (hl)	4 × 100	7 × 5300
Area requirement (m^2)	150	700
Height requirement (m)	7	25
Building costs (million €)	0.2	0.5
Process and tank costs (million €)	1.4	2.0
Total investment costs (million €)	1.6	2.6
Total savings (million €)	–	0.9[2]

[1] Equal to 2 hours.
[2] If maturation tanks were 17 × 2650 hectolitres (hl), the total savings would be 3.0 million euros.
Source: Pajunen (1995).

Besides direct economic advantages some operational advantages are reported, too. The application of centrifuging leads to less beer loss in tank bottoms. All the treatments lead to reduced kieselguhr usage in beer filtration as well as long filtration cycles. Production planning is easier to manage because of predictable maturation time.

The most critical factor in applying any new technology is, of course, the beer quality. Pajunen (1995) reported changes in foam stability (improved consistently) and SO_2 content between (lower) beers produced by the continuous maturation process and beers produced by the traditional process. Mensour *et al.* (1995), too, reported differences observed in lager beers produced by the continuous maturation process and traditional process. The reported values of both sources are expressed as per cent changes in Table 12.3. From the table it can be seen that the changes are small and not always in the same direction for the two different beers. Foam stability has improved and the concentration of 3-methyl butyl acetate has decreased in both cases.

The quality of beer produced through this continuous maturation process has been reported to be excellent. The Finnish lager beer so produced won The Brewing Industry International Award in 1996 in the bottled lager beer category.

Alpha-acetolactate decarboxylase

Solving the diacetyl problem by adding an enzyme was proposed as early as 1964 (Sandine *et al.*, 1964) but it was discovered that the enzyme used, diacetyl reductase, did not catalyse the rate-limiting step (Godtfredsen and Ottesen, 1982; Sone *et al.*, 1987a) rather only the reduction of diacetyl (and 2,3-pentanedione) to acetoin (and 2,3-pentanediol). The rate-limiting step is the decarboxylation of the precursor, alpha-acetolactate, to diacetyl. This reaction is catalysed by alpha-acetolactate decarboxylase. This enzyme was used by Godtfredsen and Ottesen (1982) to decrease the levels of alpha-acetolactate and acetohydroxy butyrate in freshly fermented beer. In 24 hours at 10°C levels were

Table 12.3 Comparison between continuously ('immo.') and traditionally ('trad.') matured beers

	Pajunen			Nothaft		
	Immo.	Trad.	Immo. (%)	Immo.	Trad.	Immo. (%)
Original gravity	9.9	10.2	−2.9	11.63	11.7	−0.6
Alcohol	4.2	4.2	0.0	3.8	3.98	−4.5
Real extract	3.4	3.7	−8.1	4.25	3.98	6.8
Real attenuation	65.5	63.5	3.1	82.1	83.1	−1.2
pH	4.2	4.2	0.0	4.25	4.1	3.7
Bitterness	17	17	0.0	17	15.5	9.7
Colour	9.4	8.5	10.6	4	5	−20.0
Haze	0.7	0.6	16.7			
Foam stability	247	176	40.3	134	108	24.1
Flavonoids	19	19	0.0			
SO_2	4.2	5.2	−19.2	7.3	6.1	19.7
Acetaldehyde	10.3	11.1	−7.2			
Ethyl acetate	12.6	12.1	4.1	14.4	14	2.9
3-methyl butyl acetate	0.9	1.1	−18.2	0.32	0.59	−45.8
Propanol	7.6	7.8	−2.6			
2-methyl propanol	8.3	8	3.8			
2-methyl butanol	11.7	11.4	2.6			
3-methyl ethyl butanol	33.5	32.7	2.4	41.9	48.6	−13.8

Source: data from Pajunen (1995) and Nothaft (1995).

reduced below the taste thresholds of the corresponding diketone (diacetyl and pentanedione). The enzyme used in the experiment was isolated from *Enterobacter aerogenes*. Subsequently alpha-acetolactate decarboxylases with more appropriate characteristics have been isolated (Godtfredsen *et al.*, 1984, 1987) from *Bacillus* and *Lactobacillus* strains. They are produced commercially by some bacteria (Rostgaard-Jensen *et al.*, 1987).

The US Food and Drug Administration approved the use of alpha-acetolactate decarboxylase for brewing applications in the United States on 16 May 2001 (Hannemann, 2002). It is commercially available. This enzyme is produced by a *Bacillus subtilis* strain into which a gene coding alpha-acetolactate decarboxylase is inserted from a *Bacillus brevis* strain (Jeppsen, 1991). The enzyme was shown to reduce diacetyl below taste threshold levels in both all-malt and adjunct beer at the dosage level of 4 ml/hl (Hannemann, 1996). It is claimed to be widely used in the brewing industry. The same enzymatic activity as above was used in the continuous immobilized process, to some extent in a rather similar way, as described in Scetion 12.5 (Laurent *et al.*, 1989).

An interesting solution to substitute for the heating step in the continuous maturation process (see above) was found by simulation (Dulieu *et al.*, 2000). In the proposed process immobilized alpha-acetolactate decarboxylase (Dulieu *et al.*, 1997) was used together with immobilized yeast to reduce alpha-acetolactate to acetoin.

Zeolites
Zeolites are reported to convert alpha-acetolactate at cold temperatures directly into acetoin. The conversion varied between 50% and 80% in buffer solutions, the best results being achieved at 5°C (Andries *et al.*, 1997, 1998). In beer conversion patterns were reported to be similar.

12.3.3 Techniques using genetic engineering

Modifying brewers' yeast
As genetic engineering became possible, brewers' yeast became a target for many modification trials. Brewers' yeast is much harder to modify successfully than laboratory yeast as it has multiple chromosomes, though this has not prevented researchers from producing yeasts with better characteristics.

The enzymatic reactions and genes involved have been elucidated by Holmberg *et al.* (1987), Gjermansen *et al.* (1988) and Holmberg and Litske-Petersen (1988). In Fig. 12.5 for VDK metabolism, relevant genes are presented together with the reactions which the genes' encoded enzymes catalyse.

Concerning maturation of beer, three different strategies can easily be applied. The first is to prevent the formation of alpha-acetolactate and thus prevent its leakage out of the yeast cell and subsequent formation of diacetyl. The second is to increase the utilization of alpha-acetolactate inside the yeast cell and minimize the leakage by keeping the internal pool of alpha-acetolactate low. The third option is to incorporate the gene encoding alpha-acetolactate decarboxylase enzyme into brewers' yeast and so speed up the decarboxylation of alpha-acetolactate outside the yeast.

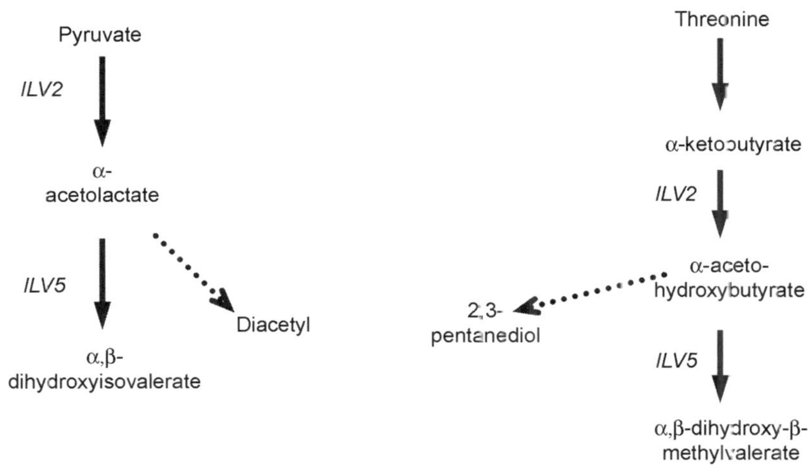

Fig. 12.5 Genes involved in vicinal diketone formation and reduction.

The gene ILV2 codes the enzymes leading to the formation of alpha-acetolactate. Blocking this step was proposed very early by Ramos-Jeunehomme and Masschelein (1977). A different method to achieve the same effect, complete deletion of ILV2 gene, was used by Gjermansen *et al.* (1988). The full or partial deletion of ILV2 can be achieved also by *in vivo* mutation and recombination (Hansen and Kielland-Brandt, 1997). One downside of this strategy is impairing the valine biosynthesis in yeast which may – and probably will – lead to flavour changes in beer other than diacetyl reduction.

The second strategy employs the amplification of the gene, ILV5, that inside the yeast leads to the formation of valine from alpha-acetolactate. This enzyme converts alpha-acetolactate into α,β-dihydroxy isovalerate. Gjermansen *et al.* (1988) achieved 60% reduction of diacetyl production by inserting a plasmid containing ILV5 into brewers' yeast. Although the brewing performance was good, the stability was not (Villanueba *et al.*, 1990).

To overcome the instability problem, the ILV5 gene was incorporated into the yeast genome by Goossens *et al.* (1987). This strain produced 57–63% less diacetyl than the parent strain. The same reduction was achieved by Mithieux and Weiss (1995) by using a different genetic technique. This strategy will not impair valine production so its effects on the flavour of beer can be assumed to be less significant than the blocking strategy. In some mutants based on this strategy the maturation step has been reported to be unnecessary.

The third alternative is to bring the gene encoding alpha-acetolactate decarboxylase into yeast. This enzyme makes a short-cut in the diacetyl metabolism directly to acetoin from alpha-acetolactate (see above). Instead of adding the enzyme into wort or beer, a more advanced solution is to make the yeast to produce the enzyme.

Alpha-acetolactate decarboxylase is produced by many bacteria, and the choice of bacteria has differed between different researchers. Sone *et al.* (1987a,b) used the gene from *Enterobacter aerogenes* under alcohol dehydrogenase I promoter (ADHI) in a high copy number plasmid. The diacetyl production was only 14% of that of the unmodified strain. On a one-litre scale, amino acid utilization and flavour compounds were in the normal range.

Suihko *et al.* (1990) took the genes from *Enterobacter aerogenes* and from *Klebsiella terrigena* under two different promoters in a plasmid. In some combinations the diacetyl production was so low that no further diacetyl reduction was necessary after the main fermentation. This approach was patented in 1987 (Enari *et al.*, 1987). Blomqvist *et al.* (1991) used the same gene and promoters but incorporated it into one of the brewers' yeast chromosomes. With these strains diacetyl reduction was unnecessary and no adverse quality effects were seen. Fujii *et al.* (1990) incorporated the alpha-acetolactate gene in high numbers into yeast genome so that no bacterial DNA other than that encoding alpha-acetolactate was left in the yeast.

Genes from different bacteria have been used, for example *Klebsiella terrigena*, *Enterobacter aerogenes* (Blomqvist, 1992; Blomqvist *et al.*, 1991; Suihko *et al.*, 1989, 1990) and *Acetobacter xylinum* (Yamano *et al.*, 1994, 1995).

The promoter used has been PGK or ADH or modified alcohol dehydrogenase promoter (Onnela *et al.*, 1996).

These new types of brewers' yeast can be used in conventional batch fermentation or in a continuous maturation process (Kronlöf and Linko, 1992). Continuous fermentation with immobilized genetically modified yeast without any subsequent maturation is a realistic option. The total time, the residence time in the bioreactor, needed for the fermentation would then be about one day compared with weeks or months in traditional brewing (Linko *et al.*, 1993). This radical change is possible without affecting the flavour of the final beer.

The use of genetic engineering, of course, is under debate, but as some genetically modified products are on the market (e.g., corn, soy) and a few beers produced using genetically modified materials have been accepted by authorities to be released on the market, it is probably only a matter of time until modified brewers' yeast is used on a larger scale. Consumer acceptance of different manufacturing processes has been evaluated (Caporale and Monteleone, 2004). The findings indicate that both sensory properties of beer and information on the manufacturing properties affect consumer liking.

12.4 Warm stabilization

Ales were traditionally matured by storing the beer in casks or in a tank at around 14°C (Lloyd Hind, 1940), but today's current practice is to mature ales in the absence of yeast at −1°C for a week (*EBC Manual*, 1999). A long shelf-life is required for today's market. This means for beer that no haze may form in the customer package (bottle) and the flavour characteristics must not change. The latter is dealt with elsewhere in this book. The haze in beer is caused by formation of protein tannin complexes which are large enough to be seen. If one of the components is removed from beer, the haze formation is prevented.

Phenols are potent antioxidants. The most important polyphenols of beer with regard to haze formation are flavonoids and more precisely flavan-3-ols and their oligomeric derivates.

Beer contains proteins and some proteins are essential for foam and foam stability. Therefore, when increasing colloidal stability (and shelf-life) by removing proteins from beer, care must be taken in order to keep the foam-active proteins in beer. Current practices to remove proteins from beer include adding silica gel before filtering the beer, addition of proteases before filtering, and adding gallotannin or bentonite in the lagering.

Adding proteases, e.g., papain, bromelin or ficine, will convert protein into amino acids which cannot form large enough complexes with polyphenols, thus preventing the haze formation. A drawback with enzymes is the residual protease activity in the product which may cause some unwanted effects: free –SH groups, foam killing, etc.

12.5 Cold stabilization

Beer proteins and polyphenols can slowly react during shelf storage to form a colloidal complex. This complex becomes insoluble. The result is formation of an undesirable cloudiness: chill haze. Haze formation is accelerated by oxidation of beer during processing, by the introduction of headspace air during packaging, by elevated storage temperatures and by the presence of heavy metals such as copper and iron. The formation of the hydrogen bonds in protein–polyphenol complexes is not very rapid, and the temperature is more important a factor than time. The ideal conditions for haze formation are −1 to −2°C for 2–3 days. At higher temperatures the breakdown rate of hydrogen bonds increases rapidly. As the haze is a product of protein and polyphenols, removing or reducing the concentration of either component from beer will prevent haze formation. Several techniques are used for this purpose.

12.5.1 Current practice

Because the chill haze precursors cannot totally be removed from beer, breweries use stabilizers to prevent the formation of haze. Commonly used stabilizers have included enzymes, tannic acid, polyvinylpolypyrrolidone (PVPP) and silica gels. Silica gels have become the most used method over the last two decades.

After the storage time suspended material, possible yeast cells and haze particles must be removed from beer. Current practice relies on filtration, usually kieselguhr filtration, or centrifugation. A combination of both methods is also used (see Chapter 13).

12.5.2 New aspects

There are no totally new acceleration techniques for cold stabilization, but there exist – and have long done so – many ways to accelerate the sedimentation: collagen from the swim bladder of fish (isinglass), alginate (which can be combined with collagen) and wood chips. The acceleration relies mostly on better equipment design and on improved understanding of the chemical reaction taking place during cold stabilization (see, e.g., Bamforth, 1999). Some of these methods are dealt with in the next chapter in this book, but brief notes are given here when their use is more directly related to accelerating the stabilization, not filtration itself.

Perhaps the easiest new way to shorten the cold stabilization time is just to shorten it. The increased knowledge about haze-forming reactions can justify testing how long is really needed for the stabilization of the beer in question. Many of the reactions need only a few seconds to take place, and the use of heat exchangers installed in-line speeds the temperature reduction considerably. Miedl and Bamforth (2004) found that a short period at low temperature is more efficient in forming insoluble material than a prolonged period at warmer temperatures. Their results showed that 6 hours at −2.5°C produced the maximum amount of haze. The beer used (6.3% v/v ethanol) did not freeze at this temperature, nor was it expected to. The authors were cautious in generalizing the results to beers other than the one tested. In the continuous secondary fermentation process, beer is cooled to −1.5°C in a heat exchanger and led to post-fermentation treatment (Pajunen, 1995), though the stabilization time is not mentioned. The use of this very cold, but rapid stabilization procedure requires the beer to be filtered at the same temperature as it is stabilized (Miedl and Bamforth, 2004).

Agarose system

Simultaneous adsorption of both protein and polyphenolic precursors of haze by insoluble agarose particles is a new idea for beer stabilization (Jany and Katzke, 2002). One unit is installed in Russia with a capacity of 300 hl/h (Jany and Katzke, 2002) and several other units are reported to be in industrial use (Jany and Katzke, 2003).

The system is installed after kieselguhr filter in-line. The agarose is placed in a small vessel through which the beer flows with a contact time of only a few seconds. The beer flows through this vessel and the proteins and polyphenols are removed to the preset concentration. The adsorption system is regenerable by rock salt and caustic soda (Jany and Katzke, 2002).

Enzymatic treatment

Proteolytic enzymes, papain, bromelin, ficin and acid protease from *Bacillus subtilis* are used to hydrolyse beer protein to peptides which cannot form haze complexes with polyphenols. The enzymatic treatment is performed by adding the enzyme in maturing beer. The dosage is enzyme dependent, but a normal dosage is a few grams per hectolitre. Alternatively the enzyme can be added to final beer. The danger of enzymatic treatment is in reduction of foam stability. This is especially so if the enzyme is still as active in the final product, for example in non-pasteurized beer.

Silica treatment

Silica gel is an efficient protein adsorbent and is widely used in brewing. Silica gels can be added into maturation, but more often they are added into the beer stream leading to kieselguhr filtration.

Gallotannin treatment

Other protein absorbants are gallotannins which are extracted from gall nuts or sumach tree leaves. Gallotannins react with high molecular weight acid proteins and polypeptides by reacting –SH and $–NH_2$ groups. Gallotannins are added to the maturing beer, but preferably during the transfer into the maturing tank, as addition directly into the tank may lead to unfilterable complexes (*EBC Manual*, 1999). After 5 days at 0°C the beer is easy to filter. Acceleration of the process may be achieved by adding gallotannins before the cold stabilization process.

Polyvinylpolypyrrolidone treatment
Polyvinylpolypyrrolidone (PVPP) is a polyphenol adsorbant. The PVPP treatment can complement a protein adsorption treatment but it can also be used as the only treatment. PVPP can be added to the maturation tank or dosed online to beer going to filtration.

A way to improve the process is to treat the wort prior to fermentation with a carrageenan-micronized PVPP composite (US Patent 2005019447; Rehmanji *et al.*, 2002). The composite is added about 10 minutes before the end of the boil, in the wort kettle. Carrageenan absorbs positively charged proteins, carbohydrates and metal ions. Benefits of this treatment are claimed to be higher wort yield, improved wort clarity, faster fermentation, improved yeast vitality and crop, longer filter runs, lower beer haze and longer stability of the packaged beer.

Barley breeding
By mutation the amount of haze-forming anthocyanogens was reduced in barley (US Patent 4165387). Chill haze stability in beers produced from the mutant barley was excellent (Figueroa *et al.*, 1989). However, yield was lower in some cases and a darker colour was also observed.

12.6 Two maturation philosophies

The 'no yeast needed philosophy' is that the understanding of beer chemistry and development of refrigeration, carbonation and filtration equipment have rendered the long cold-maturation process obsolete (Munroe, 1995). Advocates of long maturation with yeast cells at cold temperatures for longish times insist, on the other hand, that the yeast cells excrete flavour-active compounds into beer and improve the palate fullness and mouthfeel (Masschelein and Van de Meersche, 1977; Masschelein, 1986). The two philosophies may coexist in the same company for different brands of beer. A premium brand is matured for 6 weeks by a Canadian brewing company, and even one European beer is reported to be matured for nine months (Steward, 2002).

The major metabolic sequences responsible for improvements in sensory quality are breakdown of cell constituents and release of amino acids, peptides, nucleotides, phosphates and other cell material into the beer (Masschelein and Van de Meersche, 1977). Other changes which are reported to occur during maturation include changes in sulphur compounds, reduction of aldehydes, changes in volatile fatty acids and release of proteinases.

As mentioned earlier, the higher the temperature the faster the release of fatty acids into beer (Masschelein, 1986), the rate at 20°C being 20 times that in 0°C in horizontal 400 hl tanks. To the advocates of long maturing times these fatty acids are needed in certain amounts to give the proper mouthfeel (Delvaux, 1996). As caprylic flavour is an off-flavour (yeasty) at high concentrations, its release imposes strict control requirements. Another issue long debated is whether the shape of the tank (cylindroconical or horizontal) has an effect on the sensory quality of beer.

Masschelein and Van de Meersche (1977) found that also amino acid content increased in beer over 8 weeks of maturation and the amino acid profile showed increases in lysine and glutamate proportions. Sodium glutamate is a known flavour enhancer. In Japan hydrolysis products of RNA (inosinic acid and guanylic acid) have been used as food seasonings. These are produced industrially by hydrolysis of yeast RNA (Nakao, 1979). These are compounds which may have an effect on beer flavour.

Perpète and Collin (2000) have reported that fresh yeast (aerobically propagated) has enzymatic activity to produce aldehydes in cold temperatures in non-alcoholic beer fermentation. If the capability still exists after the main fermentation of beer, it may lead to flavour changes.

Although dimethyl sulphide (DMS) is a very volatile compound, due to the low CO_2 evolution during maturation the stripping of H_2S and DMS from beer is very low (Zangrando and Girini, 1969).

Details of the precise changes in flavour (or in flavour compounds) are still somewhat cloudy at the time of writing and the debate over the maturation time has been going on for a long time. A Japanese group found that 'no single flavour component affects flavour changes during lagering and no benefit in taste was derived from prolonged lagering' (Hashimoto *et al.*, 1969), while other researchers concluded that 'the minimum maturation times for a 12 degree Plato light beer are: 70 days for all malt beers, 70 days for beers containing unmalted barley, 50 days for beers brewed using sugar' (Ferkl and Curin, 1979).

The increased knowledge, faster and more precise control will undoubtedly speed the maturation of beer, but where lies the limit? There will be no single solution, but a spectrum of different maturation times, temperatures and processes.

12.7 Sources of further information

Basics of brewing

KUNZE, W., 2004, *Technology Brewing and Malting* (3rd edn), Verlagsabteilung der VLB Berlin, Berlin.

HARDWICK, W.A., 1994, *Handbook of Brewing*, Food Science and Technology, Vol. 64, Marcel Dekker, New York.

Current research

The American Society of Brewing Chemists:
Website: http://www.asbcnet.org
Journal: *The Journal of the American Society of Brewing Chemists*

Master Brewers' Association of the Americas:
Website: http://www.mbaa.com
Journal: *Technical Quarterly of Master Brewers' Association of the Americas*

European Brewery Convention (EBC):
Proceedings
Manuals of good practice
Both available from http://www.hanscarl.com

Especially for this chapter

HANSEN, J. and KIELLAND-BRANDT, M.C., 2003, Brewer's yeast: genetic structure and targets for improvement, in *Topics in Current Genetics*, ed. J.H. de Winde, *Vol. 2, Functional Genetics of Industrial Yeasts*, Springer Verlag, Berlin.

Proceedings of J. de Clerck Chair VII, 15–19 September 1996, Leuven, Belgium.

12.8 References

ANDRIES, M., DERDELINCKX, G., IONE, K.G., DELVAUX, F., VAN BEVEREN, P.C. and MASSCHELEIN, C.A., 1997, Zeolites as catalysts for the cold and direct conversion of α-acetolactate into acetoin, *Proc. Eur. Brew. Conv., Maastricht*, 477–484.

ANDRIES, M., JANSSENS, B. and MASSCHELEIN, C.A., 1998, Zeolites as catalysts for the cold and direct conversion of α-acetolactate into acetoin, *Cerevisiae, Belg. J. Brew. Biotechnol.*, 23(1), 3–6.

BAKER, D.A. and KIRSOP, B.H., 1973, Rapid beer production and conditioning using a plug fermenter, *J. Inst. Brew.*, 79, 487– 494.

BAMFORTH, C.W., 1999, Beer haze, *J. Am. Soc. Brew. Chem.*, 57, 81–90.

BLOMQVIST, K., 1992, *The bacterial budABC operon; isolation and application for accelerated brewing*, PhD thesis, VTT and University of Umeå, 74 pp.

BLOMQVIST, K., SUIHKO, M.-L., KNOWLES, J. and PENTTILÄ, M., 1991, Chromosomal integration and expression of two bacterial α-acetolactate decarboxylase genes in brewer's yeast, *Appl. Envir. Microbiol.*, 57, 2796–2803.

CAPORALE, G. and MONTELEONE, E., 2004, Influence of information about manufacturing process on beer acceptability, *Food Quality and Preference*, 15, 271–278.

CHRISTENSEN, C.M., 1997, *The Innovators Dilemma When New Technologies Cause Great Firms to Fail*, Harvard Business School Press, Boston, MA, 225 pp.

DELVAUX, F., 1996, Why there is a need for post fermentation period?, *J. de Clerck Chair VII*, 15–19 September 1996, Leuven, 1–6.

DILLEMANS, M., GOOSSENS, E., GOFFIN, O. and MASSCHELEIN, C.A., 1987, The amplification effect of the ILV5 gene on the production of vicinal diketones in *Saccharomyces cerevisiae*, *J. Am. Soc. Brew. Chem.*, 45(3), 81–84.

DULIEU, C., BOIVIN, P., MALADA, M., DAUZENBERG, H. and PONCELET, D., 1997, Encapsulation of α-acetolactate decarboxylase to avoid diacetyl formation, *Proc. Eur. Brew, Conv., Maastricht*, 455–460.

DULIEU, C. MOLL, M., BOUDRANT, J. and PONCELET, D., 2000, Improved performance and control of beer fermentation using encapsulated α-acetolactate decarboxylase and modeling, *Biotechnol. Prog.*, 16, 958–965.

EBC MANUAL OF GOOD PRACTICE FERMENTATION AND MATURATION, 1999, Fachverlag Hans-Carl, Nürnberg, Germany, 187 pp.

ENARI, T.-M., NIKKOLA, E., SUIHKO, M.L., PENTTILÄ, M. and KNOWLES, J., 1987, *Yeast strains suitable for accelerated brewing*, US Patent Application 07/044 244.

FERKL, P. and CURIN, J., 1979, Change of organoleptic parameters of beer in relation to the use of adjuncts and shortening of maturing times, *Tech. Q. Master Brew. Assoc. Am.*, 16(4), 214–217.

FIGUEROA, J.D.C., MADSON, M.A. and D'APPOLONIA, B.L., 1989, The malting and brewing quality of crosses of barley anthocyanogen-free mutants, *J. Am. Soc. Brew. Chem.*, 47, 44–48.

FUJII, T., KONDO, K., SHIMIZU, F., SONE, H., TANAKA, J.-I. and INOUE, T., 1990, Application of a ribosomal DNA integration vector in the construction of a brewer's yeast having α-acetolactate decarboxylase activity, *Appl. Envir. Microbiol.*, 56, 997–1003.

GJERMANSEN, C., NILSSON-TILLGREN, T., PETERSEN J.G.L., KIELLAND-BRANDT, M.C., SIGSGAARD, P. and HOLMBERG, S., 1988, Towards diacetyl-less brewer's yeast. Influence of ilv2 and ilv5 mutations, *J. Basic Microbiol.*, 28, 175–183.

GODTFREDSEN, S.E. and OTTESEN, M., 1982, Maturation of beer with α-acetolactate decarboxylase, *Carlsbeg Res. Comm.*, 47, 93–102.

GODTFREDSEN, S.E., RASMUSSEN, A.M., OTTESEN, M., RAFN, P. and PEITERSEN, N., 1984, Occcurence of α-acetolactate decarboxylase among lactic acid bacteria and their utilization for maturation of beer, *Appl. Microbiol. Biotechnol.*, 20, 23–28.

GODTFREDSEN, S.E., OTTESEN, M., SIGSGAARD, P., ERDAL, K., MATHIASEN, T. and AHRENST-LARSEN, B., 1987, Use of α-acetolactate decarboxylase for accelerated maturation of beer, *Proc. 21st EBC Congress, Madrid*, IRL Press, Oxford, UK, 161–168.

GOOSSENS, E., DILLEMANS, M., DEBOURG, A. and MASSCHELEIN, C.A., 1987, *Proc. 21st EBC Congress, Madrid*, 563.

GRONIECK, E., GROPPE, H., DILLENHOFER, W. and RONN, D., 1997, Secondary fermentation of beer with immobilized yeast (SFIY). A practical experience, *Brauwelt Int.*, 15(3), 264–266.

HANNEMANN, W., 1996, α-Acetolactate decarboxylase for diacetyl control, *J. De Clerck Chair VII*, 15–19 September, Leuven.

HANNEMANN, W., 2002, Reducing beer maturation time and retaining quality, *Tech. Q. Master Brew. Assoc. Am.*, 39(3), 149–155.

HANSEN, J. and KIELLAND-BRANDT, M.C., 1997, Brewer's yeast, in *Yeast Sugar Metabolism: Biochemistry, Genetics, Biotechnology and Applications*, eds Zimmermann, F.K. and Entian, K.-D., Technomic Publishing, Lancaster, UK, 519.

HASHIMOTO, N., KUROIWA, Y. and SOGAWA, H., 1969, Maturation of beer flavor during lagering, *Report of the Research Laboratories of Kirin Brewery Company No. 12*, 57–68.

HOLMBERG, S. and LITSKE-PETERSEN, J.G., 1988, Regulation of isoleucine-valine biosynthesis in *Saccharomyces cerevisiae*, *Curr. Genet.*, 13, 207–217.

HOLMBERG, S., PETERSEN, J.G.L., NILSSON-TILLGREN, T., GJERMANSEN, C. and KIELLAND-BRANDT, M.C., 1987, Biosynthesis of isoleucine and valine in *Saccharomyces*. Gene structure and regulation, in *Biological Research on Industrial Yeasts, Vol. 1*, CRC Press, Boca Raton, FL, 177–187.

HYTTINEN, I., KRONLÖF, J. and HARTWALL, P., 1995, Use of porous glass at Hartwall Brewery in the maturation of beer with immobilized yeast. *Eur. Brew. Conv. Monograph XXIV, EBC Symposium on Immobilized Yeast Applications in the Brewing Industry, Espoo, Finland*, Verlag Hans Carl Getränke-Fachverlag, Nürnberg, 55–65.

INOUE, T., MURAYAMA, H., KAJUNO, K., KAMIYA, T., MITSUI, S and MAWATARI, M., 1991, *Proc. 23rd EBC Congr., Lisbon*, 369–376.

JANY, A. and KATZKE, M., 2002, CSS – a new beer stabilization process, *Tech. Q. Master Brew. Assoc. Am.*, 39(1), 96–98.

JANY, A. and KATZKE, M., 2003, CSS – a new beer stabilization process, *Proc. EBC Congr., Dublin*, CD-ROM, poster 26.

JEPPSEN, S., 1991, A new enzyme for diacetyl control during beer fermentation, *4th Meeting Industrial Appl. Enzymes, IQS, Barcelona.*

KAMIYA, T., KAJINO, K., HIRATSU, H., MAWATARI, M. and INOUE, T., 1993, A new schema for the conversion of acetolactate in young beer, *Tech. Q. Master Brew. Assoc. Am.*, 30(1), 14–15.

KRONLÖF, J. and LINKO, M., 1992, Production of beer using immobilized yeast encoding α-acetolactate decarboxylase, *J. Inst. Brew.*, 98, 479–491.

LAURENT, M., GOOSSENS, E. and MASSCHELEIN, C.A., 1989, Accelerated beer maturation with permeabilized and immobilized *Lactobacillus casei* cells, *Proc. 22nd Congr. Eur. Brew. Conv., Zürich*, IRL Press, Oxford, 363–370.

LEVIN, Y., PECHT, M., GOLDSTEIN, I. and KATCHALSKI, E., 1964, A water-insoluble polyanionic derivate of trypsin. I Preparation and properties, *Biochemistry*, 3, 1905–1913.

LINKO, M. and ENARI, T.-M., 1967, Lagering of beer. I, Shortening of the lagering period with the aid of raised temperature, *Mallasjuomat*, 199–218.

LINKO, M., SUIHKO, M.-L., KRONLÖF, J. and HOME, S., 1993, Use of brewer's yeast expressing α-acetolactate decarboxylase in conventional and immobilized fermentations, *Tech. Q. Master Brew. Assoc. Am.*, 30(3), 93–97.

LLOYD HIND, H., 1940, Bulk carbonated ales, in *Brewing Science and Practice, Vol. II*, Chapman & Hall, London, 894.

LOURENÇO, J., CRUZ, J.M. and GONÇALVES, M., 1971, Contribution à l'étude du lavage de la bière en garde après l'acide carbonique, *Proc. 13th Congr. Eur. Brew. Conv., Estoril*, Elsevier, Amsterdam, 294–305.

MASSCHELEIN, C.A., 1986, The biochemistry of maturation, *J. Inst. Brew.*, 92(3), 213–219.

MASSCHELEIN, C.A. and VAN DE MEERSCHE, J., 1977, Flavour maturation of beer, *Tech. Q. Master. Brew. Assoc. Am.*, 13(4), 240–250.

MENSOUR, N.A., MARGARITIS, A., BRIENS, C.L., PILKINGTON, H. and RUSSELL, I., 1995, 'Gas lift' systems for immobilized fermentations. *Eur. Brew. Conv. Monograph XXIV. EBC Symposium Immobilized Yeast Applications in the Brewing Industry, Espoo, Finland.* Verlag Hans Carl Getränke-Fachverlag, Nürnberg, pp. 125–132.

MENSOUR, N.A., MARGARITIS, A., BRIENS, C.L., PILKINGTON, H. and RUSSELL, I., 1997, New developments in the brewing industry using immobilized yeast cell bioreactor systems, *Journal of the Institute of Brewing*, 103(6), 363–370.

MIEDL, M. and BAMFORTH, C.W., 2004, The relative importance of temperature and time in the cold conditioning of beer, *J. Am. Soc. Brew. Chem.*, 62(2), 75–78.

MITCHELL, C.A., 1926, *Vinegar: Its Manufacture and Examination* (2nd edn), Griffin, London.

MITHIEUX, S.M. and WEISS, A.S., 1995, Tandem integration of multiple ILV5 copies and elevated transcription in polyploid yeast, *Yeast*, 11(4), 311–316.

MUNROE, J.H., 1995, Aging and finishing, in *Handbook of Brewing*, ed. Hardwick, W., Marcel Dekker Inc., N.Y., pp. 355–381.

NAKANISHI, K., ONAKA, T., INOUE, T. and KUBO, S., 1985, A new immobilized yeast reactor system for rapid production of beer, *Proc. 20th Congr. Eur. Brew. Conv., Helsinki*, IRL Press, Oxford, UK, 152–157.

NAKAO, Y., 1979, Microbial production of nucleosides and nucleotides, in *Microbial Technology. Vol. I, Microbial Processes*, eds. Peppler, H.J. and Perlman, D., Academic Press, New York, 312.

NARZISS, L. and HELLICH, P., 1971, Ein Beitrag zur wesentlichen Beschleunigung der

Gärung und Reifung des Bieres, *Brauwelt*, 111, 1491–1500.

NARZISS, L. and HELLICH, P., 1972, Rapid fermentation and maturing of beer by means of bio-reactor, *Brewer's Digest*, September, 106–118.

NOTHAFT, A., 1995, The start-up of an immobilized yeast system for secondary fermentation at Brama, *Eur. Brew. Conv. Monograph XXIV, EBC Symposium Immobilized Yeast Applications in the Brewing Industry, Espoo, Finland*, Verlag Hans Carl Getränke-Fachverlag, Nürnberg, 41–54.

ONNELA, M.-L., SUIHKO, M.-L., PENTTILÄ, M. and KERÄNEN, S., 1996, Use of a modified alcohol dehydrogenase, ADH1, promoter in construction of diacetyl non-producing brewer's yeast, *J. Biotechnol.*, 49, 101–109.

PAJUNEN, E. (1995), Immobilized lager beer maturation: DEAE-cellulose at Sinebrychoff, *Eur. Brew. Conv. Monograph XXIV, EBC Symposium Immobilized Yeast Applications in the Brewing Industry, Espoo, Finland*, Verlag Hans Carl Getränke-Fachverlag, Nürnberg, 24–40.

PAJUNEN, E. and MÄKINEN, V., 1975, The effect of pH and agitation on beer fermentation, *Proc. 15th Congr. Eur. Brew. Conv., Nice*, Elsevier, Oxford, UK, 525–538.

PAJUNEN, E. and ENARI, T.-M., 1978, *Accelerated Lagering and Maturation, EBC Fermentation and Storage Symposium Monograph V*, Verlag Hans Carl Getränke-Fachverlag, Nürnberg, 181–187.

PAJUNEN, E. and GRÖNQUIST, A., 1994, Immobilized yeast fermenters for continuous lager beer maturation, *Proc. 23rd Conv. Institute of Brewing (Asia-Pacific Section)*, 101–103.

PAJUNEN, E., MÄKINEN, V. and GISLER, R., 1987, *Proceedings 21 EBC, Madrid*, 441–448.

PAJUNEN, E., GRÖNQVIST, A. and LOMMI, H., 1989, Continuous secondary fermentation and maturation of beer in an immobilized yeast reactor, *Tech. Q. Master Brew. Assoc. Am.*, 26(4), 147–151.

PAJUNEN, E., GRÖNQUIST, A. and RANTA, B., 1991, Immobilized yeast reactor application in continuous secondary fermentation in industrial scale operation, *Proc. 27th Congr. Eur. Brew. Conv., Lisbon*, IRL Press, Oxford, UK, 361–368.

PENTTILÄ, M. and ENARI, T.-M., 1991, Genetic engineering of industrial yeasts, in *Biotechnology – Current Progress*, eds Cheremisinoff, P.N. and Ferrente, L.M., Technomic Publishing, Lancaster, UK, Vol. 1, 173–202.

PERPÈTE, P. and COLLIN, S., 2000, Evidence of strecker aldehyde excretion by yeast in cold contact fermentations, *J. Agric. Food Chem.*, 48, 2384–2386

PIERCE, J.S., 1987, The role of nitrogen in brewing, *J. Inst. Brew.*, 93, 378–381.

RAMOS-JEUNEHOMME, C.L. and MASSCHELEIN, C.A., 1977, Contrôle génétique de la fermentation des diketones vicinales chez *Saccharomyces cerevisiae*, *Proc. 16th EBC Congress, Amsterdam*, 267–283.

REHMANJI, M., GOPAL, C. and MOLA, A., 2002, A novel stabilization of beer with Polycar Brewbrite, *Tech. Q. Master Brew. Assoc. Am.*, 39(1), 24–28.

ROSTGAARD-JENSEN, B., SVENDSEN, I. and OTTESEN, M., 1987, Isolation and characterization of an α-acetolactate decarboxylase useful for accelerated beer maturation, *Proc. 21st Congr. Eur. Brew. Conv., Madrid*, IRL Press, Oxford, UK, 393–400.

SANDINE, W.E., ELLIKER, P.R., PACK, M.Y. and BAVISCTTO, V.S., 1964, Use of diacetyl reductase to remove diacetyl from beer, *Tech. Q. Master Brew. Assoc. Am.*, 155–159.

SATO, T. and TOSA, T., 1993, Optical resolution of racemic amino acids by aminoacylase, in *Industrial Application of Immobilized Biocatalysts*, eds Tanaka, A., Tosa, T. and Kobayashi, T., Marcel Dekker, New York, 3–24.

SONE, H., FUJII, T., KONDO, K. and TANAKA, J., 1987a, Molecular cloning of the gene encoding α-acetolactete decarboxylase from *Enterbacter aerogenes*, *J. Biotech.*, 5, 87–91.

SONE, H., KONDO, K., FUJII, T., SHIMIZU, F., TANAKA, J. and INOUE, T., 1987b, Fermentation properties of brewer's yeast having α-acetolactate decarboxylase gene, *Proc. 21st EBC Congress, Madrid*, 545–552.

STEWARD, G.G., 2002, Brewing – the evolution of a tradition into a technology, *Ingenia*, 11 February, 31–35.

SUIHKO, M.-L., PENTTILÄ, M., SONE, H., HOME, S., BLOMQVIST, K., TANAKA, J., INOUE, T. and KNOWLES, J., 1989, Pilot-brewing with α-acetolactate decarboxylase active yeasts, *Proc. 22nd Congr. Eur. Brew. Conv., Zürich*, IRL Press, Oxford, UK, 483–490.

SUIHKO, M.-L., BLOMQVIST, K., PENTIILÄ, M., GISLER, R. and KNOWLES, J., 1990, Recombinant brewer's yeast strains suitable for accelerated brewing, *J. Biotechnol.*, 14, 285–300.

VILLANUEBA, K.D., GOOSSENS, E. and MASSCHELEIN, C.A., 1990, Subthreshold vicinal diketone levels in lager brewing yeast fermentations by means of ILV5 gene amplification, *J. Am. Soc. Brew. Chem.*, 48(3), 111–114.

VINH, T., SCHWARTZ, G. and MOLL, M., 1981, Identification of 2-ethyl-hexanoic (2-ethyl-caproic) acid in beer, *J. Am. Soc. Brew. Chem.*, 39, 44–50.

YAMANO, S., KONDO, K., TANAKA, J. and INOUE, T., 1994, Construction of a brewer's yeast having alpha-acetolactate decarboxylase gene from *Acetobacter aceti* ssp. *xylinum* integrated in the genome, *J. Biotechnol.*, 32(2), 173–178.

YAMANO, S., TOMIZUKA, K., SONE, H., IMURA, M., TAKEUCHI, T., TANAKA, J. and INOUE, T., 1995, Brewing preformance of a brewer's yeast having α-acetolactate decarboxylase from *Acetobacter aceti* subsp. *xylinum*, *J. Biotechnol.*, 39, 21–26.

YAMAUCHI, Y., OKAMOTO, T., MURAYAMA, H., NAGARA, A., KASHIHARA, T. and NAKANISHI, K., J., 1994, Beer brewing using an immobilized yeast bioreactor. Design of an immobilized yeast bioreactor for rapid beer brewing system, *Ferment. Bioeng.*, 78(6), 443–449.

ZANGRANDO, T. and GIRINI, G., 1969, *Proc. 12th Congr. Eur. Brew. Conv., Interlaken*, 445.

13

Filtration and stabilisation of beer

G. Freeman, Brewing Research International, UK

13.1 Introduction

Many of the beer products now produced are aimed at a global market. This sits alongside a distinct pressure in the brewing industry for centralisation of brewing capacity into fewer, larger breweries. Production costs are thereby minimised according to the principles of economy of scale. However, both these trends necessitate extended distribution networks for products. The subsequent increase in the interval between production and consumption has put pressure on brewing companies to provide beers with longer and longer shelf lives. This chapter will outline the processing operations currently used to render beer stable for subsequent packaging. Subsequently, the major part of the chapter will review newer or developmental technologies that may enable benefits in environmental health and safety, product quality and purity, and of course process economics.

13.2 Current filtration practice

Filtration in breweries is most commonly accomplished by the use of filter aids. These substances, used as slurried powders, form incompressible and highly porous filter beds, thus allowing the relatively free flow of beer. The most common filter aid used in breweries is kieselguhr or diatomaceous earth. These materials comprise fossils or skeletons of microscopic salt or freshwater life known as diatoms. When they die they sink and form deposits that are mined, processed and size-classified to give kieselguhr of various grades. The disadvantages of kieselguhr are that it is a health hazard (by dust inhalation)

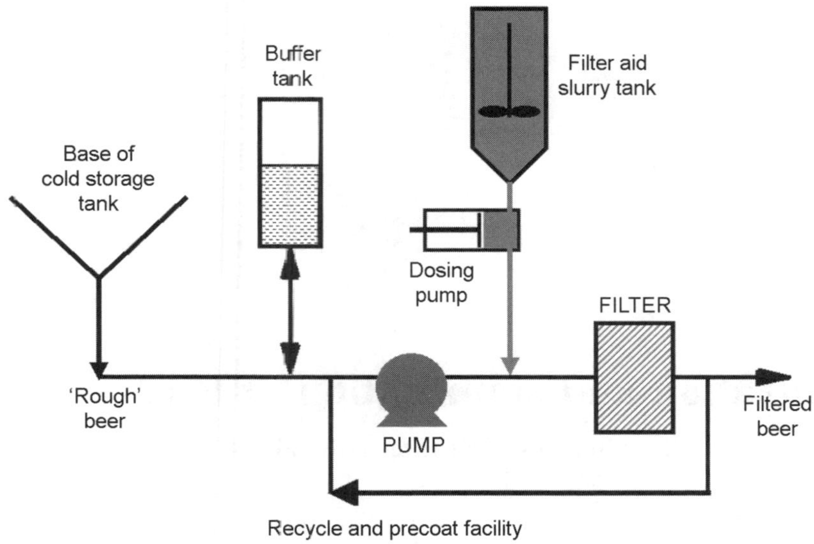

Fig. 13.1 A schematic of a typical system for filter aid filtration.

in its dry form as delivered to the brewery and that it is in itself non-biodegradable, but with a concentration of organic solids, and thus expensive to dispose of in landfill sites.

The configuration of a filter aid filtration system is shown in Fig. 13.1. Before processing occurs a precoat of filter aid is deposited onto the filtration surface. This is achieved by recycling of a water/filter aid slurry around the filter. After several minutes the precoat will be deposited completely onto the filtration surface and the recycling water is clean. The precoat is necessary to ensure efficient filtration of the early part of the beer run, to guarantee the integrity of the filter throughout the run and to aid removal of the filter cake after the process cycle. After precoating, the filter is smoothly put into 'forward flow' mode. The filter aid slurry is added continuously to the flowing beer stream. Thus the filtration surface is constantly being regenerated. In this way the filtration run time is extended, causing the process to be commercially viable. With regard to the actual filter unit, options may be divided into plate and frame type, leaf type and candle type. Plate and frame filters have been a workhorse of the brewing industry since the inception of filter aid filtration. They are known to enable excellent filtrate clarity. However, they are not amenable to full automation, causing long downtimes between filter runs and an increased manpower requirement. Most brewers would now choose leaf or candle filters for which beer recovery, cleaning and restarting may be automated by process control systems. Selection of leaf or candle is a brewery-specific decision. Leaf filters are mechanically more complex and need higher maintenance, but have more flexibility in flowrate and are not as vulnerable to process interruption.

The bulk filtration duty in a brewery is a demanding unit operation. It is essential for product clarity, and also for colloidal stability. It should significantly lower the quantity of contaminant micro-organisms presented to the pasteuriser, since heat should be used sparingly if flavour impairment is to be avoided. If sterile filtration is employed the bulk filtration stage must still give a high degree of clarity since the majority of sterile filtration systems have very limited dirt-holding capacities.

In this chapter alternative filtration technologies and filter aids will be considered. These alternatives are challenged by the intrinsic difficulties of beer as a product for filtration. The low temperature (0°C) and presence of dissolved solids and alcohol mean that viscosity is quite high (at least 2 mPa.s). Of even more significance is the nature of the suspended solids. These may be present in very high levels, perhaps up to 0.2% by volume or even higher over short periods during tank run-off. Practically all of the suspended beer solids are compressible, which causes them to form filter cakes impermeable to beer flow. Filtration may also be impaired by colloidal substances such as β-glucan gels (Waiblinger, 2002).

Alternative filter aids (and alternative technologies) are currently in use. Perlites consist of thermally expanded volcanic glass (Davies, 2004), crushed to form microscopic flat particles. Perlites are less efficient filter aids than are kieselguhrs but are perceived as being safer than kieselguhr. However, perlite lacks the remarkable skeletal structures of the diatoms that comprise kieselguhr. As a consequence its filtration performance is not as good. In order to achieve the required filtration performance, secondary filtration (e.g. sheet or cartridge filters) is required.

It is interesting to compare the filtration performance of kieselguhr and perlite in some detail. Figure 13.2 shows the particle sizes remaining in beer after filtration with kieselguhr and perlite (of similar permeability). Note that the kieselguhr exhibits a very exact particle size at which almost all particles smaller will pass through and almost all particles larger will be retained. This is not the case with perlite which, although it removes more very small particles than kieselguhr, allows particles in excess of 1.5 microns to pass into filtered beer. The two filter aids operate in a very different manner. The kieselguhr behaves somewhat like a sieve. The perlite, however, filters more like a depth filter (mass filter). The perlite performance in Fig. 13.2 is of less desirable performance than the kieselguhr filtration. The larger particles will manifest themselves as a more obvious, visible haze in the beer. Also, a higher proportion of any spoilage bacteria will pass through the perlite filter. Hence the need for secondary filtration or an increase in pasteurisation intensity. The latter may impinge adversely on beer quality.

It is likely that many of the novel, alternative filter aids now on the market will not filter beer to the same standard of clarity that is possible with kieselguhr. This is because of the unique structures in kieselguhr that cannot be artificially reproduced economically. The advantages of the alternative filter aids must be one or more of a stabilisation effect, the ability to regenerate and reuse, health and safety or environmental.

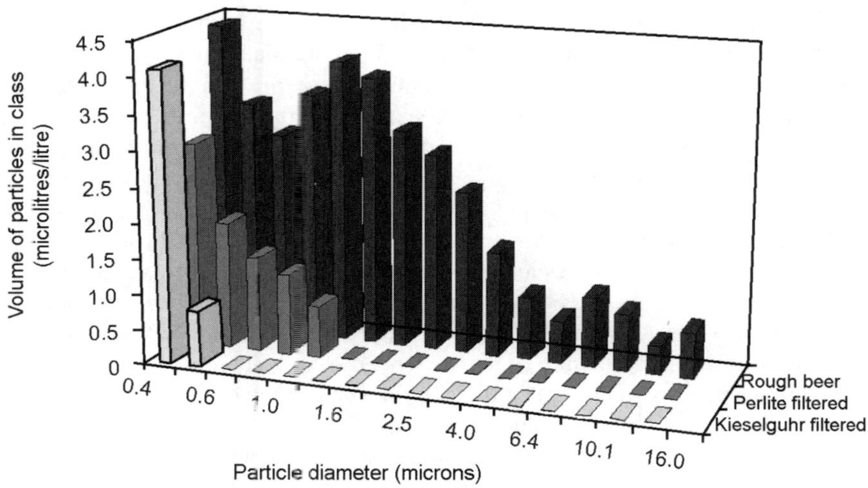

Fig. 13.2 A comparison of particle counts remaining in beer after perlite and kieselguhr filtration.

After the main filtration operation many breweries employ secondary filtration operations. Often the purpose is to provide sparkling clarity and enhanced shelf life before the beer becomes turbid in-pack. Some products are sterile filtered. This means that essentially all microbes with the potential to spoil beer are removed. The most common technologies installed for these purposes today are sheet filters (Brunner, 1987) and cartridge filters (Tubbs, 1998). Sheet filters consist mainly of cellulosic fibres that have been compressed into a thin mat and arranged in a plate and frame filter press. Cartridge filters comprise small units enclosing a filter element, commonly of polymeric fibre sheets that are pleated. The demands on beer quality today, especially if sterile filtration is required, mean that on occasions three or four filtration steps are performed.

13.3 Possibilities in recycling of kieselguhr

It is clear that kieselguhr would become much more environmentally acceptable if it were to be reused in the process. Alternatively, it could become a by-product of the brewing process. It has been suggested that spent kieselguhr could be employed for enhancing the nutrient value and structure of agricultural soil either directly or by composting first (Russ, 1993). However, in recent years this use has become unfashionable because of the perceived health risk.

It is possible to add spent kieselguhr to construction materials such as bricks and tiles. The problem here is economic in that individual breweries do not produce enough to justify the transportation to central processing.

Hence, there is still scope for engineers to develop processes that enable kieselguhr to be reused in the brewery. There are established technologies involving sodium hydroxide and also furnacing (Russ, 1993). In both cases, economy of scale suggests the need for removal of spent kieselguhr from the brewery to a centralised processing station. Recently, studies have been performed on the use of hydrocyclones to separate the filter aid particles from organic material (Poku, 2004). Hydrocyclones operate by the conversion of pressure energy into vortex flow, thus enabling separation of relatively dense particles from liquid suspension and less dense particles. Filter aid particles may be concentrated by a factor of up to 25 (depending on particle size) in one pass. Yeast cells do not get concentrated and therefore are effectively being washed out from the spent filter aid. There is the possibility of utilising several in series to effect a good separation, though there will still be a need for chemical washing of some stubborn organic residues.

With all technologies the skeletal structure of kieselguhr means that only partial recycle will most likely be possible. There will be loss of material due to particle attrition.

13.4 Crossflow filtration

Researchers have been trying to optimise this process for beer processing for over two decades. The technique has been successfully applied to cider filtration, beer recovery and wine filtration. Now, however, it has become a viable process for bulk beer filtration. This is because the membrane technologies that are employed in many aspects of process engineering plants have caused membranes to become more effective and also cheaper.

The mechanism of crossflow filtration is to employ a flow over the membrane to scour particles away from the surface, minimising fouling. With time, however, there is fouling of the surface and the inside of the membrane pores (Fig. 13.3). This fouling effect is retarded in some crossflow processes by backflushing. This entails periodic backflow of filtrate in order to lift filter cake from the membrane. Backpulsing can be employed with some membrane formats. This is frequent backflushing over very short time intervals (e.g. every few seconds for a few milliseconds). Another alternative is to replace the crossflow movement of the beer with movement by oscillation of the membrane itself (Sellick and Pratt, 1998). This has been achieved by a torsion bar vibrating a stack of PTFE membrane discs. This technology has been successful in the recovery of beer from yeast slurries. A major advantage of crossflow filtration over filter aid filtration is that it is amenable to automation, so much so that in some circumstances it can be operated with minimal supervision for hours (Anon., 2005).

The membranes may be either ceramic or polymeric. Ceramic membranes are more expensive but very long lasting (Fillaudeau and Lalande, 1999). With care they may last in excess of 10 years. Polymeric membranes are significantly cheaper, perhaps a quarter of the price per unit area, but have a working life of perhaps two years maximum.

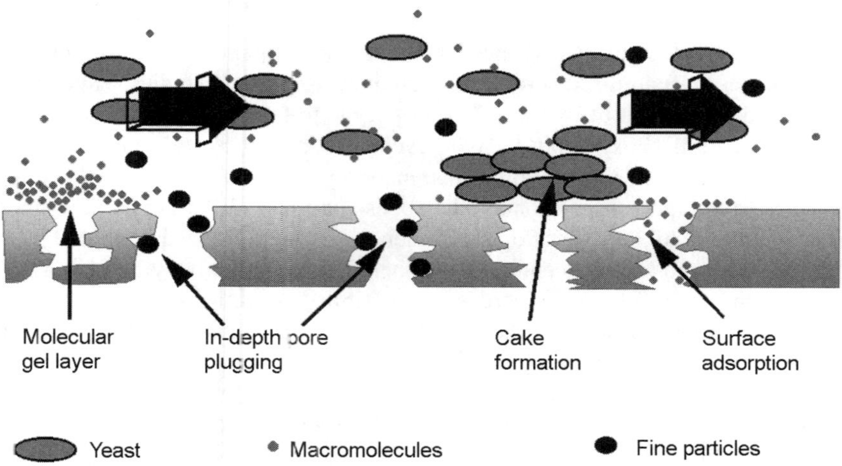

Fig. 13.3 A diagram of the fouling mechanisms that occur in membrane filtration.

Perhaps the main incentive to develop commercially viable crossflow filtration systems is the prospect of elimination of kieselguhr usage. This has increased the pace of development due to close collaborations between filtration companies and environmentally aware brewers. One system comprises tubular, polymeric membranes arranged in modular housings (Noordman *et al.*, 1999). The modules comprise membrane fibres 1 m long and of internal diameter 1.5 mm. Each module comprises just under 10 m^2 of filtration area. The system is pressurised on the 'retentate' (recirculating unfiltrate) side by a feed pump. Crossflow is provided by a separate recirculation pump. It is this pumping power that makes crossflow a relatively energy-intensive process, a fact that is compounded by the need to remove much of the pumping energy by increased cooling. There are at least four full-size plants in beer production at time of writing. Filtered beer quality is equivalent to that from kieselguhr filtration. The most recent developments in the process have been optimisation of backflushing frequency and also the cleaning regime. Options on the cleaning regime include caustic, acid, oxidative and enzymatic rinses. Another option that may be applicable to many plant formats is to drain the membrane housing and refill. The interface effects between air, liquid and membrane may enable removal of fouling material. Similarly, it has been suggested that gas formation may be one mechanism that increases the effectiveness of oxidative cleans. Total costs of the process are stated to match those of filter aid filtration (Schuurman *et al.*, 2003).

Two other collaborations have employed crossflow microfiltration and supplemented its performance through centrifugation immediately upstream. This has the advantage of improving the membrane performance. Also, the reduced particle loading onto the membrane means that a greater 'concentration factor' is possible. Thus beer recovery is required from only a relatively small volume compared to a crossflow unit alone. Of course, these benefits must be offset against the cost of purchase, operating and maintenance of the centrifuge(s).

Some time ago some experiments were performed with crossflow microfiltration of unfiltered beers of varying size distributions and numbers of particles (Burrell *et al.*, 1994). They demonstrated that the majority of the fouling was being caused by small particles (Fig. 13.3), which were not significantly removed by the centrifuge. However, the success of the new systems reflects the improved performance of modern disc stack centrifuges. For example, centrifuges may now be designed with very small sedimentation distances through the incorporation of a large number of discs in the stack. Employment of very high performance centrifuges may allow economic processing, in some cases, with sheet or pulp filters rather than crossflow filtration. Another application being utilised by some brewers is to replace isinglass finings, which have been subjected to labelling requirements in some parts of the world. They lack some of the stabilising effects of finings, however.

Two suppliers of the new centrifuge–crossflow systems have claimed cost benefits (mostly on variable costs) over kieselguhr filtration (Gaub, 2004; Borremans and Modrok, 2003). However, some brewers have reported that they calculate the costs of crossflow filtration to be higher (Stenhouse *et al.*, 2003). This must be placed in the context of a need for full operating optimisation and also a background of increasing costs for kieselguhr filtration.

Both systems employ polyethersulphone membranes in the range 0.45–0.65 micron rating, allowing for the pore size to be altered to optimise the brewery-specific conditions. One system (Gaub, 2004) employs the membranes in tubular format, the other (Borremans and Modrok, 2003) arranged in cassette format rather like a plate and frame filter. This format requires relatively clear pre-filter beer, as is forthcoming from the centrifuge, otherwise the flow channels will become blocked with solids. Similarly attractive filtrate flowrates are obtained with both systems. The narrow flow channels of the cassette format can be engineered to increase turbulence and thus membrane scouring. One method is to place plastic 'gauze' separators between the membranes that cause vortices in the flow.

Perhaps the most advanced technology targeted at the brewing industry comprises thin ceramic discs that have very precise pore sizes chemically etched into them (Lommi *et al.*, 2003). The high porosity and thinness of the discs mean that the filtrate flow rates are 100 times that achieved with most other processes (per unit membrane area). However, the system is not fully commercialised, requiring development of cleaning cycles and mass production of the discs in order to reduce price (Lommi *et al.*, 2003).

It is a feature of all filtration systems that performance is very dependent on the brand of beer being processed. This is especially true of membrane systems. The only means of evaluating performance, thus enabling plant design, is to perform laboratory or pilot studies.

Potentially crossflow microfiltration can produce sterile beer ready for package. However, the emphasis of the described systems is on filtration alone.

This is because of many brewers' requirement for stabilisation downstream, although there are several methods for effecting stabilisation upstream.

13.5 Single-pass membrane filtration

This technology has been an important part of the beer processing industry for many years. The technology options, normally based on a cartridge-type design, are compact and hygienic. Many modern systems are modular, thus enabling cleaning, maintenance or removal or changing of cartridges whilst production continues in the other modules. Some membrane formats enable backflushing and mild chemical cleaning, enabling longer membrane life.

Cartridge filters are commonly employed for sterile filtration. Sterile filtration is often presented as 'cold filtration' to the public, a misnomer that is used to avoid unfortunate alternative connotations of the word 'sterile'. Sterile filtration is also possible by crossflow microfiltration or fine sheet filtration. However, the compact size of cartridge filtration plants is an advantage because they may be installed very close to the package filling head. This minimises the risk of recontamination after sterilisation. This compactness is maximised by the arrangement of the filtration medium. Thin layers are often pleated or it is possible to employ multi-layers of media in one cartridge. The relatively low capital expenditure of cartridge filter housings makes them a viable option when only a small percentage of a brewery's products require such processing. However, sheet filters, normally employed in a plate and frame format, enable processing alternatives. It is possible to partition the unit and perform multi-stage filtration in one filter. Some sheets are available that are infused with PVPP to enable stabilisation.

There are broadly two types of media employed in the cartridges: depth and membrane. Through fine membrane filtration the absence of beer spoilage organisms is guaranteed. This is because a membrane removes all particles above a certain size; it is an absolute filter (Fig. 13.4). However, without the scouring effect of a crossflow of beer, the capacity of a membrane filter for retaining solids whilst still allowing filtrate flow is very small. Thus the membrane filter is the last in a filtration sequence illustrated in Fig. 13.5. After filter aid filtration there is most usually a coarse filter to capture any leakage of filter aid. The membrane filter needs to have an absolute cut-off of 0.65 microns at most, preferably 0.45 microns. The depth filter has an optimum specification to maximise run length. If it has too fine a rating its run time will be too short, and too coarse a rating will cause the subsequent membrane filter to blind more quickly. The optimum specification will depend on characteristics of the beer being filtered. Thus the brewer should employ laboratory or pilot tests to optimise the process. Samples of the membrane material and rating options may be used to filter the beer, and the maximum beer volume that may be processed through each is compared. This is known as an Esser test (Freeman, 1996).

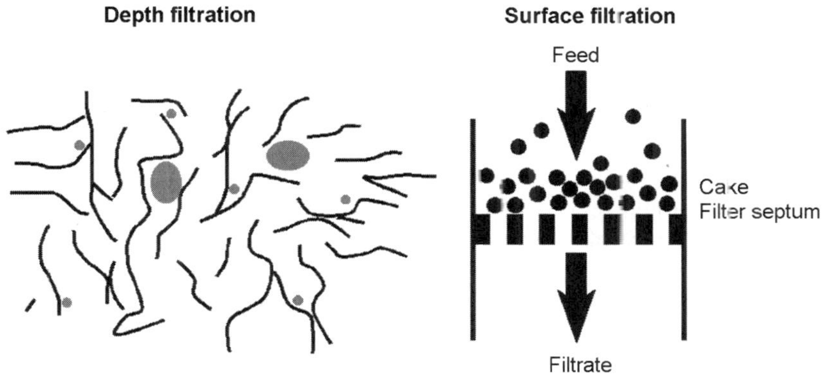

Fig. 13.4 A comparison of the modes of operation of depth filters and surface filters.

Just as has been the case with crossflow membranes, the chemical compositions of the filtration materials have been improved. Relatively new materials such as polyether sulphone (Riddell, 2002) have been demonstrably effective.

In recent times there has been a trend towards reducing the amount of heat used in pasteurisation processes. This is because the heating action has been blamed for off flavours and increased staling of the beer. Although a consequence is reduced security in microbiological stabilisation, this is offset by the improved hygienic standard in modern breweries. Thus there is often in reality little differentiation in quality between pasteurised and membrane-filtered beer.

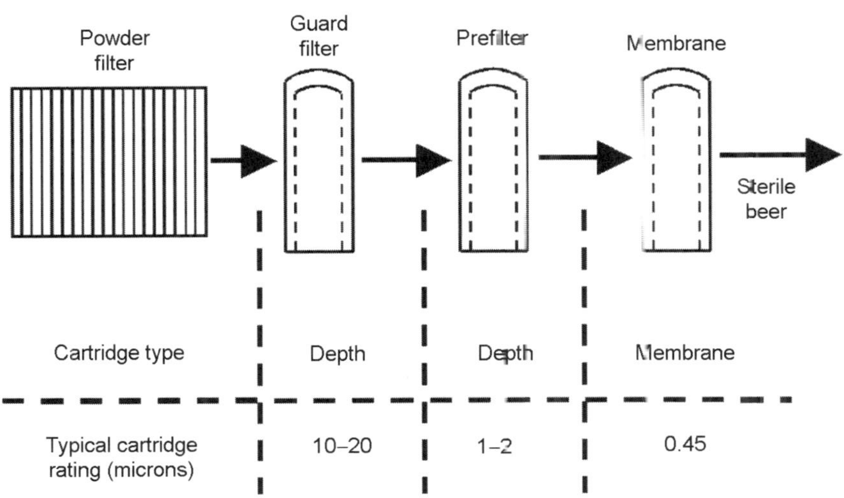

Fig. 13.5 A typical filtration sequence for the production of sterile beer with cartridge filters.

There are, however, some economic aspects to the selection of the microbiological stabilisation technology. Pasteurisation, either in pack or process, requires a large initial capital expenditure but subsequent operating costs are relatively low. The reverse is true of cartridge filtration. Initial expenditure is low but operating costs are relatively high, largely due to the cost of replacement membranes. Thus, in many cases there may not be much to choose between the technologies for new process plant (Dunn *et al.*, 1996). Pasteurisation may have the advantage where plant exists or can be acquired inexpensively. Membrane filtration may be useful for small production runs, when a minority of the production of the brewery is specified for sterile filtration.

13.6 Novel filter aids

Some alternative filter aids would best be employed as partial replacements for kieselguhr. In other words, they are blended in with the kieselguhr slurry. Cellulose fibres may be employed as coarse precoats or with the bodyfeed slurry. Complete replacement of kieselguhr might be achievable by blending the cellulose with silica gels for employment of the bodyfeed. This would offset disposal costs by enabling disposal with spent grains. Cellulose fibres have a natural elasticity that resists 'pressure shocks' on the filter, possibly contributing to filter run reliability. There are products on the market (Blümelhuber *et al.*, 2003) that comprise mainly cellulose and allow for complete kieselguhr replacement. The purchase price is significantly higher than for kieselguhr but this will be increasingly offset by lower disposal costs.

Another alternative as a partial kieselguhr is rice hull ash, a by-product of the food industry (Villar *et al.*, 2004). Hulls are removed from rice during milling and are incinerated in the drying process. It is possible to treat the resulting particles so that leaching of contaminating ions into beer is minimised. Partial (50%) replacement is recommended, although it is unlikely that the same standard of clarity of filtered beer would be achieved with all beers. There is scope for development of this product to commercial viability by controlled milling and size classification to optimise its performance.

A variety of synthetic, regenerable filter aid filtration systems have been proposed. One system is based on a synthetic polymer (Brocheton *et al.*, 1995). It is granular with a typical particle size of 35 microns, which is larger than that of the kieselguhr grades normally employed. However, the particles are claimed to be hydrodynamically 'lighter' than kieselguhr, which will assist in the development of smooth, even filtration cakes. Regeneration is by hot caustic solutions, meaning that the filter aid may be blended with PVPP and regenerated together. Stabilisation (of polyphenolic sources of instability) and filtration are achieved in one unit operation. Another system employs a mixture of synthetic microballs (for filtrate clarity) and fibres (for cake flexibility) (Harmegnies *et al.*, 1997). Similarly to the previous system, the main regeneration process is with hot caustic and therefore PVPP may be incorporated into the filter aid to provide stabilisation plus filtration. With both systems, filtrate quality and an evaluation of economic worth were both positive. Disadvantages of regenerable filter aid systems are the need for precoating and bodyfeed mixtures to be identical, and also there is very limited flexibility over the dosage rate of stabiliser if it is employed.

For many years, mainly in the US, pulp filters have been employed for secondary and sometimes primary filtration. However, a new supplier (Miller, 2001) claims an improvement to this technology, enabling reliable single-stage bulk filtration. The fibrous 'pulp' mass is filtered on to a belt with the beer, and the belt continuously moves the mass to a regeneration tank. The 'pulp' medium comprises positively charged micro-spheres that efficiently remove the suspended particles in the beer, most of which exhibit a negative surface charge. A feature of this technology is the potential for a high degree of automation.

13.7 Current stabilisation processes

There are many forms of haze, an unsightly suspension of particles that can occur in beer. Hazes may be microbiological, yeasts, bacteria or even moulds. Some hazes are inorganic, such as calcium oxalate (beer stone). Hazes can comprise carbohydrate material such as starch, beta-glucan or pentosan. Beer is intrinsically susceptible to one form of haze, however. This is caused by cross-linkage of certain polyphenolic constituents with certain protein/polypeptide fractions. The 'sensitive protein' fraction is characterised by a significant presence in the molecules of the imino functional group (–NH). Polyphenols exhibit more haze-forming potential when they form more oxidised states and polymerise into dimers, trimers, etc. Such a haze manifests itself at first when the beer is cold (a 'chill haze'). The proteins and polyphenols cross-link in this state with weak hydrogen bonds, which break down when the temperature is raised, and thus the haze disappears. This feature is one of the main reasons for cold storage of beers in breweries. Removal of chill haze by filtration lengthens shelf life. Chill hazes will eventually form stronger covalent cross-linkages that make the haze permanent.

It should be noted that the two main contributors to haze instability for a given beer composition are dissolved oxygen in pack and initial haze. Thus, before stabilisation treatments are considered it is worth reviewing packaging and process options. Finings or centrifugation enable finer filtration techniques to be employed. Membrane filtration techniques or sheet filters enable very bright product, but care must be taken not to over-filter or there may be unacceptable loss of body, bitterness or foam

Current technologies for stabilisation include PVPP, silica gels, tannic acid and papain. Optimisation is possible with shelf-life tests and collaboration with the suppliers. The longest shelf lives are achieved with employment of more than one stabiliser. For example, both silica gel and PVPP may be optimised together by testing a 'matrix' of different loadings of each stabiliser.

Optimisation of tests can be accelerated by employing forcing tests of cycling temperatures and possibly agitation. However, forcing tests are notoriously unreliable indicators of true shelf life. The longer the shelf-life requirement, the less reliable are forcing tests as a predictor.

PVPP (polyvinylpolypyrrolidone) is an insoluble polyamide that contains the same –NH functional group as haze-sensitive proteins. Thus it adsorbs the polyphenolic fraction that contributes to haze. It is possible to employ PVPP (or the less cross-linked PVP) as a one-off addition to beer storage or filtration. More commonly, however, it is used and regenerated. A dedicated vessel (similar to a leaf or candle filter) is employed to catch the PVPP after it is dosed into the beer stream. Regeneration is possible with caustic.

Silica gels are sold as xerogels or hydrogels, the latter having more water in the particles. They comprise particles with a highly porous structure enabling a high adsorption area. They selectively adsorb haze-active proteins and polypeptides. At the dosage rates employed they have no perceptible effect on beer foam (which relies on protein for its structure). Silica gels are not regenerated; they are employed in beer storage or as an addition (at 0.3–0.8 g/litre of beer) to the filter aid.

Given that brewers often employ both silica gel and PVP(P), some suppliers have taken to combining the two stabilisers in order to facilitate simpler processing. One utilises a tightly bound coating of PVP on the silica (McKeown and Thompson, 2003). This has some process advantages over PVPP in that a swelling time is not required in the process. Another employs a simpler mixture (Rehmanji *et al.*, 2000). In this case the attractive forces between the xerogel and PVPP cause the mixture to disperse more easily than would the two components on their own and the hydrated particles to be less compressible. This means that it performs quite well as an addition to the beer filter.

Tannic acid (or gallotannin) precipitates haze-forming protein. It also precipitates significant proportions of transition metal ions and lipids. These components adversely affect beer stability, the former through formation of free radicals and the latter through oxidation to stale-flavoured compounds. A large dosage of tannic acid may cause a large amount of sediment, resulting in loss of high quality product. Some imaginative processing ideas have been conceived by the suppliers of tannic acid. For example, one involves the use of a temperature cycle following the addition of tannic acid. Large flocs are formed that are readily removed by centrifuge. The remaining beer is very amenable to cross-flow filtration (Mussche and De Pauw, 1998).

Papain is a proteolytic enzyme that stabilises by reducing the chain length of haze-forming proteins (polypeptides) (Lemaire, 2000). As a stabiliser it is cost effective. However, there is some concern that it will also reduce foam potential in the beer. It is added to filtered beer, and its activity reduces with storage and pasteurisation. The brewer must take care that little residual activity remains in the packaged beer. An alternative to papain has recently been developed in prolyl endoproteases (Edens and Lopez, 2003). It is recommended to add the enzyme immediately after wort boiling, although it can be added to the beer.

Finings have a significant impact on colloidal stability. Copper finings increase the amount of cold trub produced when the hot wort is cooled ready for fermentation. Copper finings comprise carageenan derived from seaweed. Other types of fining agent are employed for beer clarification. Isinglass is collagen derived from the swim bladders of fish. They form a positive charge in beer, thus greatly enhancing the precipitation of the predominant negatively charged particles in beer. Auxiliary finings comprise polysaccharides, polysilicates or a mixture of both. They are applied before isinglass to enhance its performance. They help with the precipitation of positively or neutrally charged particles, most notably dead yeast cells that lose their surface charge. As well as precipitating haze particles large enough to be visible haze, they also work on smaller particles that may polymerise or nucleate visible hazes during the shelf life of the beer. They also precipitate lipid material, enhancing flavour and foam stability.

Some brewers are not able to employ fining agents because of regional 'beer purity' laws or commercially unviable labelling restrictions. The significant effect of fining agents on beer stability should not be overlooked if the beer processing is changed with their exclusion.

13.8 Novel stabilising systems

Alternative stabilising systems are now available. One employs a mixture of cellulose fibres and PVPP that is completely regenerable (Wackerbauer and Evers, 1998). The regeneration is by hot caustic rinse, similar to that employed with conventional PVPP processing. In order to achieve the required filtrate clarity, the researchers found that very fine fibrils of cellulose were required. In addition to this there was a need to have a low loading of yeast on to the filter. This could be achieved either via a centrifuge upstream or by a suitably long settlement period in cold storage tank. The reduction in haze-active polyphenols was similar to that achieved by conventional PVPP processing, and shelf-life measurements were satisfactory. Given the complete replacement of kieselguhr filtration by this technology, benefits are seen in environmental terms and perhaps in process economics, depending on the disposal costs of spent kieselguhr in the appropriate region.

Another system employs a regenerable agarose medium that adsorbs both haze-active proteins and polyphenols (Jany and Katzke, 2002). It is claimed to be more cost-effective than other stabilisation techniques. The medium is derived from algae and is treated in a confidential process that enhances the adsorption of the protein fraction. It is non-swelling. The vessel design is compact and low volume, meaning that beer brand changeover is rapid and also it is easy to retrofit. Regeneration of the material is in the same vessel and employs sodium chloride and caustic solutions. Losses of the agarose are very small because the adsorbing particles are quite large at 100–300 microns. Process and performance flexibility is achieved by adjusting residence time of the beer in the vessel. This is most easily achieved by incorporating a bypass into the system, so that typically not all of the beer passes through the gel. This means that different beer brands are easily treated optimally, and over-treatment of some brands is avoided. The technology has been successfully commercialised, firstly in Russia.

In the future there may be possibilities for the high pressure treatment of beer. At the moment, the very high capital costs of the process vessels are prohibitive. After all, beer is a relatively high volume product. In order to effect pasteurisation of the beer, 4000 atmospheres are required (Herdegen *et al.*, 1998), otherwise some bacteria such as the remarkably resilient *Lactobacillus brevis* will survive. Another study (Perez-Lamela *et al.*, 2004) found that high pressure treatment had a stabilising effect. A haze is immediately formed and thus can be removed by filtration, resulting in a more stable beer. An improvement in foam stability was also measured. The reason for this is unclear; perhaps there was a partition of materials into the haze that are detrimental to foam in solution (e.g. lipids).

13.9 The effect of modern processing technologies on stability

It is certain that there are enormous incentives for brewers to select optimal processing to secure the futures of their brands. Today the brewer has several options for major processing operations and an optimal choice requires a look into possible future scenarios. There are increasing requirements for high performance end-processing operations. Many brewing companies of all sizes have long distribution chains. These can involve temperature changes (storage and retail), flavour damage from ultraviolet light (notably retail) and agitation (transportation). In those countries such as Australia that serve beer extremely cold, the need for colloidal stability is greater. The need to achieve the required, labelled shelf life is absolutely of fundamental importance to the company. No amount of brand recognition or marketing expense will make up for a customer receiving a cloudy or poorly flavoured beer.

For many years now many brewing companies have employed high gravity brewing. This entails brewing beer in a concentrated form and diluting back to the correct alcoholic strength at the end of the process. Because of the freezing point depressing effect of ethanol, this enables cold storage and filtration to be effected at lower temperatures. In this event superior colloidal stabilisation is achieved. Therefore, any technology advancement that enables beer to be brewed satisfactorily at a higher strength could potentially aid stability. Modern mash filters in the brewhouse enable high strength worts. There may also be problems in fermentation, however, and these problems need to be solved on the yeast technology and handling fronts.

When the beer is in package, colloidal stability is most affected by the initial clarity, oxygen minimisation in-pack and the chemical stabilisation processes. The more thorough the filtration, the better the prospects for a long period before

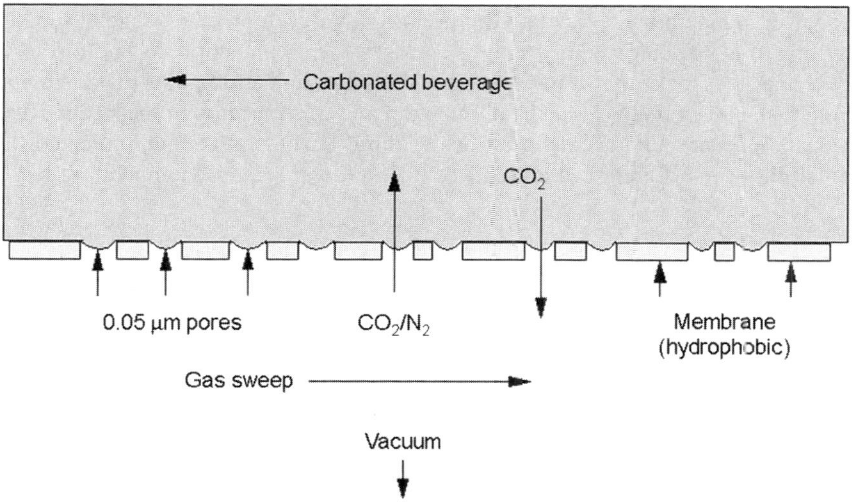

Fig. 13.6 A schematic of the mode of operation of hydrophobic membranes for dissolved gas concentration adjustment in beer.

haze formation. Oxygen minimisation is an increasing issue and many breweries now achieve 100 parts per billion or less. A technology that may assist in minimising both haze and oxygen levels is hydrophobic membrane gas control technology (Gill and Rogers, 1998). This mechanism is illustrated in Fig. 13.6. Most processes require addition of carbon dioxide (and in some niche products nitrogen) after filtration. If it is bubbled into the beer stream, there is the potential for particle formation from the bubble 'skins' as the gas dissolves. However, the hydrophobic membranes allow gas addition without bubbling. Furthermore, as the desired gas is added to the beer there is a simultaneous removal of oxygen, which diffuses from the beer stream to the gas stream that contains negligible oxygen. This may have a very significant shelf-life benefit.

13.10 Pasteurisation

Microbiological stabilisation of beer by thermal treatment is still the most common method. Alternatives are sterile filtration as described above and inclusion of a culture of brewing yeast in the pack (e.g. cask-conditioned or bottle-conditioned beers). The issue with pasteurisation is that oxidative damage occurs. Staling and ageing occur very rapidly in the short pasteurisation time. There are incentives for brewers to reduce the number of 'Pasteurisation Units' employed, otherwise the microbiological stabilisation is at the expense of colloidal and flavour stability. This reduction is achieved through lower pasteurisation temperature and/or time. Thus there is an incentive to develop a new technology that employs another mechanism, which supplements the effect of raised temperature. It is at a very preliminary developmental stage at time of writing, but this technology may exist as a steam injection technology that generates a 'shockzone' downstream of the steam injection point (Todman and Freeman, 2004). In this zone the momentum and condensation of the steam cause a decompression and acceleration of the fluid. Early results have indicated the possibility of sufficient bacteria kill rates at a lower required temperature rise.

13.11 Future trends

Environmental issues will become more and more important in the brewing industry. Brewing companies generally have to demonstrate (especially at the larger sites) that they are performing within acceptable parameters for environmental damage, within the constraints of product quality, process safety and economic viability. Currently in the European Union these constraints are set by the determination of Best Available Techniques (BAT), industry by industry.

For filtration, the most obvious issue is the disposal of kieselguhr. Filtration with this material has been the technology of choice for decades; however, its future does not look secure given the competitive technologies now available as described above. BAT allows for differentiation between existing plant and new installations. Perhaps within a few years new filtration plant will not be designed around kieselguhr, although some existing filtration plants may continue to operate for as long as 20 years.

The brewing industry is a large user of water per unit volume of product. This fact will increase the pressure on brewers from water treatment and legislative bodies. It also utilises a lot of thermal energy early in the process and in most cases even more cooling energy later. In the future, some technologies may fall by the wayside by being less water or energy efficient than the alternatives.

Brewers will consider package designs and stabilisation issues that are driven by long distribution chains and increasing delays in stock of ever-larger retail outlets. It is likely that plastic packaging will increase in the brewing industry worldwide. Beer is likely to follow to some extent other beverages such as carbonated soft drinks, for which plastic bottles are now the dominant primary package. This will be supported by the relatively low weight of a plastic bottle compared to a glass one, and the availability of colourants that will protect beer against the damaging effects of ultraviolet light. However, plastic bottles are permeable to gases (both loss of fizz and ingress of destabilising oxygen). Technologies such as multi-layer and coated plastic bottle and oxygen scavenging caps will assist, but there will be greater onus on the brewing process to provide increasingly colloidally- and flavour-stable beer. Emphasis may also move towards flavour stability from colloidal stability. The latter is generally used as the marker for shelf life because it is easier to measure. However, new analytical methods may provide means to predict and/or measure flavour stability more easily. The brewing process may change to prolong acceptable beer flavour

13.12 Acknowledgement

The author would like to thank the Chief Executive of Brewing Research International for permission to publish this chapter.

13.13 References

ANON. (2005), 'Beer filtration without kieselguhr', *Filtration and Separation International Edition*, 2005(5), 29–31.

BLÜMELHUBER G, BLEIER B and MEYER-PITTROFF R (2003), 'Trials with an alternative filter aid based on cellulose fibre', *Brauwelt International*, 21(6), 370–373.

BORREMANS E and MODROK A (2003), 'Membrane beer filtration by Alfa Laval and Sartorius', *Brewing and Distilling International*, 34(4), 10–11.

BROCHETON S, HERMIA J, RAHIER G and VAN DEN EYNDE E (1995), 'The basic principles of a new beer filtration process', *Proceedings of the European Brewery Convention, Brussels*, Nürnberg, Fachverlag Hans Carl, 427–436.

BRUNNER M (1987), 'Filter sheets', *International Bottler and Packer*, 61(6), 27–32.

BURRELL K, GILL C, McKECHNIE M and MURRAY J (1994), 'Advances in separations technology for the brewer: ceramic crossflow microfiltration of rough beer', *MBAA Technical Quarterly*, 31(2), 42–50.

DAVIES K (2004), 'Filter aids', *Brewer International* 4(2), 14–18.

DUNN A E, LEEDER G I, MOLLOY F and WALL R (1996), 'Sterile beer filtration', *Ferment*, 9(3), 155–161.

EDENS L and LOPEZ M (2003), 'Improved method for the prevention or reduction of haze in beverages', *World International Property Organisation Patent Application WO 03/ 104382 A1*.

FILLAUDEAU L and LALANDE M (1999), 'Crossflow microfiltration of rough beer with tubular ceramic membranes – qualitative and quantitative performances of the process', *Proceedings of the European Brewery Convention, Cannes*, Nürnberg, Fachverlag Hans Carl, 823–830.

FREEMAN G (1996), 'Liquid and solid separation – in beer', *Brewers'* Guardian, 125(12), 27–32.

GAUB R (2004), 'The PROFi membrane filtration system for clarifying beer', *Brewing and Distilling International*, 35(4), 14.

GILL C and ROGERS T (1998), 'Trouble free control of dissolved gases in beer', *Proceedings of the Fifth Aviemore Conference on Malting, Brewing and Distilling*, London, Institute of Brewing, 366–369.

HARMEGNIES F, BONACCHELLI B and TIGEL R (1997), 'Beer filtration with regenerable filter aid', *Proceedings of the European Brewery Convention, Maastricht*, Nürnberg, Fachverlag Hans Carl, 517–524.

HERDEGEN V, MEZGER R, RUSS W, VOGEL R F and MEYER-PITTROFF R (1998), 'Die Hochdruckbehandlung – neue Wege für die Getränkeindustrie', *Brauindustrie*, 83(2), 105–107.

JANY A and KATZKE M (2002), 'CSS – a new beer stabilisation process', *Technical Quarterly of the Master Brewers Association of the Americas*, 39(2), 96–98.

LEMAIRE M (2000), 'Beer stabilisation with papain', *Brewers' Guardian*, 129(8), I–VII.

LOMMI H, RASPE O J A, VAN RIJN C J M and VOS J (2003), 'New filter and method for beer clarification', *Proceedings of the European Brewery Convention, Dublin*, Zoeterwoude, European Brewery Convention, 9 pp.

MCKEOWN I and THOMPSON N (2003), 'Lucilite TR silica – a cost effective alternative to PVPP', *The Brewer International*, 3(5), 54–57.

MILLER P (2001), 'The nutsche filter comes of age', *Filtration*, 2(1), 5–10.

MUSSCHE R A and DE PAUW C (1998), 'Total stabilisation of beer in a single operation', *Proceedings of the 25th Convention of the Institute of Brewing – Asia Pacific Section*, East Brisbane, Institute of Brewing and Distilling Asia Pacific Section, 125–130.

NOORDMAN T R, BERGHUIS O A E, MOL M N M, PEET C J, MULLER J L M, BROENS L and VAN HOOF S (1999), 'Membrane filtration for bright beer, an alternative to kieselguhr filtration', *Proceedings of the European Brewery Convention, Cannes*, Nürnberg, Fachverlag Hans Carl, 815–822.

PEREZ-LAMELA C, REED R J R and SIMAL-GANDARA J (2004), 'High pressure application to wort and beer', *Deutsche Lebensmittel-Rundschau*, 100(2), 53–56.

POKU M (2004), 'An investigation into the recycling of filter aids for the brewing industry', Ph.D. Thesis, University of Essex, Department of Biological Sciences.

REHMANJI M, MOLA A, NARAYANAN K and GOPAL C (2000), 'Superior colloidal stabilization of beer by combined treatment with silica (xerogel) and PVPP, Polyclar plus 730', *MBAA Technical Quarterly*, 37(1), 113–118.

RIDDELL P (2002), 'Sterile filtration of beer. The PES story', *Brewer International*, 2(11), 31–35.

RUSS W (1993), 'Disposal of kieselguhr – kieselguhr recycling', *Brauwelt International*, 11(1), 51–55.

SCHUURMAN R, BROENS L and MEPSCHEN A (2003), 'Membrane beer filtration – an alternative way of beer filtration', *Technical Quartely of the Master Brewers Association of the Americas*, 40(3), 189–192.

SELLICK I and PRATT J (1998), 'Advances in the recovery of beer from tank bottoms', *Brewing and Distilling International*, 29(2), 26–27.

STENHOUSE R, PARAMINSEE S, DENNINGER H and BOCK M (2003), 'Membrane filtration for beer clarification', *Proceedings of the IGB Africa Section Convention, Victoria Falls*, Hout Bay, Institute of Brewing and Distilling Africa Section, 123–130.

TODMAN M and FREEMAN G (2004), 'Opportunities brewing for process improvement', *Chemical Engineer*, 762/763, 34–36.

TUBBS J (1998), 'Cartridge filtration – part 2', *New Brewer*, 15(3), 56–58.

VILLAR J C, CANETE R E and MANGANELLY E A (2004), 'Why adding rice hull ash can benefit beer clarification', *Filtration and Separation*, 41(6), 32–33.

WACKERBAUER K and EVERS H (1998), 'Kieselguhr-free filtration (F & S) in industrial use', *Brauwelt International*, 16(2), 120–125.

WAIBLINGER R (2002), 'Beer filterability', *Brewer International*, 2(1), 15–18.

14

Packaging of beer

J. Browne, Technical Management Support Ltd, UK

14.1 Introduction

The last 10 years has seen many changes in beer packaging, from both the packagers' and the suppliers' perspective. There has been increased marketing enthusiasm for the introduction of new products. As one marketing company put it, 'packaging is becoming the emotional interface with the consumer and a tangible manifestation of a brand and its values' (Tutssel, 2005). Globalisation is happening and is being accelerated through the consolidation of the major brewers. There are now some bold moves by brewers into China and Russia, and this is creating yet another dimension.

On the supply side, with more sophisticated technology and materials available, there is an even greater choice from a huge range of possibilities. It is not only the brewers that have been consolidating. The suppliers of materials and machines have also become bigger. Brewing companies will normally have partnership agreements for the purchase of their materials; and when purchasing a new packaging line, it is more likely that a turnkey arrangement is made with a single supplier.

14.2 Trends in packaging formats

14.2.1 Glass bottles

The early challenge in packaging was to apply a good foil to a bottle. It was a delight to see this significantly reduced; in fact, very few bottles are now foiled. This change effectively came about due to the modern trend in drinking, which is to drink from the neck. The glue under the foil never properly dried out and the taste of glue did nothing to compliment the beer. Although the change was good, the trend of drinking from the neck meant that no glass fragment of any kind could be allowed to be present in the bottle, as it could be swallowed by the consumer. This has led to one brewer employing empty and full glass inspection. The same brewer in fact suffered a very costly recall which was caused by glass inclusion in the bottle. Full bottle inspection is costly. Not only does it mean the introduction of another machine with the penalty of lower efficiency, but such machines are also expensive to buy.

Marketing were never satisfied with the champagne-necked bottle, and the long-necked bottle was eventually introduced for 275 and 330 ml bottles. This shape of bottle, with slight differences, has been adopted by virtually everyone. This is not good news for packagers. If the height of the bottle is more than three times its diameter, the bottle becomes unstable. Unfortunately, all long-necked bottles break this rule.

14.2.2 PET bottle

Over the last 10 years a lot of suppliers of plastic bottles have been trying hard to change our beer-drinking habits, but there is still no enthusiastic shift to plastic. This is mainly due to the traditional values of our noble drink where glass still has a premium quality image. If beer is in plastic the feeling is that it should be cheaper, but this is not possible as plastic has not had the price advantage. PET (polyethylene terephthalate) is like a sieve: it allows the passage of CO_2 and O_2 out of and into the beer respectively, giving a shelf life of less than two weeks under normal conditions. PEN (polyethylene naphthalate) is much better but is more expensive. This is used by one brewer for returnable bottles.

The way of overcoming the loss of CO_2 and the ingress of O_2 is to introduce a barrier. This can be done by coating the bottle internally or externally, usually with silicon oxide (known as SiO_x), or introducing another one or two layers into the PET giving a built-in barrier, and creating a three-layer or five-layer bottle respectively. Multilayer technology has proved to be the most popular method, and the three-layer technology has been the most favoured. The barrier is normally nylon (MXD6), and to enhance it an oxygen scavenger is added, making it an active rather than a passive barrier. Any plastic bottle is formed in two stages. First the preforms are injection moulded, and then these are stretch blow moulded to form the shape of bottle that is desired (see Fig. 14.1). In order to introduce the barrier layer, a unique injection nozzle has to be used and this requires a special injection moulding unit to be built. These are much more expensive than the standard moulding machine, but they will also produce standard PET preforms and can also produce layered preforms at the same speed, which was not previously the case.

The most recent and exciting developments have been with blended materials, and one plastics company has recently succeeded in blending the nylon and scavenger into PET, and producing preforms on a standard injection moulding machine. When the preform is blown, the bottle has excellent clarity,

Fig. 14.1 Shows a preform and a stretch blow moulded bottle. Courtesy Krones UK.

which is something that previous blends have not achieved. The biggest market at present for beer in PET is Russia and in many instances straight PET is used. One brewer uses a blended material but it has poor clarity. The turnaround time for beer is fast and generally consumers are not as discerning as their Western counterparts, but they will be, and the blended material does seem to be the answer. The huge advantage is that such bottles can be produced on a standard PET injection moulding unit.

For PET, one of the outstanding matters is what happens to the bottle afterwards. Recycling facilities for PET in Europe are woefully inadequate. The recycling rate for most individual materials in Europe is above 50%, but for plastics it is around 15%. Add to this the fact that the beer bottle will most likely be green or amber, and will not be monopolymer due to the barrier material, and it is likely that the bottle will not be recycled at all. It is wrong to say that plastic is totally bad. It is much lighter than glass, with a weight ratio of 1:6, which will give transport savings, and also there will be no breakage during transportation and handling. Interestingly, a 330 ml plastic bottle is the same size as a 275 ml glass bottle. The author's view is that we should be talking about 'reuse' and 'recover'. If possible PET bottles should be reused, though this is not always practicable, and one day these bottles will become waste as well. The best way is to incinerate the bottles, with other plastics, and recover the energy. It is appalling to think how much energy is being thrown away each day!

14.2.3 Cans

Cans have always had their place in the market. They are easy to package, handy, easy to transport, easy to chill and can be recycled. Furthermore, the beer does not become light-struck! However, the downside is that they are dull – they all have the same shape. This has always been a problem with the can. As a result all sorts of ideas have been developed – thermo-chromic ink which changes colour with temperature; scratch and sniff which gives the aroma of the fruit drink; tactile to give the feel of, perhaps, an orange; embossed cans to give the look of, for example, a football; shaped cans, and so on.

Out of this list, thermo-chromic inks, embossing and shapes have the best synergy with the brewing industry. However, in order for a can to be shaped or embossed it needs to be made from aluminium, with more aluminium in the can walls. This, ultimately, with the extra stage in the manufacturing process, makes such cans much more expensive. Also, a shaped can will not be of the same diameter as a standard can (known as '211', i.e. $2\frac{11}{16}$ inches). This means that the extra expense of changeovers is necessary. One brewery was contract-packaging shaped cans, then decided to stop due to the expense of changeovers. Another brewery has succeeded, but has built a dedicated filler/seamer for their own shaped can. The can has a distinctive barrel shape, and is targeted to the US market. It has been very successful and the extra cost has been more than offset by the sales volume achieved. The embossed cans will be the same diameter as a standard can and are usually used for special events. One brewery has used embossing to highlight their brand name with some success. Recently two beers have appeared in the sleek and the slim can respectively. The sleek can has a 206 body and a 202 end and the slim a 200 body and a 200 end. As the sleek has the same end as a standard can, manufacturers of can fillers for the standard 211 can are already talking about manufacturing fillers with the capability of handling the 206 can. It looks good and is an exciting development.

Self-chilling and self-heating cans often come up in discussion. The self-heating can has been tried for coffee and has been withdrawn, because of the resulting poor coffee and insufficient rise in temperature, known as Δt (instructions say 'set off at 20°C'). Also, the device occupies about 120 ml of space in a 330 ml can. For beer, self-cooling would be good, but the same difficulties of Δt and space apply. Also, they are not very environmentally friendly. Many companies have patented their inventions.

It should be added that printing on the can has vastly improved. A new technique has recently been introduced giving near-photographic quality.

14.2.4 Labelling

Another trend that is taking place is to replace the existing paper label by a smarter plastic self-adhesive label to give the product an up-market image (these are normally known as pressure sensitive labels, abbreviated to PSL). The bottle also could be fully sleeved, giving marketing the chance of giving the bottle an extra classy design. For more information see Section 14.3.

14.2.5 Final packaging

Final packaging has also changed with the introduction of multi-packs in all styles, giving marketing all sorts of scope and creating many more SKUs (Stock Keeping Units) for the packager. Coupled with this, lines have become faster, requiring flexibility and reliability.

One point worth making is that a standard size for a pack in the UK is the 24 unit. This covers multiples of that figure, i.e. 2×12, 3×8 or 4×6. However, an 18 or a 20 pack, for example, could slow the line down, since the palletiser is built to handle the number of cycles per minute for the 24 unit. For 18 or 20 the palletiser would need to go faster.

With legislation requiring the packager to reduce packaging, and with the normal need to reduce costs, the end package is changing. The first example of this was the removal of cardboard dividers that separated the bottles. They are now packed much more tightly, so eliminating the bottle to bottle impact. In other areas, tray wall heights have been reduced, replaced by a pad, or, if the machine can cope, eliminated completely. Polyethylene film is being used to replace cardboard wraps to collate cans and bottles. Film printing is now excellent, allowing a high quality image to be produced. It is both lighter and cheaper, although there is the same environmental argument for plastic as discussed under the PET bottle. However, marketing departments still seem to have a preference for board due to its more premium image.

Something which cannot go unmentioned is the power of the supermarkets in all of this. They demand their own pack sizes and price points, and also how the product is to be delivered, such as the use of merchandisable units (MUs). Display pallets remove the need for an extra layer of packaging such as trays and shrinkwrap, as the pallet will contain all packs ready for sale. Another example of an MU is the Dolly. The size is 600 mm × 400 mm (half a Euro pallet) and it is on wheels. The product is packed on these in trays and is then stretch-wrapped to give it stability. It then ends up in a supermarket aisle and all that needs to be done is to remove the stretch film.

14.2.6 Draught beer in keg

The draught beer market in the UK has been decreasing year on year, with the trend moving rapidly towards lager at the rate of 1–2% per year. Ales have been competing with the introduction of nitrogenated beers which have a much smoother texture. In order to give the drinker an added experience, extra cold beers have also been introduced.

Trends for kegs have been to go smaller due to the health and safety regulations, and to be manufactured in stainless steel rather than aluminium. An issue with kegs has always been their traceability and information on their condition.

More people are now tending to drink at home so there is a lot of interest in the 5-litre mini-keg market. Self-cooling kegs are now available with a capacity range of 5 to 20 litres and the result is impressive. They work using latent heat of evaporation and can generate a Δt of 25°C. Due to their insulation they can keep cool for up to 12 hours. Their great advantage is that they can be regenerated after use by subjecting the surface of the keg to an external temperature of 400°C for a short period of time.

14.3 Developments in canning and bottling

As can be seen from the introduction and trends, the consumer is being given much greater choice. Also, with initiatives like ECR (Economic Consumer Response) the consumer is receiving the beer fresher. An added advantage is that the stock or inventory of products carried in the supply chain is substantially reduced, so reducing the cost of warehousing and the value of stock. For packaging this means more flexibility and more changeovers. The challenge is therefore to be more flexible, comply with new health and safety regulations, lower the packaging costs and meet stringent quality parameters, especially concerning oxygen.

14.3.1 Servo motors and line philosophy

So what changes will one see on a modern canning or bottling line? Perhaps the most remarkable change is the use of servo motor technology. This allows all movements in a machine to be controlled electronically. This would include vertical and horizontal drives on a depalletiser or palletiser, and separate drives in dry end machines, which used to be full of chains and shafts. It would also include the synchronisation of a can filler and seamer, where the seamer drive is presently connected to the filler via a shaft. Servo technology has also allowed changeovers to be faster and more accurate, making machines more flexible with less settling down time.

Another development has been with the control systems, making the setting-up of line philosophy much easier. Frequency drives are now fitted locally and communication cables fed back via PLCs to a central point using a profi-bus system. An asi-bus system is also used to transmit data from all the sensors on the line. This substantially reduces the cost of cabling and allows programming from one central point. It also makes it possible, via Ethernet, for diagnostics, and perhaps adjustment, to be carried out by the supplier from their home base. This avoids costly journeys and time delays.

14.3.2 Beer fillers

All fillers for beer are barometric (counter-pressure) fillers. When a filler is being purchased, there is a choice of mechanical or electro-pneumatic control, short- or long-tube filling, or volumetric filling. If the line is packaging non-returnable bottles – PET or glass – the filler is often combined with a rinser, known as a monobloc.

The mechanical filler is the most common form of filler and will normally have a short vent tube. For glass bottles, most short-tube fillers will now be fitted with a double pre-evacuation which will be operated before filling commences. With a 90% vacuum this reduces the air content in the bottle to 1%, whereas a single evacuation would reduce the air content to only 10%. Before the double pre-evacuation was introduced, a long-tube filler was often preferred to lessen the oxygen uptake. For PET it could still be the preferred route as the PET bottle can only be flushed with CO_2. However, for cans only a short-tube filler can be used due to the way they are introduced to the filler.

The mechanical filler needs to be balanced to the line, as the filling cycle is governed by the speed of the filler. However, if the filling cycle is automatic, it will take place independently of the rotation. This method of filling is now more popular, as it allows a variable speed at the filler and ensures a complete fill when the filler stops without being emptied out. The automatic filler is electro-pneumatically controlled and is now in common use. Apart from variable speed control, a small fill height adjustment is also possible without changing the tubes. One would readily recognise such a filler from the control hub located in the centre of the filler. The electronic signals are converted into air signals and these open and close the appropriate diaphragm which in turn opens the port on each head in the correct sequence.

The early fillers were extremely difficult to clean due to all the tubes being fed to each head. On the newer fillers this has been tidied up, and on the most recent model, shown at Drinktec 2005, the central hub has been removed and replaced by the product tank. The electronic signal now goes to each head and the channels are opened and closed by solenoid valves in the filling head. Also the filling tube retracts, making it unnecessary to have bottle lifts. This particular filler is impressive in that not only is the filler very open and easy to clean, but each moving part is driven by its own servo motor. This makes the filler extremely hygienic and excellent for sterile filling. See Fig. 14.2, in which the product tank is in the centre, though this can be a conventional annular tank shaped like a doughnut. Again the shape of the tank is changed to a round tube to facilitate cleaning. Although this filler is a volumetric filler designed for PET, this technology will soon be used for short-tube beer fillers.

Volumetric fillers measure the quantity of beer into each container. This is done either by using a measuring chamber which is fitted to each head and fills to a pre-programmed volume, or by using a magnetic flow meter. Volumetric fillers are often used for the filling of carbonated beverages into PET or for can filling where variable volumes are required, for example for widgetted cans versus standard cans. For bottles it is better to fill to a level with a short vent tube so all levels look the same.

14.3.3 Labelling

Labelling has now become much more demanding. Most packaging lines have only wet glue labellers, so with the trend moving to self-adhesive labelling

Fig. 14.2 Shows the openness of a servo driven filler. Courtesy Krones UK.

(known as pressure sensitive labels or PSL), the only way to label the bottle is for the bottle supplier to pre-label them. Fortunately it is possible to pass a PSL pre-labelled bottle through a tunnel pasteuriser. However, always check the colours first to ensure that they are not affected by the heat!

Pre-labelling is, of course, expensive, and this is added to the cost of an already expensive label. In most cases only the alcopops and premium beers are being labelled in this way. The trouble is that to put in an extra labeller means an additional conveyor and space which many lines do not have. Suppliers of labellers, seeing the opportunity, have introduced a modular labeller which allows the wet glue label stations to be removed from the carousel and to be replaced by PSL labelling stations. A trolley device was designed to allow the changeover procedure to be as easy as possible. This concept was introduced four years ago and has had teething problems with the lining-up of the stations due to the differing floor levels. These problems have now been overcome. A labeller has also been developed with all the stations in place, allowing the user to switch over from one type of label to another. So the choice is to have either all the modules in place – say six stations, three for wet glue and three for PSL – or just three stations which are interchangeable. The latter arrangement will allow a smaller carousel to be used at a lower cost. See Fig. 14.3.

For those used to wet glue labelling, it should be said that one can be fooled into thinking that applying a self-adhesive label is a simple business. The application of the plastic label to a bottle can be compared to applying a sticker to a glass window, which is not easy when it is wet and often ends up with an air bubble! So the bottles must be dry, and the wipe-down time must be sufficient!

Presently PSL label suppliers are working on PSL-type labels that can be

Fig. 14.3 Shows an interchangeable labelling station on a wet glue/PSL labeller. Courtesy Krones UK

applied by wet glue stations. A PSL label that can be removed from a returnable bottle in a bottle washer was launched at Drinktec 2005.

The plastic sleeve is not so easy to apply. This has to be carried out as a separate operation on the line and will most likely reduce line efficiencies. Often the bottles will be delivered by the glass manufacturer with the sleeve already applied. Sleeves are purchased in a roll and fed to the applicator which opens the sleeve and then cuts them at the register point, so leaving each sleeve identical. The sleeve is then placed over the bottle, and as it leaves the applicator, there is usually a small flail that assists in assuring that the sleeve has reached the bottom of the bottle. This is the easy part. Achieving even shrinkage is the most difficult, as the heat must be directed to the correct part of the bottle, usually the base, in order to tighten the sleeve onto the bottle in the correct position. The heat can be dry or wet. As it is easier to shrink film onto an empty bottle dry heat is often used, but it is accepted that wet heat (steam) gives the most satisfactory and consistent result. For a full bottle only steam will work satisfactorily. This is because the steam penetrates the plastic material and as a result gives a better result. The setting up of the tunnel in which shrinking takes place is vital, especially for a curved bottle.

Maximum filler outputs for cans and bottles tend to have stabilised to 1500 cpm (cans per minute) and 1000 bpm (bottles per minute) respectively; 2000 cpm can fillers have been tried but the availability figure is not so good. A recent bottling line installation has seen the sharing of a single tunnel pasteuriser by two 50 000 bph (bottles per hour) lines, giving a total output of 100 000 bph (see Fig. 14.4). This approach is clearly a good way to save costs, with an

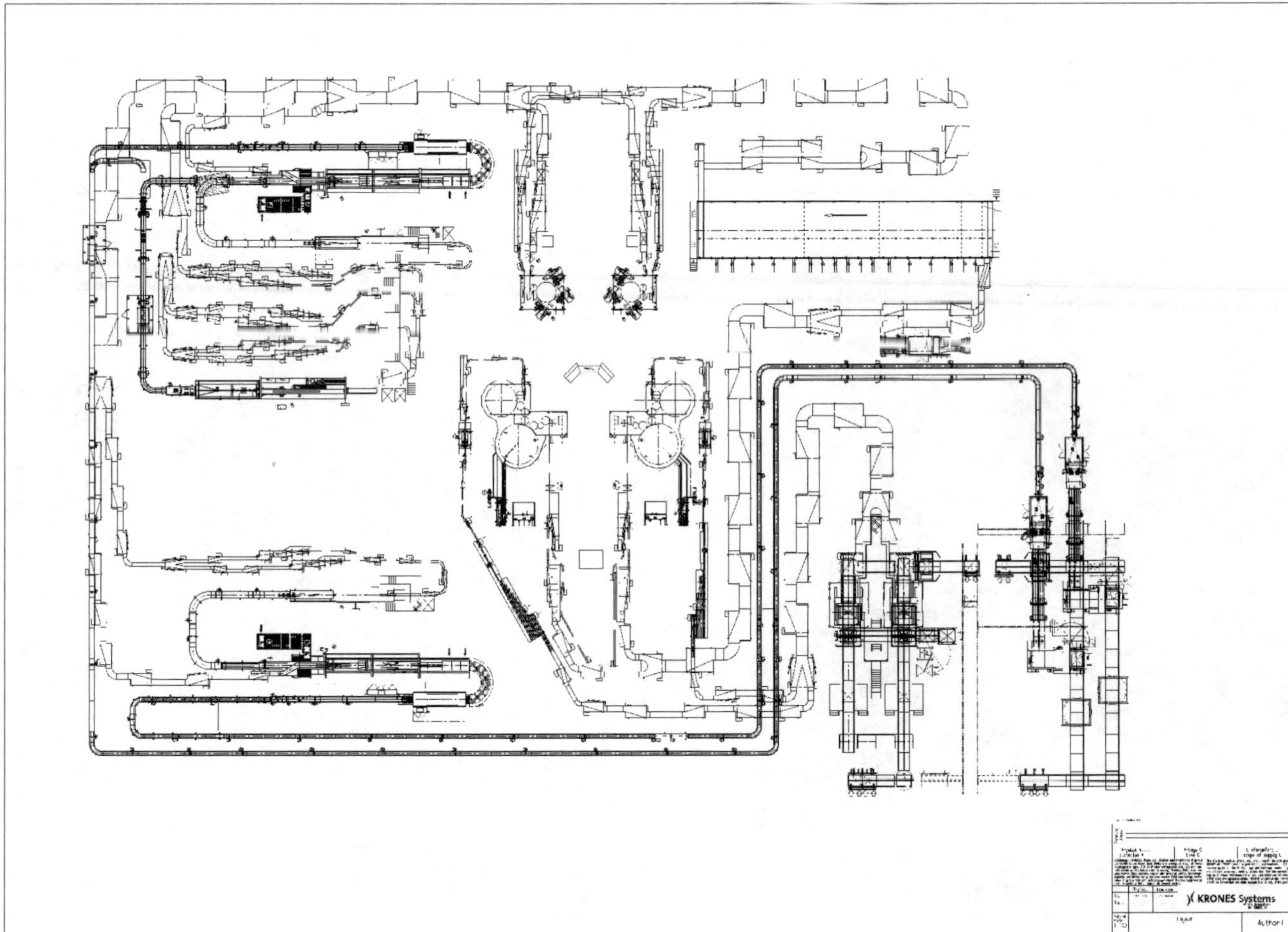

Fig. 14.4 Shows a layout for two 50,000 bph lines using a single tunnel pasteuriser (Courtesy Krones UK).

installed pasteuriser costing in excess of £1m. In the UK, unlike in Germany, it is not normal to run a beer line without a tunnel pasteuriser. However, the temptations are there, with significant savings in capital and revenue costs, and a much simpler layout. Another recent installation has been carried out in the UK using a flash pasteuriser for a 50 000 bph line and the company has no regrets (see Fig. 14.5).

Although robotic depalletisation and palletisation is now freely available, the traditional form of operation is still favoured. What is being seen more often is live-bed delivery of cans and bottles (cans and bottles being transferred from the lorry directly onto the conveyor at the packaging plant), and these being automatically delivered to the depalletiser by conveyor or shuttle. Alternatively a laser guided vehicle (LGV) is being used for all movements into and out of the warehouse. These vehicles now have a much greater built-in intelligence, making this form of material movement very reliable.

14.4 Developments in handling kegs

Technology has also moved forward for kegs, with some excellent machinery now being manufactured. One supply company has the monopoly for rotary fillers and is now producing fillers which have less waste and are more hygienic. The arguments in favour of rotary have moved on, as the maintenance becomes easier, changeovers more straightforward and central distributors more reliable. The next stage is to develop a filler which does not need change parts, i.e. the infeed and discharge starwheels, and the plough.

In smaller installations, lane systems are probably best. However, there is a cross-over point where a rotary installation will give the best financial return, and this is recognised at being around 450 kegs per hour (see Fig. 14.6). For lane systems, as the output goes up extra lanes are required. For example, with the output per lane being 65 kegs/hour for 30 or 50 litre kegs, seven lanes would be required to produce 450 kegs per hour.

When deciding which system to purchase, the following issues need to be considered:

- The 36 gallon and the 100 litre are not so popular, as they are heavy. However, some breweries still use them and a lane system handles them better.
- Slim kegs (20 litre) run better on rotary lines, and these are becoming more popular.
- If one lane should break down it can be worked on while the others are still running. With rotary, if a head fails, the filler (or washer) can still be run with lower output and the repair is carried out later.
- Rotary lines take up less space and are suited to a 'U'-shaped layout.

Another interesting development is in the outsourcing of keg management. This has now become popular in the UK. Increasingly, an active radio frequency

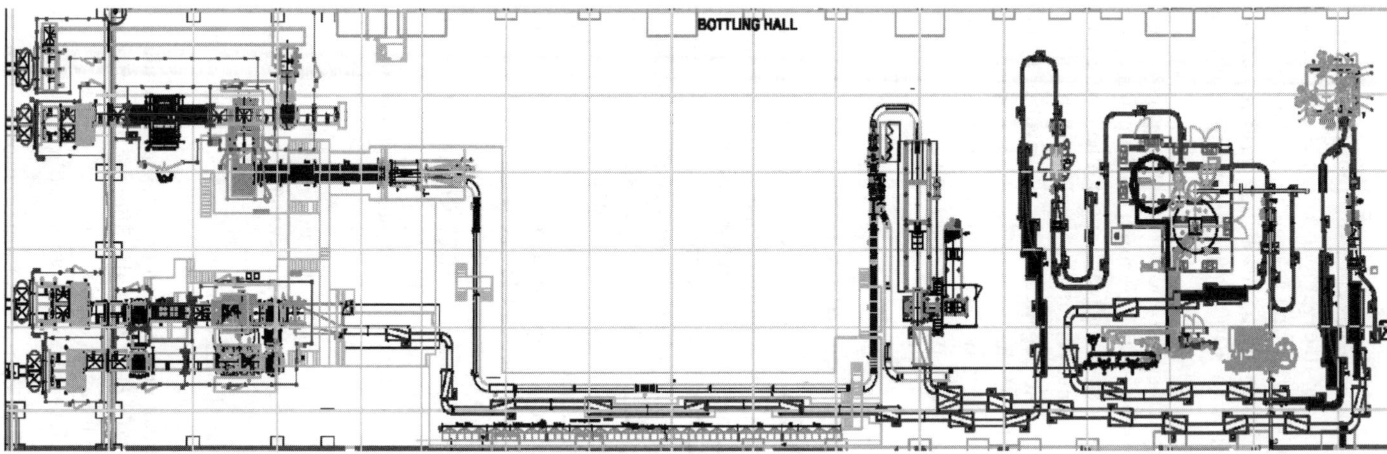

Fig. 14.5 Shows the layout of the 50 000 bph bottling line installed at Inbev, Magor, Wales, UK. The beer is flash pasteurised and sterile filled (Source: courtesy KHS).

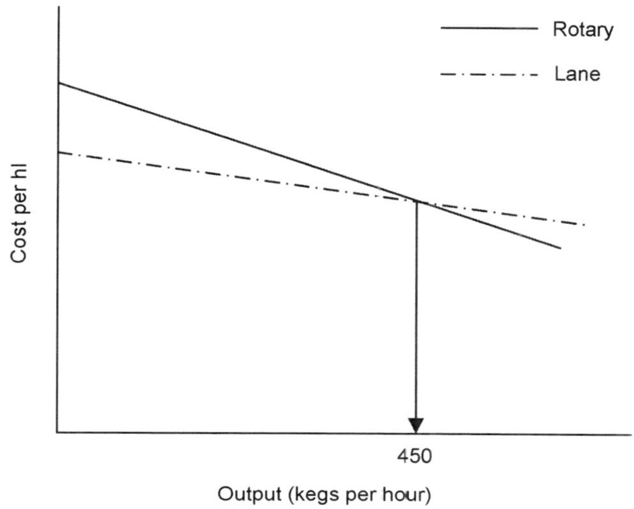

Fig. 14.6 Shows how the costs favour a rotary filler above 450 kegs/hour.

identification (RFID) tag, known as a transponder, is being fixed to each keg to enable information such as tare weight, type of metal, last service and last date filled to be captured.

14.5 Future trends

There has been a lot of consolidation in the industry, with both brewers and suppliers. Major pressure is already being applied by brewers on suppliers to deliver at a competitive cost, and this is now extending to life cycle costs. This requires the supplier to provide costs for servicing the equipment for periods from five to ten years. Some breweries are also asking for consignment spares which give the brewery a spares consignment and spares that are paid for only when they are used. This puts an extra financial burden on the supplier and could lead to higher costs for the equipment and possibly further consolidation in the industry.

As regards small pack trends, the sleek and slim cans look like a new generation of package for premium products such as beer. They look smart and trendy, and two breweries have started using them. New can lines can be designed to handle the sleek can which has a 206 body and a 202 end. The slim can is 200/200 so the can line is likely to be dedicated for this size. For bottles the same complexity of packaging will continue unabated in order to continuously capture the premium market. Glass will continue as the favourite material for beer in the West, and plastic will continue to be used more and more in the East.

A recently developed machine applies sleeves to bottles from plastic film, making sleeving 30% cheaper. With present technology, the sleeve is supplied in tubular form. It is opened up, cut to length and placed over the bottle. This new machine takes the film (not in tubular form) and cuts it to length before transferring it to a vacuumed mandrel on the bottle carousel. The film on the mandrel is then ultrasonically welded before the mandrel drops down, allowing the bottle, which is sitting on top of the mandrel, to follow it down into the sleeve. This development could make sleeving a more attractive proposition. Not only does it look good, but it also makes the glass bottle safer.

The PET bottle will continue to struggle in developed markets due to people's discerning taste for beer. However, in Russia and Asia the existing surge in demand will continue.

For kegs the trend will be for cask beers (UK) to become niche products for the regional brewers. The bright beers (Keg) will be packaged on high-speed rotary lines (1000–1200 kegs per hour) and the 20-litre slim keg could be the next development. The Japanese use this keg extensively. Another pack that could become interesting is the 5-litre keg. Once the presentation and delivery of this pack in the home is solved, it should become popular, especially with the swing to home consumption.

One topic that should be mentioned is RFID (Radio Frequency Identification). Barcodes have been in use for the last 30 years so it is not surprising, with the advances in technology, that smarter propositions are being made. The RFID tag is perhaps the most talked about, but it potentially adds to the cost of packaging and therefore must be justified. The dream is to be able to identify everything in the warehouse instantly. The analogy is a room full of people all shouting out their names to one individual in the room, and that individual being able to hear each name separately! The transmitting of data over distances has, however, been difficult due to the power of the antenna in the tag that is required to do this. As mentioned in Section 14.4, RFID technology is starting to be used for kegs, but the tags are large and are being used only on high value items. For small packs, the cost needs to be less than €0.05 per tag. The present cost is around €0.20 to €0.30 and, with the added costs of label conversion, losses in application and encoding, these costs could triple. The cost of manufacture will reduce significantly when billions of tags are produced, but that would mean a wholesale change to this technology. High cost items such as perfumes and spirits could tolerate higher costs but these have to be balanced against the cost savings in the supply chain.

The barcode is not dead yet. The traditional 1-D barcode is simple and once originated is easily printed onto the package. However, the 1-D barcode is unable to carry the 96-bit information required by electronic product coding, and this has led to the development of 2-D barcodes which will be able to compete with RFID technology (White, 2005).

Another topic that receives a lot of press is nanotechnology. This technology has been in use for some time in the printing, paper and packaging industries, in some instances for up to 20 years. Nanotechnology has gained momentum with the use of atomic force microscopy (AFM) and scanning probe microscopy (SPM), which allow technologists to know what is happening on a nanoscale.

This allows them to manipulate materials to give desired properties like better heat resistance, greater strength and improved electrical conductivity. More recent examples of what is being worked on are more intelligent RFID, nanobarcodes, intelligent packaging (e.g. information on freshness of package contents), antistatic properties, better UV protection, gas barrier protection in PET, and so on. The applications are without limitation (Pira International, 2005)!

14.6 Sources of further information

BROWNE J and CANDY E, *Excellence in the Packaging of Beverages*, Hook, Binsted, 2001.

HANLON J F, KELSEY R J and FORCINIO H E, *The Handbook of Package Engineering* (3rd edn), Lancaster, PA, Technomic Publishing, 1998.

SOROKA W and EMBLEM A, *Fundamentals of Packaging Technology* (3rd edn), Institute of Packaging, 1999.

The Future of Intelligent Retail Packaging, Leatherhead, Pira International, 2004.

The Future of Nanotechnology in Printing and Packaging, Leatherhead, Pira International, 2005.

14.7 Acknowledgements

KHS Kisters Ltd, Unit 6 Monkspath Business Park, Highlands Road, Shirley, Solihull B90 4NY.

Krones UK Ltd, Westregan House, Great Bank Road, Wingates Industrial Park, Westhoughton, Bolton BL5 3XB.

Simonazzi UK Ltd, Unit 2, The IO Centre, Hearle Way, Hatfield Business Park, Hatfield, Hertfordshire AL10 9EW.

14.8 References

PIRA INTERNATIONAL, 2005, 'A new frontier', *Flexible*, 4(1), 27–29.

TUTSSEL G, 2005, 'Enterprise IG', *Packaging Magazine*, 19 May, 6.

WHITE N, 2005, 'Smart pricing', *Flexible (Pira Publication)*, 3(6), 33–37.

15

Modern brewery sanitation

D. Loeffler, Loeffler Chemical Corporation, USA

15.1 Introduction

Brewing is an old tradition, and a romantic flair is usually associated with the 'art' of brewing. Creativity, experimentation and individuality generally lead to the creation of new beer styles or sometimes simply a great new beer recipe. Cleaning and sanitizing, on the other hand, is a very scientific process, and should be understood and approached as such. Technology advancements in many areas of brewing have made quantum leaps over the past 10 years, and the chemical industry is no exception. With the development of more versatile products containing ever more powerful ingredients, cleaning is quickly becoming a very complex and scientific area of brewing. It would take the entire book here just to discuss new surfactant technology alone. Therefore, I will focus on providing the knowledge and background information necessary to make intelligent choices when selecting the cleaners, additives and sanitizers/disinfectants for a successful and comprehensive sanitation program.

One of the most important factors is to develop a comprehensive sanitation program that is applied throughout the entire production process. It is extremely important to incorporate all areas in the sanitation program. What good does a great sanitation program in the brewhouse and cellar do, if the packaging department contaminates the beer during the filling process? Why spend millions of dollars on low air if the beer is already spoiled when it is packaged? Every brewery should have a designated 'brewery sanitarian' who directly reports to the plant manager or owner and supervises the sanitation protocols in every area of the brewery in close cooperation with the individual department heads. A successful sanitation program incorporates all operational aspects of a brewery, from raw material storage all the way into the package. Only then can a brewery maintain its quality over time, assuming that the quality control for the raw materials is equally strong and enforced.

But let's come back to cleaning and sanitizing, also commonly referred to in Europe as R&D (German for '*R*einigung & *D*esinfektion' – Cleaning & Sanitizing/Disinfecting). Before we discuss how to clean and sanitize, we should define a few key terms.

15.2 Sanitation terminology

As time passes, phrases and terms tend to be used for areas that far exceed their original meaning and definition. Today, almost any type of cleaning not directly involving hand scrubbing is commonly referred to as CIP. However, for the purpose of our discussion, we shall define the terms CIP, COP and foam cleaning as follows.

15.2.1 CIP – Cleaning in place

Cleaning in place shall be defined as a method of cleaning equipment without the need to disassemble the equipment prior to cleaning. It is commonly used for the interior cleaning of brewing equipment such as tanks, brewhouse vessels, heat-exchangers, pipe-systems and hoses, and other production equipment such as fillers, etc. The equipment is cleaned without any major modifications to the equipment itself. Cleaning solutions are generally transported via pumps or applied via special spraying devices. Automated cleaning via CIP systems has the following benefits:

- Consistent and repeatable cleaning results
- Safe operation without much exposure to employees
- Very economical operation.

15.2.2 COP – Cleaning out of place

Cleaning out of place is essentially the opposite of CIP, and refers to most manual cleaning applications. Either the equipment must be broken down into pieces or major modifications must be performed before the cleaning can take place. Some equipment that is normally cleaned via CIP should be cleaned periodically in a COP mode, such as heat-exchangers. Therefore, COP cannot and should not be completely avoided. Some of the drawbacks and benefits of COP are:

- Cleaning results may vary with operator/employee
- Exposure of employee to cleaning solutions
- Time-consuming process
- Easily verifiable through visual inspection
- May expose residuals left by CIP cleaning.

15.2.3 Foam cleaning

Foam cleaning is a method for cleaning large, exterior surfaces. Special chemicals with foaming agents or gels are applied via low pressure foam units. Foaming is a very economical method for cleaning large exterior surfaces that cannot be cleaned via CIP. Here is an example. One liter of chemical and 49 liters of water combined with 450 liters of air produces 500 liters of foam, which is enough foam to cover an area of approximately 200 m^2. Before applying foam, each surface should first be rinsed with ambient water. Pressure up to 30 bars may be used for the rinse. The foam itself can be applied with either low or high pressure foamers. However, low pressure foam cleaning has clear advantages over high pressure foam cleaning and is today the preferred method of use. The surface to be cleaned is applied with a thin film of foam, and then left for 10–20 minutes to react with the soil on the surface. After the 10–20 minutes, the foam is then rinsed off at 10–30 bars with warm water (the temperature should be above the melting point of any fat or grease present).

Foam application with low pressure units has displaced high pressure for the following reasons:

- No physical/mechanical impact or damage on equipment
- No aerosol formation
- No cross-contamination from deflected cleaning solutions into already cleaned areas
- Foam is visible during application and rinsing, making it easier to cover all areas.

The importance of a well-balanced mixture of air, water and chemical must be stressed. The most desirable foam is wet foam; however, if the foam is too wet, sufficient contact times cannot be achieved. A perfect blend of water, air and chemical can yield very effective and highly cost-efficient cleaning results. So-called hose-end foamers do a mediocre job at best, and usually do not allow for adjustments of air. Commercial-grade foam units yield the best results and their slightly higher cost will pay off quickly due to their ability to perfectly blend water, chemical and air. Dry foam clings well to the equipment, but does not clean very well. Very wet foam cleans well, but typically lacks the required contact times since it quickly runs off the equipment.

Figure 15.1 shows the mechanics behind foam cleaning. In Fig. 15.1(a), we can see how the foam adheres to the surface to be cleaned. The cleaning solution is encapsulated in the air bubbles. In Fig. 15.1(b), we see how the bubbles immediately contacting the surface are bursting, thereby releasing the encapsulated cleaning solution onto the surface. A cleaning film has been formed (Fig. 15.1(c)). This cleaning film is constantly being renewed by additional bursting bubbles, resulting ultimately in an extended contact time. The cleaning solution inside the foam bubbles does the cleaning, which is why wet foam will result in better cleaning results.

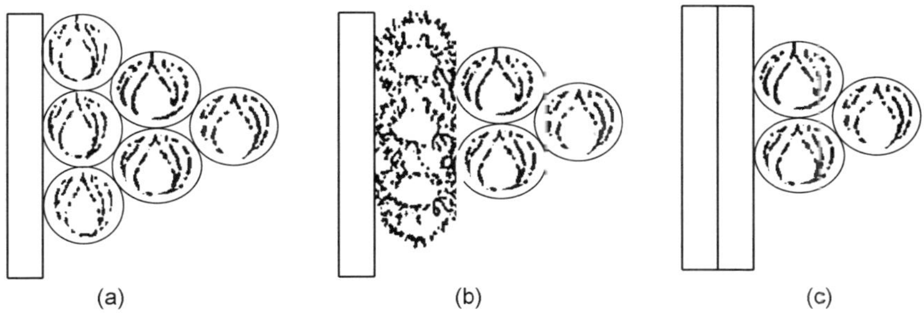

Fig. 15.1 The mechanics of foam cleaning.

15.3 Theoretical aspects of cleaning

Soil adheres to surfaces in very complex ways. It can be trapped mechanically in pores, cracks or other inclusions, which explains the choice of hard-surface materials such as finished stainless steel. We also see electrostatic binding forces, both between the surfaces and the soil as well as between different types of soils such as protein and mineral salts. The sum of all these binding forces combined can be expressed as the adhesion energy, which is the energy that has to be achieved during the cleaning process to remove the soil. During cleaning, the adhesion energy is derived by combining the energy from chemicals, mechanics, and temperature, whereas the energy from these three components is interchangeable within certain limits.

The process of soil removal can be divided into four major steps:

1. Transport of the cleaning solution to the soil with complete wetting of the soil.
2. Chemical reactions and physical processes during the cleaning process:
   - Reaction of the cleaning solution with hard water constituents and/or suspended soil
   - Convective and diffusive transport of the cleaning agents from the cleaning solution to the soil
   - Wetting or transport of the cleaning agents within the soil itself
   - Cleaning reaction with the soil, both chemically and physically
   - Diffusive transport of soil particles removed during the cleaning process.
3. Removal of the soil from the surface and transfer into the cleaning solution via dispergation and/or emulsification.
4. Prevention of re-depositing removed soil through stabilization in the cleaning solution and transport of removed soil away from the surface.

The cleaning result is always determined by the following four factors:

- Chemical [C]

- Mechanical energy [M]
- Temperature [T]
- Time [Z]

These four factors are interchangeable within certain limits.

The following formula is desirable:

$$C_R = C + M + T + Z$$
$$C_R = \text{cleaning result} = \text{constant}$$

We also have certain parameters influencing our cleaning results. The three major groups of parameters influencing cleaning results can be divided as follows.

Equipment parameters
Equipment parameters are generally determined during the construction or the purchase of the production equipment. Important is the proper design of the equipment, especially with respect to eliminating equipment parts not contacted by cleaning solutions. Equally important is the proper choice of materials with respect to cleaning and sanitizing. Materials with high chemical compatibility and smooth surfaces as well as minimal electrostatic binding forces for industry-relevant soil particles should be chosen. It is therefore important to start planning for optimal sanitation during the planning and construction phase of the brewery.

System parameters
System parameters can be defined as factors that are determined through the regular operation of the brewery such as the chemical composition and quantity of the soil as well as the quality of the process water, particularly the water hardness.

Operational parameters
Operational parameters are the parameters that influence the cleaning result during the cleaning process itself. Here, we have again the four classic parameters of the *Synergistic Circle* (Fig. 15.2):

- [C] – Chemical properties of the cleaning solution, such as composition, concentration, surface tension, activity, etc.
- [M] – Mechanical properties of the cleaning solution, such as flow velocity, Reynolds number, etc.
- [T] – Temperature of the cleaning solution
- [Z] – Contact time of the cleaning solution.

Not all parameters have to be constant during the cleaning process. An optimal cleaning process may be obtained through carefully combining and/or interchanging selected parameters.

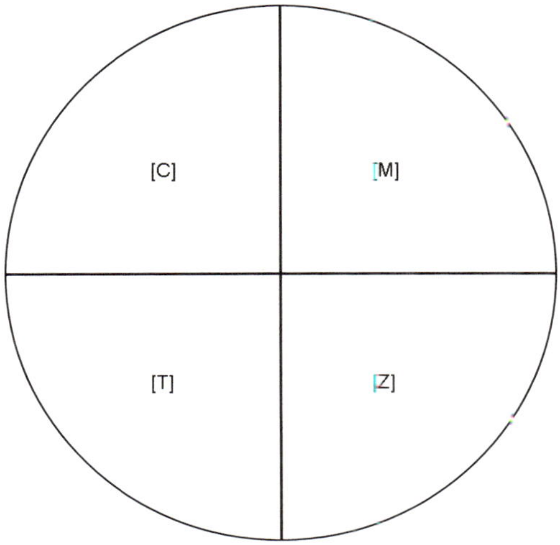

Fig. 15.2 The four parameters of the *Synergistic Circle*.

15.4 The use of water in cleaning

Except for a few special cases, such as disinfecting gases, all cleaners, detergents and sanitizers/disinfectants are used diluted. The exclusive agent for diluting chemical agents is almost always water. In addition to providing a medium to obtain the correct solution strength, water is also the main instrument to provide temperature and mechanical energy to the cleaning process. Considering the fact that over 95 percent of a chemical cleaning solution is made up from water, we can safely say that water is of central importance to the cleaning and sanitizing process. With water often being the only 'chemical agent' used in the important first step of cleaning, the prerinse, we can also say that water is truly a part of our fleet of cleaning chemicals.

Some of the physically important factors influencing cleaning results are the boiling point, surface tension, kinetic resistance, electrical conductivity and pH value. The water available to breweries always contains salts in solution. Among all the salts in water, calcium and magnesium are of primary importance. It is these two salts that determine the hardness of the water. The hardness present in the form of carbonate or hydrogen carbonate forms the carbonate hardness. However, earth alkalis can also be bound to Cl^-, SO_4^{2-}, and PO_4^{3-}, which make up the non-carbonate hardness. The sum of carbonate and non-carbonate hardness equals the total hardness of the water. General guidelines by the US Department of the Interior, US Geological Survey, for classification of waters in the United States are that 0 to 60 mg/l (milligrams per liter) calcium carbonate is classified as soft water, 61 to 120 mg/l as moderately hard water, 121 to 180 mg/l as hard water, and more than 180 mg/l as very hard water.

In addition to these earth alkalis, several heavy metals, particularly iron and manganese, should also be mentioned, as they can react with the complexing agents in the cleaners. They are also relevant with respect to corrosion as they are important components of the water's redox system. Next to the already mentioned anionic components, chloride is especially important with respect to the corrosive properties of the water. Chloride content in municipal and/or process water can be as high as 100–600 mg/l, but should be kept to a minimum, if necessary through treatment, as 40–50 mg/l are sufficient to cause severe pitting under the right circumstances.

In addition to the salts, we also find gases in solution, such as O_2 and CO_2. Carbon dioxide is especially important, since it can significantly influence the amount of carbonate hardness:

$$CaCO_3 + CO_2 + H_2O \rightleftarrows H_2CO_3 + CaCO_3 \rightleftarrows Ca(HCO_3)_2$$

As the temperature of the water rises, increasing amounts of CO_2 are being released. Consequently, the balance of the water shifts towards $CaCO_3$. Due to its much lower solubility (1.4×10^{-2} g/l) as opposed to that of $Ca(HCO_3)_2$ (166 g/l), the carbonate in hard water precipitates out during heating as 'limescale' or hard water scale.

The importance of good quality water cannot be stressed enough. I frequently come across hot liquor tanks that are corroded beyond repair and covered in rust and limescale several centimeters thick. Today's technologies offer solutions for every type of water, from elaborate filter systems through simple water softeners to ion exchangers and reverse osmosis systems. With water being such an important part of every cleaning and sanitizing solution, it should get the attention it deserves.

15.5 Detergents/cleaners

15.1.1 Commodity chemicals

Commodity chemicals such as sodium hydroxide, nitric acid, phosphoric acid, sodium hypochlorite (bleach), etc., cannot be considered sanitation products. Due to the broad use of these chemicals throughout the chemical industry, many of them often do not even satisfy the quality requirements set forth by most responsible producers of sanitation products. For example, sodium hydroxide should have a low chloride content to minimize the potential for corrosion during cleaning. Nitric acid must contain inhibitors to prevent the formation of nitrous gases. Many of these characteristics are not present in commercial-grade commodity chemicals. But it is primarily the lack of performance of these commodity chemicals that renders them unsuitable for use as cleaners and/or sanitizers. The use of complex formulated products, also known as 'built' products, is far more economical than the use of commodity chemicals.

15.5.2 Built cleaners/detergents

The performance requirements for modern cleaners are high due to the number of chemical and physical processes that are taking place during the cleaning process. The minimum requirements for a modern cleaner/detergent are:

- Fast and complete solubility in water
- Good wetting properties
- Good dispersing and suspension properties
- Good fat emulgating properties
- High sequestration properties
- Good defoaming properties in CIP applications
- Non-corrosive on applicable material
- Fast and complete rinsability
- Biodegradable
- Low toxicity.

Without a single chemical available that would satisfy even half of these requirements, it is necessary to combine several different chemical agents in order to formulate a product with all or most of these before-mentioned properties. In these often complex products, each individual component fulfills a special requirement.

Cleaners are classified by pH into three major categories:

- Acid cleaners
- Alkaline or caustic cleaners
- Neutral cleaners.

We also differentiate cleaners by their method of application into subcategories:

- CIP cleaners
- Foam cleaners
- COP or manual cleaners.

Finally, we differentiate between 'built' cleaners and additives. Built cleaners are generally products that are formulated for a specific application and merely require dilution with water for use. Additives on the other hand are added to either 'built' cleaners or even commodity chemicals such as sodium hydroxide in order to enhance the performance of these products in certain applications or to provide special characteristics, such as oxidation. Additives are usually very complex products that are designed to work in a wide variety of applications and to boost the performance of the products they are added to

The most important components in cleaners and their properties are as follows:

- Alkalis such as sodium hydroxide and/or potassium hydroxide, soda ash, sodium hydrogen carbonate, sodium metasilicate, trisodium phosphate, etc., which act as the major components for removing organic soil.

- Complexing agents (also know as chelating agents), which react with the calcium and magnesium ions of the water and prevent the precipitation of water hardness.
- Sequestering agents, which are important in alkaline formulations in addition to the complexing agents.
- Surface active ingredients (surfactants, wetting agents) such as anionic, cationic, amphoteric and nonionic surfactants, depending upon the electrical charge. Surfactants are used for the removal of water-insoluble soil as well as to reduce the surface tension of the water.
- Acids such as phosphoric acid, nitric acid, amidosulfonic acid, and urea nitrate, and urea phosphate, as well as other organic acids, which are primarily used for the removal of inorganic deposits.
- Oxidizers such as sodium hypochlorite, hydrogen peroxide, sodium percarbonate or other oxygen donators, which are used to chemically modify deposits through oxidation.
- Defoaming agents such as special alkalene oxide derivates may be used to suppress foaming.
- Corrosion inhibitors, often of a complex organic nature, are used to protect materials from chemical attack.
- Enzymes, such as proteases, may be used for special cleaning applications such as membrane cleaning, etc.

The chemical ingredients in cleaners and detergents are commonly referred to as 'builders', thus the definition of the 'built' cleaner.

Alkalis

The composition of alkaline components in a cleaner determines its alkalinity. Acidic product residues as well as carbon dioxide can partially neutralize and reduce the original free alkalinity. The hardness constituents of the water can also reduce the free alkalinity in a cleaner, as shown in the following equations:

$$Ca(HCO_3)_2 + 2NaOH \rightarrow CaCO_3 \downarrow + Na_2CO_3 + 2H_2O$$
$$CaSO_4 + Na_2CO_3 \rightarrow CaCO_3 \downarrow + Na_2SO_4$$

Sodium hydroxide provides the most alkalinity, which is one of the key factors in removing organic soil in brewery cleaning. Potassium hydroxide is sometimes used in combination with sodium hydroxide to improve rinsability and/or to reduce the freezing point of the chemical concentrate. Potassium also contributes to the cleaning effectiveness of the cleaner. In order to accommodate automated dosing equipment in modern brewery CIP systems, most products today are sold in liquid form. Solid products based on sodium carbonate, soda ash and sodium metasilicate are slowly being phased out. Lately, products based on sodium metasilicate have become increasingly popular in the US and are marketed as so-called 'non-caustic' alternatives to traditional NaOH-based cleaners. While compatible with a variety of soft metals such as copper and brass, these products generally require much higher use concentrations in order to remove protein deposits and thus are not very economical to use. Another negative aspect is the problematic rinsability of these products with the potential for permanent silicate deposits in the equipment if allowed to dry.

Acids

- *Mineral acids*. The corrosivity as well as the incompatibility of most mineral acids with other active components commonly used in cleaning products limits their use to primarily phosphoric acid and nitric acid. Sulfuric acid may be used at temperatures not exceeding 30°C. Hydrochloric acid should be avoided at all cost.
- *Organic acids*. The most important criteria for organic acids are their odor (short-chained carbonic acids), solubility and strength. Commonly found products from this group are formic acid, oxalic acid, citric acid and lactic acid.

Beerstone and mineral deposits are primarily based on mineral components. These types of deposits are virtually impossible to remove with alkaline products alone. However, acids will take water-insoluble salts and chemically transform them into a soluble, rinseable form. For example:

$$Ca_3(PO_4)_2 + 4HNO_3 \rightarrow Ca(H_2PO_4)_2 + 2Ca(NO_3)_2$$

Complexing agents

Complexing agents are commonly used in alkaline products to prevent the precipitation of water hardness as well as to remove very small films of mineral deposits. The most commonly used complexing agents are EDTA (ethylene-diaminetetraacetate), NTA (nitrilotriacetate), gluconate, phosphonate, and polymerphosphates. Occasionally, sodium glucoheptonate is still used in selected formulations, but it is being more and more phased out due to its strong odor, tar-like properties and tendency to discolor plastics.

Sequestering agents

Sequestering agents are essentially complexing agents. However, they are special as they are able to prevent the crystallization of calcium and magnesium salts by disturbing their crystal structures (the threshold effect) at extremely low concentrations in the under-stoichiometric environment. Sequestering agents are extremely important for the rinsability of alkaline cleaning solutions. Complexing agents are here effective usually only in a stoichiometric environment and cause problems during the increasing dilution of the cleaning solution with the rinse water. A rapidly decreasing concentration of complexing agents is faced with an increasing amount of hard water constituents from the rinse water. Thus, hard water precipitation can occur with alkaline solutions and/or on hot surfaces, resulting in hard water scale deposits.

Surfactants

Surfactants reduce the surface tension of cleaning solutions, thereby providing the solution with better wetting properties. Each surfactant combines in its molecule a hydrophilic and a hydrophobic group. Therefore, surfactants are vertically adsorbed at the water interface, causing a compressive force to act on the surface. This reaction reduces the relatively high surface tension of the water (72 dyn/cm) to a minimum. A low surface tension allows the solution to easily penetrate the soil both from behind at the equipment interface as well as through cracks and inclusions in the soil, resulting in much faster soil removal (capillary activity). Surfactants are also able to emulgate hydrophobic, and therefore water-insoluble, deposits such as fat. Most foam cleaners use anionic surfactants such as Alkyl Sulfonates due to their foaming properties. Most CIP cleaners use nonionic surfactants such as ethoxylated fatty alcohols because of their low foaming properties. It has been primarily improvements in the surfactant area that have led to constant performance improvements and have helped to shape the modern cleaners of today.

Defoamers
Defoamers play an important role in many CIP products, as foaming can render a product useless in a CIP application. Mostly nonionic surfactants such as ethoxylated or propoxylated fatty alcohols with low cloud points are used here. These types of surfactants are able to suppress both foam caused by other surfactants as well as foam caused by cleaning byproducts such as saponified proteins as they reach or exceed their cloud point (beginning of water insolubility). The great benefit of this type of defoamer is that as soon as the temperature of the solution falls below the cloud point again, as is the case during an ambient rinse, the surfactant becomes water soluble again and can easily be rinsed off. It is important to note that products containing silicone and paraffin defoamers must not be used for food and beverage applications.

Builders
Builders are substances that synergistically improve the detergency and efficiency of surfactants. They also significantly promote the dispergation of many soil components, such as protein. Products worth mentioning here are phosphonates, polyphosphonates, polyacrylates, NTA, gluconates and citrates.

Soil dispergators
Soil dispergators are used to increase the soil load tolerance of cleaning solutions, thereby preventing redeposition of removed soil to the cleaned surfaces. Silicates, polyacrylates, phosphonates and polyphosphonates are commonly used. We can see here that some materials have multiple properties (see builders above).

Oxidizers
Oxidative additives can be found in 'built' cleaners as well as in additives. Oxidative additives using oxygen donators such as hypochlorites and peroxides can assist in the removal of burnt-on deposits (protein, hop-resin) and insoluble, high-molecular-weight compositions (carbon) as well as destroy color pigments

(bleaching). Hypochlorites are today being replaced more and more by oxygen donators from the peroxide groups such as hydrogen peroxide, sodium percarbonate and similar products, due to the high corrosivity of hypochlorites on stainless steel and the well-known wastewater problem of hypochlorites (AOX, TMH, etc.).

Corrosion inhibitors
Corrosion inhibitors have lost some of their importance, as modern breweries have turned to stainless steel as the predominant material for production equipment. However, corrosion inhibitors are still used whenever soft metals such as aluminum, brass, copper or iron are present. Most modern corrosion inhibitors are of a complex organic nature and formulated to protect a specific metal in either acid or alkaline environments. However, traditional products such as silicates and nitrogen compounds can still be found in some products.

Solubilizers
Solubilizers are used in some formulations to 'force' certain ingredients into solution (e.g. hydrotopes). These products do not contribute to the cleaning performance of the product and their use is generally restricted to an absolute minimum to keep cost low.

15.6 Mechanical aspects of cleaning

The mechanical aspects of cleaning are an important part of the cleaning program, and often are not completely understood. Resulting problems are often blamed on chemical performance. While well-formulated cleaners can help in overcoming mechanical shortcomings, a truly cost-efficient and cost-effective cleaning program relies heavily on properly applied mechanical energy.

As detailed in the previous section, cleaners work chemically through dispergation, emulsifying, saponification, etc. However, once the soil has been chemically altered to the point where it is soluble, mechanical energy is always required to physically remove it from the equipment surface and transport it via the cleaning solution away from the surface.

From a mechanical standpoint, soil removal takes place in several steps:

- Prerinse removes loose soil not chemically or physically attached to the surfaces.
- Chemicals are transported to the soil to chemically alter the deposits, allowing mechanical forces (water) to remove it.
- Chemically altered deposits are removed via fluid mechanics.

Going back to the Synergistic Circle, four criteria influence a successful sanitation program: chemicals, mechanics, temperature and time. The mechanical aspects affecting our cleaning results are primarily pressure, volume and flow velocity.

In manual cleaning applications, mechanical energy is applied via brushing, scrubbing or other physical means. However, once CIP cleaning is utilized, we are relying almost exclusively on the cleaning solution and the mechanical forces it exerts to remove deposits from the equipment. Pressure, flow volume and flow velocity are determined and affected by the CIP supply and return pumps, installed CIP sprayballs, pipe diameters and lengths, etc.

15.6.1 Tank cleaning mechanics

Two different types of technologies are used for cleaning tanks and other closed vessels in breweries:

- High pressure cleaning (impingement cleaning)
- Low pressure cleaning (film or chemical cleaning).

High pressure cleaning, also commonly referred to as impingement cleaning or hydrodynamic cleaning, relies on the direct impact of a cleaning solution to provide the mechanical energy necessary to remove the soil deposits. The deposits are literally blasted away from the surfaces. Using pressures generally ranging from 3–12 bars, water is sprayed through small openings of a mechanically operated spray system, also sometimes referred to as a 'cleaning machine'. The theoretical concept behind these impingement cleaning systems relies on a mechanically determined, time-related, increasingly smaller spray pattern where in theory all areas of the tank are directly impacted by the spray, thereby blasting the deposits off the tank surface (see Figs 15.3(a) and (b)).

Usually these cleaning machines must be run for a predetermined amount of time in order to contact the entire surface of the tank. Experience has shown that

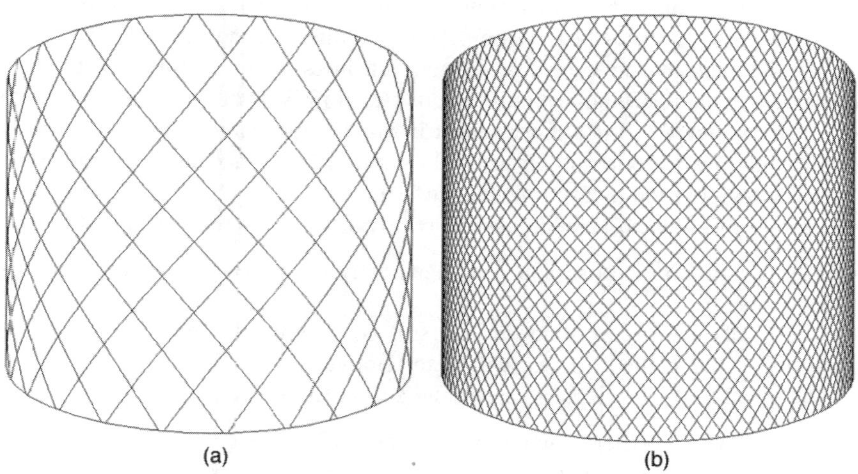

Fig. 15.3 (a) Large spray pattern (b) small spray pattern (Images of Butterworth C&D Inc. Houston, USA).

this type of technology is successful in only a limited number of applications, due in part to the following reasons:

- Tank cleaning cannot be easily combined with line cleaning.
- The equipment suffers from high mechanical wear and tear (needing increased maintenance).
- High pressure is not applicable for all surfaces.
- Any mechanical malfunction will leave soil residuals as a result.

For these and other reasons, many breweries rely on low pressure cleaning technology for cleaning tanks. As opposed to impingement cleaning, low pressure cleaning can be considered true chemical cleaning.

A few criteria must be fulfilled to ensure consistently good cleaning results using low pressure systems:

- The sprayballs must be properly sized and engineered for the tank diameter, volume and height.
- The sprayballs must be self-cleaning due to the potential for the sprayball being immersed in the product.
- The tank walls (including any installed instruments, etc.) must be evenly applied with cleaning solution.
- The applied cleaning film must have a sufficient thickness and volume in order to produce turbulent flow on the tank walls.
- As the applied cleaning film runs down the walls of the tank, the flow velocity and turbulence at the tank's wall surface supplies the mechanical energy to carry away any deposits that have been emulgated, dispersed or otherwise chemically modified.

As we can see in Fig. 15.4, low pressure CIP sprayballs come in many different variations and configurations. We differentiate between static sprayballs and rotating sprayballs. Rotating sprayballs can sometimes be beneficial when isolated thick soil deposits are present and direct impingement energy is needed in a small area. This method of cleaning should not be mistaken for true impingement cleaning. While rotating sprayballs can provide additional mechanical energy, they are also known to occasionally jam due to material trapped in the sprayball.

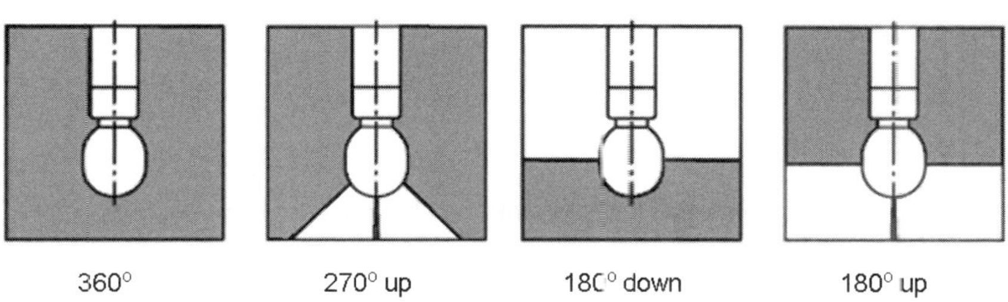

Fig. 15.4 Types of CIP sprayball (Reproduced with permission of Alfa Laval Inc., Richmond, USA).

It is important not to switch sprayballs between different-sized tanks without consulting the manufacturer of the sprayball or calculating the required spray volume and spray pattern for the tank and sprayball first.

For pipes and other closed systems such as heat-exchangers, filling equipment, etc., different mechanical requirements apply. Usually, pipes and other closed systems are cleaned by filling the entire system with cleaning solution followed by circulating the solution. Once the deposits have been solubilized and dispergated by the cleaning agents, the particles need to be transported away from the surfaces by the moving solution. However, the mechanical energy necessary to transport the particles away from the surface occurs only when turbulent flow is present.

By definition of fluid dynamics, turbulent flow is a flow regime characterized by semi-random, stochastic property changes. This includes low momentum diffusion, high momentum convection, and rapid variation of pressure and velocity in space and time. Flow that is not turbulent is called laminar flow. The (dimensionless) Reynolds number characterizes whether flow conditions lead to laminar or turbulent flow. Imagine the flow of water over a simple, smooth object, such as a sphere. At very low speeds the flow is laminar, i.e., smooth (though it may involve vortices on a large scale). As the speed increases, at some point the transition is made to turbulent ('chaotic') flow. In turbulent flow, unsteady vortices appear on many scales and interact with each other. Drag due to boundary layer skin friction increases. The structure and location of boundary layer separation often changes, sometimes resulting in a reduction of overall drag. Because the laminar–turbulent transition is governed by the Reynolds number, the same transition occurs if the size of the object is gradually increased, or the viscosity of the fluid is decreased, or the density of the fluid is increased (see equation below). When designing piping systems, turbulent flow requires a higher input of energy from a pump than laminar flow. However, for applications such as heat-exchangers, pipes and other closed systems, turbulent flow is essential for obtaining good cleaning results. For the praxis, this means that we need to know the Reynolds number present in our system.

The Reynolds number is the most important dimensionless number in fluid dynamics and provides a criterion for determining dynamic similitude. Where two similar objects in perhaps different fluids with possibly different flow rates have similar fluid flow around them, they are said to be dynamically similar. The Reynolds number is generally given as follows:

$$\mathrm{Re} = \frac{\rho v_s L}{\mu} \quad \text{or} \quad \mathrm{Re} = \frac{v_s L}{\nu}$$

where v_s = mean fluid velocity
L = characteristic length (equal to diameter $2r$ if a cross-section is circular)
μ = (absolute) dynamic fluid viscosity
ν = kinematic fluid viscosity: $\nu = \mu/\rho$
ρ = fluid density.

The Reynolds number is the ratio of inertial forces ($v_s\rho$) to viscous forces (μ/L) and is used for determining whether a flow will be laminar or turbulent. Laminar flow occurs at low Reynolds numbers, where viscous forces are dominant, and is characterized by smooth, constant fluid motion, while turbulent flow, on the other hand, occurs at high Reynolds numbers and is dominated by inertial forces, producing random eddies, vortices and other flow fluctuations. The transition between laminar and turbulent flow is often indicated by a critical Reynolds number (Re_{crit}), which depends on the exact flow configuration and must often be determined through experimentation. Within a certain range around the critical Reynolds number, we find a region of gradual transition where the flow is neither fully laminar nor fully turbulent, and predictions of fluid behavior can be very difficult. For example, within circular pipes the critical Reynolds number is generally accepted to be 2300, where the Re is based on the pipe diameter and the mean velocity v_s within the pipe. However, engineers generally avoid any pipe configuration that falls within the range of Reynolds numbers from about 2000 to 4000 to ensure that the flow is either laminar or turbulent.

In an existing pipe system, both pipe diameter and product viscosity of the cleaning solution are generally constant. This means that the Reynolds number is primarily determined by the flow velocity of the solution. The resulting conclusion from this is that an aqueous solution must have a flow velocity of at least 2 m/s. The volume that has to be circulated within a system to reach the critical flow velocity of 2 m/s consequently is dependent upon the diameter of the pipe. Table 15.1 gives some examples.

The flow velocity can be calculated as follows:

$$v = \frac{4Q}{3.600\pi d^2} \text{ m/s}$$

where Q = flow volume (m^3/h)
d = pipe diameter (m).

The required flow volume and the required or existing pipe diameter determine the size of the Vorlauf pump, or CIP supply pump. When sizing the pump it is important to consider that the pump has to overcome pressure loss in the pipe system resulting from friction loss in the pipe itself, elbows and tees as well as any installed equipment or instruments. When pipes with different diameters are combined in a system, it is important to observe and calculate the flow velocity for each pipe as well as to compensate for varying pressure losses.

Table 15.1 Calculation of flow volume dependent upon pipe diameter

Pipe diameter (mm)	40	50	65	80	100	125
Flow volume (m^3/h)	9	14	24	36	54	86

Example 1: Pipe diameter changes from DN 80 to DN 65
As we can see in Table 15.1, a flow volume of 36 m^3/h is required to achieve a flow velocity of 2 m/s in a DN 80 pipe. The pressure loss here is 0.55 bar per 100 m pipe. If the DN 80 pipe is reduced to a DN 65 pipe and the flow volume is 36m^3/h, the result will be a flow velocity of 3 m/s and a pressure loss of 1.8 bar per 100 m.

Example 2: Pipe diameter changes from DN 80 to DN 100
If a pipe diameter is increased from DN 80 to DN 100 and the flow volume is maintained, the flow velocity will drop to 1.25 m/s. Not only will the flow velocity drop below the number required for turbulent flow, but air may accumulate in the larger pipe as hollow space in the larger pipe may allow air to be drawn in from vertical pipes. Increases and decreases in pipe diameters should be avoided whenever possible when designing pipe systems. If a change in pipe diameter cannot be avoided, the increase or decrease should never exceed more than one DN or more than $\frac{1}{2}$ inch (US).

15.6.2 CIP systems

While not directly a part of mechanical energy, CIP systems play a critical role for sanitation in modern breweries. It is here where chemical solutions are prepared, originate in the CIP process and return to after use. We can differentiate between two major design concepts:

- Recovery systems
- Non-recovery systems.

We can differentiate further by separating both designs into centralized CIP systems and decentralized CIP systems. Centralized CIP systems have been largely abandoned due to the many restrictions they come with and the expense of their design. However, they are still useful when space is at a premium and several CIP systems cannot be installed.

CIP systems are usually engineered and designed to meet each brewery's specific requirements. Therefore, the concept drawings in Figs 15.5 and 15.6 are designed only to show the basic operation of a CIP system. Figure 15.5 shows a simplified typical design of a system designed to recover cleaning solutions. This system design allows both recovered as well as non-recovered cleaning. Systems that are not designed to recover cleaning solutions generally lack the recovery tanks as well as most of the process valves. A simple system may look like that shown in Fig. 15.6.

The benefits of a non-recovery CIP system usually outweigh the benefits of a recovery system, if the total volume of a solution is less than approximately 800 liters. Automated CIP recovery systems are expensive and require a great deal of planning and engineering. Therefore, the expense should be primarily viewed as a quality control instrument and not on an ROI basis. Even simple systems such

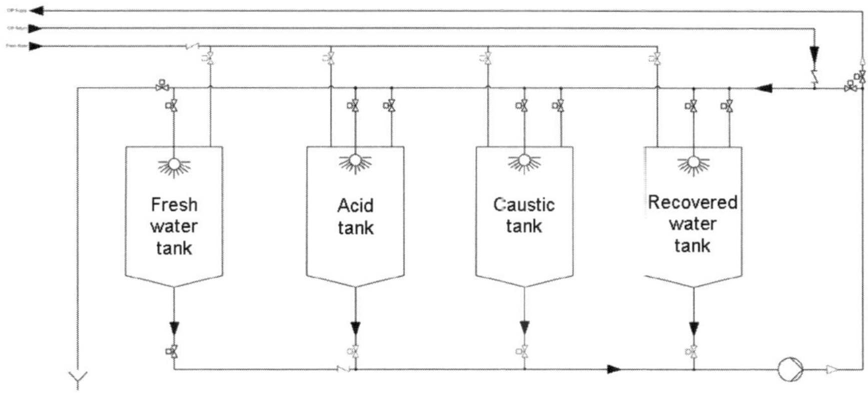

Fig. 15.5 Typical design of a CIP system designed to recover cleansing solutions (Loeffler Chemical Corp.).

as that shown in Fig. 15.6 provide the great advantage of automated chemical dosing and can be installed at minimal cost.

Most modern CIP systems rely heavily on conductivity to separate process cycles and to accurately dose and monitor chemical concentrations. However, a few important aspects must be understood about conductivity systems in CIP systems. We differentiate between two methods of measuring conductivity:

- Conductive measuring (contacting)
- Inductive measuring (non-contacting).

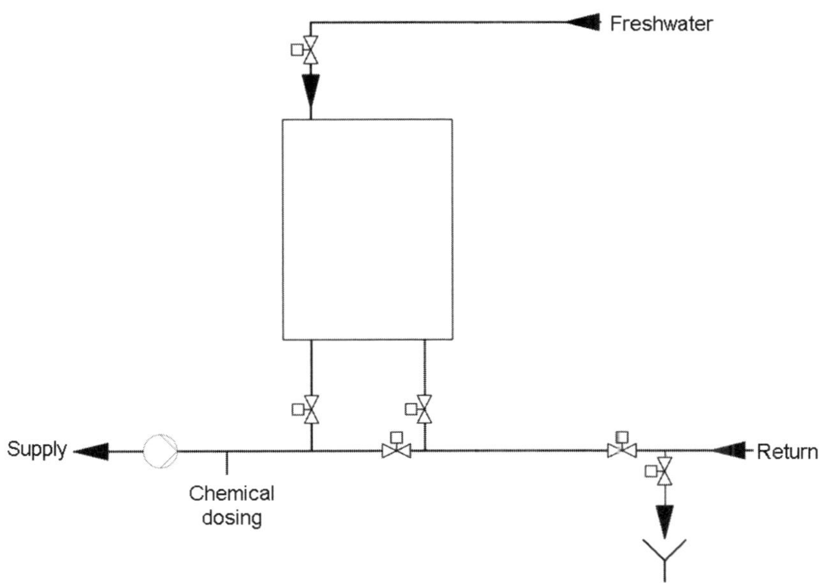

Fig. 15.6 Typical design of a non-recovery CIP system (Loeffler Chemical Corp.).

As a basic rule, the conductivity of the finished chemical solution must be considerably different than that of the water it is diluted with in order to control chemical solutions via conductivity. The following equation applies:

$$R = \frac{C}{x} \text{ ohms } (\Omega)$$

where R = resistance of solution ($\Omega = 1/S$)
C = electrode constant (1/cm)
x = specific conductivity (S/cm)

The measured resistance R (the inverse value is conductivity) in a solution containing electrolytes at a certain temperature is determined by the electrode constant C (which varies with material and construction) and the specific conductivity x, which is determined by the composition and concentration of the chemical in the solution. The conductivity probe measures the resistance R; however, what is displayed is normally its inverse factor, the conductivity. In modern systems, inductive or contactless conductivity probes are used due to their higher accuracy and lower maintenance requirements. If installed inline, these probes are virtually maintenance free as the flow velocity and various chemicals constantly clean the probe. Modern materials such as PTFE allow these probes to be used virtually anywhere and at any temperature. It is important to choose a controller or analyzer with a built-in temperature compensation, or to manually compensate for temperature differences. Conductivity changes greatly with varying temperatures, as Table 15.2 shows.

Table 15.2 Temperature dependence of conductivity

Solution	20°C	70°C	85°C
1% NaOH	47.5 mS	93 mS	107 mS
2% NaOH	90 mS	177 mS	203 mS
3% NaOH	127 mS	254 mS	293 mS

Many chemicals used in industrial cleaners such as sodium hydroxide, nitric acid or sulfuric acid have an almost linear relationship between concentration and conductivity. It is recommended to verify conductivity tables provided by chemical companies and to calibrate the equipment with accurately prepared reference solutions on a regular basis.

15.7 Temperature

The temperature of a cleaning solution is primarily determined by the type of deposit, the composition of the deposits and the chemical makeup of the detergent and the application. The development and use of high-performance

Table 15.3 Commonly found temperature ranges

Temperature range	Applications
Ambient to 40°C	Fermentation, storage and bright beer tanks, fillers
70°C to 90°C	Brewhouse vessels, heat-exchangers, wort lines, pipes, etc.

surfactants enables breweries today to operate with ambient temperatures in applications that would have required high temperatures only 5–10 years ago.

In breweries, prerinses should always be performed at ambient or at temperatures below 35°C. Protein and starch, two deposits commonly found throughout breweries, will be chemically modified (denatured) above 38°C, making them significantly more difficult to remove. Some common temperature ranges are shown in Table 15.3.

Acids are usually used at ambient temperatures, unless heat contributes to disinfection, such as in cleaning kegs. Caustic solutions should reach temperatures of at least 40°C during the cleaning cycle. However, large cooled tanks may not permit such a temperature without sophisticated safety equipment such as pressure switches and temperature probes. The pressure differential created by only a 4–6°C temperature drop in a large tank may be sufficient to implode the tank. The optimal temperature should be carefully selected in close cooperation with equipment manufacturers and the chemical provider.

Since we know that the parameters of the Synergistic Circle are interchangeable within limits, the addition of oxidative additives and/or additional time and mechanical energy may be used to compensate for temperature restrictions set by the equipment.

15.8 Time

The chemical processes during cleaning and sanitizing such as wetting of soil (dried-on protein or starch), removal of mineral scale or stone by acids, saponification of protein and dispergation (e.g. hop-resins) undergo certain time laws. Even the improved performance chemicals of today can only shorten but not eliminate the time requirement for soil removal.

Contact times are generally defined by the reaction of the cleaning solution with the soil at a certain concentration, temperature and mechanical conditions. Time is a costly commodity in a modern production brewery. Consequently, the turnaround time and thus the contact time during cleaning is always pushed to its limits. However, chemical reactions are time dependent, and even modern cleaning products using the latest performance products require certain contact times to do their magic. To better understand the limitations, we have to picture deposits as multiple layers on top of the equipment surface. As the chemical solution is reacting with the soil, it first reacts only with the outermost layers of the soil. While the use of high quality surfactants may result in more angles of attack such as from the sides, back and front of the soil (surface tension, capillary activity), the chemical solution is still removing the soil layer by layer. Furthermore, we need to take into consideration that the full concentration of the cleaning solution is only available after a few minutes once the mixing phase of water and cleaner has been pushed out of the system.

The following aspects should be taken into consideration when choosing contact times:

- *Prerinse.* The contact time for the prerinse is generally determined by the amount of time needed to flush out all loose soil deposits in the system. Once the water starts running out clear, the prerinse is completed.
- *Cleaning.* The contact times for cleaning solutions vary dramatically from application to application. It is recommended to establish the cleaning times required to remove all deposits under normal operating conditions by visually inspecting critical system components (heat-exchanger plates, pipe bends, interior tank surfaces or instruments, doorways, etc.). Keep in mind that the cleaning times for the same piece of equipment may vary with different beer styles. If only one CIP program is available and a certain product requires additional cleaning time, the system should be programmed with the longest necessary time. In addition, a safety buffer of 5–10 minutes is generally recommended and should be added to the predetermined contact time.
- *Intermediate rinse.* The completion of an intermediate rinse is usually determined by measuring the pH or conductivity of the rinse water. The rinse is considered complete once the conductivity or the pH has returned to the starting values of the process water. A safety buffer of 1–2 minutes is recommended when any chemical compound of the following cycle is negatively affected by a compound of the preceding cycle (e.g. alkalinity/iodine). On the other hand, the intermediate rinse may be shortened or completely eliminated if no undesirable reactions occur with the following cleaning products (e.g. acid cleaners and sanitizers).
- *Final rinse.* The complete absence of one or several characteristic elements (pH, conductivity, etc.) of the cleaner or sanitizer determines the length of the final rinse. A safety buffer of 1–2 minutes is recommended.

15.9 Sanitizers/disinfectants

The single most important precondition for successful sanitizing is an effective cleaning program. The old rule still applies: You cannot sanitize a dirty surface! Residual deposits can protect microorganisms from contact with the sanitizer. At the same time, the effectiveness of many sanitizers is directly negatively affected by organic soil. Many organisms not killed during the sanitizing cycle will grow in the moist atmosphere left by the sanitation cycle and may lead to product spoilage. Although it is virtually impossible to grow pathogens in beer, a brewery may never recover from the impact that a spoiled batch of beer has on its consumers and retailers.

Due to the fact that more and more countries impose very high registration fees on sanitizers, very little advancements have been made with respect to new sanitizers and disinfectants in comparison with cleaners. For the purpose of this chapter, the words sanitizer and disinfectant shall be interchangeable. The brewery's choice for a sanitizer should not be driven by the price of the product, but primarily by its efficacy and impact on the product it is intended to protect, in this case beer. In addition, consideration should be given to the compatibility with any special environments, such as carbon dioxide or low pH (beer).

We differentiate between two methods of sanitizing: thermal and chemical sanitation.

15.9.1 Thermal sanitation

The effectiveness of thermal sanitizing is highly dependent upon the temperature and the time during which a specific temperature is maintained. A system can only be sanitized if the *entire* system can be heated and the heat can be maintained throughout the entire system for the required length of time. The temperature during thermal sanitizing should always be measured at the end of the system or at the point where the lowest temperature is expected during the cycle. Today, thermal sanitizing (water or steam) is mostly used for sanitizing pipes and closed systems such as filters and heat-exchangers.

If steam is used, the quality of the steam largely determines the success and quality of the sanitizing cycle. Wet or saturated steam produces the best results. However, it is also difficult to produce and to maintain. Temperatures of 70–80°C must be maintained over periods of 15–20 minutes. Even at these temperatures, certain molds and bacteriophages may not be completely killed. Temperatures as high as 130–140°C and contact times of 20 minutes and more are needed, which can only be realized in an autoclave environment (pressurized pipes), which is generally not practical for a brewery. Another important aspect is the complete elimination of potential recontamination through outside air during the cooling phase. Thermal disinfection exerts enormous stress on metal surfaces that may result in hair fractures and weld failures. For some equipment such as glycol jacketed tanks, tanks in cooled cellars (condensation) and other cold or sensitive surfaces, thermal disinfection is completely unsuitable.

In certain applications such as keg sanitizing, steam and chemical sanitizers may be combined to achieve optimal results in the short times allotted.

15.9.2 Chemical sanitizing

As opposed to thermal sanitizing, chemical sanitizers do not require heat to kill microorganisms. However, some chemical sanitizers can react with either the environment in the equipment such as residual carbon dioxide or the beer itself. While a final rinse is generally performed to remove chemical residues from the equipment, small residues may remain due to equipment malfunction or poor system design. In some countries such as the United States, a rinse is not always required after sanitizing. While I believe that every responsible brewery should perform a final rinse or at least a series of quick burst rinses followed by a pressure evacuation (leaves trace residues behind, which sterilizes the rinse water), many US breweries actually just allow their equipment to air dry after sanitizing. This practice requires a particularly careful selection of sanitizers with respect to compatibility with the equipment, carbon dioxide and most importantly the beer.

Even though a cleaning regime using both alkaline and acid cleaners provides some antimicrobial activity based on the significantly different pH values alone, a separate sanitizing cycle is always required to ensure a 100 percent clean surface. Some of the commonly used sanitizers today are chlorine, chlorine dioxide, peroxide compounds, iodine, and quaternary ammonium compounds (QACs).

Chlorine

Most liquid products derive the active chlorine from sodium hypochlorite (bleach) while powdered products generally use chlorinated trisodium phosphate (TSP) or sodium chloroisocyanurate. Chlorine reacts negatively with beer and may cause protein haze, head retention problems, or, worse, the formation of *ortho*-chloro-phenolics. Chlorine based products must never be used in the presence of carbon dioxide as poisonous chlorine gas is released. Chlorine's antimicrobial activity increases as its pH decreases; however, a pH environment of 9 or higher should always be maintained in order to minimize the danger of corrosion. While inexpensive and offering a broad antimicrobial spectrum, chlorine is more and more being phased out due to its incompatibility with beer, its corrosivity and its wastewater problems (AOX, COD, THM, etc.). Chlorine's activity is based on irreversible oxidative action on the cells, in which both the cell structure as well as enzymatic proteins and nucleic acids are so strongly changed that the microorganism is destroyed and cell regeneration is arrested.

Chlorine dioxide

In recent years, chlorine dioxide has displaced chlorine in many applications where chlorine has been traditionally used. Chlorine dioxide provides chlorine's benefits without its detriments and does not chlorinate organic material. This eliminates or significantly lowers trihalomethanes (THMs), haloacetic acids (HAAs) and other chlorinated organic compounds. However, the application of chlorine dioxide has its limitations in breweries. First of all, chlorine dioxide is a gas in solution and must be generated on site by either generation or activation. Chlorine dioxide can be activated at a relatively low yield of 25–35 percent active ClO_2 from dilute sodium chlorite solutions. By lowering the pH of the sodium chlorite solution using either phosphoric or citric acid, chlorine dioxide is generated. However, the amount of active chlorine dioxide released during activation can vary and is extremely difficult to monitor and control. This method is also rather expensive once we consider the activation ratio, the cost for the activator, labor and the precursor.

Chlorine dioxide can be generated on site via a variety of chemical reactions, using chemical precursors. The most common generation methods are as follows.

The conventional chlorine dioxide process uses sodium chlorite, chlorine, or sodium hypochlorite and hydrochloric acid to convert sodium chlorite to chlorine dioxide:

$$NaClO_2 + \tfrac{1}{2}Cl_2 \rightarrow ClO_2 + NaCl$$
$$5NaClO_2 + 4HCl \rightarrow 4ClO_2 + 5NaCl + 2H_2O$$
$$2NaClO_2 + HOCl + HCl \rightarrow 2ClO_2 + H_2O + 2NaCl$$

These systems are fairly reliable and efficient but they require attention for continuous operation at high efficiency with low levels of by-products and residual chlorine. They are also difficult to monitor for yield, efficiency and product purity in operations that require frequent capacity and/or concentration changes.

These conventional processes require an excess of chlorine or acid to maximize sodium chlorite conversion. The amount of excess is dependent upon the generator design but can be 10–15 percent of the stoichiometric amount. The control and proportioning of two or three chemical feeds in these 'on-demand' systems is difficult. If not carefully monitored, they can lead to untreated chlorite or excessive amounts of chlorine, which in turn can lead to the formation of sodium chlorate via other side-reactions or even to the formation of chlorine related by-products in the finished water, which is what we are trying to avoid.

Newer technologies utilize electrochemical generation using an anode/cathode system and/or membrane technology. These systems utilize the electrochemical oxidation of sodium chlorite and water. By-products are usually dilute sodium hydroxide solutions (10–20% NaOH) as well as small amounts of hydrogen (0.1–0.3%). The predominant anode reaction is the rapid and kinetically favored 'one electron transfer oxidation' of the chlorite ion as shown below:

$$ClO^{2-} \rightarrow ClO_2 + e^-$$

The reaction at the cell cathode is the reduction of water, producing hydrogen gas (H_2) and hydroxide ions (OH^-) as shown below:

$$2H_2O + 2e^- \rightarrow H_2 + 2OH^-$$

While chlorine dioxide offers a broad killing spectrum similar to those of chlorine or peracetic acid, its generation and monitoring technology is still in baby shoes. Chlorine dioxide works essentially on the same principle as chlorine. Most generators are expensive, and monitoring and controlling accurate concentrations is still difficult. Chlorine dioxide should not be used on beer contact surfaces as it can cause head retention problems and protein haze. Chlorine dioxide breakdown products may react with the carbon dioxide in the tank, causing severe headspace corrosion in tanks and other closed systems. Its use should be limited to water treatment applications such as final rinse compartments in bottle washing machines and bottle rinsers. Lately, it has also shown promising results as an additive to makeup water for conveyor belt lubricants.

Peroxide compounds

Hydrogen peroxide and peracetic acid are the two most commonly found sanitizers from this category. Hydrogen peroxide is a fairly non-corrosive chemical which is often used in aseptic filling systems as well as in foam cleaners as a non-corrosive and effective replacement for chlorine. The oxygen radical released oxidizes the biologically active system of the cell, which is then destroyed. Hydrogen peroxide is an excellent heat sanitizer.

Peracetic acid, although discovered in the 1920s, has only recently become a favorite in many breweries. As opposed to chlorine, chlorine dioxide and thermal disinfection, its killing spectrum includes viruses, phages and spores (including endospores). Peracetic acid reacts not only with the protein content of the microorganisms' cell wall, but through wall penetration. The weakly dissociated acid enters the inner parts of the cell itself where all proteinous components of the cell including all enzyme systems are destroyed through oxidative–destructive action. The cell dies.

Peracetic acid is a chemical equilibrium between acetic acid and hydrogen peroxide and therefore requires special stabilization by the manufacturer. It breaks down into oxygen, acetate and water, making it one of the few sanitizers that are truly 'compatible' with beer. However, larger amounts of Peracetic acid residuals may cause oxidation of the beer. It is one of the strongest oxidizers commercially available and its oxidation potential is higher than those of chlorine, chlorine dioxide and hydrogen peroxide. It is also extremely effective at cold temperatures, making it a typical and true cold sanitizer.

Iodine

Iodine-based sanitizers are generally sold mixed with phosphoric acid as iodophors. Iodine exhibits similar oxidative qualities as chlorine and reacts with the molecular structure of the microorganisms. Like chlorine, it also reacts with any organic deposits remaining in the system, inhibiting the killing effect. Iodophors are effective against most types of bacteria, viruses, fungi and spores. The antimicrobial effect is negatively affected by alkalinity (pH>7). The antimicrobial effectiveness of iodine and its oxidative power is decreased in the following order: I_2, HOI, OI^- and IO_3^-. As we can see in the following equations, the solution must be acidic to obtain the highest efficacy:

$$\text{pH 3–5: } I_2 + H_2O + H^+ \rightleftharpoons I^- + HOI + 2H^+$$
$$\text{pH} \sim 7\text{: } I_2 + H_2O \rightleftharpoons I^- + HOI + H^+$$
$$\text{pH} >7\text{: } 3I_2 + 6OH^- \rightleftharpoons 3I^- + 3OI^- + 3H_2O \underset{\text{while standing}}{\rightleftharpoons} 2I^- + IO_3^-$$

Iodine by itself is not soluble in water and is therefore solubilized using nonionic surfactants. The surfactant encapsules the iodine in micelles. As the iodophor is diluted, the micelles separate and release the iodine. Due to its higher effectiveness in acidic pH environments, iodine complexes are generally mixed with phosphoric acid or other acids. As we can see in the above equations, alkaline residues in the system would negatively affect the antimicrobial properties of iodophors. It is therefore important to ensure that no alkalinity is present in the system prior to sanitizing with Iodine. Iodophors contain 'tamed' iodine. The encapsulation in micelles reduces the vapor pressure of the iodine and dramatically reduces the danger of corrosion through sublimation. However, the practical use of iodophors should be limited to temperatures below 40°C at all times. Even without direct corrosion, brown discoloration may occur above the liquid level of the solution.

Iodophors are fairly stable unless contaminated with organic soil, and are often used to soak small parts and hoses. Iodophors may negatively affect head retention and can produce a metallic off-flavor in beer. Due to the corrosivity of iodine, similar to that of chlorine, and some undesirable breakdown by-products in the wastewater stream, iodine-based sanitizers are more and more being replaced by other alternative sanitizers such as peracetic acid.

Quaternary ammonium compounds
Quaternary ammonium compounds (QACs) are cationic surfactants that exhibit broad antimicrobial properties and are commonly neutral in pH. They kill through the inactivation of enzymes and the denaturation of cell proteins. By adsorption of the cationic QAC into the cell surface, the permeability of the cell walls of the microorganism is impaired. The cell contents are released, thereby killing the cell.

The effectiveness of QACs is negatively affected by protein contamination, cellulose and other polymers. QACs tend to foam and can cause protein haze in beer. They are difficult to rinse due to their high surface activity; however, this can be beneficial if a residual activity such as on exterior surfaces is desired. QACs are generally impractical for use in modern CIP systems due to their foaming characteristics and their lack of conductivity or oxidative properties, making automatic monitoring and dosing impossible.

15.10 Future trends

With environmental concerns mounting and new laws being signed in effect almost every year, the efforts of the chemical industry have been primarily towards conservation and optimization. Water conservation through combining cleaning cycles, water recovery and using recovered solutions in less critical applications have been successfully tested and implemented. Enzyme technology has failed to fulfill its promises as the problems of allergic reactions and limited heat tolerance (denaturation) remain. However, enzyme technology is being successfully used for membrane cleaning and as it moves forward we will probably see its use expanding into other areas of the brewery.

Many mechanical improvements with respect to chemical automation, cleaning equipment and surface finishing have helped to make cleaning more effective and efficient. Chlorine is being phased out of breweries and replaced by peroxide technologies which offer many advantages over chlorine. As the chemical industry develops ever more powerful and versatile performance products, the cleaners we use will become even more effective and more universally usable. The dream of cleaning all tanks under CO_2 atmosphere with acid products alone is still some way off, but we see what the future will hold for us. New product technologies already allow us to clean storage and bright beer tanks *almost* exclusively with acid cleaners and sanitizers. Then again, who would have thought in 1900 that humans would walk on the moon in 1969? The limit is our imagination.

15.11 Sources of further information

Unfortunately, there are very few books and publications available on cleaning and sanitizing. Of the few published, many are so old and outdated that I cannot recommend them here with good conscience. A few good and recent publications are listed below:

Praxis der Sterilisation Desinfektion – Konservierung by Karl Heinz Wallhaeuser, ISBN 3-13-416305-5.

Brewing Microbiology (second edition) by F.G. Priest and I. Campbell, ISBN 0-412-59150-2.

Farbatlas und Handbuch der Getraenkebiologie, Teil II by Werner Back, ISBN 3-418-00760-0.

Chemische und Technologische Grundlagen in der Flaschenreinigung by Hans Peter Jungmann, ISBN 3-418-00754-6.

Some helpful websites:

http://www.gcisolutions.com/flow.html – turbulent flow calculator
http://en.wikipedia.org/wiki/Main_Page – free encyclopedia
http://www.efunda.com/formulae/fluids/calc_pipe_friction.cfm – pipe pressure loss calculator

16

Waste handling in the brewing industry

R. Reed, Black & Veatch, UK

16.1 Introduction

The volume of wastewater being discharged from a brewery may vary between 2.5 and 10 times the volume of beer being produced. If this wastewater is discharged to sewer then trade effluent charges in the UK, the combined cost of water and wastewater disposal are similar to the energy costs for a brewery. To put this into context, a reduction of 10% of these costs is equivalent to improving malt extract by 1.5%. Hence any opportunity to reduce these costs should be seriously considered. There are a number of potential routes to achieve a reduction in the water/wastewater bill:

- Improved water management
- Recovery and reuse
- Partial on-site treatment of the wastewater.

Where no discharge to sewer is practical, such as in the case of new breweries or those in rural localities, purpose-built treatment will be required before discharge to a waterway to deliver a water quality specified by the Environment Agency.

This chapter will review the sources of wastewater and other wastes and the technologies available for their treatment. Additionally other environmentally sensitive issues, including the energy conservation and carbon dioxide emissions that are the subject of environmental legislation, will be reviewed.

16.2 Assessment of the character and strength of wastewaters

Prior to discussion on wastewater it is necessary to define those parameters generally used in the wastewater treatment industry. Some such as pH are in common use but others are specific to or have specific meanings in wastewater technology.

pH

pH is a measurement of the alkalinity or acidity of a wastewater. High and low pH values can result from CIP operations and are critical in that sewers can be corroded and biological treatment processes disrupted if uncorrected. Numerically,

$$\text{pH} = \log_{10} (1/\text{free hydrogen ion concentration as g/litre})$$

In neutral water pH is 7, with acid giving low values and alkali giving high values; generally wastewater sources are mildly acidic except when alkaline CIP liquors are discharged.

Suspended solids

Suspended solids are generally defined as that material which does not settle after 30 minutes of quiescent settling. It is measured by filtration on filter paper and its units are generally expressed as mg/litre after drying at 105°C. Suspended solids can smother bottom-living aquatic and plant life if discharged direct to waterways as well as impose the need to collect and treat the sludge they generate in treatment works.

Biological Oxygen Demand (BOD)

BOD (or more accurately BOD_5) is a measure originally devised to measure the impact of a wastewater on a river as a consequence of the growth of micro-organisms on waste products and associated uptake of oxygen from the water; it is also a good indicator of the concentration of biodegradable matter in a waste stream. It is measured as the oxygen uptake by bacteria mixed with the wastewater over a five-day period, over which time most of the waste is degraded. It is not generally used to measure the strength of industrial or trade wastes due to the likelihood that the waste is devoid of viable bacteria or that components within the waste are inhibiting the development of bacteria.

Chemical Oxygen Demand (COD)

COD was developed as a rapid means of measuring BOD and also as a means of measuring industrial or trade wastewaters. The is the unit generally used in the definition of Mogden charges, as trade wastewaters may be toxic to treatment bacteria when undiluted and in any event frequently lack the bacteria necessary for the measurement of BOD. A sample of the wastewater is refluxed with a highly oxidising mix of sulphuric acid and potassium dichromate for 2 hours and the extent of reduction in the dichromate assessed. This process can oxidise matter that is generally inert to bacterial action, with the result that its value for a wastewater is greater than that for BOD.

Other measurements
Measurements such as Total Organic Carbon (TOC) may be used in some areas as a means of assessing strength of industrial wastewater. Given that the BOD and COD tests register the oxidation of one carbon atom as two oxygen atoms, TOC values tend to be far smaller than BOD values, except when inhibitory action may be expected. The advantage of TOC is that it is a test that can be fully automated to operate on-line.

16.3 Sources and nature of wastewater

The Chemical Oxygen Demand (COD) in wastewater from maltings, breweries and cider manufacturers is generally biodegradable. There is some evidence that the COD in brewery CIP wastewater may be partly 'intractable' or 'bio-refractory', i.e. resistant to conventional treatment as a result of the CIP liquid reacting with sugars and proteins. This could have a cumulative effect with other industrial discharges to cause a receiving wastewater treatment authority to fall foul of the Urban Waste Directive (see Section 16.7). In such a case the authority is likely to call on the industrialist to reduce the discharge of intractable COD.

A breakdown of the sources and nature of wastewaters for maltings and breweries is given in Table 16.1. The flows and loads emanating from malting and brewing processes are given in Table 16.2. The values given in the table are a combination of observed measurements and theoretical prediction. The points to note are:

- A high proportion of beer being packaged small will increase the wastewater ratio, but even so, the spread of values indicates that some breweries are very inefficient in terms of water usage. The actual volumetric ratio of wastewater to product, as collated by the British Beer and Pub Association,[1] ranges between a minimum of 2.4:1 up to 19:1.
- The load of material being discharged is greatly affected by the presence of yeasty heads in shallow top fermentations and conditioning tank sediments.
- COD or BOD in the wastewater generally represents wort and beer losses and can usefully be used as a measure of brewery operating efficiency.

The figures generally relate to a modern, well-managed brewery. It should be noted that there is no detailed breakdown of the consumption of water by canning. The bottling and canning figures are excluded from the totalised values at the bottom of the table. The average wastewater generation by the UK industry is 4.5 hl/hl product (specific water consumption is 5.5 hl/hl product).

Low usage of a process and aged plant tend to go hand in hand with inefficient water usage, with water:product ratios up to 30:1 in some old bottling

Table 16.1 Sources of wastewater

Wastewater source	Nature of waste
Maltings	
Steepwater	Neutral to slightly acidic Treatable COD in solution Some suspended solids
Brewery	
Water treatment	Saline, may be both acidic and alkaline
Vessel washings – first rinses	Neutral to slightly acidic Generally treatable COD Some suspended solids
Vessel rinses – CIP	Strongly acidic/alkaline Some COD may be intractable
Weak process streams (wort and beer last runnings)	Neutral to slightly acidic Treatable COD Some suspended solids
Trub (wort kettle residue)	High suspended solids and COD
Fermenter and conditioning tank sediments	High suspended solids and COD
Beer filter cake discharges	High suspended solids, comparatively low COD
PVPP regeneration	Hot alkaline High COD
Keg and bottle washing	Hot, alkaline (if bottle washing) High COD

plants where short runs (the result of low demand for the products) result in disproportionate cleaning time. Normally, surplus yeast is sold off as food or animal feed (for further processing) and conditioning tank sediments are pressed and the filtrates returned to the product. When this does not happen there is a great increase in the quantity of COD discharged to drain.

16.3.1 Spent grains and yeast

Spent grains and spent yeast are generally regarded as by-products from the brewing process. It is becoming critical from the aspects of environmental legislation that the markets for these biodegradable products remain open, as disposal of these materials to landfill sites is prevented under the EU Landfill Directive.[2]

- Quantities of spent grains are typically some 30% of the dry mass of malt used in the process, but at some 80% moisture correspond to 150% of the weight of malt or 21 kg/hl final product, assuming no adjunct and depending on the strength of the final beer.

Table 16.2 Brewery wastewater loads

Discharging process	Flow (litre/hl beer)	COD (kg/hl beer)
Mashing	3	0.003
Lautering		
• Last runnings	8	0.06
• Wash	4	0.005
Boiling		
• Clean	14	0.01
• Rinse	3	0.005
Whirlpool/hopback	8	0.12
CIP caustic brew	10	0.01
Fermentation		
• Rinse	8	0.06–0.5
• Clean	2	0.01–0.05
Surplus yeast		
• If discharged direct to drain	2	0.35
• If pressings discharged to drain	1.5	0.15
• If sold for food or animal feed	0	0
Conditioning		
• Sediments		
– Discharged direct to drain	3	0.3
– Pressings discharged to drain	2.5	0.25
– Pressings to product	0	0
• Washing	10	0.07
Filtration		
• Last runnings	2	0.06
• Washing	50	0.01
• Cake		0.02
Stabilisation		
• Regeneration	50	0.01
Kegging		
• Washing	30	0.08
• Pasteuriser	12	0.005
Cask filling		
• Returns	1	0.1
• Washing	up to 100	0.1
Bottle washing	<100 up to 3000	0.05–0.2
Bottle/can filling	<100 up to 2000	0.05–0.3
Total (assuming 100% keg beers)	**223–228**	**0.538–1.668**
Total (assuming 80% cask beers, 20% bottled beers)	**190–1150**	**0.553–1.523**

- The quantity of spent yeast is low, some 0.2 kg/hl beer, based on 75% moisture.

The current dependency on the animal feed market for spent grains disposal has led to attempts to reprocess this material into a form that can be utilised within the brewing process; unfortunately there has as yet been no successful outcome. Spent grains could presumably be composted along with other 'green' waste if an animal feed route is not locally available.

Yeast is currently used as a raw feedstock for food ingredients, generally in a hydrolysed form. Some of the products are popular brands and there is no reason to foresee any decline in this market.

16.3.2 Filter-aid

Spent filter-aid is largely mineral in nature and can be disposed to landfill. Most beer filters discharge filter aid as a cake that can be readily handled during disposal, though some of the older styles of candle filter depend on flushing the spent filter-aid from the filters to drain. This imposes an additional cost for wastewater discharge that can readily be costed using the Mogden formula.

As in the case of spent grains, attempts have been made to process the spent filter-aid for reuse in filtration operations. These processes tend to break the structure of the filter-aid, limiting its capacity for beer solids and limiting the extent of reuse to coarse precoating or small admixture to body feed.

16.3.3 Carbon dioxide

For each kilogram of alcohol produced in beer fermentation, 1 kg of carbon dioxide is generated; allowing for 1.5 volumes of gas per volume of product ex-fermenter, some 900 g of CO_2 will be released. The gas is generally released to the atmosphere, but is not deemed to be a polluting gas nor as an addition to greenhouse gases as the source (the malt) is renewable.

Carbon dioxide can be recovered for use elsewhere in the brewing process, as opposed to buying in carbon dioxide. However, the gas must be carefully collected to ensure that air is totally excluded and scrubbed to remove organic components; generally gas will also need to be compressed to enable its utilisation. The economics of gas recovery are generally open to question and in any case are site specific.

16.3.4 VOCs (volatile organic compounds)

Under the EU Solvent Directive there are three categories of VOCs according to the harm that the VOCs will have on health and the environment. The Directive defines a volatile organic compound as a substance having a vapour pressure of 0.01 kPa or more at 293.15 K (20ºC), or having a corresponding volatility under the conditions of its use. Hence, while many of the organic compounds being released to the atmosphere during wort boiling or fermentation, such as acetate esters, are VOCs, they fall into a low risk category which, combined with the very low concentration and associated partial vapour pressure in the process off-gases, results in the brewery vapours and gases as being considered to be unharmful.

16.4 Extent of treatment

The extent of treatment required for a brewery wastewater must be carefully assessed:

- In principle, on-site treatment of a strong wastewater such as from a brewery should be more efficient than treatment in combination with weaker municipal streams, particularly if partial treatment is being considered. However, return on investment and management time generally render on-site treatment unattractive if there is a convenient disposal route to a municipal treatment works. It should be noted that brewery wastewater is frequently of benefit to the treatment of municipal wastewater in that the latter is rich in nutrients (nitrogen and phosphorus) which are increasingly required to be removed prior to discharge to watercourses; the nutrient-deficient brewery water results in improved take-up of the nutrients into the biomass.
- Partial or roughing treatment enables the use of relatively small-footprint technologies utilising high F:M (short sludge age aerobic processes) or high-rate anaerobic processes, increasing the likelihood of achieving a satisfactory payback relative to disposal costs to sewer.
- Partial treatment may be required if the local municipal works has a limited capacity.
- Full treatment (down to BODs as low as 20 mg/l, COD 100 mg/l) will be required if there is no other option than disposal to river or soakaway.

Partial treatment may be attractive if it is feasible to collect the strongest waste streams and treat these, but this is not easy to realise.

16.4.1 Costs of discharge to sewer

In the UK industrialists are always charged when a wastewater is discharged, even if it is fully treated and is discharged to a river, completely meeting its specified consent; in this case it is the Environment Agency which levies a comparatively small charge. Partly or untreated wastewaters being discharged to sewer are subject to much higher charges to cover the cost of conveying the water to the treatment works and to cover the cost of treatment itself. The charge is calculated from the *Mogden formula* in the UK:

$$P = C + V + \frac{S_T}{S_S}S + \frac{O_T}{O_S}O$$

where P = cost (pence) per m^3
C = conveyance charge, pence per m^3

V = volumetric charge, pence per m^3
S = suspended solids charge, pence per m^3
O = COD charge, pence per m^3
S_T = suspended solids content of trade waste, mg/l
S_S = suspended solids content of sewage, mg/l
O_T = COD of trade waste, mg/l
O_S = COD of sewage, mg/l.

All the above parameters vary between authorities, depending on local treatment costs and the extent of treatment required, e.g. should a biological treatment stage not be included then O equals zero. Even if the wastewater is simply pumped out to sea (an increasingly rare situation) there will be a conveyance charge. The suspended solids charge can be considerable due to the extensive treatment required to treat sludge, as can be the oxidation charges.

16.5 Outline of processes and technology

The traditional methods of treating wastewater are facultative lagoons (where naturally generated aerobic and anaerobic bacteria effect treatment) and low rate trickling filters (where aerobic bacteria in a slime layer on inert media effect treatment). However, these methods demand too high a land area to be practical for most trade waste situations, except for facultative lagoons in tropical areas where there may be more space available and reaction rates are faster as a result of higher temperatures. There is a wide range of water/wastewater treatment technologies that could be applied to water recovery operations in breweries and cider plants.

16.5.1 Chemical and physical treatment

Generally there will be the need for the physical separation of suspended solids and perhaps the correction of pH. The cleaning wastewater flow from CIP (brewery and cider production) and rinsing operations is generally the largest. This stream is usually caustic, is high in COD and sometimes contains oxidising agents. Consequently the wastewater may need to be held in balancing tanks to allow the oxidising agents time to react and hence prevent them from inhibiting biological processes (some of which, in the case of some membrane-related processes, can deliver bacteria-free treated water – see below).

Unless both acid and alkaline CIP can be used to generate a neutral effluent after flow balancing, either acid or alkali may have to be purchased and stored on site if on-site biological treatment is provided. An alkaline wastewater may be acceptable to a treatment authority in that the sewer system is unlikely to be damaged (except at very high pH), provided that there is sufficient buffering within the general municipal wastewater.

Ideally, alkaline pH may be corrected using the CO_2 that is generally wasted from the brewing process; alternatively the alkaline stream is also ideal for scrubbing biogas (from anaerobic treatment processes) to remove the CO_2 and hence enhance the calorific value of the biogas. However, the problems of coinciding operations and long costly piping runs generally negate these theoretical options.

Suspended solids should be removed by gravity settlement (in tanks or through tilted/lamella plates) or Dissolved Air Flotation (DAF) plant. The supernatant can then be treated by biological oxidation or anaerobic digestion.

16.5.2 Biological treatment

This operates by using the capability of micro-organisms to use waste components as food. The trick is to load the organisms at such a rate that they can manage to give the correct degree of treatment. If the load, measured as the mass of BOD per mass of biological solids per day (the F:M ratio), is too high, then the process will be overloaded and only a low efficiency of removal will be achieved. As the load is reduced the efficiency of treatment becomes higher. A similar concept is the sludge age where a more mature sludge, generated at low sludge loadings, gives better treatment. Both concepts are used in design, the selection depending on the application. Biological processes can proceed either with oxygen (aerobic) or without oxygen (anaerobic), Fig. 16.1. The former is generally faster but generates more biomass, which can be problematic to dispose of.

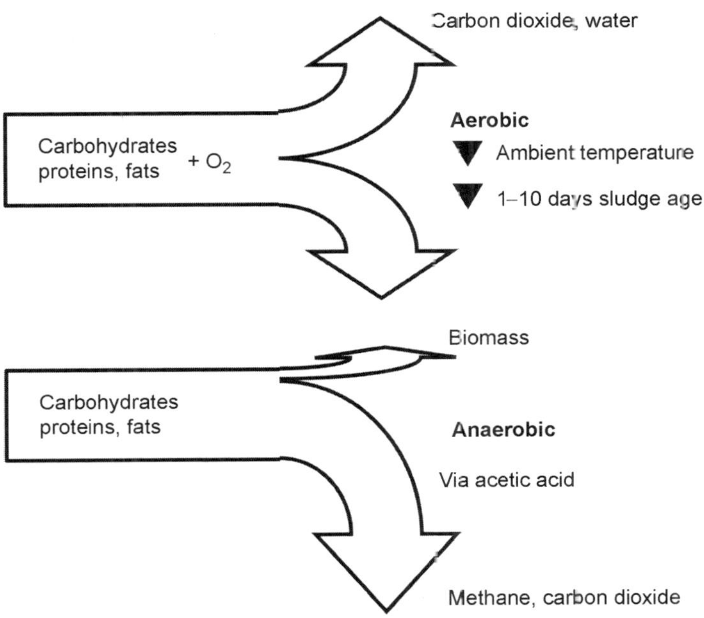

Fig. 16.1 Outline of wastewater biological processes.

16.5.3 Aerobic oxidation

Aerobic processes can be via the bacteria and protozoa growing on a slimy 'zoological' film growth or as suspended flocs, where materials exuded by the bacteria and filamentous fungi act as a glue or flocculation aid; this allows the biomass to be settled from the treated wastewater. Organic matter is oxidised to carbon dioxide and water. Organically bound nitrogen is assimilated into the biomass and released as nitrogen gas or converted to nitrate. Ammonia is oxidised to nitrate. Generally, though, brewery effluent is deficient in nitrogen and phosphorus and these nutrients will have to be added to the wastewater for on-site treatment, particularly aerobic. Aerobic processes are a good source of energy for the bacteria and result in a high conversion of the waste organic material into biomass. Extending the age of the sludge has the effect of breaking down bacteria and encouraging the growth of higher organisms to assist in the breakdown of the biomass to reduce its net production.

Suspended growth

The wastewater is treated in concrete or glass-lined steel tanks using a variety of methods to dissolve oxygen. These may be submerged air diffusers or surface aeration devices. There are a number of forms of the 'Activated Sludge Process', but all have one thing in common in the use of settlement tanks to recover the biomass from the treated wastewater and recirculate it back to the process, thereby enabling the concentration and the sludge age to be increased (Fig. 16.2).

In simple aerated lagoons, there is no settlement tank and biomass is allowed to overflow with the treated wastewater. If high-rate treatment is the objective,

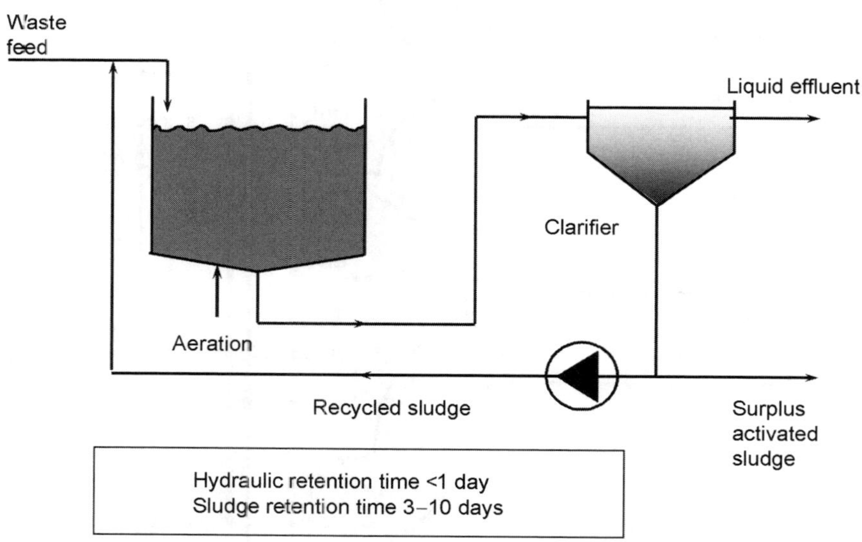

Fig. 16.2 Schematic of the activated sludge process.

then this is not an issue because the bacteria growth rate is so fast that settleable flocs do not form because the sludge age is too low for suitable flocculating organisms to develop. However, higher sludge ages can achieved by sequentially switching off the aerators for controlled periods allowing settlement, similar to the activated sludge process, with the treated wastewater being decanted off.

Film growth

Conventional trickling filters take up far too much space to be of use on industrial sites, but high rate systems built in a tower format using plastic support media offer an effective means of achieving high rate treatment (50–80% reduction in BOD) in a small footprint. The same principles apply as in the low rate system, with the liquid trickling down through the media in constant contact with air. With particularly concentrated wastes it is necessary to recirculate the effluent to dilute the feed to the biological film. The film is constantly sloughing off the media and must be removed in a settlement tank referred to as a 'humus tank'. Although there is no aeration equipment to power, the energy involved in pumping is equivalent.

16.5.4 Anaerobic reduction

Anaerobic reduction of organic material is achieved by a relatively small number of species of bacteria in the absence of oxygen. The process comprises the liquefaction of insoluble organic material, followed by 'acetogenesis', i.e. the production of acetic and other fatty acids, then 'methogenesis', i.e. the conversion of the acids into methane and carbon dioxide. These biological processes are poor generators of energy for the bacteria, resulting in slow growth and a low conversion of the organic material into biomass. Conventionally the process is operated at 37°C, which is the optimum temperature to speed up the process. The low generation of sludge is beneficial in that it presents the minimum of a disposal issue. The generation of the flammable biogas can be used to raise steam or generate electricity, although some may well have to be used to maintain temperature at the optimum. The value of the surplus biogas and the fact that there is no need for energy-intensive aeration systems mean that anaerobic systems are cheaper to operate than aerobic plants. Furthermore, the low growth of biomass means that the demand for nutrients such as nitrogen and phosphorus is also less. This is of particular relevance to brewery wastewater.

Conventional anaerobic systems have a poor reputation in terms of the degree of treatment that can be attained, with 60% removal of BOD being typical. Modern developments that ensure good separation of the biomass from the treated wastewater are more on a par with aerobic systems, with up to 97% reduction in COD being achieved.

16.6 Treatment processes

16.6.1 Activated sludge

For a typical effluent generated by an efficient modern brewery (Table 16.2), the wastewater per hectolitre of beer would comprise a minimum of 2.5 hl containing 0.55 kg COD or roughly 0.3 kg BOD. To achieve some 90% removal of BOD a typical F:M loading of 0.5 would be required, needing an aeration tank volume equivalent to 2 hl/hl beer per day at an MLSS concentration of 3 g/l. For a brewery generating say 4000 hl/day (1 million hl/annum) this volume translates to an area of 200 m^2 at an operating depth of 4 m. This area is for the aeration tank alone and excludes a 10 m diameter settlement tank, a balancing tank and an area for ancillary equipment.

To make the process more compact two things are necessary: good oxygen transfer and a high concentration of biomass. Depth or the use of pure oxygen rather than air can be used to achieve the former. The only conventional ways to achieve a higher concentration of biomass are to dose in a polyelectrolyte to thicken the flocs or to use Dissolved Air Flotation; the former is the more common. However, systems incorporating membrane filtration as the method of separating biomass from the treated water have recently been developed. This delivers a higher concentration of biomass than is otherwise possible whilst also making settlement redundant, further reducing the footprint of the plant.

16.6.2 Sequencing batch reactors (SBRs)

Sequencing batch reactors[3] provide a well-established alternative to conventional activated sludge treatment; the biology is the same and only the format varies. No clarifiers are used to settle and recover the biomass, rather settlement is achieved in the aeration chamber itself by switching off the aeration device (Fig. 16.3); hence two or more reactors are required to operate in sequence. Generally this provides a far more compact plant. SBRs are best applied when the discharge consent is not too tight, as the method of collecting the supernatant leads to a relatively poor removal of biomass, resulting in higher BODs and suspended solids when compared to conventional activated sludge systems.

16.6.3 High rate filters

High rate filters are a development of conventional biological filters except that the growth support media comprise thin, light plastic media with a high specific volume (some twice that of mineral media) that enables a large media volume to be stacked high (e.g. 10 m), thereby giving a small footprint. These filters are generally loaded at high rates designed to achieve say only 70% removal of BOD that further reduces the plant size. The plants are well suited to roughing treatment, to reduce the discharged load, but still requiring further treatment or as a preliminary treatment for activated sludge-based processes.

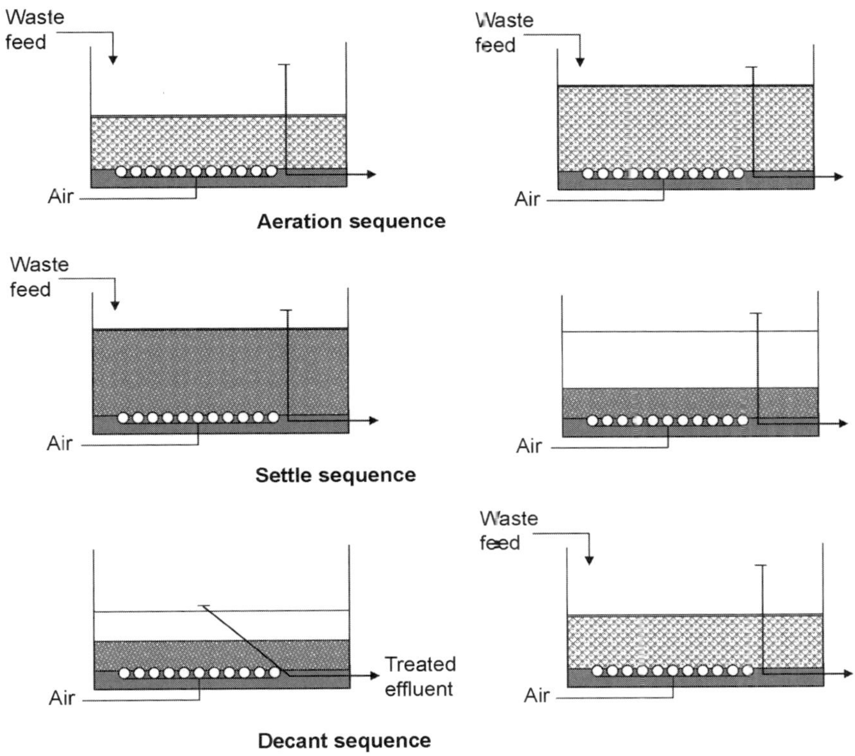

Fig. 16.3 Sequence of sequencing batch reactor operations.

The medium is at risk of drying out due to poor flow distribution over its surface, especially with high concentration wastewaters such as from a brewery. Stacking the medium high helps to reduce the risk, but pumped recirculation of treated effluent back to the top of the tower is also required. The pumping costs are considerable but are still less than the power costs associated with activated sludge systems.

16.6.4 The HCR process

This is a variant of the activated sludge process but operates at a much higher rate.[3] It is a small-footprint system that utilises a tower reactor (Fig. 16.4), to minimise the footprint, and also provides an aeration system that has very high oxygen transfer rates that appear to benefit the development of high rate bacteria above other organisms. A 100-fold increase in viable bacteria population is claimed, enabling very intensive treatment. The secondary settlement stage has to be preceded by a deaeration stage to drive out the excess air and reflocculate the suspended matter. The process favours strong wastewaters in that the area required for settlement does not predominate over land utilisation and costs. The

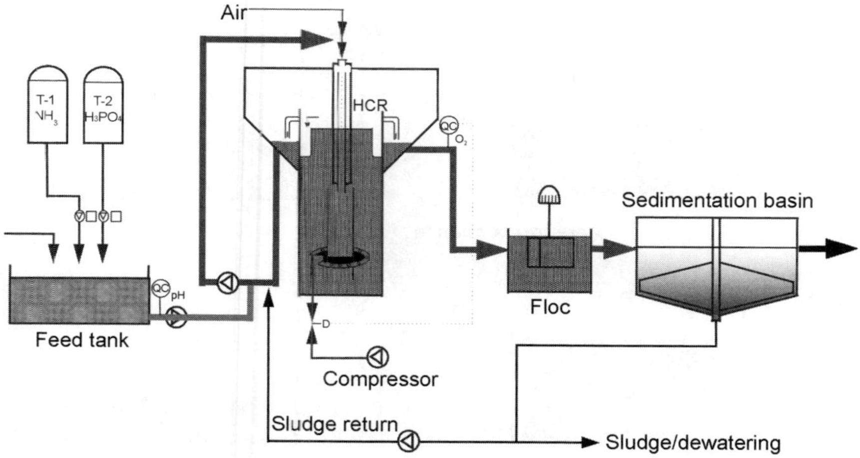

Fig. 16.4 The HCR process (Source: courtesy of Aker Kvaerner).

process has been applied only to high strength paper and pulping wastewaters with pilot-scale operation for brewery and high strength food wastewaters. The gains in energy efficiency and small-plant footprint justify consideration of this process.

16.6.5 Biological aerated flooded filters

Again there have been developments in this field in recent years, resulting in the 'Biological Aerated Flooded Filter (BAFF)' sold under trade names including Biobead,[4] Biofor[5] and Biopur[3] (Fig. 16.5), amongst others. These processes have generally been developed for the production of high quality effluents. In this case air is blown through the submerged 'filter', where the medium may be a fixed bed structure, similar to that used in high rate filters, supporting a submerged film growth, or free-moving porous ceramic or plastic beads. This format facilitates more prolonged contact between the growth and the waste stream compared to conventional biological filters. In this way the volumetric capacity of the process is enhanced; the high capacity of the medium for solids growth results in greater biomass densities compared to activated sludge which enables a 3–5 times more intensive operation. The filters need regular backwashing, typically once a day, to prevent the medium from becoming blocked with growth. The backwash wash water is generally treated to remove the suspended matter by recirculating to a preliminary settlement stage; this avoids the need for a dedicated settlement facility, but storage tanks are required to collect water for backwashing and to provide balancing of the backwashed water. This effectively precludes the process from industrial wastewaters of low suspended solids content that do not require the primary settlement stage. The effluent quality from the process is generally excellent with very low suspended solids content.

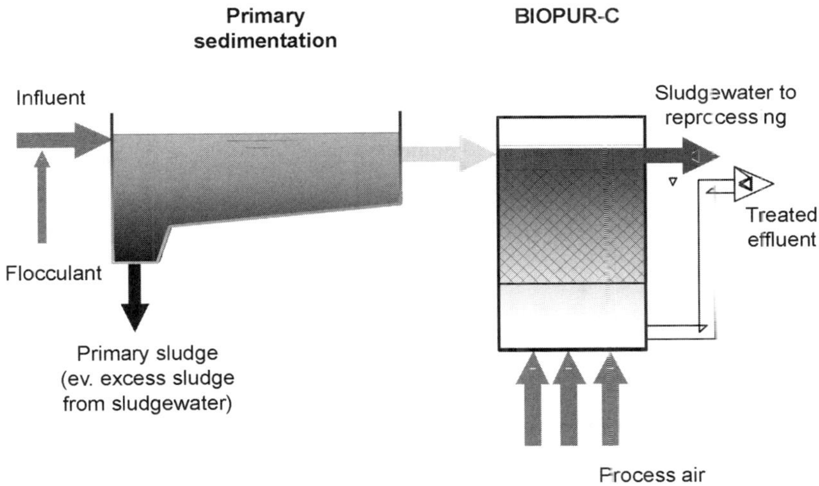

Fig. 16.5 Schematic of biological aerated flooded filter (Source: courtesy of Aker Kvaerner).

The media used in BAFFs and the control systems are specific to the suppliers and no general design equations can be supplied; to this extent the process equipment is proprietary and the responsibility for designs rests with the suppliers.

16.6.6 Submerged aerated flooded filters

This is a lower cost system than the BAFF,[6,7] of similar concept but avoiding the need for complex backwashing and control. The medium is generally fixed in place. Biomass is continuously washed off the medium by means of intensive aeration, and the effluent must pass through a settlement tank if effective treatment is to be realised.

16.6.7 Moving Bed Biological Reactors (MBBR)

This system is intermediate between the BAFFs and SAFFs, where free-moving plastic medium is utilised, such as in some BAFF processes, but excess growth is sloughed off the medium and must be removed in clarifiers. An example is sold under the trade name of Kaldnes.[8] A threefold intensification compared to conventional activated sludge is achieved, while the process is suited to retrofitting into existing activated sludge tanks, with the media being retained in place using open mesh baffles. The process can be used for high rate roughing operations or for producing high quality effluents.

16.6.8 Membrane Biological Reactors (MBR)

MBRs are wastewater treatment systems utilising micro-crossflow filtration membranes[6,9] (Fig. 16.6). The body of the reactor containing the activated sludge

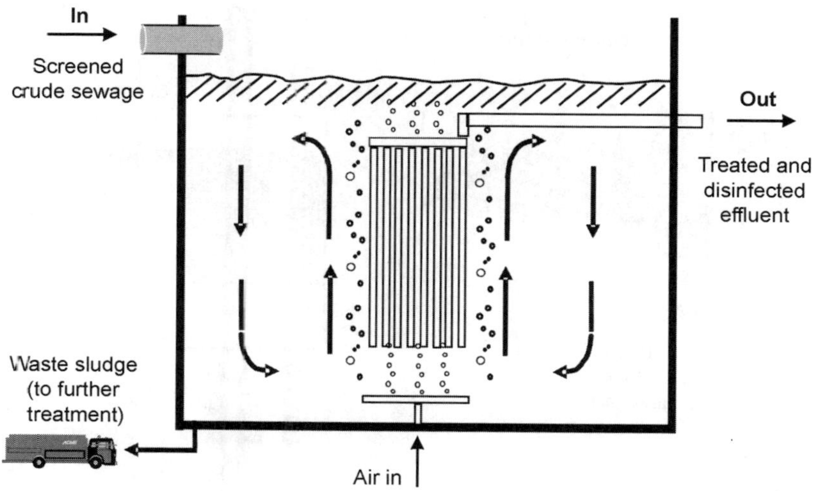

Fig. 16.6 Membrane biological reactor (Source: courtesy of COPA Ltd).

also contains membrane envelopes which are generally held under a small negative pressure to withdraw the treated wastewater, rather than using settlement of the biomass. This enables very high concentrations of activated sludge to be maintained in the reactor, say 2% solids or so, which results in a very small footprint for the process. An additional advantage is that the membranes have a particle cutoff size of typically 0.2 microns, which combined with the sludge ages results in a very high quality effluent suitable for immediate reuse. High levels of aeration are used to maintain the surface of the membrane clean. The process is sensitive to contaminants such as oils that will blind the membranes, and generally the membranes must be removed periodically for cleaning.

During the early stages of development the membranes were very costly and their life was unknown, resulting in the potential for very high operating costs. However, falling membrane costs have resulted in the process being cost-effective even for the treatment of municipal sewage in sensitive areas and where a bacteria-free discharge is required, as on bathing beaches. If reuse of treated effluent is being considered, then MBRs should be considered for the treatment of brewery effluent; otherwise this is likely to be too expensive a technique.

16.6.9 Reed beds

Reed beds[10] have been developed over the past few decades. While the reeds do participate in removing polluting materials, the majority of the treatment is effected by bacteria which grow around the root mass. Oxygen is transferred to the bacteria via the reed leaves and roots. Design of reed beds is a very specialised skill. They are generally used for polishing operations but can be designed for full-strength wastewater.

16.6.10 Anaerobic systems

The conventional process as used in sewage sludge treatment is aimed at the breakdown of solid matter and could be applied to the treatment of biological sludges grown in a system treating brewery or maltings wastewater. It is generally a batch process, taking place in large vessels of 5–10 m height. The hydraulic retention time is long, and while it could be used to treat liquid wastes, modification of the process is required to reduce the land area and capital cost. The approach is along similar lines to the aerobic process.

The processes are designed on the basic of volumetric COD loadings, i.e. in terms of kg COD/m^3.day. Using the guide loadings given in Section 16.6.13, a crude guide to the size of the wastewater treatment reactor can be calculated. However, design of anaerobic systems is by no means easy, and well-established suppliers should be approached when considering the size and cost of on-site anaerobic treatment.

16.6.11 Anaerobic contact process

This is the direct analogue of the activated sludge process and is shown in Fig. 16.7. A number of methods are used to recycle the biomass. These include cooling to stop the generation of gas followed by gravity settlement, centrifugation, and dissolved gas flotation. These separation processes can involve a fair degree of maintenance. The volumetric loading that this type of system can manage is in the range of 1.5–5 kg COD/m^3.day. A system along these lines is in operation at the Hall and Woodhouse brewery in Dorset, operating at a volumetric loading of approximately 1.5 kg COD/m^3.day and a retention time of 3 days plus. An interesting aspect of this reactor is that it accumulates volume during the working week to maintain treatment over the weekend.

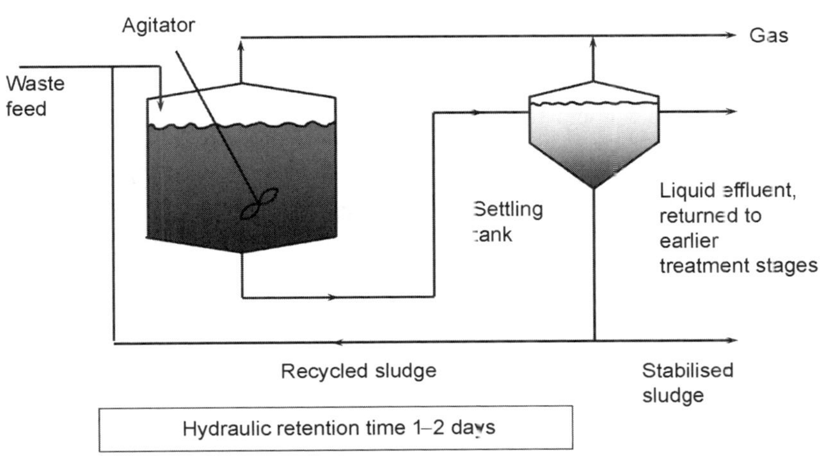

Fig. 16.7 Schematic of anaerobic contact process.

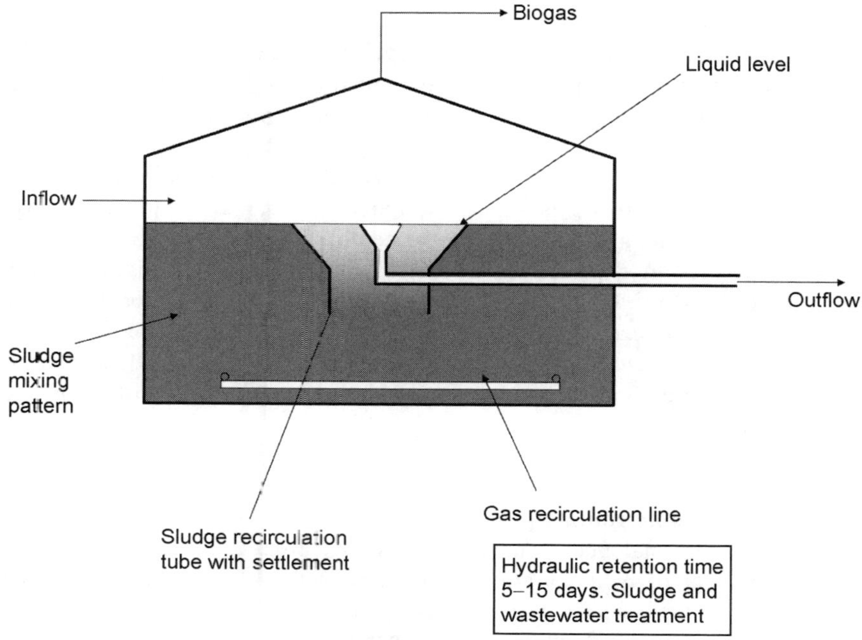

Fig. 16.8 Schematic of simple digester.

16.6.12 Digester with internal biomass recovery

This system, shown in Fig. 16.8, represents a lower-cost alternative to the above processes, although the efficiency of the internal settlement system is not particularly great (50–70%). Its simplicity makes it worth considering, especially if concentrated waste streams can be isolated for treatment.

16.6.13 Upflow anaerobic sludge blanket (UASB)

It has been found that if anaerobic biomass is grown under the correct conditions, it grows in a granular form which has excellent settling properties, with the sludge developing concentrations of up to 70 g/l in the settlement hoppers. This makes the treatment of low concentrations of wastewaters a feasibility provided that the stream is warm. The liquid waste is pumped up through the dense sludge where it is rapidly treated. A greatly simplified form of the plant is shown in Fig. 16.9. This enables the use of very high loadings within the reactor, such as 10–15 kg COD/m^3.day, and short retention times, typically less than 48 hours and as low as 12 hours. Generally a 'pre-acidification' stage incorporating flow balancing is provided upstream of the reactor, providing approximately a 12-hour retention time for the wastewater. Good flow distribution within the reactor vessel and biomass/gas separation provide the key to the success of this type of treatment. The addition of pH correction is also critical to satisfactory operation. A number of suppliers have gained recognition for the success of

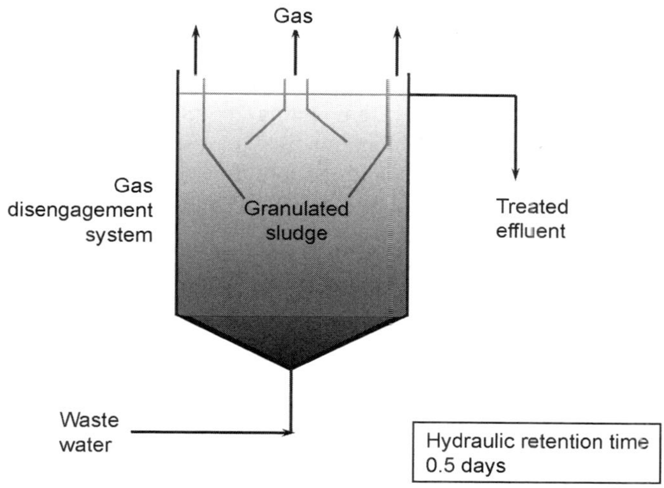

Fig. 16.9 Simplified schematic of up-flow sludge blanket reactor.

these designs, including Biothane[11] (based in the USA) and Paques[12] (based in the Netherlands). Despite the low hydraulic retention times, the complexity of the engineering of the flow distribution and gas disengagement still results in a high capital cost.

This technology has now been further developed for 25 m tower-type steel reactors with low footprint; low retention times of a few hours are claimed for the weaker wastewaters.[12] Internal recirculation of the treated wastewater plays a major role in these high rate designs.

16.7 Summary of the new legislation

The generation of wastewater is coming under increasing pressure not only in terms of the charges imposed by the receiving authority, but also legislation, namely the European Urban Waste Directive[13,14] and, of greater immediacy, the European Directive on Integrated Pollution Prevention and Control (IPPC). Linked with the IPPC regulations is the Climate Change Levy[15] (and its rebate) on fuel usage which is being imposed by the Department of the Environment, Transport and the Regions. It is the pollution 'load' (the product of strength and volume) which is the concern of the European legislation, and more specifically minimisation of pollution in the case of the Climate Change Levy and the associated IPPC legislation. In general, for maltings and breweries, the polluting load is no longer simply the mass flow of Biological (or Chemical) Oxygen Demand (BOD/COD) but the salinity and metal content, e.g. copper or chromium from plant. Any future limits on salinity and metal content will present an entirely new range of water management problems, especially to brewers, with implications for raw water treatment and CIP.

16.7.1 Urban Waste Directive

The UK interpretation of the Urban Waste Directive is defined in the Statutory Instrument No. 2841, 1994. This stipulates that any industrial discharge or geographical group of discharges amounting to more than 120 kg BOD/day (equivalent to roughly 100 000 hl/annum in the case of a brewery or 15 000 tonnes/annum for a maltings) discharging into a waterway requires 'secondary' treatment by the end of 2005. The limits on BOD and COD entering a waterway are 25 mg/l and 125 mg/l respectively or a 70% reduction in BOD and COD, whichever is the greater. The extent of treatment into estuaries depends on the dispersal characteristic of the estuary. Generally most plants in England or Wales already comply with this Directive, either treating on site or discharging to a sewer for subsequent treatment, apart from perhaps the COD in the treated wastewater which can be regarded as 'biorefractory' (see Section 16.3). Some Scotch whisky distillers may still not be in compliance. Of greater significance, sites close to the coast discharging more than 600 kg/day BOD to sea, either directly or via a municipal sewer, must ensure treatment of the wastewater, but to a standard less stringent than above. The above are the minimum requirements and more stringent discharge consents may be imposed.

16.7.2 Climate Change Levy and IPPC

The voluntary Climate Change Agreement with the UK government secures a reduction in the Climate Change Levy in exchange for meeting agreed biennial targets through to 2010. This runs alongside the mandatory EU's Emission Trading Scheme which is aimed primarily at reducing fuel consumption and associated carbon dioxide emissions; it is targeted at all industrial users.

A good dataset collated by the Brewers' Society and now the British Beer and Pub Association dating back to the 1970s has demonstrated the determined and continuous drive by the brewing industry to reduce water and energy use. Since 1975 the reduction in specific water usage is 38% per unit of beer production. The reduced volumes of water have resulted in lower pumping and energy-loads, which in association with other energy saving technologies has resulted in the reduction in specific energy usage by 47%. This is despite the swing towards small-pack beers (and away from kegged beers) demanding more water, and the swing towards lager from ale production with the demand for more refrigeration. Carbon dioxide is the key measure of environmental emissions: electricity results in twice the amount of CO_2 compared to natural gas over the point of generation to end use; oil and coal are also inefficient compared to gas. Gas has completely replaced the role of coal as an energy source for the industry and much of the role of oil is now much diminished. The result is that between 1990 and 2003 the specific CO_2 emissions by the industry have reduced by 44%. These factors have enabled the industry to obtain an 80% reduction in the Climate Change Levy, worth a saving of some £4 million per year to the UK industry. It should be noted that CO_2 release during fermentation is excluded from the calculations as this gas is recovered by the next harvest.

The EU Energy Trading Scheme applies to any facility that totals more than 20 MW oil- or gas-fired combustion plant.[1] This infers that 19 or so breweries need to register for the scheme, which will operate alongside the UK Climate Change Levy scheme. Under the EU scheme any facility using more than their allocated emissions allowance will have to purchase additional allowances to continue to operate. The EU scheme has completed baseline verification and all eligible facilities will be allocated their emission allowances under the national plan. By December 2004 the UK industry had 56 000 tonnes of 'over-achievement' in CO_2 reduction, enabling a reduction in the tightening of the brewing sector target to a reduction in CO_2 emissions of 2% by 2006 and an aspirational reduction of 3.5% by 2010.

The EU Integrated Pollution and Prevention Control (IPPC) requires sites with a design capacity broadly more than 2000 hl/day, 0.5 million hl/annum in the case of breweries and cider manufacturers, and 300 tonnes malt/day, to obtain a Pollution Prevention Control permit. This demands that the site will have to be operated in a way that prevents or reduces to acceptable levels emissions to air, land and water. These sites must install 'Best Available Technology' (BAT) to minimise both power consumption and 'emissions to the environment' (solid, liquid, gaseous, noise and vibration), minimising waste and recycling water wherever possible. There was a tendency by the regulatory authorities to proscribe exactly what BAT is, but this approach has been modified to reflect plant location, age and product quality. Gaining a permit is eased by having an Environmental Management System (e.g. ISO 14001) in place and by being part of the Climate Change Agreement or a similar scheme. There are clearly conflicts within the objectives of reducing waste discharges and minimising the use of energy, particularly if aerobic waste treatment systems are used; the newer anaerobic systems clearly have IPPC and environmental advantages.

Under a recent legal ruling regarding IPPC, spent grains and yeast are not considered waste provided that these materials are passed directly to another undertaking for processing into food or drink; this is critical in the determination of the threshold calculations for the Pollution Prevention Control permit.

It can be seen from the above that there are both increasing legislative and economic pressures to improve performance regarding simultaneous power and water usage, and waste minimisation.

16.8 Management approaches to water and waste minimisation

Management of water use in breweries is seldom an easy task, seeing that water is used in most process stages and adequate metering is seldom installed. Knowledge of water usage for each processing area, along with the strength of wastewater, is the critical primary step to the minimisation of water usage and waste. Ideally meters should be installed to assist in achieving this end, but surprisingly good information can often be acquired by personnel skilled in the estimation of water usage even without meters. Once the water usage has been mapped, areas of excessive usage and waste generation can be targeted. It is worth noting that waste in the context of a brewery or cider plant represents the loss of wort, juice or final product.

A typical approach to water management prioritises the following steps:

- Elimination or reduction of waste generation/water usage
- Direct reuse/recovery of water/product at source
- Treatment at source enabling recovery/reuse of water/product
- Finally, treatment of groups or the combined waste streams if appropriate to reduce overall disposal costs.

Reduction in beer losses, from a conditioning tank for example, can be effected by allowing more time for tank drainage, collecting first tank rinses and improving the layout of associated pipework. The layout of pipework is particularly important when attempting to reduce the quantity of deaerated liquor used to flush pipework prior to the pumping of beer, where contorted runs and inverted U-bends can result in the hold-up of air in the pipes. There are doubtless many possibilities for the reduction of water use and beer losses, but many will be site specific.

16.9 Future trends

There is likely to be continuing pressure from EU Directives and associated UK legislation on reducing water consumption and energy-related CO_2 emissions. While much of the EU legislation relates to the larger production facilities, experience in the water industry is that there is a continuing trend to lower the limits of production capacity for bringing facilities within the legislation net. Current climate change legislation does not appear to take into account the range of products that any particular site is generating, but the pollution control legislation does allow for the location, age and product quality on the definition of best available technology restricting emissions to the environment.

Where on-site treatment of wastewater is necessary, the demands to limit the consumption of fossil fuel-derived energy and associated financial benefits are likely to increase the development of anaerobic digestion as the preferred method of treatment.

Energy-saving techniques such as combined heat and power (CHP) systems are actively encouraged in legislation for government grants/allowances but a full process/engineering/financial assessment should be made to assure that the energy released during power generation can be usefully captured by the brewing facility. Operations based on one operating shift per day are unlikely to be benefited by CHP systems.

The British Beer and Pub Association is acting as the interface between the industry and government and has an extensive database on water and energy utilisation.

16.10 References

1. Long D, Tighe A, McCrimmon E: An update on EU and UK pollution regulations, *Brewing and Distilling International* 2005(2) 14–16.
2. www.defra.gov.uk/environment/waste/topics/landfill-dir
3. www.akerkvaerner.com
4. www.brightwater.uk.co.uk
5. www.ondeo-degremont.com
6. www.copa.co.uk
7. www.wpl.co.uk
8. www.purac.net/index.cfm
9. www.naston.co.uk
10. www.armreedbeds.co.uk
11. www.biothane.com
12. www.paques.nl
13. www.defra.gov.uk/environment/water quality
14. www.environment-agency.gov.uk/business/directives/IPPC
15. A business guide to climate change projects: www.dti.gov uk/ccpo

17

Quality assurance in brewing

G. Jackson, Brewing Research International, UK

17.1 Introduction

There has been increased pressure worldwide for breweries to implement systems to assure the safety, quality and legality of their products. These pressures have derived in some cases from legislative requirements and in other cases from commercial requirements. There has also been an increase in the number of standards and guidelines to assist breweries,[1] and other food producers, to comply with requirements and, where necessary, to demonstrate the effectiveness of the systems implemented by achieving certification. This chapter outlines the systems in place in breweries to achieve product safety, quality and legality, how these are implemented and, where applicable, the process for certification.

17.2 Hazard Analysis Critical Control Points (HACCP)

HACCP is a management system designed to assure the safety of food products. It was developed in the 1960s and is now recognised by the World Health Authority and other international bodies. It has now been implemented in breweries worldwide. Several countries have legal requirements for HACCP. For example, in Europe, European Council Directive 93/43/EEC[2] and Regulation 178/2002[3] made it mandatory for all food producers, including breweries, to carry out risk assessment based on HACCP principles. This applies to all stages of beer production, including processing, packaging, storing, transportation, distribution, handling and sale.

Before implementing HACCP in a brewery there are a number of requirements and systems that must be in place. These so called prerequisites are activities that reduce some food safety hazards (and hence reduce the number of Critical Control Points) or that are required to operate the HACCP system effectively. Prerequisites for breweries include the following aspects:

- There should be a food safety policy outlining the brewery's commitment to producing safe products. This should be disseminated to all staff.
- Potential contamination from surrounding industries should be assessed and suitable precautions taken. Site boundaries should be sufficiently protected to avoid accidental or malicious contamination.
- High risk areas should be identified; this includes areas where there is open product, package, process or raw material (e.g. open fermentations, bottling and canning areas). These areas require more stringent controls and should be protected from the outside. Eating, drinking and smoking is not permitted in these areas.
- Buildings used for storage or production should be fit for their purpose, adequately maintained and cleaned.
- Equipment should be suitable for its intended purpose, easily cleaned and subject to a planned maintenance programme.
- Suppliers of raw materials and packaging materials should be controlled. Goods should be checked on receipt to ensure that the correct grade has been delivered and the packaging is intact.
- Housekeeping and hygiene should be maintained to a high standard to prevent contamination from the surroundings.
- Suitable toilets and hand-washing facilities should be available; these must not open directly into production areas.
- A pest control programme should be in place to minimise pests. Precautions should be taken to prevent the materials used from contaminating product.
- The use of glass in production areas should be minimised to prevent product contamination. There should be a glass policy which states the glass breakage procedure. A glass register should also be available.
- Vehicles used for raw materials or products shall be suitable and should avoid deterioration of loads carried. Vehicles should be loaded and unloaded so as to avoid contamination or deterioration.
- All staff, including temporary staff, shall be adequately trained before carrying out a task.
- A product recall procedure should be in place. Where a product has been withdrawn because of an immediate health hazard, other products which are produced under similar conditions, and which may present a similar hazard to public health, should be evaluated and may need to be withdrawn.

Once the prerequisites are in place then HACCP can be implemented. Typical steps in implementing HACCP in breweries are shown below:

- Senior management commitment is essential to sanction staff time in setting up the plan and maintaining it.
- The scope of HACCP should be defined; this normally includes all process

steps from receipt of raw materials to beer in package on-site. It also normally includes all hazards (chemical, physical and microbial) that may occur.

- A HACCP team should be used to ensure that all potential hazards are considered. Large, multi-site breweries may operate more than one HACCP team. A Team Leader should be appointed to coordinate activities and ensure that a systematic approach is followed.
- A process flow chart is prepared to include all process steps within the scope of HACCP. This helps with the analysis of hazards.
- The HACCP team should identify all potential hazards to the consumer and appropriate control measures for each hazard. Since beer does not support the growth of microorganisms, the main hazards are from chemical sources (e.g. caustic, nitrosamines, heavy metals and mycotoxins) and physical sources (e.g. glass or metal fragments, filling tubes, etc.). However, *Salmonella* and *Escherichia coli* can survive for some time in non-alcoholic beers and the protozoan *Cryptosporidium* can contaminate beer through the use of contaminated water. The bacteria *Obesumbacterium proteus* and *Bacillus* species may harm the consumer as they are implicated in the production of nitrosamines. These are normally also considered in HACCP in breweries. Table 17.1 gives some typical examples of hazards in breweries.
- The Critical Control Points are identified; these are the stages in processing where controls must be in place to eliminate or reduce hazards to an acceptable level. A decision tree can be used to identify the CCPs. Table 17.2 gives an example of a typical CCP in a brewery.
- Critical limits are set for each control measure at each CCP; if the critical limit is exceeded then the process is out of control and unsafe product is produced. Critical limits, such as time, temperature or pH, need to be easily measured, so that they can be monitored routinely. Table 17.2 gives a typical example of critical limits.
- The process is then monitored at the CCP. This checks that the critical limits are not exceeded and that CCPs are under control. When critical limits have been exceeded then corrective actions should be taken. Records of monitoring should be retained as evidence that the process was under control. Table 17.2 gives a typical example of monitoring in a brewery.
- Corrective actions must be taken when target values or critical limits are exceeded (Table 17.2). They include actions to rectify the immediate situation plus actions to prevent the problem recurring. Records of corrective actions should be retained.

Once the hazard assessment has been completed it is implemented in the brewery. Implementation includes:

- Setting up records for monitoring critical limits at each CCP. Clear accept/reject criteria, based on the critical limits, should be defined.
- Amending operating procedures, as needed, to include HACCP requirements.
- Training operators to make them aware of the monitoring required, its

Table 17.1 Examples of hazards in breweries

Operation	Examples of hazards
Raw materials, processing aids, additives and all food contact materials procurement	• Agricultural residues such as pesticides and herbicides, heavy metals • Chemical contamination
Raw material storage	• Chemical contamination from hazardous chemicals stored in close proximity
Materials intake	• Oil from delivery vehicles
Water intake	• Potential contaminants of brewing liquor
Malt conveying/all points of lubrication	• Chemical contamination, e.g. oil from conveyor motor oil
Sieving/dust removal/destoning	• Foreign bodies in malt, e.g. pests, stones, metal
Addition of salts to grist case	• Addition of potentially hazardous material • Over-addition of material with a legal limit
Liquor heating	• Chemical contamination, e.g. from boiler treatments (only if direct steam injection is used)
Wort mashing and separation	• Damage to vessel that may allow bacteria to grow in cracks and lead to ATNC production
Wort boiling	• Chemical contamination from boiler treatment (only if direct steam injected)
Copper additions	• Over-addition/addition of hazardous material
Trub separation	• ATNC formation due to microbiological growth
Wort cooling	• Hazardous coolant leakage into product due to damaged plate heat exchanger, e.g. IMS, glycol, methanol
Addition of yeast nutrients	• Over-addition above safe limits of zinc sulphate
Fermentation	• Over-addition of antifoam above the legal limit • Operator cleaning a full fermentation vessel • Chemical contamination – propylene glycol, coolant from coolant jacket due to damaged vessel wall
Post-fermentation hopping	• Addition of potentially hazardous material
Chilling	• Chemical contaminant from secondary coolant due to damaged plate heat exchanger plate
Filtration	• Foreign bodies introduced from previous process steps
Tanker loading	• Chemical contamination from cleaning agents • Chemical contamination from previous tanker load(s) • Physical contamination from flexible hoses

Table 17.2 Critical Control Point example

Process step	Chill beer during transfer from BBT to filler bowl
Hazard and cause	Chemical contamination from secondary refrigerant due to leaking heat exchanger
Control measure	Product pressure higher than coolant pressure during beer transfer
Critical limits	Pressure differential = x bar
Monitoring	Check coolant inlet pressure and product outlet pressure (frequency: once per hour; responsibility: to be specified)
Corrective action	Stop beer forward flow; examine heat exchanger and repair (responsibility: to be specified). Isolate product produced since last check and analyse for presence of secondary coolant (responsibility: to be specified)

importance and significance and what corrective actions are required if critical limits are exceeded.

The HACCP system should be reviewed whenever there are changes to raw materials, equipment, process or packaging or as a result of recommendations from an audit.

Typical records and documents which are retained in breweries include the HACCP plan (process flow diagrams, hazards identified and CCPs), records of monitoring of CCPs, deviations at CCPs and the actions taken, audit reports, non-compliance sheets, minutes of review meetings, operating instructions, procedures for non-conforming product and training records.

Some breweries have applied for certification for their HACCP systems. At present, this is based on compliance with requirements of the Codex Alimentarius. However, the draft ISO22000 Standard 'Food Safety Management Systems – Requirements for Organisations throughout the Food Chain'[4] has been published. This will enable breweries to apply for certification of their HACCP to this Standard. The Standard sets out the requirements for a food safety management system that will ensure food safety along the food chain up to the point of final consumption. The key elements of this are interactive communication, system management, process control, HACCP principles and prerequisite programmes. The Standard can be applied on its own or integrated into another management system. It is compatible with ISO9001:2000[5] to enable food safety systems to operate within the framework of a structured management system.

To comply with this Standard, breweries will need to implement HACCP as described above plus some additional requirements, as detailed below:

- There should be communications with external bodies (suppliers and contractors, customers, food authorities) to provide information on the brewery's products. The aim is to agree the level of food safety required along the food chain so that hazards are controlled along the chain. There should also be communication within the brewery to ensure that the food safety team is informed of any changes such as new products, raw materials, production, packaging, changes in legislation, complaints involving health hazards, etc.

- The food safety management system should be reviewed at intervals to ensure it is still effective (in the same way that the quality management system is reviewed). The review should consider verification results, changes that might affect food safety, product recalls and results of reviews and updates to the system. The meeting should be committed to improving the effectiveness of the food safety management system and assuring the safety of the products.
- There should be adequate resources for effective operation of the food safety management system. This includes training for the food safety team and other staff whose job affects product safety. The skills needed should be identified and provided; this includes training for monitoring at CCPs and taking corrective actions when there is loss of control of processes.
- The Standard identifies two different Prerequisite Programmes (PRPs). The first is the Infrastructure and Maintenance Programmes; this includes ensuring that the infrastructure minimises hazards and includes the layout and design of buildings, air, water and energy supplies, equipment maintenance and removal of waste. The second PRP is the Operational programme; this includes personnel hygiene, cleaning and sanitising, pest control, preventing cross-contamination, management of purchasing and packaging procedures. The brewery can decide whether to classify PRPs into Infrastructure/ Maintenance or Operational PRPs. The difference between them is that Operational PRPs are subject to hazard analysis and identification of CCPs, whereas the Infrastructure/Maintenance PRPs are not based on hazard analysis nor subjected to validation.
- Traceability must operate throughout the brewery. This should identify the source of incoming materials and distribution of product.
- Records of monitoring at CCPs should be reviewed by a designated person with authority to initiate corrective actions. All non-conformities at CCPs must be actioned to prevent recurrence and bring the process back in control. End product affected by non-conforming CCPs or operational PRPs should be assessed and appropriate actions taken.

17.3 Quality management systems and ISO9001:2000[5]

The most common standard for quality management systems for breweries is ISO9001:2000. This replaced the ISO9000:1994 series of standards that had been in operation in many breweries. This Standard specifies the quality system that the brewery should implement to prove its ability to manufacture and supply product to an established specification. It relates only to the management system for controlling quality. It defines activities for which the brewery must provide appropriate controls but it does not define how they are to be controlled. It outlines the key elements of a quality management system that should be in place to ensure the quality of the products produced. The following sections consider the key elements of a quality management system (based on ISO9001:2000) and how these are implemented in breweries:

- A quality policy should be available, usually signed by the Chief Executive Officer. This informs employees and customers of the brewery's commitment to quality and its commitment to meet the needs of customers. All staff in the brewery are made aware of the quality policy and its implications; this can be achieved by training sessions or by placing the policy on notice boards.
- A quality manual should be available that outlines the scope of the quality management system and the interaction between the processes of the quality management system. It provides a map of the quality system and its links to other systems, such as HACCP. It can be cross-referenced to relevant procedures and work instructions.
- Written procedures should be available for key operations affecting product quality. This ensures that there is no variation in operations.
- Documents relating to quality must be controlled. Controlled documentation can include specifications (e.g. for raw materials, processing and products), drawings (e.g. artwork for packaging), legislation and codes of practice, equipment manuals, the HACCP plan and HACCP documentation. Systems for document approval and document changes should be in place. All staff must use the latest version of documents relating to quality.
- The brewery should keep sufficient records to show that the quality management system works effectively and to meet legal and contractual requirements. They are stored so as to avoid deterioration and to enable retrieval of archived documents as needed. The retention time of documents (such as internal audit reports, management review minutes, HACCP documentation) must be defined; claims from customers can be made after the shelf life has expired, so the retention time may need to be set accordingly. A blanket retention period of several years for all records generated may not be appropriate.
- Review meetings should be held at defined intervals (at least annually and preferably more frequently) to review the quality system. The aim is to assess its continuing suitability and effectiveness and to decide opportunities for improvement and the need for changes. This meeting provides the overview of the company and ensures that it is moving in line with long-term goals or aspirations. The review considers evidence of the functioning of the quality system. The agenda may include the quality policy, results from audits, customer complaints, performance of suppliers, effects of changes in plant, specifications, staffing or legislation.
- Training should be provided for staff whose jobs affect product quality; no member of staff should be asked to do a job for which they are untrained. There should be procedures for identifying training needs, providing training

and reviewing the effectiveness of training. Training is vital to new employees. Even if they have previously worked in a brewery they will need to be trained in the particular procedures of the new company. Additional training may be required as a result of changes in plant, staff or procedures. Training might include in-house, on-the-job or external training courses.

- Purchasing should be controlled for all key materials and services that could affect product quality, for example ingredients, processing aids, water and water treatment, equipment, packaging, contract brewing and packaging operations, sub-contract laboratories, training, transport and warehousing. Materials or ingredients that come into direct contact with the product will probably need tighter controls than other materials. Most breweries have a list of approved suppliers. Suppliers can be evaluated on the basis of their ability to meet specifications or by doing audits of suppliers. The performance of suppliers should be monitored on an on-going basis, by analysis or by checking the goods delivered. Suppliers may also be encouraged to implement a formal quality management system, preferably registered by an accredited body, and preference may be given to these suppliers.
- Equipment that is used to control the process or to check conformance to specifications should be calibrated at specified intervals. Calibrations should be traceable to international or national standards. This also includes newly purchased equipment which should undergo an initial calibration, where necessary, before being released for use. It excludes equipment used to monitor the condition of the plant (e.g. power consumption gauges and some pressure gauges). Most breweries have a calibration schedule, defining, for example, the equipment, its location, calibration frequency, calibration method, accept/reject limits, and the actions to be taken if calibration is out of specification. Recalibration should be performed sufficiently frequently so that it occurs before equipment goes out of calibration; the interval between calibrations can be lengthened or shortened as necessary. Equipment should have its calibration status clearly identified. If the equipment is used for indication only, then this must be clearly indicated on it to avoid the possibility of it being inadvertently used for critical measurements. Equipment that is found to be out of calibration, or equipment which has gone over its calibration date and is awaiting calibration, should be labelled to indicate that it should not be used.
- Internal audits are used to ensure that the quality management system conforms to planned arrangements and is effectively implemented and maintained. An audit schedule is used to define the scope, frequency and methods defined. All aspects of the quality system should be audited within one year; critical processes and areas where problems have occurred in the past should be audited more frequently. In the brewery audits are done by staff who are suitably qualified, experienced and competent and who are independent of the area being audited; staff should not audit their own areas of responsibilities. In multi-site breweries staff can be used to audit other breweries in the group. A report is issued at the end of the audit giving a summary of audit events and findings. This could, typically, give details of

deficiences found, the extent of the problem, persons responsible for corrective action, proposed corrective action and date for completion of corrective action. Deficiencies are brought to the attention of management; it is then the responsibility of the appropriate manager to decide upon and implement the necessary corrective action. Follow-up audits may be used to ensure that the necessary corrective actions have been done and are effective in correcting the problem.

- Systems should be in place to handle non-conforming product. It must be clearly labelled to ensure that it is not used until a decision has been made as to how to deal with it. In the case of contract brewing or packaging products, the customer should normally be informed and product is released only on the authorisation of the customer; records should then be kept of the authorised concession. Written procedures should describe who is authorised to decide the fate of non-conforming product, how the decision is communicated and how subsequent operations are controlled.
- Complaints from customers should be recorded by authorised persons and an acknowledge sent to the complainant. Complaints should be allocating to a named person to ensure they are dealt with; this person will validate that the complaint is genuine and respond to the complainant. Corrective actions should include actions to resolve the immediate problem and actions to prevent the same problem recurring. Customer complaints and the actions taken should be reviewed at the management review meetings.
- Companies should take action to eliminate the cause of non-conformities in order to prevent recurrence. Corrective actions shall be appropriate to the effects of the non-conformities encountered. This includes reviewing non-conformities, determining the causes, carrying out actions and reviewing the actions taken to ensure that they are effective.
- In addition, breweries should identify potential non-conformities in order to prevent their occurrence. Preventive actions shall be appropriate to the effects of the potential problems. Current information on the causes of problems can be useful in predicting areas of potential problems; this enables actions or amending working procedures to be done to prevent problems occurring.

17.4 Integrated management systems – the BRC Global Standard – Food[6]

Traditionally, breweries in the UK that provided beers to retailers (particularly own-label brands) were audited by the retailer to assure the safety, quality and legality of goods supplied. Each retailer had its own standard and breweries supplying multiple retailers were audited against different standards – one for each retailer.

In 1998 the retailers, under the auspices of the British Retail Consortium (BRC), produced a new standard for food producers, including breweries. This is now named the BRC Global Standard – Food. This combined the best elements from each of the individual retailer standards and provided a single standard that breweries, and other food producers, could be audited against. The Standard has been revised on three occasions; Issue 4 was published in January 2005. Its objective is to specify safety, quality and operational criteria that must be in place to supply food products to UK retailers. Although the Standard was originally developed for the supply of retailer branded products, it is now widely used for products supplied to retailers, pub groups and contract clients for brewing and packaging. The Standard requires:

- The adoption and implementation of Hazard Analysis and Critical Control Point (HACCP)
- A documented and effective quality management system
- The control of factory environment standards, products, processes and personnel.

Some of the requirements are designated 'Fundamental'. This includes Quality Management System, Internal Audit, Corrective Action, Traceability, Layout, Product Flow and Segregation, Housekeeping and Hygiene, Handling Requirements for Specific Materials, Control of Operations and Training. These requirements must be fully implemented if the brewery is to gain certification against the Standard.

To comply with the sections on HACCP and quality management systems, breweries must implement these systems as described above. Some additional specific requirements of the Standard are listed below:

- The site should be securely maintained, with access to production and storage areas being limited to authorised staff.
- The site must be located and maintained so as to prevent contamination of product. Local activities which might have an adverse impact on the brewery must be considered and protective measures put in place where necessary and regularly reviewed.
- The brewery grounds must be finished and maintained to an appropriate standard. External parts of the brewery must be kept in good condition with grassy/planted areas being well maintained. External walls of buildings should be provided with a clean and unobstructed area.
- Brewery buildings must be suitable for the intended purpose. Walls should be designed and maintained to prevent dirt, condensation and mould growth. Floors should be designed to meet the demands of production and kept in good repair. Drainage should be adequate.
- There should be systems to minimise contamination from glass and brittle plastic. Most breweries eliminate all unnecessary glass and brittle plastic from production areas. Items that cannot be removed are listed on a glass register and these are checked at defined intervals to ensure that they are still intact.
- Most breweries now have a written hygiene policy to minimise contamination of product by staff; this documents the personal hygiene standards to

be adopted by all staff and visitors. This includes a ban on smoking, eating and drinking in production or storage areas; these are permitted only in designated areas. The site must also have a procedure for notification by employees of any relevant infectious diseases which they may be suffering from or may have been in contact with; if necessary, staff must be moved to a non-production area. Suitable protective clothing must be issued for staff, visitors and contractors in production areas. In addition, further controls are needed in bottling and canning areas. For example, in most breweries jewellery is not permitted with the exception of a plain wedding ring and sleeper ear-rings; where jewellery is worn for ethnic, medical or religious reasons it must be one-piece and controlled to avoid contamination. Rings and studs in exposed parts of the body, such as nose and eyebrow, are not permitted. Also, for bottling and canning areas, hair must be covered to prevent product contamination; staff and visitors with beards must wear a snood. Some sites now also operate these policies in any area where there are open vessels, e.g. open fermentation vessels.

Breweries can be audited against the requirements of the Standard to gain certification. Audits normally take a minimum of $1\frac{1}{2}$ days and may take longer depending on the size of the brewery and the complexity of its operations. There are three levels of non-conformity:

- Critical – There is a failure to comply with a food safety or legal issue.
- Major – There is doubt as to the conformity of the product being supplied.
- Minor – Where absolute compliance has not been met, but the conformity of the product is not in doubt.

Where a critical or major non-conformity has been raised against a 'Fundamental' clause, the company's certification is suspended or withdrawn.

17.5 Feed Materials Assurance Scheme (FEMAS)

As a result of food safety incidents in the United Kingdom, there followed two courses of action. Firstly, the European Union issued a series of Directives and Regulations[7] aimed at improving the safety and traceability of material used to feed animals that are then eaten by humans. Secondly, retailers and dairy producers insisted on traceability of feed materials for animals. This means that breweries that produce co-products (such as brewers' grains, malt dust, trub, surplus yeast and surplus beer) which are used to feed animals that are subsequently used for human consumption must ensure that these materials are safe for the animals and also safe for humans who eat the meat. Brewery co-products are therefore no longer considered to be a waste material to be removed from the site but are now considered to be a food grade material produced by the site.

To assist breweries and other feed material producers to comply with the new requirements, the Feed Materials Assurance Scheme (FEMAS)[8] was developed by various industry bodies. FEMAS is based on HACCP, management and quality assurance, food safety/quality policies, quality management structure, risk assessment, documented procedures, technical support, internal auditing procedures, personnel skills and training, product liability insurance, recall and crisis management procedures, and maintenance of complaints procedure records.

Some of these systems already operate in breweries, especially those with HACCP, quality management systems and BRC certification. However, the new requirements mean that systems previously applied to beer need now be applied to feed materials as well. In addition, the new scheme has meant changes in the way that feed materials are handled in breweries. Some of these changes are discussed below:

- FEMAS requires that feed materials are sold in accordance with a contract. Contracts must clearly state the name, quantity, specification, price and collection/delivery period of the feed materials sold. Breweries supplying co-products have not traditionally had a contract in place.
- Each feed material must have a written specification. Specifications must include analytical and nutritional characteristics of the feed and must be updated when any changes take place. Typical specifications include moisture and crude protein in order to set nutritional requirements. Where surplus beer is supplied then it may be necessary for the brewery to indicate whether the beer has a higher than normal alcohol content so that the processor or farmer can dilute it with other materials.
- All feed materials must be sampled; every load leaving the brewery must be sampled. Sampling techniques must ensure a true representation of the feed materials. Most breweries arrange for samples to be taken by the driver. The samples of feed materials must be retained for a minimum period of six months, unless risk assessment studies show that shorter periods are sufficient or longer periods are required. If the feed is to be used quickly as animal feed then a retention period of less than six months may be appropriate; if it is to be used as silage then a longer retention period may be required. Samples of feed must be kept in appropriate, sealed containers and labelled to ensure traceability. They must be stored so as to minimise deterioration; for brewers' grains and yeast this means storage in a freezer.
- All feed materials must be inspected physically; inspections include colour, physical form, odour, freedom from contamination by insect pests, droppings and other extraneous matter, freedom from mould and freedom from excessive damage.
- Sufficient microbiological testing must be carried out to ensure the safety and critical quality of the feed materials. Most breweries carry out *Salmonella* tests on each feed material at intervals of up to six months.
- Sufficient chemical tests must also be carried out. Typical tests are those for crude protein (for nutritional requirements) but additional tests (such as for pesticide residues, mycotoxins and heavy metals) may be done to ensure that these are not concentrated in feed materials. Where breweries are supplying

waste beer that has been returned from trade (e.g. unbroached casks or broached/unbroached kegs), risk assessment is essential to consider the hazards involved in this practice and additional analyses may be required.
- Breweries must be able to demonstrate traceability of all feed materials supplied. A traceability trail is required for each batch of feed materials. For brewers' grains, most breweries use records of malt deliveries, malt silos used, brewing numbers and dates, and dates of collection of the brewers' grains. For surplus yeast (derived from fermenters, tank bottoms, processing, etc.) traceability is often more difficult. Where breweries are supplying surplus beer derived from a warehouse there is usually adequate traceability.
- The brewery must have written procedures to ensure that feed material production, storage and transport facilities are cleaned so as maintain quality and safety of feed materials. Traditionally, feed material handling systems have had a minimal programme of cleaning. However, breweries must now carry out cleaning on a regular basis and records must be kept. For brewers' grains, this might involve cleaning of the dump tank under the mash tun or lauter, the Pondorff valve, lines to the brewers' grain silo(s) and the silos. Several breweries have needed to install additional sprayballs in the silos.
- All transport used to carry feed materials must be controlled with regard to hygiene and potential contamination. All transport must be inspected by an authorised person just prior to loading to ensure it is clean. Records of these checks must be maintained. In addition, the previous three loads carried must be recorded and assessed for compatibility with feed materials. Transport must be covered en-route to the brewery, uncovered just before loading and then re-covered immediately after loading to prevent contamination of feed materials.
- A formal risk assessment, based on HACCP principles, must be carried out to identify and control any hazards that might affect the quality and safety of feed materials supplied. Risk assessment in breweries has traditionally been concerned with the production of beer and hence risks to the consumer; the presence of physical contaminants early in processing is not traditionally considered a high risk since the beer stream is filtered. For feed materials the HACCP plan will need to be extended to consider the production, processing and storage of feed materials; this will need to consider the risks to the animals as well as to humans who eat the meat produced. Hence risks from physical contamination such as glass fragments need to be considered, since they may harm the animals.

To date, most of the breweries in the United Kingdom have been assessed against the FEMAS Standard and have received certification. Without this certification they will not be permitted to send co-products for animal feed.

The Standard has now been launched as an International Standard. It is expected that it will be implemented worldwide.

17.6 Future trends

The increase in legislation worldwide and increased commercial pressures to assure product safety, quality and legal requirements seem likely to result in increased pressure for breweries to implement quality systems to fulfil these demands. This, in turn, seems likely to result in increased pressure for certification of quality systems to obtain external validation of their effectiveness.

17.7 References

1. *Managing Food Safety in the European Brewing Industry through the Application of HACCP Principles*, published by the Brewers of Europe, 2003.
2. Council Directive 93/43/EEC on the hygiene of foodstuffs.
3. Regulation (EC) No. 178/2002 of the European Parliament and of the Council of 28 January 2002 laying down the general principles and requirements of food law, establishing the European Food Safety Authority and laying down the procedures in matters of food safety.
4. ISO22000 Food Safety Management Systems – Requirements for Organisations throughout the Food Chain.
5. ISO9001:2000 Quality Management Systems – Requirements.
6. BRC Global Standard – Food (Issue 4).
7. Regulation (EC) No. 183/2005 of the European Parliament and of the Council of 12 January 2005 laying down requirements for feed hygiene.
8. FEMAS, fourth edition – July 2005 International Core Standard.

18

Brewing control systems: chemical analytes

K. J. Siebert, Cornell University, USA

18.1 Introduction

Many chemical transformations take place in the brewing process; as a result relatively few of the compounds found in beer are in the same chemical forms as they were in the raw materials. For example, of the compounds most important to beer flavor (those typically present at more than twice their flavor threshold) (Meilgaard, 1981, 1982), the hop bitter compounds in beer (isoalpha acids) are derived from the alpha acids in hops during kettle boiling, and ethanol and CO_2 are formed during fermentation from simple sugars produced from starch breakdown during mashing. Many other compounds found in beer are similarly altered during the process. Measurements of chemical substances are thus needed to maintain consistent quality of products, to control process economics, and in some cases to comply with government regulations.

Successful analytical methods employ a means of detection that is sufficiently sensitive to determine a substance of interest in relevant samples with reasonable precision and accuracy. The analyte must be adequately separated from interferences prior to detection; the degree of separation required is determined by the specificity of the detector and the nature of the sample. In some cases the separation may be achieved mathematically rather than physically. Methods are increasingly likely to be automated or moved on line; both offer advantages. New technologies (including the use of biological recognition systems, microfluidics and laboratory information management systems) should lead to improvements in speed, precision and throughput.

18.2 Brewing analytical methods

A number of scientific organizations carry out collaborative tests of brewing analytical methods and confer official status upon those that are deemed to perform satisfactorily. These organizations include the American Society of Brewing Chemists (ASBC), the Brewery Convention of Japan (BCOJ), the European Brewery Convention (EBC), the Institute of Brewing and Distilling (IBD), and the Mitteleuropaeische Brautechnische Analysenkommission [the Central European Brewing Technology Analysis Commission] (MEBAK). Each of these organizations publishes its official methods and periodically adds new ones. Most of the examples of analytical methods mentioned in this chapter are drawn from the ASBC *Methods of Analysis* (ASBC, 1992).

18.3 Detection, separation and measurement techniques

Analytical methods are intended to measure the concentration of either a species of interest, called the analyte, or a chemically similar group of compounds. An example of the former is measurement of diacetyl in beer; examples of the latter are methods for measuring total protein, total acidity or total polyphenols. Every method employs some form of analyte detection. In successful methods the detection approach must have adequate sensitivity to respond to the analyte of interest in relevant samples with reasonable accuracy and precision (Siebert, 1983), and the sample must be free of interferences that lead to significant measurement errors.

Interferences occur when some other component of a sample matrix besides the analyte produces a response with the detector used. This can cause results that are erroneous. If an interferent causes a positive response in the detector, results will be erroneously high. Some interferences can lead to lower responses (e.g. by competition, inhibition or quenching). The use of appropriate sample blanks can correct for the effects of some interferences.

Some detectors are highly specific for an analyte, which minimizes responses from possible interfering substances. Such detectors can produce good results even in very complicated mixtures (Siebert, 1983). Some sample matrices may lack interferences that cause a response in a particular detector; in such cases no separation is needed before detection. However, the same detector may fail to work properly with other samples if interferences are present. This is why procedures that successfully determine compounds in water often fail to work in wort or beer.

One frequently used approach to remove interferences is to carry out a separation of the analyte from the interference before detection. Depending on the nature of a sample matrix and the specificity of a detector, the adequacy of separation needed ranges from none to modest to high (Siebert, 1983). Separation can be achieved in two main ways: by a physical separation, such as distillation or chromatography; or by a mathematical separation, most commonly applied with detection means that produce multiple channel results.

So a successful analytical method combines some degree of separation with an appropriate degree of detector specificity for a particular type of sample.

18.3.1 Detection

Detection can be accomplished either directly by an instrument or after carrying out a chemical reaction. Physical and physicochemical measurements can often be carried out directly. These include determinations such as specific gravity, refractive index, turbidity, and viscosity.

A few chemical detectors are sufficiently specific that no separation before measurement is needed even with very complicated mixtures. The glass electrode measurement of hydrogen ion concentration under most pH conditions is a good example: see Wort-8 and Beer-9 (ASBC, 1992).

Some detectors are very non-specific and respond to most compounds, at least to some extent. The thermal conductivity detector (TCD) in gas chromatography (GC) is a good example of this; it responds to permanent gases and volatile inorganic as well as organic compounds. The flame ionization detector (FID), also used in GC, is slightly more specific, as it responds to all compounds that can be combusted in a hydrogen flame to produce ions. As a result, it responds to most organic compounds, but not to inorganics. Use of detectors with little or no specificity requires either samples that contain few constituents or a high degree of separation prior to detection. In examination of brewery or purchased CO_2 (where oxygen content is of concern), only a few gases are present in significant concentration (CO_2, O_2, N_2 and Ar), and only a modest degree of separation (low resolution packed column GC) before a TCD detector is needed. On the other hand, there are hundreds of organic compounds in beer (Meilgaard, 1981, 1982), so a high resolution separation (capillary GC) is needed in order to employ an FID for analysis of a beer solvent extract.

Some detectors are specific for particular elements. These include flame photometry, atomic absorption spectroscopy (AAS) and inductively coupled argon plasma (ICAP). In all these cases a flame or furnace is used to heat samples, and the extent of absorbance at element-specific spectral lines is measured with considerable sensitivity. With any of these techniques the amount of a particular (and with AAS and ICAP usually a metallic) element is determined regardless of its oxidation state or molecular form (e.g. free or in a chelation complex).

Measurement of 'protein' in brewing samples is often made either by Kjeldahl nitrogen determination or by a combustion method. Either approach is specific for nitrogen, and as such produces responses not only to protein but also to free amino acids, peptides, nucleotides, and to some extent nitrate and nitrite ions. Multiplication by a factor leads to a number that assumes that all the nitrogen found in a sample was contained in proteins. Obviously, this produces only an approximation of true protein in wort and beer.

Some detectors applied in gas chromatography are element-specific. Nitrogen specific detectors are reportedly useful for detecting pyrazines and thiazoles in beer (Herent and Collin, 1998). The thermal energy analyzer is a chemiluminescent detector that can operate in several modes (nitrogen, nitro and nitroso). In the last of these it has been widely used to measure the carcinogen dimethyl nitrosamine in malt and beer, as in Beer-40 (ASBC, 1992).

Sulfur-specific detectors are useful in brewing because of the presence of potent sulfur-containing flavor compounds such as H_2S, methyl mercaptan, ethyl mercaptan, isopentenyl mercaptan, dimethyl sulfide and others. The sulfur flame photometric detector was used extensively at one time, but its non-linear response was problematic. The sulfur chemiluminescent detector (SCD), which is simpler to use, has largely replaced this, as in the method for dimethyl sulfide in beer; see Beer-44 (ASBC, 1992).

The electron capture detector (ECD) responds to compounds that capture 'slow' electrons produced by a radioactive source and thus reduce the baseline current seen in their absence. This makes it quite specific for compounds that contain halogens (not many of these are found in brewing samples), but it also responds well to relatively few organic structures, including the vicinal diketone function found in diacetyl and 2,3-pentanedione. As a result of this specificity, a sample of beer headspace needs only a moderate resolution GC separation prior to ECD, as in Beer-25E (ASBC, 1992).

Refractive index detectors respond to solutes that differ in refractive index from a solvent. As such, they respond to many different compounds and have little specificity. This detector is sometimes used in liquid chromatography (LC).

Spectroscopy is useful for compounds that absorb in the electromagnetic spectrum. Commonly used detectors operate in the ultraviolet (UV) and/or visible bands or in the infrared (IR) region. Such detectors can employ a fixed wavelength or may be tunable or multichannel. A fixed-wavelength UV detector, for example, typically operates only at 254 nm (where the vast majority of the energy of a mercury vapor lamp is concentrated); as a result, it is only useful for analytes that absorb at least somewhat at this wavelength. UV absorbances are often fairly broad, however, so compounds that have maximum absorbance at nearby wavelengths can often be measured. Variable wavelength UV or UV-visible detectors can be set to a particular wavelength anywhere within their range and are thus 'tunable'. Some detectors take data at multiple wavelengths either by scanning or by simultaneous observation (as with a diode array detector, DAD). These both produce multichannel results. IR absorbances at particular wavelengths are characteristic of certain structures in organic molecules; IR detectors are often multichannel.

Fluorescence detectors can be separately set to employ particular wavelengths or bands for both excitation and emission. Since relatively few compounds fluoresce, and those that do have very different patterns of excitation and emission, fluorescence detectors are both specific and tunable.

Electrochemical (amperometric) detectors apply a potential that either oxidizes or reduces compounds in solution and can be employed directly to a sample or, more often, after separation by high performance LC (HPLC). The current flow is measured and used to indicate the magnitude of the response.

Since relatively few compounds can be induced to undergo electrochemical reactions at modest potentials, their detection is fairly specific. The direction (positive or negative) and the magnitude of the potential determine whether and which compounds will be oxidized or reduced. If the potential applied is low, only compounds that are readily electroactive are affected. Higher potentials affect a broader spectrum of compounds. So this detector is to some degree tunable. Versions of electrochemical detectors that employ electrode arrays with increasing potentials give multichannel results.

18.3.2 Separation

If the detection mechanism to be used for a measurement is sufficiently specific that no component of a sample other than the analyte produces a response, then no separation is needed. Obviously this depends on both the specificity of the detector and the nature of the sample. If, on the other hand, one or more components of the sample other than the analyte produces a detector response, this constitutes interference and the analyte must be separated from the interference before detection. This separation may be achieved physically or mathematically.

Physical separation

The simplest degree of physical separation is into two different phases. This can be accomplished by distillation, solvent extraction, filtration, adsorption or phase partitioning. In some cases this can also lead to concentration of an analyte.

With distillation, components of a sample with boiling points at or below the boiling temperature are separated from the non-volatile material. This is the traditional method used in determining alcohol in beer. Both alcohol and the dissolved solids in beer affect density. However, if the volatile fraction is separated, its density (measured with a hydrometer, a pycnometer or a density meter) is almost entirely a function of alcohol content; this is the basis of Beer-4A and Beer-4B (ASBC, 1992). The density of the residue (which leads to real extract) is almost entirely due to dissolved solids, as in Beer-5A and Beer-5B (ASBC, 1992).

Another way to separate volatile from non-volatile compounds is by enclosing a liquid or solid sample in a vial together with some headspace. Compounds that are volatile under the conditions used will partition into the headspace and can then be sampled by removing a portion of the gas phase or by solid phase microextraction (SPME) of the gas phase with a suitable SPME fiber. The temperatures used for equilibration are typically quite a bit lower than beer boiling temperature, so the compounds available are generally more volatile than ethanol. The samples are typically analyzed by gas chromatography. An example is headspace sampling for VDK analysis by GC with ECD, where both diacetyl and 2,3-pentanedione can be measured and, if a high enough equilibration temperature is used, the precursors are converted into the VDKs, giving a VDK potential indication: see Beer-25E (ASBC, 1992).

Solvent extraction of a solid sample generally dissolves substances of similar polarity to the solvent. Solvent extraction of liquids requires the use of a solvent that is not miscible with the sample. If the volume of solvent used is smaller than that of the sample, concentration of the analyte occurs. For an aqueous sample like wort or beer, a non-polar solvent can be used to extract non-polar compounds. An example of this is the use of isooctane to extract hop resins from beer or wort in the determination of bittering units: see Beer-23A (ASBC, 1992).

Two-phase physical separations can also be achieved by solid–liquid separations like filtration or centrifugation. Filtration can separate material ranging from large solids down to the molecular scale (in the case of dialysis or ultrafiltration).

Another kind of two-phase separation can be achieved by the use of solid-phase adsorbents; these are widely used to prepare samples for chromatography or other procedures by removing polar, non-polar or ionic compounds from solution (depending on the nature of the adsorbent used). They can be used either to remove an interferent from a sample or to extract (and possibly concentrate) the analyte, provided it can then be quantitatively released from the adsorbent. This approach has been accepted by ASBC for extracting hop compounds from beer (Murphey *et al.*, 1993).

Finer separation or separation into more fractions can be achieved through some form of chromatography or electrophoresis. This can range from a modest degree of separation, with a low number of theoretical plates (a packed column in GC or large particle size packings in LC), through very high resolution with capillary GC or very efficient HPLC columns.

Mathematical separation

Mathematical separation can be used in some cases where actual physical separation has not been achieved. This falls into two main areas: mathematical peak resolution (of chromatographic, electrophoretic or spectroscopic peaks) and multivariate resolution of multiple channel data.

Procedures for mathematically calculating the size of peaks that are incompletely resolved have been developed (van den Bogaert *et al.*, 1994; Shao and Sun, 2001; Ford *et al.*, 2003). These use knowledge about the shape of chromatographic or electrophoretic peaks to permit better quantitation than would otherwise be possible (the simple assumptions made by many integrators).

In the case of multichannel data, mathematical calculations can be used to resolve analytes that are not physically separated at all, provided they have different patterns of responses (Martens and Naes, 1989). This approach is most often applied with spectral data. Knowing the spectra of each of the constituents of a mixture and Beer's Law, it is possible to deconvolute mixtures and estimate the amounts of each species present. Often, however, the identities and spectra of all components of a mixture are not known. It is still possible to use multivariate calibration (typically with partial least squares regression) to determine an analyte in a difficult matrix or even to determine several different analytes in the presence of multiple interferences (Martens and Naes, 1989). This is often used with near-infrared (NIR) data to rapidly determine protein, as in Barley 7D, and moisture, as in Barley 5C, content when barley is delivered to a grain elevator (ASBC, 1992). Similar approaches have been developed for determination of alpha and beta acids and hop storage index in hop materials (Garden *et al.*, 2000), and for alcohol and OG in beer (Coventry, 1994).

18.3.3 Measurements

Making an analytical measurement requires a suitable combination of detector specificity and separation as well as sufficient sensitivity to detect the analyte at the concentrations that normally occur in samples of interest. Highly specific detectors can be applied directly to a sample without any separation. Very non-specific detectors, on the other hand, require either essentially complete separation of the analyte from all interferences, as for example with high resolution in chromatography, or mathematical separation using multiple spectral measurements and multivariate calibration to 'model around' interferences.

Often the specificity of a detector for a particular analyte can be improved by using some sort of 'recognition' element. This can be done either by carrying out a reaction that is specific to a molecular feature, or by employing biological recognition.

Chemical reagents are often designed to react with specific compounds or functional groups. As a result, they produce a response that is more specific for a particular detector, at least under the measurement condition used, than was the case in the original sample. Depending on the nature of the method and the spectrum of compounds in the sample, this may produce a satisfactory method. Most commonly such measurements are made by photometry, although other detection means are employed in some methods.

Great specificity can often be achieved by employing biological recognition approaches such as enzyme reactions or antibodies. Antibodies can distinguish between different proteins, for example between haze proteins and foam proteins (Ishibashi *et al.*, 1996). They can respond not only to different functional groups but in some cases to particular molecular species, and may even discriminate between stereochemical isomers. Monoclonal antibodies, for example, can be highly specific in attaching to target compounds. A variety of test formats can be used to provide quantitation. A widely used format is enzyme-linked immunosorbent assay (ELISA), in which an enzyme is bound to an antibody; after the analyte is bound by one antibody, another antibody attached to an enzyme binds to either the analyte or the first antibody and then catalyzes a reaction, which serves to amplify the response. Antibody-based procedures have been widely used to detect and quantitate mycotoxins in beer and brewing materials; many of these are small molecules (Papadopoulou-Bouraoui *et al.*, 2004).

With enzyme reactions, changes in the product concentration after a defined time under defined conditions are typically used to estimate either the amount of substrate or the amount of enzyme present, depending on which is the analyte. Examples include diastatic power (Malt-6) and alpha-amylase (Malt-7) in malt (ASBC, 1992).

Biological detection systems are often less stable than instrumental methods, so particular care in storage and use is required to keep them free of biological contamination and to detect changes in their response.

18.3.4 Measurement quality

The quality of a measurement depends on two main things, the accuracy with which a method measures the true concentration of an analyte and the reproducibility (precision) of the measurement (Massart *et al.*, 1997). Precise methods with poor accuracy are essentially useless. If a method has reasonable accuracy but poor reproducibility, it would be possible, though inconvenient, to average replicated measurements to arrive at a correct result. Of course, a method must have adequate sensitivity to determine an analyte in the concentration range normally encountered in samples of interest. Linearity of response is desirable, but not absolutely necessary as either a non-linear or a piecewise calibration can be applied.

18.4 Combining different techniques

Some examples will be given to illustrate how different combinations of detector selectivity and degree of separation result in successful analytical methods.

Direct methods

Physical measurements, such as density, turbidity or viscosity, or chemical measurements with highly specific detectors, like a pH electrode, can usually be made directly on a sample. Apparent extract in wort can be found from a direct observation of density by a hydrometer (Wort-4) or a density meter (Wort-2) (ASBC, 1992).

Atomic absorption employs spectral lines that are highly specific for elements and so can be used directly to determine the total amount of a particular element in a sample, but not its chemical form or oxidation state. This is used in official methods for Mg and Zn in wort (Wort-15 and Wort-16, respectively), and for Fe (Beer-18B), Cu (Beer-19C), Ca (Beer-20C), Na (Beer-36), K (Beer-37), Mg (Beer-38A), and Al (Beer-42) in beer (ASBC, 1992).

Wet chemical methods

These typically involve the addition of a reagent, which leads to a reaction that results in some observable property. This is most frequently light absorption at some wavelength and can be observed with a photometer of some sort. Examples of this are the total polyphenols (Beer-35) and SO_2 (Beer-21) in beer methods (ASBC, 1992). In some cases detection may be by turbidimetry. Quite often wet chemical methods are group methods; that is, they respond to all members of a class of compounds present in the sample. Free amino nitrogen in wort (Wort-12) is an example (ASBC, 1992).

Titration involves the gradual addition of some reagent to a sample until some change is observed either visually (as with a color indicator) or with some sensor (such as a pH or conductivity electrode or a probe colorimeter). With the conductometric lead acetate titration for hop alpha acid content (Hops-6B), the end point is determined by the intersection of two lines (ASBC, 1992). Automated titrators are commercially available and can carry out titrations with greater reproducibility than human analysts.

Methods with limited separation

When separation is needed, the least degree of separation is into two phases. The traditional method for measuring alcohol in beer is to separate a beer sample into two phases by distillation, as described previously. Also previously mentioned was the bitterness determination where an isooctane extraction separates isoalpha acids from other substances in beer with UV absorbances in a particular wavelength region. Polarographic oxygen electrodes employ a membrane that permits gas but not liquid to diffuse into a solution containing an electrode polarized to reduce O_2; the current flow indicates the amount of oxygen in a liquid or gas sample in contact with the membrane. The Embra CarboCheck analyzer also uses a gas-permeable membrane that permits diffusion of gases through it; the resulting pressure on the gas side of the membrane is observed and, after correcting for nitrogen, gives the amount of CO_2 in a sample. In the determination of hop essential oils, a steam distillation is used to collect and concentrate the oil from ground hops: see Hops-13 (ASBC, 1992).

Methods with modest resolution (physical separation) and low specificity detection

In special cases only modest resolution may be needed before low specificity detection. An example is the use of an RI detector with low resolution HPLC for the determination of fermentable saccharides in wort: see Wort-14B (ASBC, 1992). In this case the sugars in question are in vastly greater concentration than all of the other substances present. So a small response to other compounds in wort (e.g. amino acids), which undoubtedly occurs, causes little interference.

Methods with high resolution (physical separation) and low specificity detection

When a detector is relatively non-specific (e.g. GC with thermal conductivity or flame ionization detection, or HPLC with refractive index detection) and the analytes are in low to modest concentration, many compounds can produce responses and so act as interferences. In order to make accurate measurements it is then necessary to employ powerful means to separate the analyte(s) from interferences before detection. With GC this can often be achieved with capillary columns. With HPLC, columns with high resolution (many theoretical plates) can be employed.

Methods with modest resolution (physical separation) and high specificity detection

When detectors with great specificity (GC with ECD or SCD; HPLC with fluorescence detection) are used, the number of potential interferences is greatly reduced and a less rigorous separation may suffice. This may be possible with packed (non-capillary) GC columns or low resolution HPLC columns.

Methods with high resolution (physical separation) and high specificity detection

The most powerful measurement approaches are when a high specificity or a multichannel or tunable detector is combined with high resolution separation. Examples include the UV–visible diode array detector (this observes absorbances at multiple wavelengths simultaneously) in HPLC, and GC-mass spectrometry, either full scan (multichannel) or with selected ion monitoring (tunable). Compounds would need to elute very close to the analyte and to have a similar response to a specific detector to interfere with measurement.

Methods with multiple measurements and mathematical resolution

When two or more measurements are made on a sample, either using separate procedures or by making two or more observations with the same instrument (e.g. absorbances at two or more wavelengths in spectroscopy), it is sometimes possible to calculate concentrations of two or more analytes. For many years measurements of refractive index and density in beer and a series of equations were used to determine alcohol, original gravity (OG), real extract (RE), apparent extract (AE), real degree of fermentation (RDF) and calories: see Beer-4C, Beer-6A, Beer-5C, Beer-6B and Beer-33, respectively (ASBC, 1992).

In recent years instruments that perform two measurements and carry out the calculations have been offered. The Anton Paar beer analyzer employs a density meter and a sound velocity instrument and uses their outputs to determine alcohol, OG, RE, and RDF. The SCABA beer analyzer measures specific gravity with a density meter, measures alcohol with a ceramic sensor that oxidizes evaporated vapor, and also measures pH and color The instrument then calculates specific gravity, OG, RE, AE, RDF, and caloric content and reports these along with color and pH. These instruments have been tested and approved by the major brewing organizations (EBC, ASBC, BCOJ, MEBAK) – see, for example, Beer-4E (ASBC, 1992) – and are used by many major brewers.

The ASBC spectrophotometric method for alpha and beta acids in hops (Hops-6A) uses absorbance measurements at three wavelengths to calculate the amounts of alpha and beta acids in a solvent extract (ASBC, 1992). The hop storage index method (Hops-12) uses a ratio of the absorbances at two of those wavelengths to assess hop condition. The spectrophotometric determination of protein in unhopped wort and in beer (recently accepted as an official method by ASBC) employs a similar approach (Carruthers *et al.*, 2005). Calculation of color by the tristimulus method is more complicated (many more wavelengths are used), but the calculation is sufficiently straightforward that it can be carried out by a spreadsheet program or in a dedicated instrument (Cornell, 2002).

More complicated applications need additional data (more complete spectra) and more powerful calibration and data analysis tools. We are often faced with the situation that we don't know the identities of all the compounds present, let alone their full spectra. It is still possible to make measurements after using a reasonably large set of samples on which determinations of the analyte of interest have been carried out with a primary method as a calibration data set (Martens and Naes, 1989). Calibration, even in the presence of unknown amounts of unknown interferences, can often be accomplished with partial least squares (PLS) regression. In this case the calibration 'models around' the interferences. Previously mentioned examples of this are determinations by NIR of moisture and protein in barley; alpha-acids, beta-acids and HSI in baled hops; and alcohol and gravity in beer.

When separation is combined with a multichannel detector, three-way data results. Chromatographic data, for example, can be considered to have time as the x-axis, the measurement channel as the y-axis, and response intensity as the z-axis. In the case of HPLC-DAD, plotting z against x at some value of y appears as a chromatogram, while plotting z against y at some value of x (here retention time) appears as a spectrum. With GC/MS, the spectrum is a mass fragmentogram. The spectral information can be used to identify unknown peaks. If a change in spectra occurs at different time slices within a peak, it indicates a two or more component peak comprised of compounds with slightly different retention times. If a measurement channel or a ratio of measurement channels that is unique for a component can be discovered, it is possible to separately quantitate that compound (Brereton and Dunkerley, 1999).

18.5 What and why do we measure?

Analytical procedures are carried out to assess and control product quality and to satisfy regulatory requirements. In the latter case the necessary measurements are specified by a government agency and often so is the methodology to be employed. Generally, regulations are designed to protect the consumer (e.g. nitrosamines, which are carcinogenic; or SO_2, to which some individuals are hypersensitive) or to assure collection of appropriate taxes (e.g. alcohol in countries where the beer tax rate is a function of alcohol content). Measurements other than those required by government are made in the brewer's interest in controlling product quality and process economics.

The product quality attributes that matter most are those that consumers can directly perceive: flavor (odor, taste and chemesthesis) and appearance (color, haze, and foam). While these attributes could be, and frequently are, assessed by sensory panels, it is often advantageous to employ analytical measurements because they are typically more precise, accurate and reproducible than panel results, and fewer man-hours are required to obtain results.

Producing a consistent quality product requires either consistent raw materials and processing conditions or, if raw materials are variable in some predictable way (i.e. following broad trends or making step changes between crop years, suppliers, or shipments), making appropriate adjustments during processing to achieve consistent quality.

Where knowledge and analytical methodology are sufficient, it is of course preferable to directly measure the molecular species that are perceived (Siebert and Acree, 1993). However, because many transformations occur during the brewing process and many compounds contribute to beer flavor and appearance, this is often not possible or practical. It may, however, be possible to measure factors related to the important qualities. For example, measuring both enzyme levels (alpha- and beta-amylase) and the amount of substrate present in malt (lab mash and adjunct extract) is generally sufficient to predict and control wort gravity. In some cases particular analytical methodology is not sufficiently sensitive, robust or affordable for routine application. Observations may then depend on indirect factors, which in combination constrain important properties within a suitable range. In a sense the strategy of developing a list of specifications for a raw material results in constraints on suppliers that are likely to limit variations in a number of properties other than those directly measured.

Achieving a consistent quality fresh product is not sufficient; the product must also maintain its quality during the intended product shelf life. Often this is approached by measuring factors related to aging (most notably package oxygen content, which is known to impact both flavor and colloidal stability). It is also normal practice to employ forcing tests followed by turbidimetry, as in Beer-27.II (ASBC, 1992), and/or sensory analysis to predict product aging behavior.

Appearance factors should, in principle, generally be measured without much difficulty. Color can certainly be measured by spectrophotometry, but measurement at a single wavelength, as called for in the official method for beer color used for many years, Beer-10A (ASBC, 1992), implicitly assumes that all beers have the same hue and differ only in the intensity of brown color, which is obviously not entirely correct. The tristimulus approach to color uses three scales (dark–light, red–green and blue–yellow), and collaborative trials with this method in brewing have been carried out successfully and led to its adoption as an official method for beer (Cornell, 2002). Turbidity can readily be measured by light scattering, but the angle to the incident beam at which the observation is made has a large effect on results; 90° scattering is commonly, but not universally, used in lab instruments. It is specified in Beer 27.I.B (ASBC, 1992). Most turbidimeters use white light, but some employ monochromatic light, which produces somewhat different results. Foam measurement is problematic due to sensitivity to product composition (CO_2, protein and alcohol content, etc.), cleanliness of glassware, and the foaming technique used. Methods based on manual pouring typically have poor reproducibility; this has led to development of more controlled pouring and foam sensing technologies: see, for example, Beer-22 (ASBC, 1992).

Flavor assessment by chemical analysis is problematic, in part because there are numerous taste, odor and chemesthetic sensations and individual perceptions of these vary widely (Siebert and Acree, 1993). In some cases measuring the responsible compound is difficult or impossible due to the low concentrations at which it occurs in beer and to interfering substances. It appears, for example, that some aroma compounds have flavor thresholds below the sensitivity of analytical instruments.

Analytical error is clearly a function of analyte concentration, as shown by many years of experience with collaborative testing of methods (Horwitz *et al.*, 1980). The coefficient of variation increases exponentially as the analyte concentration decreases. Presumably this is largely due to the increasing number of interferences as analyte concentration decreases. Even a very modest response of a detector toward a compound present in much larger concentration could become a significant interference.

18.6 Where and how do we measure?

Traditionally, analytical measurements have been performed on samples brought to the brewery laboratory. When justified by the numbers of samples, simple automation of wet chemical methods (especially those where a reagent is added, time to react is provided and a spectrophotometric observation is made) has sometimes been achieved by segmented flow chemistry (as in the Technicon Autoanalyzer) (Siebert, 1982; Crawshaw, 1988). More recent flow chemistry approaches have tended toward unsegmented flow using smaller-volume apparatus, such as in flow injection analysis (FIA) (Stockwell, 1990). FIA methods have employed a large range of detectors including photometers, fluorometers and electrochemical instruments, among others. Methods for β-glucans in barley, malt, wort and beer have been developed based on addition of Calcofluor in an FIA fluorescence instrument (Munck *et al.*, 1989). In general FIA methods are more rapid (leading to higher throughput) and use considerably smaller reagent volumes than conventional methods.

A very simple approach to automation is to use an autosampler to deliver samples to an instrument. This minimizes waiting time between samples, permits operation around the clock, and eliminates variability in sample introduction. Chromatographic instruments (both GC and HPLC), spectrophotometers of various types, and many other kinds of instruments can employ automatic samplers.

A more recent development that has not as yet been adopted by many brewing labs is the use of laboratory robots, which can perform analyses with greater precision than a human analyst. Commercially available laboratory robots can weigh, pipette, vortex, centrifuge, carry out solid phase treatments, make spectrophotometric measurements, and carry out numerous other operations in various sequences without human involvement other than programming. By changing the programming it is possible to use one robotic system to perform many different analytical methods. This flexibility may make a robot justifiable where automation of a single analysis would not.

A number of routine measurements can successfully be made by instruments that directly monitor some property at a location in the brewery (Siebert, 1984b; Besford, 1990; Daoud, 1991; Forrest, 1996). These may be described as on-line, at line or in-line depending on the nature of the sampling arrangement. The advantage over grab sampling is a much better appreciation of the extent of process drift, noise and other variations that occur over time. Further, it is often possible to use the output of such an instrument to control a process. A prime example is the use of outputs from two on-line instruments (e.g. sound velocity and specific gravity) to control blending of high gravity beers to desired package strength; results for calories, alcohol or OG can then be used to produce light, low alcohol or regular beers, respectively. Another example is the use of on-line CO_2 analysis to control touch-up carbonation. While on-line analyzers usually require laboratory involvement in calibration, they also involve instrument engineers in their selection, installation design, and maintenance.

A relatively recent development is movement toward supplier certification. In the past the same lot of material was likely to have been examined by both the supplier and the brewer; this was clearly duplication of effort. Where agreements can be reached to rely on supplier measurements, it is possible to reduce lab work at the brewery.

Recent trends are to carry out more analyses in the brewery lab with automated instruments and increasingly to use more on-line observations and supplier certification. As a result, brewery labs may be performing more analyses that were previously not regarded as routine on a routine or semi-routine basis.

18.7 Impact of brewery operation scale

The scale of operation of a brewery has a large impact on the measurements carried out and the location of the analysis. Brewpubs and microbreweries generally have much less laboratory equipment and rely to a much greater extent on sensory perceptions. As the size of a brewery increases, it becomes increasingly more practical to carry out analytical work and eventually to employ automated procedures in the laboratory or to move observations on-line. In part this is necessitated by production rates: when something goes awry in a larger plant, greater volumes of out-of-specification product will be produced per minute. So providing rapid analysis and fault detection is clearly more valuable.

18.8 Changes over time

Over time, instruments and methodologies tend to improve in quality and robustness. This enables some methods originally used only in research or for studying particular problems to be employed semi-routinely or even routinely. Improvements can occur in separation, detection or data analysis.

Sample preparation methodologies have improved greatly, perhaps most notably in compact, easy to use separation cartridges that can be applied before detection in order to remove interfering substances from the analyte or to separate and concentrate analytes. Compact devices with membranes that carry out molecular-size separations can be used to separate macromolecules from small molecules. One such format uses a centrifuge to provide the driving force. Solid-phase cartridges have gained wide acceptance for separating analytes or interferences from liquid samples. These often function as adsorbents that bind to polar, non-polar or ionic species. In cases where the analyte is bound, it is then necessary to elute it from the adsorbent before analysis. A number of these approaches can be applied to modest or large numbers of samples in parallel, and they facilitate automation. This approach has in many cases replaced solvent extraction and has the advantage of avoiding work with sometimes flammable materials and those that are difficult to dispose of.

18.9 Traditional, emerging and future methods

Methods that were originally used only in research due to their expense, limitations in stability, technique requirements or complexity gradually improve in these respects and become available for semi-routine or even routine use. Methods that are used frequently may be automated. Other methods are adapted for on-line use, resulting in much more complete information (due to their continuous nature), and may be used for process control.

An obvious trend that will increase in the future is the use of computers; this was foreseen over 20 years ago (Siebert, 1984a). Computers are already commonplace components of many instruments where they add features that improve measurement quality such as signal averaging, provide programmability that permits storage and recall of methods and calibrations, and generally contribute to ease of use. Adding computers to instruments also enables communications among them and with networks such as laboratory information management systems (LIMS). These permit automatic printing of barcoded labels for sample aliquots when they are logged into the lab. By using barcode readers at each instrument to identify samples as they pass through procedures, transfer of results to a LIMS can occur directly; this reduces the likelihood of introducing errors through mis-keyboarding and also makes results available more rapidly, not only to lab managers but also to ultimate clients once results have been released by the lab. The clients may be local or remote (either in different locations in a plant or at different sites in the case of multi-plant organizations).

On-line instruments are also likely to feed directly into data reporting systems, making results conveniently available both locally and remotely. Combined with good graphical displays and statistical analysis tools, these can detect trends, aberrations, or out-of-specification products quickly.

18.10 New technologies

In capillary electrophoresis, voltage is applied across buffer-filled capillaries. This leads to the separation of ions, which move at different speeds depending on their size and charge. As they pass through a detector, solutes are seen as peaks, and the area of each peak is proportional to the concentration of the substance. Analysis times are in the region of 1–30 minutes. Capillary electrophoresis has been applied in brewing to both organic and inorganic anions (Klampfl, 1999), carbohydrates by an indirect approach (Cortacero-Ramirez *et al.*, 2005), and alcohols, amines, amino acids, flavonoids and nucleosides (Cortacero-Ramirez *et al.*, 2004)

Since many brewing methods involve addition of a reagent and development of a response that is sensed either photometrically or by some other detector, they are particularly amenable to FIA. The ASBC has accepted FIA methods for bitter compounds in beer (Hassinger *et al.*, 1995) and diastatic power in malt (Laycock *et al.*, 1997), among others. The EBC has accepted the FIA fluorescence method for beta-glucans in barley, malt, wort and beer (Munck *et al.*, 1989). And FIA methods have been published for ascorbates (Luque-Perez *et al.*, 2000); alcohol (Worsfold *et al.* 1981); chloride (Ferreira *et al.*, 1994); copper (Fernandes *et al.*, 1998b); glycerol (Prodromidis *et al.*, 1995); glucose (Ruz *et al.*, 1988); iron (Fernandes *et al.*, 1995); lactate (Perez-Ruiz *et al.*, 1999); sodium, potassium, calcium and magnesium (Fernandes *et al.*, 1997); total carbohydrate (Larew *et al.*, 1988); total phosphorus (Fernandes *et al.*, 2000); total SO_2 (Fernandes *et al.*, 1998a); total phenols (Peris *et al.*, 1991); and tannins (Tomas *et al.*, 1993) in beer. Further miniaturization of FIA systems has led to very small fluidic analytical systems; depending on their size these have been described as 'microfluidic' and even 'nanofluidic'. Often an entire system is built on a single substrate; this has been called a 'lab-on-a-chip'. It is only a matter of time until these are used for routine brewing analyses.

Many claims have been made for 'electronic nose' instruments (eNoses); these are essentially gas sensor arrays that give characteristic response patterns to different compounds, mixtures or substances. Some of the claims made appear questionable, as the demonstrated sensitivity of most eNoses is insufficient to measure some of the beer constituents they are claimed to determine. They clearly do respond differently to different samples and may have value in detecting cultivars of barley or hops or brands of beer, but their use as quantitative instruments appears in question, at least to date.

Mathematical approaches are likely to be applied to a much greater extent in the future.

18.11 References

ASBC (1992), *Methods of Analysis*, 8th edn, St. Paul, MN, American Society of Brewing Chemists.

BESFORD R P (1990), 'In-line and on-line analysis', *Ferment* 3(1), 46–47.

BRERETON R G and DUNKERLEY S (1999), 'Deconvolution of partially overlapping tailing peaks in diode array high performance liquid chromatography using purity ratios', *Analyst* 124, 705–711.

CARRUTHERS A, BERG K, FELSKE J, FOSTER A, GUALDONI P, LE Q, VANHOECKE H, ZINANTI J and MARADYN D (2005), 'Spectrophotometric analysis of protein in beer', *J. Amer. Soc. Brew. Chem.* 62(4), 218–219.

CORNELL J (2002), 'Beer color using tristimulus analysis', *J. Amer. Soc. Brew. Chem.* 60(4), 215–219.

CORTACERO-RAMIREZ S, SEGURA-CARRETERO A, CRUCES-BLANCO C, ROMERO-ROMERO M-L and FERNANDEZ-GUTIERREZ A (2004), 'Simultaneous determination of multiple constituents in real beer samples of different origins by capillary zone electrophoresis', *Anal. Bioanal. Chem.* 380(5–6), 831–837.

CORTACERO-RAMIREZ S, SEGURA-CARRETERO A, CRUCES-BLANCO C, HERNAINZ-BERMUDEZ-DE-CASTRO M and FERNANDEZ-GUTIERREZ A (2005), 'Indirect determination of carbohydrates in wort samples and dietetic products by capillary electrophoresis', *J. Sci. Food Agric.* 85(3), 517–521.

COVENTRY A G (1994), 'In line measurement of alcohol and original gravity using fiber optic sensors', *Cerevisia & Biotechnol.* 19(4), 48–52.

CRAWSHAW J R (1988), 'Automatic analysis – a practical approach', *Ferment* 1(6), 48–51.

DAOUD I S (1991), 'New techniques for in-line measurement', *Ferment* 4(1), 40–46.

FERNANDES S M V, RANGEL A O S S and LIMA J L F C (1995), 'Colorimetric determination of iron in beer by flow injection analysis using the merging zones technique', *J. Inst. Brew.* 101(4), 281–284.

FERNANDES S M V, RANGEL A O S S and LIMA J L F C (1997), 'Flow injection determination of sodium, potassium, calcium, and magnesium in beer by flame emission and atomic absorption spectrometry', *J. Agric. Food Chem.* 45(4), 1269–1272.

FERNANDES S M V, RANGEL A O S S and LIMA J L F C (1998a), 'Determination of total sulphur dioxide in beer by flow injection spectrophotometry using gas-diffusion and the merging zones technique', *J. Inst. Brew.* 104(4), 203–205.

FERNANDES S M V, RANGEL A O S S and LIMA J L F C (1998b), 'Merging zones standard addition technique for determination of copper in beer by flow injection atomic absorption spectrophotometry', *J. AOAC Int.* 81(3), 645–647.

FERNANDES S M V, LIMA J L F C and RANGEL A O S S (2000), 'Spectrophotometric flow injection determination of total phosphorus in beer using on-line UV/thermal induced digestion', *Fresenius J. Anal. Chem.* 366(1), 112–115.

FERREIRA I M P L V O, LIMA J L F C and RANGEL A O S S (1994), 'Flow injection titration of chloride in food products with a silver tubular electrode based on an homogeneous crystalline membrane', *Food Chem.* 50(4), 423–428.

FORD S J, ELLIOTT M A and HALBERT G W (2003), 'Use of mathematical derivatives (time-domain differentiation) on chromatographic data to enhance the detection and quantification of an unknown "rider" peak', *J. Pharm. Biomed. Anal.* 33(4), 563–570.

FORREST I S (1996), 'In-line instruments, the way towards brewing efficiency?', *Ferment* 9(5), 273–280.

GARDEN S W, PRUNEDA T, IRBY S and HYSERT D W (2000), 'Development of near-infrared calibrations for hop analysis', *J. Amer. Soc. Brew. Chem.* 58(2), 73–82.

HASSINGER T, ACKERMANN R, BIKSACKY M, CATTANACH D, GUALDONI P, HERBST D, IZAWA I, KUSTELSKI G, MCLINN C, SAKUMA S, VENNE M, YOUNGER E and MURPHEY J (1995),

'Bitterness in beer by automated flow analysis', *J. Amer. Soc. Brew. Chem.* 53(4), 216–219.

HERENT M F and COLLIN S (1998), 'Pyrazine and thiazole structural properties and their influence on the recovery of such derivatives in aroma extraction procedures', *J. Agric. Food Chem.* 46(5), 1975–1980.

HORWITZ W, KAMPS L R and BOYER K W (1980), 'Quality assurance in the analysis of foods for trace constituents', *J. Assoc. Offic. Anal. Chem.* 63(6), 1344–1354.

ISHIBASHI Y, TERANO Y, FUKUI N, HONBOU N, KAKUI T, KAWASAKI S and NAKATANI K (1996), 'Development of a new method for determining beer foam and haze proteins by using the immunochemical method ELISA', *J. Amer. Soc. Brew. Chem.* 54(3), 177–182.

KLAMPFL C-W (1999), 'Analysis of organic acids and inorganic anions in different types of beer using capillary zone electrophoresis', *J. Agric. Food Chem.* 47(3), 987–990.

LAREW L A, MEAD D A, JR and JOHNSON D C (1988), 'Flow-injection determination of starch and total carbohydrate with an immobilized glucoamylase reactor and pulsed amperometric detection', *Anal. Chim. Acta* 204(1/2), 43–51.

LAYCOCK G, CHAN S, CHRISTOPHER D, EDNEY M, JONES B, JOYCE M, KUESTER H, KOUHI-LAVANDER E, LEWIS P, LOVE D, MCLINN D, MAURICE M, RUEL G, SIEBEN R, SMITH G, SWENSON W, SWIMS J and MUNAR M (1997), 'Alpha-amylase and diastatic power in malt by automated flow analysis', *J. Amer. Soc. Brew. Chem.* 52(4), 1999–2002.

LUQUE-PEREZ E, RIOS A and VALCARCEL M (2000), 'Flow injection spectrophotometric determination of ascorbic acid in soft drinks and beer', *Fresenius J. Anal. Chem.* 366(8), 857–862.

MARTENS H and NAES T (1989), *Multivariate Calibration,* New York, John Wiley & Sons.

MASSART D L, VANDEGINSTE B G M, BUYDENS L M C, DE JONG S, LEWI P J and SMEYERS-VERBEKE J (1997), *Handbook of Chemometrics and Qualimetrics: Part A*, Amsterdam, Elsevier Science.

MEILGAARD M (1981), 'Beer flavor', *Dissertation Abstracts International, B* 42(6).

MEILGAARD M (1982), 'Prediction of flavor differences between beers from their chemical composition', *J. Agric. Food Chem.* 30(6), 1009–1017.

MUNCK L, JORGENSEN K G, RUUD-HANSEN J and HANSEN K T (1989), 'The EBC methods for determination of high molecular weight beta-glucan in barley, malt, wort and beer', *J. Inst. Brew.* 95(2), 79–82.

MURPHEY J, BURKHARDT R, CASTANE J, DONLEY J, GUZINSKI G, KAKUDO Y, KOSTELECKY T, MCMURROUGH I, ONO M, VERHAGEN L C and GRIGSBY J (1993), 'Iso-alpha-acids in beer by solid-phase extraction and high-performance liquid chromatography', *J. Amer. Soc. Brew. Chem.* 51(4), 173–174.

PAPADOPOULOU-BOURAOUI A, VRABCHEVA T, VALZACCHI S, STROKA J and ANKLAM E (2004), 'Screening survey of deoxynivalenol in beer from the European market by an enzyme-linked immunosorbent assay', *Food Addit. Contam.* 21(6), 607–617.

PEREZ-RUIZ T, MARTINEZ-LOZANO C, TOMAS V and MARTIN J (1999), 'Flow injection determination of lactate based on a photochemical reaction using photometric and chemiluminescence detection', *Analyst* 124(10), 1517–1521.

PERIS M, MUELLER D and MAQUIEIRA A (1991), 'Determination of total polyphenols in beers by flow injection analysis', *Food Chem.* 40(1), 1–8.

PRODROMIDIS M I, STALIKAS C D, TZOUWARA-KARAYANNI S M and KARAYANNIS M I (1995), 'Determination of glycerol in alcoholic beverages using packed bed reactors with immobilized glycerol dehydrogenase and an amperometric FIA system', *Talanta* 43(1), 27–33.

RUZ J, RIOS A, LUQUE-DE-CASTRO M D and VALCARCEL M (1988), 'Determination of glucose in alcoholic beverages by flow injection with two internally coupled injection valves and an enzyme reactor', *Anal. Chim. Acta* 211(1/2), 281–285.

SHAO X and SUN L (2001), 'An application of the continuous wavelet transform to resolution of multicomponent overlapping analytical signals', *Anal. Lett.* 34(2), 267–280.

SIEBERT K J (1982), 'Brewing laboratory automation. A review with special emphasis on the use of microcomputers', *J. Amer. Soc. Brew. Chem.* 40(1), 9–14.

SIEBERT K J (1983), 'Application of instrumental techniques in brewing chemistry – past, present, and future', *J. Amer. Soc. Brew. Chem.* 41(1), 4–9.

SIEBERT K J (1984a), 'Applications of computers in brewing research and quality assurance laboratories', *The Institute of Brewing Australia & New Zealand Section, 18th Convention*, Adelaide, The Institute of Brewing Australia & New Zealand Section, 196–210.

SIEBERT K J (1984b), 'Process, on-stream, and chromatographic measurements in brewing', in Manka, D (ed.), *Automated Stream Analysis for Process Control, Vol. 2*, New York, Academic Press, 139–167.

SIEBERT K J and ACREE T E (1993), 'The sensory and chemical basis of beer flavor', *Cerevisia & Biotechnol.* 18(2), 29–39.

STOCKWELL P B (1990), 'The role of flow injection analysis within the framework of an automated laboratory', *J. Autom. Methods Manage. Chem.* 12(3), 95–103.

TOMAS C, CELESTE M, CLADERA A, GOMEZ E, ESTELA J M and CERDA V (1993), 'A new flow-injection spectrophotometric method for the determination of tannins in tea and beer using iron(III) and 1,10-phenanthroline', *Food Chem.* 47(2), 201–204.

VAN DEN BOGAERT B, BOELENS H F M and SMIT H C (1994), 'Quantification of overlapping chromatic peaks using a matched filter', *Chemom. Intell. Lab. Syst.* 25, 297–311.

WORSFOLD P J, RUZICKA J and HANSEN E H (1981), 'Rapid automated enzymatic method for the determination of alcohol in blood and beverages using flow injection analysis', *Analyst* 106(1269), 1309–1317.

19

Brewing control systems: microbiological analysis

E. Storgårds, A. Haikara and R. Juvonen, VTT Technical Research Centre of Finland

19.1 Introduction

The presence of inhibitors such as hop compounds, alcohol, carbon dioxide and sulphur dioxide as well as the shortage of nutrients and oxygen and the low pH all make beer resistant to microbial contamination. Moreover, processes such as filtration, storage at low temperatures and possible pasteurisation reduce contamination. The special environment in the brewing process restricts the range of microorganisms likely to be encountered to relatively few species (Ingledew 1979, Haikara 1984, Back 1994, Dowhanick 1994). Although the contaminants found may cause quality defects, pathogens have not to our knowledge been reported to grow in standard beer products. Ensuring the well-being of the production yeast strains is a fundamental part of brewing as in all processes based on fermentation technology. Thus, monitoring yeast quality and quantity is also an important part of the microbiological control carried out in breweries.

In this chapter, a summary of the microorganisms likely to be encountered in breweries and the different possibilities to detect and identify them is given. Different possibilities to quantify yeast mass and estimate the brewing performance as well as differentiate between yeast strains is also described. Both methods currently in use and emerging technologies are discussed. However, as it is not possible within the framework of this book to review all the techniques that have attracted attention among brewery microbiologists in the past, merely those methods showing most potential for brewery applications to date are reviewed here.

19.1.1 Microorganisms associated with beer production

Only very few species and strains can adapt to grow in beer. On the other hand, species adapted to the brewery environment have often not been isolated elsewhere (Back 1994, Haikara and Helander 2002). Beer spoilage organisms such as lactic acid bacteria, wild yeasts and even anaerobic bacteria are often present on the equipment, in the air or in raw materials. These organisms may survive for a long time in niches of the process, probably outside the direct product stream, without causing signs of contamination. Then suddenly, they may contaminate the entire process as a consequence of technological faults or insufficient cleaning. With the introduction of modern fingerprinting methods, such as ribotyping, into brewery microbiology, it has become evident that even the same contaminating strains can pop up after years of absence. In addition to true beer-spoiling organisms that do grow in finished beer there are a range of organisms that may grow at some stages of the process, causing off-flavours in the final product if present in sufficient numbers. There are also indicator organisms that do not cause spoilage but appear as a consequence of insufficient cleaning or errors in production. The growth of seemingly harmless microorganisms on brewery surfaces may facilitate the subsequent colonisation of beer spoilage organisms by producing a more favourable environment for their growth (Storgårds *et al.* 2006). The effects caused by different spoilage organisms during fermentation and in final beer are summarised in Table 19.1.

The microbiological safety risks involved in beer production include mycotoxin production by toxigenic fungi in raw materials, mainly barley and malt (Vanne and Haikara 2001, Flannigan 2003). Some of these fungi can proliferate on barley already in the field, such as *Fusarium*, whereas others such as *Aspergillus* and *Penicillium* grow in too humid storage conditions. Mycotoxins are often very stable compounds and can therefore survive throughout processing and enter the final product. The brewing process itself can be contaminated by *Obesumbacterium proteus* having the ability to reduce nitrate present in wort to nitrite, which in turn reacts with amines, producing nitrosamine compounds (ATNCs or apparent total N-nitroso compounds) (Prest *et al.* 1994). Various bacterial contaminants in the brewing process may also be responsible for the production of biogenic amines (Donhauser *et al.* 1993, Izquierdo-Pulido *et al.* 1996, Virkajärvi *et al.* 2001).

A wider range of microorganisms can cause problems in beer-dispensing equipment than in the brewing process or in packaged beer. This is due to the higher oxygen levels and higher temperatures at certain points in the dispensing system. These conditions favour contamination by microorganisms such as acetic acid bacteria, moderate levels of coliforms and aerobic wild yeast in addition to the oxygen-tolerant beer spoilage organisms found in the brewery environment (Ilberg *et al.* 1995, Schwill-Miedaner *et al.* 1996, Taschan 1996, Storgårds 1997). The occurrence of coliforms in beer-dispensing systems is a cause of concern due to the enteric pathogen *Escherichia coli* serotype O157:H7. This pathogen is unusually acid-resistant and has been associated with outbreaks of serious enteric infections after consumption of contaminated apple

Table 19.1 Effects of contaminants during fermentation and on final beer

Group or genera	Effects on fermentation	Turbidity	Ropiness	Off-flavours in final beer
Wild yeasts	Superattenuation	+	–	Esters, fusel alcohols, diacetyl, phenolic compounds, H_2S
Lactobacillus, Pediococcus		+	+	Lactic and acetic acids, diacetyl, acetoin
Acetobacter, Gluconobacter		+[1]	+[1]	Acetic acid
Enterobacteria	Decreased fermentation rate, formation of ATNC	–	–	DMS, acetaldehyde, fusel alcohols, VDK, acetic acid, phenolic compounds
Zymomonas		+[2]	–	H_2S, acetaldehyde
Pectinatus		+	–	H_2S, methyl mercaptan, propionic, acetic, lactic and succinic acids, acetoin
Megasphaera		+	–	H_2S, butyric, valeric, caproic and acetic acids, acetoin
Selenomonas		+	–	Acetic, lactic and propionic acids
Zymophilus		+[3]	–	Acetic and propionic acids
Brevibacillus		–	+	–
Clostridium		–	–	Butyric, caproic, propionic and valeric acids

ATNC: apparent total N-nitroso compounds; DMS: dimethyl sulphide; VDK: vicinal diketones; fusel alcohols: *n*-propanol, iso-butanol, iso-pentanol, iso-amylalcohol.
[1] In the presence of oxygen.
[2] In primed beer.
[3] At elevated pH (5–6).

cider (Semanchek and Golden 1996, Park *et al.* 1999). It is infectious at a low dose, probably due to its acid tolerance, as it can overcome the acidic barrier of gastric juice and reach the intestinal tract at low population levels (Park *et al.* 1999). Thus the possible survival in beer of acid-tolerant pathogens such as *E. coli* O157:H7 should not be overlooked.

19.1.2 Detection of microbial contaminants in breweries

Contaminations in the brewery are usually divided into primary contaminations originating from the yeast, wort, fermentation, maturation or the pressure tanks, and secondary contaminations originating from bottling, canning or kegging. About 50% of microbiological problems can be attributed to secondary contaminations in the bottling section (Back 1997), but the consequences of primary contaminations can be more comprehensive and disastrous. The spoilage character of a particular organism depends on where in the process it is found. After filtration, the brewing yeast should also be regarded as a contaminant (Haikara 1984, Eidtmann *et al.* 1998).

The concentration of process or product samples has always been a crucial step in the detection of very low numbers of contaminants in beer. Filtration of beer for the recovery of microorganisms can be improved by slightly increasing the temperature and by the use of top pressure (Hammond *et al.* 1999). A bypass-membrane filter device for continuous sampling from product lines has been developed which makes it possible to increase the sample volume up to 40-fold (Back and Pöschl 1998). Recently, the CellTrap™ device (Memteq, UK) was shown to be a useful tool for isolation and recovery of cells from contaminated beer samples prior to analysis by PCR (Whitmore and Keenan 2005).

Although breweries are still relying mainly on classical cultivation methods, a range of alternative methods has been developed for the detection of beer spoilage organisms (Barney and Kot 1992, Dowhanick 1994, Storgårds *et al.* 1998a, Quain 1999, Russell and Stewart 2003). Examples of alternative methods with brewery applications are presented in Table 19.2. Unfortunately, many of these 'rapid' techniques need a pre-enrichment step to increase the sensitivity of the method, thus still being dependent on cultivation. Reasons for the slow implementation of alternative methods in brewery quality control have been lack of the speed, sensitivity and specificity required and/or the need for advanced, expensive equipment and reagents. Therefore, classical microbiological methods remain to be the methods preferred by breweries, even though the detection of beer spoilage organisms by cultivation in laboratory media does not always provide the specificity and the sensitivity required (Jespersen and Jakobsen 1996). However, the implementation of new available technology into brewery microbiology has speeded up considerably during the twenty-first century.

Detection methods can be divided into culture-dependent and culture-independent approaches. Culture-dependent methods include traditional cultivation in combination with phenotypic (physiological and biochemical) and genotypic (species-specific PCR, DNA fingerprinting, sequencing) characterisation or identification techniques of selected, isolated strains. The advantage of this approach is that microbial cultures are available for further characterisation and exploitation. In recent years, culture-independent approaches have been developed to complement the culture-dependent ones. New powerful analytical tools enable us to investigate microbial populations in their natural environment without the need for cultivation. Direct DNA/RNA extraction approaches coupled with PCR amplification and community profiling techniques have become widely applied in microbiology (Muyzer 1999, Ercolini

Table 19.2 Microbiological detection methods with brewery applications

Method	Detection threshold	Detection time	Identification at the same time
Classical cultivation methods	Theoretically 1 cfu per sample	1 to several days or weeks	No
Fluorescence microscopy	Theoretically 1 cell per sample, depends on the application	30–60 min, if enrichment needed 1–3 days	No
Flow cytometry	10^2–10^4 yeast cells per ml	0.5–1 hour, if enrichment needed 1–3 days	Yes (in combination with fluorescent antibodies or DNA probes)
Laser scanning cytometry	1 cell per filterable sample	2–4 hours	Yes
DNA or RNA hybridisation including FISH	10^3–10^4 cells per ml	24–30 hours including pre-cultivation	Yes
PCR	10^2–10^3 cells	0.5–2 hours	Yes
ATP bioluminescence	10–100 yeast cells 10^3–10^4 bacterial cells	5 min	No
RMDS (Micro Star Rapid Microbiology System)	1 yeast cell 10^2–10^3 bacterial cells	1 day 2–3 days	No

2004). The major advantage of different culture-independent approaches is that organisms in both a cultivable and a non-cultivable state can be analysed. Moreover, a semi-quantitative picture of a microbial population can be obtained without time-consuming cultivation and isolation steps.

19.1.3 Identification and characterisation

Identification can be defined as assigning an unknown microorganism to a particular class in an existing classification (Priest 2003). Identification of microorganisms to species level is time consuming and seldom needed in brewery quality control. When identification is performed, it aims to be pragmatic, searching for key properties such as beer spoilage ability rather than for taxonomic details. This kind of characterisation of particular problem-causing strains is an important tool in the tracing of contamination sources. Identification and characterisation can be based on four levels of expression of genetic information, namely on the genome, on proteins, on cell components or morphology, and on behaviour (Gutteridge and Priest 1996). Identification, unlike specific detection, generally requires a pure culture and the use of reference

Table 19.3 Identification and characterisation methods

Method	Identification level
1. Methods based on physiological and morphological properties	
• Standard method	Genera, species
• Miniaturised and commercial systems	Genera, species
2. Chemotaxonomic methods	
• Whole cell fatty acid analysis (FAME)	Species
• Protein fingerprinting	Species, strain
• Pyrolysis mass spectrometry (Py-MS)	Species, strain
• Fourier transform infrared spectrometry (FT-IR)	Species, strain
3. Molecular genetic methods	
• Sequencing	Species
• DNA-fingerprinting	Genera, species, strain
• Karyotyping	Species, strain
• Fluorescence in situ hybridisation (FISH)	Genera, species
• Polymerase chain reaction (PCR)[1]	Genera, species, strain
4. Immunological methods	
• Monoclonal antibody techniques	Species, strain

[1] Various techniques based on PCR available.

identification libraries or keys which contain the results of those species relevant in the studied environment. The methods can be divided into genetic and phenotypic approaches, which can be further sub-divided into methods based on morphological and physiological properties and to chemotaxonomic methods (Table 19.3). Phenotypic methods study secondary properties, the expression of which depends on environmental conditions. On the other hand, genetic approaches study difference between organisms at primary (DNA/RNA) level, and are not usually affected by the physiological state of the organism. For a review see Priest (2003).

Nowadays, the standard method based on selected phenotypic and morphological properties is generally considered unreliable for identification. The most accurate method for species level identification is sequencing which reveals the differences between organisms at nucleotide level. However, high costs and labour intensity still restrict its use to specialised laboratories. Sequence heterogeneity can also be studied indirectly using a variety of DNA fingerprinting techniques. PCR fingerprinting, using specific or random primers, offers convenience and rapidity and can be used to construct identification libraries. In order to discriminate strains within a species, more complicated fingerprinting techniques, like ribotyping, pulsed field gel electrophoresis (PFGE) or amplified fragment length polymorphism (AFLP), are often required, and there is still need for a simple and rapid strain differentiation method. Ultimately, the choice of the identification method depends on the level of identification required, research question, facilities and target organisms. In general, the use of multiple techniques improves the resolution and reliability of identification.

19.2 Classical microbiological methods

19.2.1 Detection and enumeration of microorganisms by cultivation

This chapter aims at giving a brief review of classical microbiological methods, as the focus of this book is on new and developing technologies. The most widely adopted microbiological methods are still various cultivation methods that have been used for more than a century. They rely on specific media to isolate and enumerate viable bacteria, yeasts and moulds. If the right medium and cultivation conditions are chosen, the method is sensitive (theoretically one single cell can be detected from the sample) and gives both qualitative and quantitative information. A further advantage is that a sample can be simultaneously tested for the presence of various microorganisms simply by including several types of selective media in the analysis. A reason for the reluctance in adopting alternative methods in quality control may be the wealth of experience of using cultivation methods, making comparison, trend analysis and setting specifications easy.

The easiest way to detect microorganisms present and able to grow in a sample is to simply incubate it, preferably at a slightly elevated temperature, and daily inspect it for haze and possible CO_2 development. This so-called forcing test can be applied to wort as well as to bright beer samples. To speed up the method, the sample can be enriched with a suitable substrate; however, this will reduce the selectivity of the test against non-spoilage organisms. Both the forcing test and the enrichment test are qualitative and no exact number of microorganisms can be estimated. Normally, the time needed before any turbidity occurs is indicative of the amount of spoilage organisms present, although it should be kept in mind that some severe spoilage organisms are notoriously slow-growing.

The plating techniques are developed to enable better estimation of the number of organisms present. The microorganisms in the sample are fixed either on top of the surface or inside a solid substrate, i.e. a nutrient medium containing a solidifying component such as agar or gelatine. A modification for detection of low numbers of microorganisms in filterable liquid samples, such as water or bright beer, is based on filtering the sample through a membrane with a pore size small enough (normally 0.45 or 0.2 μm) to retain the microbial cells. The membrane is then transferred onto a solid medium. Irrespective of the method used to fix the cells on or into the medium, the microorganisms will eventually multiply during subsequent incubation to form visible colonies that can be counted. The methods are based on the approximation that the origin of every colony is one single cell and that all cells in the sample will grow. Unfortunately, cells are often hard to separate as they may form chains or clusters and they may also not be able to multiply in the chosen conditions. Thus the results of these methods are expressed as colony forming units or cfu's. The methods for analysing various brewery samples are described in detail in EBC *Analytica Microbiologica* (2005).

Although a considerable number of useful media have been developed to detect either all possible or only selected species or strains, none of these is able to meet all the needs. There is, for example, no single medium able to detect all lactic acid bacteria or all wild yeasts encountered in the brewery. Unfortunately, the most dreaded beer spoilage organisms, such as fastidious lactobacilli or strictly anaerobic bacteria, are often the most difficult to grow in laboratory conditions. It goes without saying that no medium will enable growth of all possible microorganisms with their various growth requirements. Thus the term 'total count' is misleading and refers mainly to mesophilic aerobic heterotrophs in case the incubation was performed aerobically at 20–40ºC.

19.2.2 Phenotypic characterisation

When unintended microorganisms are found in brewery samples, the first step is generally to group them into different categories by microscopy and by preliminary tests to be able to predict potential damages. A simplified key for the grouping of bacteria likely to be encountered in the brewery is presented in Fig. 19.1. Further identification is traditionally carried out by using various biochemical tests such as carbohydrate utilisation or enzymatic reactions.

Miniaturised commercial systems can be regarded as more reproducible than conventional methods (Gutteridge and Priest 1996). A main problem in the identification of *Lactobacillus* strains, however, is the high phenotypic similarity among species, which can be as much as 95% despite the strains being unrelated by criteria such as rRNA sequence or DNA homology (Priest 1996). The phenotypic homogeneity of the lactobacilli necessitates the use of at least 50 tests such as in the API 50 CHL system, and still the results may not be reliable. Plasmid loss may cause altered phenotypes in lactobacilli, as many plasmids code for carbohydrate utilisation pathways (Priest 1996). An additional problem associated with some beer spoilage bacteria is that they may be extremely slow growing on cultivation media such as the MRS used in API 50 CHL. For slow-growing *Lactobacillus* strains from brewery environments, prolonged incubation of API 50 CHL strips for up to 10 days (Funahashi *et al.* 1998) or even up to 18 days (Storgårds *et al.* 1998b) was needed before carbohydrate fermentation could be detected. Furthermore, beer spoilage *Lactobacillus* strains typically use only a few of the sugars available in the API 50 CHL strips (Funahashi *et al.* 1998, Storgårds *et al.* 1998b), thus making identification by phenotypic tests unsatisfactory.

Also phenotypic characterisation of yeast strains has its drawbacks. The standard methods described in yeast taxonomy (Kurtzman and Fell 1998, Barnett *et al.* 2000) are based on the ability of a strain to grow aerobically on selected carbon and nitrogen sources, to ferment different carbohydrates, to synthesise essential vitamins and to tolerate extreme conditions. In addition, characters related to sexual and asexual reproduction are studied. Depending on the yeast, 50–100 tests are required and reliable results are available only after several weeks of incubation. The standard method frequently results in unequivocal or incorrect results due to intraspecies variability of many characters and the fact that many species descriptions are based only on type strains (Kurtzman

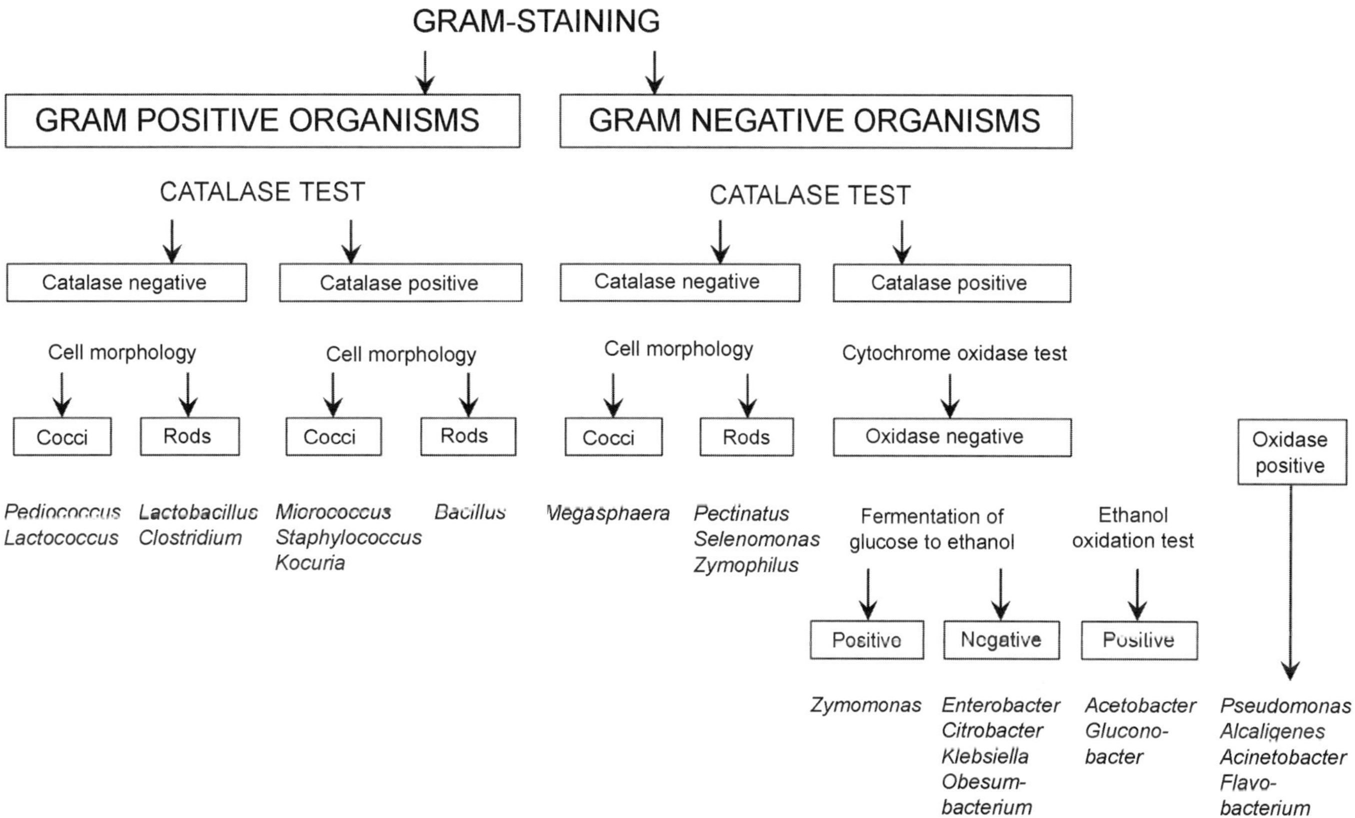

Fig. 19.1 Simplified key for the identification of bacteria likely to be encountered in the brewery.

et al. 2003). Therefore, expertise is required to interpret results. For the same reasons, atypical strains isolated from industry can be difficult to identify. The traditional approach has also limited resolution, for example *S. cerevisiae*, *S. bayanus* and *S. pastorianus* cannot be discriminated from each other using this method (Naumov *et al.* 2000).

19.2.3 Drawbacks with traditional methods

There are several drawbacks associated with the classical microbiological methods. Most importantly, the methods are too slow for rapid countermeasures. Moreover, traditional methods do not discriminate between spoilage and non-spoilage organisms (except for the forcing test) nor allow the detection of fastidious or injured microbes. Microbes in industrial environments are subjected to various stresses such as starvation, chemicals, heat, cold and desiccation, which injure the cells and may render them non-cultivable. The proportion of cultivable cells in industrial food processing premises is unknown, but in most natural environments only a small percentage of the living microbial population consists of cultivable cells (Duncan *et al.* 1994, Amann *et al.* 1995, Leriche and Carpentier 1995). In a more favourable environment these non-cultivable organisms may again become growing cells and a threat for the product. Increasing globalisation of beer markets, stricter consumer demands, avoidance of heat treatments and emergence of new microbiological safety risks put ever-increasing demands on the wholesomeness of beer. Hence, there is a need for more effective quality monitoring tools.

19.3 Optical techniques

19.3.1 Bright field and phase contrast microscopy

Studying the morphology of microorganisms has always been of vital importance for brewery microbiologists and the development of molecular techniques has basically not changed that. Preliminary microscopic examination of turbid beer samples provides a satisfactory distinction between bacteria, yeasts and moulds, subsequently enabling more specific analysis. Microscopic confirmation may also be needed at any stages of troubleshooting. While a magnification of 400 is sufficient for studying yeasts and moulds, a magnification of 800–1000 is usually needed for reliable examination of the morphology of bacteria. Phase contrast microscopy of native droplets is used, for example, for examining yeast spores or detection of bacteria in yeast suspensions. Bright-field microscopy is used for examining thin films of stained microorganisms. The most common staining procedure is Gram-staining which is used for preliminary differentiation of Gram-positive and Gram-negative bacteria. The basic techniques involved in bright field and phase contrast microscopy are described in detail in microbiological manuals such as EBC *Analytica Microbiologica* (2005) or in user guides of the microscopes.

19.3.2 Fluorescence microscopy

Direct epifluorescence microscopy

Fluorescence microscopy has been applied directly to monitor cells on filter membranes or surfaces. In the direct epifluorescence filter technique (DEFT), cells are collected from a sample by membrane filtration, stained with a fluorescent dye and counted under the fluorescence microscopy.

DEFT has been applied for the detection of beer spoilers (Haikara 1985, Rinck and Wackerbauer 1987). Using DEFT, time savings of 1–3 days were obtained in the detection of *Megasphaera cerevisiae* and *Pectinatus* spp. (Haikara 1985) and lactic acid bacteria (Rinck and Wackerbauer 1987) compared to conventional methods. Due to the tediousness of microscopic counting and the development of other alternative methods, the method was never applied on a large scale in the breweries. More recently a modification of DEFT based on enzyme-specific fluorescence was described (Jaspers *et al.* 2003). Direct epifluorescence microscopy has also been used in brewery hygiene studies (Storgårds 2000). The advantage of epifluorescence image analysis in biofilm research is that it studies cells directly on the surface, thus providing a powerful tool for detecting residual microbial cells after cleaning and disinfection treatments.

Fluorescence microscopy has also been applied to visualise small microbial colonies, so-called microcolonies, before they are visible to the naked eye. The colonies have most often been stained by adding fluorescent dyes, e.g. optical brighteners, to the growth medium, making cells fluorescent during incubation. Beer-spoilage yeasts were detected after overnight incubation (Jakobsen and Lillie 1984) and anaerobic bacteria including lactic acid bacteria after 2–3 days (Haikara and Boije-Backman 1982, Parker 1989). A more recent application based on microcolonies together with the ATP bioluminescence technique is the Micro Star Rapid Microbiology System (RMDS), which is described in Section 19.5.1.

Laser scanning cytometry

ChemScan® RDI (www.chemunex.com/literature/biblio.htm#laser) uses fluorescent markers and solid-phase laser scanning cytometry to enumerate microbial cells collected on a membrane filter. It allows quantitative detection of single cells (yeasts, bacteria, spores) in filterable samples in 2–4 hours depending on staining method. For total viable cell count, a proprietary fluorogenic enzyme substrate is used, but also fungus-specific markers are available. Fluorescent antibodies or DNA probes can be applied to detect specific microbes, but no applications have yet been developed for beer-spoiling organisms. The labelling step takes from 90 minutes to 3 hours and laser scanning less than 3 minutes. The prototype of the system was developed in the ECLAIR programme (project AGRE-0014) during 1991–92, and was evaluated by TEPRAL (Beer Research Centre of the Danone group) for microbiological quality control in the breweries. At this time, TEPRAL considered ChemScan® RDI to be a promising technique but too costly to use simply for determining the total viable count of the sample (www.nf-2000.org/secure/Fair/S1145.htm).

19.3.3 Immunoassays: immunofluorescence

Immunofluorescence is based on a specific reaction between antibodies and cell-surface antigens. Antibodies are directly or indirectly labelled with fluorochromes, and fluorescent cells are enumerated using microscopy, flow cytometry or laser scanning cytometry. Immunofluorescence was already applied early for the specific detection of wild yeasts using polyclonal antibodies (Haikara and Enari 1975), and a standard method for wild yeasts was incorporated in the Institute of Brewing *Methods of Analysis* (1997). Monoclonal antibody technology enables mass production of antibodies with defined specificity which has been applied for the detection and identification of beer spoilage bacteria. The technique has been applied for *Pectinatus cerevisiiphilus* (Gares *et al.* 1993, Ziola *et al.* 1999, 2000) and beer-spoilage strains of lactobacilli and pediococci (Whiting *et al.* 1992, 1999a, 1999b). Rapid detection and quantification of beer-spoilage lactic acid bacteria using chemiluminescent enzyme immunoassays and a CCD camera have been developed to avoid the troublesome microscopy step (Yasui and Yoda 1997, March *et al.* 2005). Immunological methods still have a lot of unexploited potential (Priest 2003). Various approaches to circumvent the tedious microscopy step have been proposed of which flow cytometry is probably the most widely used.

19.3.4 Flow cytometry

In flow cytometry (FCM), fluorescence and light scatter of individual cells is measured while cells pass in a single file through a laser beam. Hence, flow cytometers can be considered as automated cell counters, which can simultaneously analyse 2–6 fluorescence and scatter parameters at a rate of 100–1000 cells s^{-1}. Analysis of a 0.2–1 ml sample takes only a few minutes. Usually, target cells need to be first stained to discriminate them from other particles. Extensive pre-treatment of samples may also be necessary, since air bubbles and micron-sized particles interfere with the analysis (for a review see Davey and Kell 1996).

FCM has been applied to detect contaminants in food, beverages and alcoholic drinks among many other applications. FCM in combination with fluorogenic esterase substrates allowed direct detection of 10^2–10^4 yeast cells ml^{-1} in fruit juices, carbonated beverages and wine (Pettipher 1991, Malacrinó *et al.* 2001). In wine, yeast cells could be discriminated from *Oenococcus oeni* on the basis of light scatter. For the detection of contaminants in brewery samples, cells have to be pre-enriched (Jespersen *et al.* 1993) or collected by antibody-coated magnetic beads (Eger *et al.* 1995) to obtain sufficient sensitivity. By using selective pre-enrichment Jespersen *et al.* (1993) were able to detect one wild yeast among 10^6 culture yeast cells in 48–72 hours. For the specific detection of microorganisms by FCM, fluorescent antibodies or DNA probes can be used. Hutter (1994) could simultaneously identify three spoilage organisms (*Pediococcus damnosus, Lactobacillus brevis, Schizosaccharomyces pombe*) in beer when each species-specific antibody was labelled with a different-coloured fluorochrome. Bertin *et al.* (1990) showed the applicability of fluorescently labelled DNA probes for the detection of yeasts by FCM.

Automation and rapidity of analysis (0.5–1 h) are the main advantages of FCM in contaminant detection. The technique is also very versatile, allowing enumeration, identification and physiological characterisation of microorganisms. A large variety of flow cytometers is commercially available, albeit at rather high cost. However, instrument prices are continuously decreasing. In the future, FCM has potential for on-line and at-line detection of contaminants in food processes and for simultaneous analysis of multiple target organisms.

19.4 Molecular methods for detection and identification

Broadly speaking, molecular methods include the methods based on analysis of various molecular cell compounds, but generally they refer to genetic approaches based on DNA or RNA analysis. These methods, in which the primary information contained in microbial nucleotide sequences is used for detection, characterisation and/or identification of microorganisms, are rapidly becoming popular. They are very useful in solving many microbiological problems that microbiologists earlier were struggling with.

19.4.1 Hybridisation-based methods

Detection and identification can be achieved by hybridisation-based methods where DNA or RNA target molecules are hybridised with labelled capture and detection probes. For brewery use, a test kit based on this technique has been developed for detection and identification of bacteria of the genera *Lactobacillus*, *Pediococcus*, *Pectinatus* and *Megasphaera*. The test is based on immobilisation and concentration of sandwich complexes between 16S rRNA target molecules and probes by magnetic beads. The measurement is performed by absorption spectroscopy and the 96-well plate format allows high test throughput. The assay time is 24–30 hours due to precultivation, and the detection limit is 10^3–10^4 cells ml^{-1} (Bau *et al.* 2005).

Fluorescence in situ hybridisation (FISH)

Fluorescence in situ hybridisation (FISH) allows identification, quantification and localisation of single cells directly in samples. Fluorescence-labelled oligonucleotide probes are hybridised to their complementary nucleic acid targets (usually rRNA) inside the whole cells, and fluorescent cells are visualised using microscopy or flow cytometry (Moter and Göbel, 2000). The practical detection limit of FISH is around 10^3 cells ml^{-1} or higher, mainly due to the small final sample size (Kosse *et al.* 1997). Furthermore, the low rRNA content of cells often necessitates pre-incubation of samples prior to the analysis. Alternatively, the sensitivity can be enhanced by signal amplification or polyribonucleotide probes (Biegala *et al.* 2003, Zwirglmaier *et al.* 2003). Currently, commercial kits based on FISH are available for various beer-spoilage bacteria (Thelen *et al.* 2002), and a kit for spoilage yeasts in non-alcoholic beverages is being launched soon. The major drawbacks of the technique in contaminant detection are relatively poor discriminatory power and sensitivity, high instrumentation costs and lack of automation. Identification of cells by FISH may be complicated by limited accessibility of probe target sites, limited cell wall permeability, low cellular ribosome content and auto-fluorescence (Moter and Göbel 2000).

19.4.2 Polymerase chain reaction (PCR)

Even though PCR was already introduced to the scientific community in 1985, it is still today one of the most powerful and widely used approaches for detection of low levels of specific microbes, used either as such or as a preparative step to increase assay sensitivity. PCR is essentially an exponential DNA synthesis reaction in a test tube. The synthesis is primed by two short oligonucleotides, called primers, and carried out by a thermostable DNA polymerase in a three-step thermo-cyclic process. The outcome of the reaction is traditionally evaluated using gel electrophoresis. PCR allows 10^{12}-fold amplification of the target nucleic acid sequence in an hour or two (Mullis *et al.* 1986). Hence, it is a potentially very specific, sensitive and rapid technique in addition to being very simple. Several other amplification techniques (NASBA, LCR, β-Q replicase) have since been introduced, but they are not as widely applied as PCR (Cockerill and Uhl 2002).

The end-point PCR with gel electrophoresis is not an ideal method for routine diagnostics. It involves the manipulation of amplified PCR products and carcinogenic compounds, the interpretation of results is subjective, and the method is difficult to automate and quantify. Real-time PCR (also known as quantitative, on-line or kinetic PCR) provides a means to simultaneously amplify, detect and quantify nucleic acid targets (McKillip and Drake 2004). The whole analysis takes place in a closed tube and is completed in 0.5–2 hours. The accumulation of PCR products is monitored cycle by cycle using specific primers in combination with fluorescent DNA probes or dyes and a special thermocycler measuring fluorescence. During the exponential PCR phase, the onset of the fluorescence signal is inversely proportional to the initial target amount, allowing quantification. A number of different approaches can be used to generate the fluorescence signal. Most instruments allow characterisation of the PCR products by their melting point (T_m) and simultaneous measurement of two or more fluorescence parameters.

Several real-time instruments and compatible test kits for the detection and identification of beer spoilage organisms are available on the market and are already in use in many breweries (Braune and Eidtmann 2003, Hage and Wold 2003, Kiehne *et al.* 2005, Vogeser and Dahmen 2004, Wold *et al.* 2005). Recently, an EU-financed project for the development and demonstration of

PCR-based methods for process control in the brewing industry further speeded up the development of the methodology and the implementation thereof into brewery QC laboratories (Haikara *et al.* 2003, Juvonen *et al.* 2003). For the PCR detection of wild yeasts and beer spoilage bacteria in filterable and non-filterable brewery samples, simple protocols were set up. Kits were developed both for screening of chosen groups and for specific detection and identification of spoilage organisms. To date, kits cover practically all organisms of relevance in brewing microbiology, including beer-spoiling lactic acid bacteria, beer-spoiling strictly anaerobic bacteria, all lactic acid and acetic acid bacteria, enterobacteria and yeasts. Species-specific tests are available for most beer-spoiling organisms, making identification possible, and new primer sets are rapidly developed when previously unknown species emerge (Vogeser *et al.* 2005). The real benefits of the technology were recognised during the demonstration phase in the breweries (Brandl and Geiger 2003).

Discriminating beer-spoilage strains from non-spoilage strains within relevant species has always been a major concern for brewery microbiologists. Understanding the hop-resistance mechanisms has enabled development of methods aimed at dealing with this problem (Sakamoto and Konings 2003). Approaches using the PCR technique to discriminate between spoilage and non-spoilage strains of lactic acid bacteria were first described by Sami *et al.* (1997) and later by Nakakita *et al.* (2003) and Tsuchiya *et al.* (2003). These assays are based on PCR analysis of specific genes such as *horA* (Sami *et al.* 1997) or *gyrB* (Nakakita *et al.* 2003, Tsuchiya *et al.* 2003) which were observed to be related to hop resistance.

PCR offers also a sensitive and specific method for detection of toxigenic fungal species that may infest barley such as *Fusarium* (Nicholson *et al.* 2003). Recently, real-time PCR assays enabling rapid detection and quantification of *Fusarium* have been developed (Bluhm *et al.* 2004, Reischer *et al.* 2004, Waalwijk *et al.* 2004, Sarlin *et al.* 2006).

The multiplexing capability of one PCR reaction is limited to about six primer sets. PCR-ELISA (enzyme linked immunosorbent assay) offers a relatively inexpensive, high-throughput method for specific detection of multiple target organisms. PCR amplicons containing the target sequence are hybridised to species-specific probes on the wells of a microtitre plate, and thereafter visualised using an enzymatic colour reaction. PCR-ELISA was also used for detection of anaerobic *Megasphaera cerevisiae* and *Pectinatus* spp. in beer (Satokari *et al.* 1998), and later PCR-ELISA probes have been developed for many other beer-spoilage bacteria (Walker *et al.* 2003).

19.4.3 Fingerprinting methods

DNA fingerprinting

Genetic differences between organisms can also be studied indirectly using DNA fingerprinting techniques. Basically, the polymorphism in the target DNA (chromosomal and mitochondrial DNA, specific genes, plasmids) is revealed using electrophoresis, PCR, restriction enzymes (RFLP) or any combination of them. Resulting banding patterns are compared. The basic assumption is that similar patterns indicate identity and different patterns dissimilarity. Numerous DNA fingerprinting techniques are available. They differ from each other, for example, in regard to discriminatory power, ease of use, rapidity and reproducibility. In general, fingerprinting is a less expensive, quicker and easier approach for species identification than sequencing. Some techniques even allow discrimination of strains within a species, which is not usually possible by sequencing. However, no public databases are available, partly because the methods are difficult to standardise between laboratories.

In the restriction fragment length polymorphism (RFLP) analysis of rDNA, sequence heterogeneity in the selected parts of the rDNA operon is indirectly revealed using sequence-specific restriction enzymes. The target region is amplified, digested with four-base pair recognising restriction enzymes, and the fragments are size-separated using electrophoresis. This approach is also known as ARDRA (Amplified Ribosomal DNA Restriction Analysis) or PCR-ribotyping. The resolution depends on both the target area and the restriction enzymes used. RAPD-PCR (Randomly Amplified Polymorphic DNA) uses a single short primer with random sequence (van der Vossen *et al.* 2003). RAPD-PCR has been applied to distinguish beer-spoilage strains of *Lactobacillus brevis* from non-spoilage strains by identifying a genetic marker specific for spoilage strains (Hayashi *et al.* 2003). Denaturing gradient gel electrophoresis (DGGE) or temperature gradient gel electrophoresis (TGGE) are perhaps the most commonly used culture-independent fingerprinting techniques for studying microbial populations (Muyzer and Smalla 1998). DGGE has been used to study the complex microbial ecosystems in malting (Laitila *et al.* 2005).

Ribotyping

Ribotyping is used for characterisation of the restriction fragment length polymorphism (RFLP) of the ribosomal RNA genes. The total chromosomal DNA is cut with restriction enzymes, separated by gel electrophoresis and hybridised to probes for the 16S and 23S ribosomal RNA genes. Ribotyping was previously found to be quite laborious and complicated and not to show very good discrimination between strains (Prest *et al.* 1994). Automation makes the system reproducible and easy to handle, enabling the analysis of 1–8 strains to be carried out in 8 hours. The discriminatory power of ribotyping is often even better than that of partial 16S rDNA sequencing (Suihko and Stackebrandt 2003), making it a rapid and reliable tool for identification of bacteria. The RiboPrinter® system was successfully applied to differentiation and characterisation of beer-spoilage lactobacilli isolated from brewery samples (Funahashi *et al.* 1998, Storgårds *et al.* 1998b, Yansanjav *et al.* 2003), of pediococci (Satokari *et al.* 2000, Barney *et al.* 2001) of anaerobic beer spoilage bacteria (Motoyama *et al.* 1998, Suihko and Haikara 2001) and of *Obesumbacterium proteus* (Koivula *et al.* 2006).

Chemotaxonomic methods

Chemotaxonomic methods have also been found to be useful for identification and characterisation purposes. Fourier transform-infrared spectroscopy (FT-IR) can be applied directly to whole cells shortening the sample preparation required. The potential of FT-IR in the identification and typing of bacteria (Al-Holy *et al.* 2006) and yeasts (Kümmerle *et al.* 1998, Wenning *et al.* 2002) has been demonstrated. In pyrolysis mass spectrometry (Py-MS) preparation steps are also kept to a minimum as single microbial colonies can be analysed by this technique. Py-MS has proven useful for the discrimination of ale and lager yeast strains (Timmins *et al.* 1998), and has also been used for fingerprinting beer-spoilage bacteria (Beverly *et al.* 1997).

A microbial cell expresses some 2000 different proteins, which can be used as a source of information in the identification and characterisation of microorganisms. Polyacrylamide gel electrophoresis (PAGE) of cellular proteins yields complex banding patterns, which can be considered as highly specific fingerprints of the strain investigated. These electrophoregrams are highly reproducible and individual strains within a given taxon can often be recognised (Pot *et al.* 1994). Electrophoretic patterns have been found to be discriminatory at the species, subspecies or biotype level. Another advantage of the method is that a large number of strains can be compared effectively. Protein electrophoresis is regarded as particularly suitable for identification of lactic acid bacteria (Gutteridge and Priest 1996).

19.4.4 Identification by DNA sequencing

Sequencing is a direct method for comparing organisms at their DNA level. It involves the determination of the exact nucleotide sequence of a selected DNA stretch using PCR and electrophoresis. Public databases are available on the Internet to compare the obtained sequence to those of known bacteria. The large-scale use of the method is still limited by its high costs and labour intensity.

Sequencing is today the most accurate method for species identification, since it allows the analysis of genetic information at nucleotide level. Sequencing of whole genomes is impractical and instead certain diagnostic sequences are analysed. Genes and spacer regions in ribosomal DNA (rDNA) operon are the most common targets for species identification, and established protocols can be found in the literature (e.g. Kurtzman and Robnett 2003) and on the Internet for their sequencing.

19.5 Indirect methods

Indirect methods refer to those methods in which growth of microorganisms is not directly measured. However, these methods may still be dependent on microbial growth, thus they should not be confused with truly culture-independent methods.

19.5.1 ATP bioluminescence

Adenosine-5′-triphosphate (ATP) is a high-energy compound present in all living cells which can be used as an indirect measure of biomass. In the ATP-bioluminescence method, ATP is measured using a firefly luciferin–luciferase enzyme system which catalyses an ATP-specific bioluminescence reaction in which the energy contained in ATP is converted to blue-green light (Kyriakides and Patel 1994). The ATP-bioluminescence method was first applied to the rapid detection of spoilage microbes in finished beer during the 1980s. Even though good sensitivity (1–100 cells) has been obtained with yeast contaminants, the detection limit for bacteria is considerably higher (10^3–10^4 cells) (Miller and Galston 1989, Simpson *et al.* 1989). Furthermore, the method measures only total microbial ATP irrespective of its source. However, this makes the method particularly well suited for hygiene monitoring and as such it is in use in many breweries and for dispensing systems (Hammond 1996, Storgårds and Haikara 1996, Schwill-Miedaner and Eichert 1998, Werlein 1998, Quain 1999).

The Micro Star Rapid Microbiology System (RMDS) is currently the most advanced approach for the detection of microbes by ATP-bioluminescence. It was developed by Millipore in collaboration with Sapporo Breweries (Takahashi *et al.* 1999, 2000). The system consists of a reagent spray station, a luminescence detector and an image processor. Samples are filtered and the membrane is placed in the reagent spray station where ATP extraction and addition of luminescence reagents take place. The light from luminescent yeast cells or bacterial microcolonies is detected, intensified, and finally visualised on a computer display using a CCD camera. The RMDS allows quantitative detection of single viable yeast cells within one working day. Detection of beer-spoilage bacteria requires a two-day pre-incubation (Nakakita *et al.* 2002). The major drawback of RMDS is its high instrument costs. Despite this, the system is in use in brewery quality control in Japan.

19.5.2 Impedimetry

Impedance microbiology based on measurement of metabolic activity was, together with the ATP bioluminescence method, among the first alternative approaches to classical cultivation methods. Impedimetry is based on the monitoring of electrical changes caused by the breakdown of nutrient macromolecules into smaller high-charged units as a result of microbial metabolism. The bacterial population must reach a threshold level of 10^5–10^6 cfu ml^{-1} before the conductivity change can be monitored (Fung 1994). Impedimetry has been applied to some extent for detection of bacterial contamination in pitching yeast (Kilgour and Day 1983), for detection of lactic acid bacteria in bottled beer and bright beer samples (Unkel 1990) and for monitoring the effectiveness of sanitization in a brewery (Foster 1996). However, the method does not seem to be in much use in brewing microbiology, probably due to the rapid development of more specific and sensitive methods.

19.5.3 Chemical analysis of growth media

As microorganisms induce changes of a chemical nature when growing in any medium, such as in the product, looking for these changes has always been a means of microbiological QC. The most basic approaches include organoleptic evaluation and pH monitoring, but current technology also makes more advanced approaches possible. Chemical characterisation of spoilage processes can be valuable in troubleshooting, i.e. establishing the causes of spoilage (Dainty 1996). *Pectinatus* spp. can be identified based on large quantities of propionic acid and hydrogen sulphide in beer; correspondingly, *M. cerevisiae* can be identified based on butyric, valeric and caproic acids in beer (Haikara and Helander 2002). Chemical analysis of metabolised products is especially useful in the case of samples in which the bacteria are dead or non-cultivable.

19.6 Evaluation of yeast quality and quantity

19.6.1 Yeast mass determination

Yeast mass need to be determined in order to decide the optimal pitching rate. There are various methods available of which some are based on cell mass assays and others on cell count. Yeast mass can be estimated by wet weight (centrifugation), by dry weight or by modifications of these. Yeast cells are mostly counted microscopically by haemocytometrical methods, but also photometric determination or electronic counters can be used (EBC 2005).

These methods have many drawbacks as none of them indicate the viable fraction of cells. Moreover, they are often laborious and sensitive to background effects and varying measuring environments. Monitoring yeast viable biomass electrochemically overcomes many problems of the conventional methods (Kronlöf 1991). The method is based on measurement of capacitance in an electrically conducting medium, which correlates linearly with the content of viable cells (Harris *et al.* 1987, Mishima *et al.* 1991). The measurement is robust and not affected by gas bubbles or trub (Kell *et al.* 1990). A further advantage is that the method is automated, avoiding the problem of representative sampling and making on-line measurement possible. Consequently, capacitance instruments allowing automatic pitching control are nowadays common in breweries.

19.6.2 Yeast viability

Yeast viability refers to the ability of cells to grow, reproduce and interact with their environment. Thus viability is essentially a measure of living cells. In many instances accurately assessing how many cells within a population are alive is sufficient to enable decisions on yeast quality to be made (ASBC 2003). However, it should be perceived that viable cells can possess different degrees of vitality which affect the viability tests. Thus the division into viability and vitality is partly vague (vitality is dealt with in Section 19.6.3).

Viability can be assessed by growth-based methods or by various staining methods. As plate counting is slow, the more rapid slide culture technique based on microcolonies has been recommended by brewery microbiological manuals (EBC 2005, ASBC 1984, IOB 1997). However, all growth-based methods suffer from inaccuracy in assessing viability of flocculent yeasts, leading to under-estimation of viable cell count, and they are also unable to detect viable but non-cultivable cells.

Staining methods are based on either brightfield microscopy or fluorescence microscopy. The most widespread stain used with brightfield microscopy is probably methylene blue, which live yeast cells enzymatically reduce to colourless compounds inside the cell. The method is dependent both on the condition of the cell membrane and on the activity of certain oxidoreductases in the cell, thus reflecting vitality as well as viability. Dead cells are stained blue, but unfortunately the method has been found to overestimate viability in cases where the viability is below 90%. Although the method is relatively easy to use, it suffers from being rather subjective and poorly reproducible, as cells may be viable but have partly injured cell membranes and reduced enzymatic activity. Methylene violet has partly displaced methylene blue, as the dye has been reported to distinguish live and dead cells with less ambiguity (Smart *et al.* 1999). By replacing the usually citrate buffered (pH 4.6) methylene blue or violet with glycine buffered (pH 10.6) alkaline solutions, a more reliable and less subjective method with better correlation to cell proliferation and acidification power was obtained (Sami *et al.* 1994, Smart *et al.* 1999).

As brightfield stains are known to have many limitations (O'Connor-Cox *et al.* 1997, ASBC 2001), fluorescent stains have been introduced instead. These stains are alternatively based on membrane integrity such as Mg-ANS, propidium iodide, Sytox orange, Berberine and FUN-1, on membrane potential such as Oxonol and Rhodamine 123 or on esterase substrates such as fluorescein diacetate, carboxyfluorescein diacetate and calcein acetoxymethyl ester. Mg-ANS, Oxonol, Sytox orange and Berberine have been reported to be the most promising fluorescent stains for estimating viability in brewing yeast, while propidium iodide and FUN-1 were observed to overestimate or underestimate viability, respectively (Van Zandycke *et al.* 2003). Some breweries are using Mg-ANS (McCaig 1990) as the method is claimed to work well with the low viabilities that are sometimes observed in connection with increased wort strength. The advantage with fluorescent stains is that they can be used with flow cytometry, thus increasing the sensitivity and objectivity and reducing the laboriousness of the method.

19.6.3 Yeast vitality

Yeast vitality refers to yeast activity or the capacity to recover from physiological stress. It has been demonstrated that vitality actually influences fermentation performance (Bendiak 2000). The physiological condition of brewing yeast has been estimated using a range of different methods. Yeast vitality can be assessed by measuring either the concentration of intracellular components, such as storage glycogen, sterol content or ATP, or by estimating metabolic activity (Boulton 1996). Metabolic activity has been assessed in various ways including measurement of acidification power, magnesium ion release, mean cell age, intracellular pH, rates of sugar uptake, ethanol formation, CO_2 evolution, oxygen uptake and enzyme activities (Imai *et al.* 1994, Iserentant *et al.* 1996, Hutter 1997, Mochaba *et al.* 1998, Wellhoener and Geiger 2003; for a review see Heggart *et al.* 2000). It seems evident that no single method is able to define the overall physiological condition of yeast and most methods have been found to have certain limitations. All tests need to be calibrated to actual fermentation performance on production scale in order to be useful. Many authors aim at describing a simple and rapid but still reliable method that could be applied for use in breweries to decide whether a particular batch of yeast can be used and how much yeast should be pitched and to predict fermentation performance. The magnesium ion release test is a simple rapid method employed in some breweries. Monitoring the pH of yeast slurries is easy and tells how rapidly the cells are dying. Recently both vital titration, which correlates well with the acidification power test (Rodrigues *et al.* 2003), and a simplified intracellular pH measurement procedure have been proposed for this purpose (Thiele and Back 2005).

Flow cytometry in combination with fluorescent probes has been extensively used in the evaluation of yeast physiology; for a review see Edwards *et al.* (1996). Various promising staining protocols have been described for viability/vitality assessment of yeast (Breeuwer *et al.* 1994, 1995, Hutter 1997, Deere *et al.* 1998, Bouix and Leveau 2001, Boyd *et al.* 2003). These are generally based on the inability of a stain to penetrate an intact cell membrane or on some aspect of metabolic activity. The technique is less laborious and time-consuming than haemocytometry, when many samples have to be analysed simultaneously. Multicolour staining (double or triple) has been shown to identify a higher level of complexity and heterogeneity than one-parameter measurements (Attfield *et al.* 2000). Flow cytometry can also be used to follow enzyme activity of yeast populations on a single-cell level during the fermentation (Guldfeldt *et al.* 1998). Flow cytometric measurements of DNA distributions (i.e. cell cycle analysis) can be employed to monitor the growth of yeast during fermentation (Hutter 1997). DNA analysis is also applicable to the determination of the ploidy level of an unknown yeast strain (Hutter 1997).

19.6.4 Yeast characterisation

Strain differentiation

Each brewery should ideally be able to identify their culture yeast strains. Brewing yeasts belong to the *Saccharomyces sensu stricto* species complex together with other yeast strains relevant in the fermentation industry. The polyploid nature, the high genetic variability and the complexity of evolution of these yeasts make species definition troublesome (Rainieri *et al.* 2003).

According to current taxonomy, ale-brewing strains belong to the species *S. cerevisiae* while lager brewing yeasts belong to *S. pastorianus*, which in turn has been associated with the synonyms *S. carlsbergensis* and *S. monacensis* (Vaughan-Martini and Martini 1998). The problem with *S. pastorianus* is that the species is extremely heterogeneous, including strains of both hybrid and non-hybrid nature, and type strains maintained at different culture collections seem not to be identical (Rainieri *et al.* 2003). Electrophoretic karyotyping (Walmsley 1994), restriction fragment length polymorphism (RFLP) (Guillamón *et al.* 1994), PCR (de Barros Lopes *et al.* 1998), amplified fragment length polymorphism (AFLP) (de Barros Lopes *et al.* 1999) and simple sequence repeats typing (Richard *et al.* 1999) are the techniques most frequently used to differentiate yeasts at strain level.

Karyotyping involves the determination of chromosome size and number by pulsed field gel electrophoresis which allows size-separation of large DNA fragments (Lai *et al.* 1989). It is especially well suited for typing of *Saccharomyces sensu stricto* species which possess around 16 highly polymorphic chromosomes (Cardinali and Martini 1994). Karyotyping was first applied to yeasts as early as 1985, and since then it has established itself as a standard molecular method for differentiating ale, lager and wine yeast strains (Casey *et al.* 1990, Tornai-Lehoczki and Dlauchy 2000, Fernández-Espinar *et al.* 2001). Using karyotyping, Jespersen *et al.* (2000) could differentiate between almost all *S. cerevisiae* brewing contaminants and could separate them from culture yeast strains.

The patented AFLP™ technology was developed by Marc Zabeau and colleagues at KeyGene (Netherlands) (Zabeau 1992, Vos *et al.* 1995). The technique uses selective primers to amplify a subset of genomic restriction fragments which are then detected by electrophoresis. AFLP™ has been applied to the typing of brewers' yeast strains (Perpète *et al.* 2001). By using two restriction enzymes and a single primer pair, these authors could discriminate all studied lager and ale strains. Furthermore, a genetic marker specific to lager yeasts was identified.

Simple sequence repeats (SSRs) typing is based on the specific amplification of microsatellite repeats, which exhibit a substantial amount of length polymorphism due to, e.g., strand slipping during their replication (Richard *et al.* 1999). Unlike other strain differentiation techniques, it requires prior sequence information to design specific primers on open reading frames, which flank the microsatellites of interest. Microsatellite typing has been shown to be a powerful technique for typing of *S. cerevisiae* strains. In their comprehensive study, Hennequin *et al.* (2001) could differentiate 99% of 96 clinical, laboratory and industrial strains from each others and showed that genotypes were highly stable and fingerprints patterns reproducible. Rassmann and Leibhard (1999) could divide 14 lager yeast strains into five classes using six markers, whereas eight ale yeasts gave six different patterns. Results of microsatellite typing can be exchanged as quantitative data, and the analysis can be automated and multiplexed.

In mitochondrial DNA RFLP analysis (mtDNA-RFLP), differences in the sequence of mitochondrial DNA are studied using restriction enzymes (Querol *et al.* 1992, López *et al.* 2001). Wine yeast strains usually exhibit high levels of polymorphism in their mtDNA whereas brewery-related yeasts and distillers' strains are more homogeneous (Casey *et al.* 1990, Fernández-Espinar *et al.* 2001, Castrejon *et al.* 2002). PCR-fingerprinting with primers targeting repetitive elements, like intron or retrotransposon related sequences, have also proved useful in strain characterisation (Ness *et al.* 1993, de Barros Lopes *et al.* 1996, 1998, Fernández-Espinar *et al.* 2001).

The PCR method targeting Ty elements has recently been evaluated for identification of industrial *Saccharomyces* strains (Powell and Smart 2003). By using two primer combinations, all six strains could be discriminated from each other. A PCR method based on the variation in the number and position of introns in mitochondrial COX1 gene was introduced to the typing of *S. cerevisiae* strains (López *et al.* 2002). All 12 wine strains could be discriminated in 8 hours when four primers were used in a multiplex PCR. The method could be also used for typing of other yeast species.

Respiratory deficient mutants
Individual yeast cells become respiratory-deficient mutants or 'petite mutants' as a result of spontaneous mutation in their mitochondrial DNA (mtDNA). *rho*$^{-}$ refers to mutants with partly damaged mtDNA and *rho*^{0} to mutants which have completely lost their mtDNA. These mutations result in the inability of cells to respire normally and give a low yield of biomass in aerobic conditions in comparison to normal cell. The small colonies on agar plates give rise to the term 'petite'. The amount of petite mutants is normally in the range of 0.5–5 %, but this proportion is strain dependent. Factors known to induce petite mutants are high alcohol level, extreme temperatures, long storage time, recycling and starvation (Powell *et al.* 2000). The mtDNA mutations affect many cell properties such as the ability and rate of using various carbohydrates, surface structures and thus also flocculation. Cell death rate increases, causing lower overall cell age. As a result, fermentation may slow down and the beer flavour alters (Ernandes *et al.* 1993). A high percentage of respiratory-deficient cells in brewing yeast slurry poses a risk of excessive diacetyl, ester or fusel alcohol formation. The percentage of petite mutants can be easily estimated based on their inability to reduce tetrazolium dye to a coloured form (EBC 2005, IOB 1997).

19.7 Future trends

An increasing number of modern techniques allow simultaneous identification and detection of multiple target organisms in a single test. In the future, the key challenges in the methodology development are the improvement of sensitivity without using PCR amplification, incorporation of sample treatment and detection steps, lowering of reagent and instrument costs, at-line/remote detection, and standardisation and validation of the new methods. Ultimately, a low-cost, non-expert method allowing specific detection and quantification of multiple organisms at low levels in a matter of minutes is wanted. For the time being, a combination of various techniques needs to be employed (both phenotypic and genetic) for the detection and identification of harmful microbes in breweries.

19.7.1 Multiplex detection and identification

Microarrays are solid supports (such as glass) measuring a few square centimetres on which thousands or hundreds of thousands of unique probes, other DNA sequences or antibodies can be attached to a known location. They combine molecular biology, robot printing technology and matured readout systems. In a typical assay, target sequences are amplified by PCR, hybridised with specific probes on the array and detected on the basis of fluorescence or electronic signals (Call *et al* 2003). Electronic systems show more promise for simple application, since hybridisation does not require washing steps, the concept is cheap and it can be applied to remote detection (Umek *et al.* 2001). The sensitivity of the microarray assays is comparable to that of PCR, and the whole analysis can be performed in 1–2 days. Recently, an oligonucleotide microarray for the identification and differentiation of trichothecene-producing *Fusarium* species occurring on cereal grains was developed (Nicolaisen *et al.* 2005). DNA microarrays can also be applied to genotyping of microbes. Even though expectations of the microarray technology are high, there are still a lot of challenges relating to sample treatment, sensitivity, cross-reactions, probe accessibility and quantification. Luminex xMAP (www.luminexcorp.com) is a high-throughput liquid array system. It allows specific detection and quantification of up to 100 different analytes in a single well of microtitre plate, and thus around 10 000 analytes in one plate (Dunbar *et al.* 2003).

19.7.2 Transcript analysis techniques

Multiplexed and quantitative analysis of gene expression is particularly useful in monitoring the expression of a few or several dozens of genes. The methods are very powerful, meaning that various phenomena can be studied in a single assay, for example *Fusarium* toxin synthesis and barley gene expression during malting. Microarrays and chips allowing highly multiplexed DNA and RNA analysis and detection have become key tools in research, and it is only a question of time when they will be fully explored also in brewing sciences. The techniques, based on microarrays and chips, are technically demanding and produce unnecessary bioinformatics overload in routine applications. An alternative approach based on affinity capture was recently developed (Söderlund *et al.* 2004). The TRAC (transcript analysis with the aid of affinity capture) method is based on solution hybridisation between RNA and multiple fluorophore labelled probes of distinct sizes. Quantitative results can be obtained by the TRAC method (Satokari *et al.* 2005).

19.7.3 Biomarkers

Systems biology relates to the massive collection of data on all the biochemical processes in a living cell. The expression level of all the genes can be measured using DNA microarrays, a significant part of all proteins can be identified and quantified by proteomic methods, and a large number of the metabolites can be measured using combinations of chromatography and mass spectrometry. From this immense amount of data, a limited number of parameters can be identified which are predictive for a certain chain of events. Hence, in a process it is possible to follow the changes in a set of analytes which together indicate that a certain chain of events is about to start. In the future, it could be possible to better control biotechnical processes by monitoring these specific biomarkers (Söderlund *et al.* 2005).

19.7.4 Possibilities to predict microbiological spoilage

In addition to brewing yeast, there will be a certain amount of other micro-organisms present in the production stream as long as the process is not fully closed and aseptic. Of these organisms some can survive but then eventually die, whereas some grow and actually cause spoilage. The ability to grow and spoil the beer is based both on the intrinsic resistance of the beer itself and on many different characteristics of the microorganisms involved, including their genetic potential and physiological state. The need to develop a method able to predict when a microorganism present will grow and spoil the beer has been identified by major breweries (O'Sullivan and Vaughan 2005). Some of the methods described in this chapter, such as PCR and immunoassays, aim to detect and simultaneously identify those microbial species most likely to cause trouble. However, there are still many obstacles to overcome before the powerful research tools described above will allow us to predict when a particular beer-spoilage organism is actually going to spoil a particular brand of beer.

19.8 Sources of further information

The recent edition of the book *Brewing Microbiology* provides comprehensive information regarding practically all relevant microbiological aspects involved in brewing (Priest and Campbell 2003). Detailed advice on different sampling and analysis methods can be found in the recent edition of EBC *Analytica Microbiologica* (2005) or in the microbiological methods published by ASBC. For classical methods and colour photos of colonies and microscopic views, consult the manual by Back (1994), now also available in English. During the years, many reviews have been published regarding the alternative micro-biological approaches intended for microbiological control in breweries, see Barney and Kot (1992), Dowhanick (1994), Storgårds *et al.* (1998a), Quain (1999) and Russell and Stewart (2003). The various methods used to estimate brewing yeast viability and vitality are discussed in detail by Heggart *et al.* (2000).

The occurrence of mycotoxins in barley and in malt is reviewed by Flannigan (2003). The taxonomy of brewing yeast and other yeasts of industrial importance belonging to the *Saccharomyces cerevisiae sensu stricto* complex species is reviewed by Rainieri *et al.* (2003) and by Vaughan-Martini (2003). For a review on the relationship between beer-spoilage ability of bacteria and hop resistance, see Sakamoto and Konings (2003).

19.9 References

AL-HOLY M A, LIN M, CAVINATO A G and RASCO B A (2006), 'The use of Fourier transform infrared spectroscopy to differentiate *Escherichia coli* O157:H7 from other bacteria inoculated into apple juice', *Food Microbiol*, 23, 162–168.

AMANN R I, LUDWIG W and SCHLEIFER K-H (1995), 'Phylogenetic identification and *in situ* detection of individual microbial cells without cultivation', *Microbiol Rev*, 59, 143–169.

AMERICAN SOCIETY OF BREWING CHEMISTS (1984), 'Report of subcommittee on microbiology', *J Am Soc Brew Chem*, 42, 132–135.

AMERICAN SOCIETY OF BREWING CHEMISTS (2001), 'Report of subcommittee on citrate-buffered methylene violet stain for yeast viability', *J Am Soc Brew Chem*, 59, 232–233.

AMERICAN SOCIETY OF BREWING CHEMISTS (2003), 'Report of subcommittee on methods for measurement of yeast vitality', *J Am Soc Brew Chem*, 61, 239–240.

ATTFIELD P V, KLETAS S, VEAL D A, VAN ROOIJEN R and BELL P J L (2000), 'Use of flow cytometry to monitor cell damage and predict fermentation activity of dried yeast', *J Appl Microbiol*, 89, 207–214.

BACK W (1994), *Farbatlas und Handbuch der Getränkebiologie. Teil 1. Kultivierung/ Methoden, Brauerei, Winzerei.* Nürnberg, Verlag Hans Carl,165 p.

BACK W (1997), 'Technical and technological prerequisites for "cold sterile" bottling', *Brauwelt Int*, 15, 192–201.

BACK W and PÖSCHL P (1998), 'Bypass-Membranfiltration (BM-System) – Verbesserung des Spurennachweises nach der Filtration', *Brauwelt*, 138, 2312–2315.

BARNETT J A, PAYNE R W and YARROW D (eds) (2000), *Yeasts: Characteristics and Identification*, 2nd edn, Cambridge University Press, Avon.

BARNEY M and KOT E (1992), 'A comparison of rapid microbiological methods for detecting beer spoilage microorganisms', *Mast Brew Ass Amer Tech Quart*, 29, 91–95.

BARNEY M, VOLGYI A, NAVARRO A and RYDER D (2001), 'Riboprinting and 16S rRNA gene sequencing for identification of brewery *Pediococcus* isolates', *Appl Environ Microbiol*, 67, 553–560.

BAU S, MAUCHER H, ZINECKER H, LOEFFLER T and BREITENSTEIN A (2005), 'A rapid enzyme-linked test for the detection and identification of microbial brewery contaminants', *Proc. 30th EBC Congr.*, Prague, Nürnberg, Fachverlag Hans Carl [CD-ROM], 123, 1057–1058..

BENDIAK D (2000), 'Review of metabolic activity tests and their ability to predict fermentation performance', in Smart K, *Brewing Yeast Fermentation Performance*, Oxford, Blackwell Science, 34–45.

BERTIN B, BROUX O and VAN HOEGAERDEN M (1990), 'Flow cytometric detection of yeasts by *in situ* hybridization with a fluorescent ribosomal RNA probe', *J Microbiol Methods* 12, 1–12.

BEVERLY M B, BASILE F and VOORHEES K J (1997), 'Fatty acid analysis of beer spoiling microorganisms using pyrolysis mass spectrometry', *J Am Soc Brew Chem*, 55, 79–82.

BIEGALA I C, NOT F, VAULOT D and SIMON N (2003), 'Quantitative assessment of picoeukaryotes in the natural environment by using taxon-specific oligonucleotide probes in association with tyramide signal amplification-fluorescence *in situ* hybridization and flow cytometry', *Appl Environ Microbiol*, 69, 5519–5529.

BLUHM B H, COUSIN M A and WOLOSHUK C P (2004), 'Multiplex real-time PCR detection of fumonisin-producing and trichothecene-producing groups of *Fusarium* species', *J Food Prot*, 67, 536–543.

BOUIX M and LEVEAU J-Y (2001), 'Rapid assessment of yeast viability and yeast vitality during alcoholic fermentation', *J Inst Brew*, 107, 217–225.

BOULTON C A (1996), 'A critical assessment of yeast vitality testing', *Ferment*, 9, 222–226.

BOYD A R, GUNASEKERA T S, ATTFIELD P V, SIMIC K, VINCENT S F and VEAL D A (2003), 'A flow-cytometric method for determination of yeast viability and cell number in a brewery', *FEMS Yeast Research*, 3, 11–16.

BRANDL A and GEIGER E (2003), 'Microbiological quality control in breweries by PCR – a demonstration of applied methods', *Proc. 29th EBC Congr.*, Dublin, Nürnberg, Fachverlag Hans Carl [CD-ROM], 108, 1094–1104.

BRAUNE A and EIDTMANN A (2003), 'First experiences using realtime-PCR as a rapid detection method for brewery process control at Beck & Co', *Proc. 29th EBC Congr.*, Dublin, Nürnberg, Fachverlag Hans Carl [CD-ROM], 112, 1128–1135.

BREEUWER P, DROCOURT L J-L, ROMBOUTS F M and ABEE T (1994), 'Energy-dependent, carrier-mediated extrusion of carboxyfluorescein from *Saccharomyces cerevisiae* allows rapid assessment of cell viability by flow cytometry', *Appl Environ Microbiol*, 60, 1467–1472.

BREEUWER P, DROCOURT L J-L, BUNSCHOTEN N, ZWIETERING M H, ROMBOUTS F M and ABEE T (1995), 'Characterization of uptake and hydrolysis of fluorescein diacetate and carboxyfluorescein diacetate by intracellular esterases in *Saccharomyces cerevisiae*, which result in accumulation of fluorescent product', *Appl Environ Microbiol*, 61, 1614–1619.

CALL D R, BORUCKI M K and LOGE F J (2003), 'Detection of bacterial pathogens in environmental samples using DNA microarrays', *J Microbiol Methods*, 53, 235–243.

CARDINALI G and MARTINI A (1994), 'Electrophoretic karyotypes of authentic strains of the *sensu stricto* group of the genus *Saccharomyces*', *Int J Syst Bacteriol*, 44, 791–797.

CASEY G P, PRINGLE A T and ERDMANN P A (1990), 'Evaluation of recent techniques used to identify individual strains of *Saccharomyces* yeasts', *J Am Soc Brew Chem*, 48, 100–106.

CASTREJON F, CODON A C, CUBERO B and BENITEZ T (2002), 'Acetaldehyde and ethanol are responsible for mitochondrial DNA (mtDNA) restriction fragment length polymorphism (RFLP) in flor yeasts', *Syst Appl Microbiol*, 25, 462–467.

COCKERILL III F R and UHL J R (2002), 'Applications and challenges of real-time PCR for the clinical microbiology laboratory', in Reischl U, Wittwer C and Cockerill III F R, *Rapid Cycle Real-Time PCR Methods and Applications. Microbiology and Food Analysis*, Berlin and Heidelberg, Springer Verlag, 3–27.

DAINTY R H (1996), 'Chemical/biochemical detection of spoilage', *Int J Food Microbiol*, 33, 19–33.

DAVEY H M and KELL D B (1996), 'Flow cytometry and cell sorting of heterogeneous microbial populations: the importance of single-cell analyses', *Microbiol Rev* 60, 641–696.

DE BARROS LOPES M, SODEN A, HENSCHKE P A and LANGRIDGE P (1996), 'PCR differentiation of commercial yeast strains using intron splice site primers', *Appl Environ Microbiol*, 62, 4514–4520.

DE BARROS LOPES M, SODEN A, MARTENS A L, HENSCHKE P A and LANGRIDGE P (1998), 'Differentiation and species identification of yeasts using PCR', *Int J System Bacteriol*, 48, 279–286.

DE BARROS LOPES M, RAINIERI S, HENSCHKE P A and LANGRIDGE P (1999), 'AFLP fingerprinting for analysis of yeast genetic variation', *Int J System Bacteriol*, 49, 915–924.

DEERE D, SHEN J, VESEY G, BELL P, BISSINGER P and VEAL D (1998), 'Flow cytometry and cell sorting for yeast viability assessment and cell selection', *Yeast*, 14, 147–160.

DONHAUSER S, WAGNER D and GEIGER E (1993), 'Biogenic amines. Significance, occurrence and assessment', *Brauwelt Int*, 11, 100–107.

DOWHANICK T M (1994), 'Advances in yeast and contaminant determination: The future of the so called "rapid" methods', *Proc. J. De Clerck Chair VI*, Leuven, 11–15 September.

DUNBAR S A, VANDER ZEE C A V, OLIVER K G, KAREM K L and JACOBSON J W (2003), 'Quantitative, multiplexed detection of bacterial pathogens: DNA and protein applications of the Luminex LabMAP system', *J Microbiol Methods*, 53, 245–252.

DUNCAN S, GLOVER L A, KILLHAM K and PROSSER J I (1994), 'Luminescence-based detection of activity of starved and viable but nonculturable bacteria', *Appl Environ Microbiol*, 60, 1308–1316.

EBC *Analytica Microbiologica* (2005), 2nd edn, Nürnberg, Fachverlag Hans Carl, CD-ROM, ISBN 3-418-00780-5.

EDWARDS C, PORTER J and WEST M (1996), 'Fluorescent probes for measuring physiological fitness of yeast', *Ferment* 5, 288–293.

EGER C, DONHAUSER S and WINNEWISSER W (1995), 'Rapid immunoassay of beer spoilage. Possibilities and limits of flow cytometry after immunofluorescence labelling', *Brauwelt Int*, 13, 164–171.

EIDTMANN A, GROMUS J and BELLMER H G (1998), 'Mikrobiologische Qualitätssicherung: Der spezifische Nachweis von bierschädlichen Hefen im hefefreien Bereich', *Monatsschr Brauwiss*, 51, 141–148.

ERCOLINI D (2004), 'PCR-DGGE fingerprinting: novel strategies for detection of microbes in food', *J Microbiol Methods*, 56, 297–314.

ERNANDES J R, WILLIAMS J W, RUSSELL I and STEWART G G (1993), 'Respiratory deficiency in brewing yeast strain effects on fermentation, flocculation, and beer flavor components', *J Am Soc Brew Chem*, 51, 16–20.

FERNÁNDEZ-ESPINAR M T, LÓPEZ V, RAMÓN D, BARTRA E and QUEROL A (2001), 'Study of the authenticity of commercial wine yeast strains by molecular techniques', *Int J Food Microbiol*, 70, 1–10.

FLANNIGAN B (2003), 'The microbiota of barley and malt', in Priest F G and Campbell I, *Brewing Microbiology*, 3rd edn, New York, Kluwer Academic/Plenum Publishers, 113–180.

FOSTER A (1996), 'Evaluation of impedance method for monitoring microbiological quality of cleaning systems', *J Am Soc Brew Chem*, 54, 76–77.

FUNAHASHI W, SUZUKI K, OHTAKE Y and YAMASHITA H (1998), 'Two novel beer-spoilage *Lactobacillus* species isolated from breweries', *J Am Soc Brew Chem*, 56, 64–69.

FUNG D Y C (1994), 'Rapid methods and automation in food microbiology: a review', *Food Reviews Int*, 10, 357–375.

GARES S L, WHITING M S, INGLEDEW W M and ZIOLA B (1993), 'Detection and identification of *Pectinatus cerevisiiphilus* using surface-reactive monoclonal antibodies in a membrane filter-based fluoroimmunoassay', *J Am Soc Brew Chem*, 51, 158–163.

GUILLAMÓN J M, BARRIO E, HUERTA T and QUEROL A (1994), 'Rapid characterization of wild and collection strains of the genus *Zygosaccharomyces* according to mitochondrial DNA patterns', *FEMS Microbiol Letters*, 147, 267–272.

GULDFELDT L U, ARNEBORG N, SIEGUMFELDT H and JESPERSEN L (1998), 'Relationship between yeast cell proliferation and intracellular esterase activity during brewing fermentations', *J Inst Brew* 104, 333–338.

GUTTERIDGE C S and PRIEST F G (1996), 'Methods for the rapid identification of microorganisms', in Priest F G and Campbell I, *Brewing Microbiology*, 2nd edn, London, Chapman & Hall, 237–270.

HAGE T and WOLD K (2003), 'Practical experiences on the combat of a major *Pectinatus* and *Megasphera* infection with the help of TagMan realtime-PCR', *Proc. 29th EBC Congr.*, Dublin, Nürnberg, Fachverlag Hans Carl [CD-ROM], 114, 1145–1148.

HAIKARA A (1984), Beer spoilage organisms. Occurrence and detection with particular reference to a new genus *Pectinatus*, *Technical Research Centre of Finland, Publications 14*. 47 p. + app. 58 p.

HAIKARA A (1985), 'Detection of anaerobic, gram-negative bacteria in beer', *Monatsschr Brauwiss*, 38, 239–243.

HAIKARA A and BOIJE-BACKMAN S (1982), 'The use of optical brighteners in the detection of bacterial and yeast contaminants in beer', *Brauwiss*, 35, 113–117.

HAIKARA A and ENARI T-M (1975), 'The detection of wild yeast contaminants by the immunofluorescence technique', *Proc. 15th EBC Congr.*, Nice, 363–375.

HAIKARA A and HELANDER, I (2002), '*Pectinatus*, *Megasphaera* and *Zymophilus*', in Dworkin M, *The Prokaryotes: An Evolving Electronic Database for the Microbiological Community*, 3rd edn, *http://link.springer-ny.com/link/service/books/10125/.* Accessed 8 February 2005.

HAIKARA A, JUVONEN R, KOIVULA T, SCHUHBECK T, BRANDL A, VOGESER G and STORGÅRDS E (2003), 'Microbiological quality control in breweries by PCR The BREWPROC approach', Proc. 29th EBC Congr., Dublin, Nürnberg, Fachverlag Hans Carl [CD-ROM], 102, 1036–1046.

HAMMOND J (1996), 'Microbiological techniques to confirm CIP effectiveness', *Brewer*, 82, 332–338.

HAMMOND J, BRENNAN M and PRICE A (1999), 'The control of microbial spoilage of beer', *J Inst Brew*, 105, 113–120.

HARRIS C M, TODD R W, BUNGARD S J, LOVITT R W, MORRIS J G and KELL D B (1987), 'Dielectric permittivity of microbial suspensions at radio frequencies: a novel method for the real-time estimation of microbial biomass', *Enzyme Microb Technol*, 9, 181–186.

HAYASHI N, ITO M, HORI S, YOKOYAMA A, TAGUCHI H and OHKOCHI M (2003), 'An application of RAPD PCR to the identification of *Lactobacillus brevis* strains capable of growing in beer', *Proc. 29th EBC Congr.*, Dublin, Nürnberg, Fachverlag Hans Carl [CD-ROM], 105, 1065–1072.

HEGGART H, MARGARITIS A, STEWART R J, PILKINGTON H, SOBCZAK J and RUSSELL I (2000), 'Measurement of brewing yeast viability and vitality: a review of methods', *Mast Brew Ass Amer Tech Quart*, 37, 409–430.

HENNEQUIN C, THIERRY A, RICHARD G F, LECOINTRE G, NGUYEN H V, GAILLARDIN C and DUJON B (2001), 'Microsatellite typing as a new tool for identification of *Saccharomyces cerevisiae* strains', *J Clin Microbiol*, 39, 551–559.

HUTTER K-J (1994), 'Multi-colour flow cytometric analysis', *Brauwelt Int*, 12, 334–346.

HUTTER K-J (1997), 'Biomonitoring of yeasts in production. A concept for on-line monitoring of yeast during propagation, fermentation and maturation', *Brauwelt Int*, 15, 116–123.

ILBERG V, SCHWILL-MIEDANER A and SOMMER K (1995), 'Mikrobiologische Probleme an Schankanlagen', *Brauind*, 80, 992–995.

IMAI T, NAKAJIMA I and OHNO T (1994), 'Development of a new method for evaluation of yeast vitality by measuring intracellular pH', *J Am Soc Brew Chem*, 52, 5–8.

INGLEDEW W M (1979), 'Effect of bacterial contamination on beer. A review', *J Am Soc Brew Chem*, 37, 145–150.

INSTITUTE OF BREWING (1997), *Methods of Analysis, Microbiological*, vol. 2, London.

ISERENTANT D, GEENENS W and VERACHTERT H (1996), 'Titrated acidification power: a simple and sensitive method to measure yeast vitality and its relation to other vitality measurements', *J Am Soc Brew Chem*, 54, 110–114.

IZQUIERDO-PULIDO M, HERNÁNDEZ-JOVER T, MARINÉ-FONT A and VIDAL-CAROU M C (1996), 'Biogenic amines in European beers', *J Agric Food Chem*, 44, 3159–3163.

JAKOBSEN M and LILLIE A (1984), 'Rapid methods in microbiological quality control: Practical experience with epifluorescent techniques', *Proc. 18th Inst. Brew. Conv.*, Adelaide, Peacock Publ., Kent Town, 545–552.

JASPERS E, NITZSCHE F and KLUPSCH H (2003), 'Quick but not dirty: alternative methods for monitoring the hygienic state in breweries', *Proc. 29th EBC Congr.*, Dublin, Nürnberg, Fachverlag Hans Carl [CD-ROM], 118, 1181–1186.

JESPERSEN L and JAKOBSEN M (1996), 'Specific spoilage organisms in breweries and laboratory media for their detection', *Int J Food Microbiol*, 33, 139–155.

JESPERSEN L, LARSEN S and JAKOBSEN M (1993), 'Flow cytometric detection of wild yeast in lager breweries', *Int J Food Microbiol*, 17, 321–328.

JESPERSEN L, VAN DER KÜHLE A and PETERSEN K M (2000), 'Phenotypic and genetic diversity of *Saccharomyces* contaminants isolated from lager breweries and their phylogenetic relationship with brewing yeasts', *Int J Food Microbiol*, 60, 43–53.

JUVONEN R, KOIVULA T and HAIKARA A (2003), 'PCR detection of Gram-negative brewery contaminants present state', *Proc. 29th EBC Congr.*, Dublin, Nürnberg, Fachverlag Hans Carl [CD-ROM], 103, 1047–1056.

KELL D B, MARKX G H, DAVEY C L and TODD R W (1990), 'Real-time monitoring of cellular biomass: methods and applications', *Trends Anal Chem*, 9, 190–194.

KIEHNE M, BERGHOF-JÄGER K, FANDKE M and GRÖNEWALD C (2003), 'Detection of beer spoilage organisms by real-time PCR', *Proc. 29th EBC Congr.*, Dublin, Nürnberg, Fachverlag Hans Carl [CD-ROM], 106, 1073–1080.

KILGOUR W J and DAY A (1983), 'The application of new techniques for the rapid determination of microbial contamination in brewing', *Proc. 19th EBC Congr.*, London, 177–184.

KOIVULA T, JUVONEN R, HAIKARA A and SUIHKO M-L (2006), 'Characterization of the brewery spoilage bacterium *Obesumbacterium proteus* by automated ribotyping and development of PCR methods for its biotype 1', *J Appl Microbiol*, 100, 398–406.

KOSSE D, SEILER H, AMANN R, LUDWIG W and SCHERER S (1997), 'Identification of yoghurt-spoiling yeasts with 18S rRNA-targeted oligonucleotide probes', *System Appl Microbiol*, 20, 468–480.

KRONLÖF J (1991), 'Evaluation of a capacitance probe for determination of viable yeast biomass', *Proc. 23rd EBC Congr.*, Lisbon, 233–240.

KÜMMERLE M, SCHERER S and SEILER H (1998), 'Rapid and reliable identification of food-borne yeasts by Fourier-transform infrared spectroscopy', *Appl Environ Microbiol*, 64, 2207–2214.

KURTZMAN C P and FELL J W (1998), 'Definition, classification and nomenclature of the yeasts', in Kurtzman C P and Fell J W, *The Yeasts, a Taxonomic Study*, Amsterdam, Elsevier, 3–5.

KURTZMAN C P and ROBNETT C J (2003), 'Phylogenetic relationships among yeasts of the "*Saccharomyces* complex" determined from multigene sequence analysis', *FEMS Yeast Research*, 3, 417–432.

KURTZMAN C P, BOEKHOUT T, ROBERT V, FELL J W and DEAK T (2003) 'Methods to identify yeasts', in Boekhout T and Robert V, *Yeasts in Food, Beneficial and Detrimental Aspects*, Hamburg, Behr's Verlag, 69–121.

KYRIAKIDES A L and PATEL P D (1994), 'Luminescent techniques for microbiological analysis of foods', in Patel P, *Rapid Analysis Techniques in Food Microbiology*, London, Blackie Academic & Professional, 196–231.

LAI E, BIRREN B W, CLARK S M, SIMON M I and HOOD L (1989), 'Pulsed field gel electrophoresis', *Biotechniques*, 7, 34–42.

LAITILA A, WILHELMSON A, SARLIN T, JUVONEN R and STORGÅRDS E (2005), 'Novel strategies for studying the complex microbial community in barley and in malting', *Proc. 30th EBC Congr.*, Prague, Nürnberg, Fachverlag Hans Carl [CD ROM], 141, 1168–1174.

LERICHE V and CARPENTIER B (1995), 'Viable but nonculturable *Salmonella typhimurium* in single- and binary-species biofilms in response to chlorine treatment', *J Food Prot*, 58, 1186–1191.

LÓPEZ V, QUEROL A, RAMÓN D and FERNÁNDEZ-ESPINAR M T (2001), 'A simplified procedure to analyse mitochondrial DNA from industrial yeasts', *Int J Food Microbiol*, 68, 75–81.

LÓPEZ V, FERNÁNDEZ-ESPINAR M T, BARRIO E, RAMÓN D and QUEROL A (2002), 'A new PCR-based method for monitoring inoculated wine fermentations', *Int J Food Microbiol*, 81, 63–71.

MALACRINÓ P, ZAPPAROLI G, TORRIANI S and DELLAGLIO F (2001), 'Rapid detection of viable yeasts and bacteria in wine by flow cytometry', *J Microbiol Methods*, 45, 127–134.

MARCH C, MANCLÚS J J, ABAD A, NAVARRO A and MONTOYA A (2005), 'Rapid detection and counting of viable beer-spoilage lactic acid bacteria using a monoclonal chemiluminescence enzyme immunoassay and a CCD camera', *J Immunol Methods*, 303, 92–104.

McCAIG R (1990), 'Evaluation of the fluorescent dye 1-anilino-8-naphthalene sulfonic acid for yeast viability determination', *J Am Soc Brew Chem*, 48, 22–25.

McKILLIP J L and DRAKE M (2004), 'Review: Real-time nucleic acid-based detection methods for pathogenic bacteria in food', *J Food Prot*, 67, 823–832.

MILLER R and GALSTON G (1989), 'Rapid methods for the detection of yeast and *Lactobacillus* by ATP bioluminescence', *J Inst Brew*, 95, 317–319.

MISHIMA K, MIMURA A and TAKAHARA Y (1991), 'On-line monitoring of cell concentrations during yeast cultivation by dielectric measurements', *J Ferment Bioeng*, 72, 296–299.

MOCHABA F, O'CONNOR-COX E S C and AXCELL B C (1998), 'Practical procedures to measure yeast viability and vitality prior to pitching', *J Am Soc Brew Chem*, 56, 1–6.

MOTER A and GÖBEL U B (2000), 'Fluorescence *in situ* hybridization (FISH) for direct visualization of microorganisms', *J Microbiol Methods*, 41, 85–112.

MOTOYAMA Y, OGATA T and SAKAI K (1998), 'Characterization of *Pectinatus cerevisiiphilus* and *P. frisingensis* by ribotyping', *J Am Soc Brew Chem*, 56, 19–23.

MULLIS K, FALOONA F, SCHARF S, SAIKI R, HORN G and ERLICH H (1986), 'Specific enzymatic amplification of DNA *in vitro*: the polymerase chain reaction Cold Spring Harbor Symp', *Quant Biol*, 51, 263–273.

MUYZER G (1999), 'DGGE/TGGE a method for identifying genes from natural ecosystems', *Current Opinion in Microbiology*, 2, 317–322.

MUYZER G and SMALLA K (1998), 'Application of denaturing gradient gel electrophoresis (DGGE) and temperature gradient gel electrophoresis (TGGE) in microbial ecology', *Antonie van Leeuwenhoek*, 73, 127–141.

NAKAKITA Y, TAKAHASHI T, TSUCHIYA Y, WATARI J and SHINOTSUKA K (2002), 'A strategy for detection of all beer-spoilage bacteria', *J Am Soc Brew Chem*, 60, 63–67.

NAKAKITA Y, MAEBA H and TAKASHIO M (2003), 'Grouping of *Lactobacillus brevis* strains using the *gyrB* gene', *J Am Soc Brew Chem*, 61, 157–160.

NAUMOV G I, JAMES A, NAUMOVA E S, LOUIS E J and ROBERTS I N (2000) 'Three new species in the *Saccharomyces sensu stricto* complex: *Saccharomyces cariocanus*, *Saccharomyces kudriavzevii* and *Saccharomyces mikatae*', *Int J Syst Evol Microbiol*, 50, 1931–1942.

NESS F, LAVALLEE F, DUBOURDIEU D, AIGLE M and DULAU L (1993), 'Identification of yeast strains using the polymerase chain reaction', *J Sci Food Agric*, 62, 89–94.

NICHOLSON P, CHANDLER E, DRAEGER R C, GOSMAN N E, SIMPSON D R, THOMSETT M and WILSON A H (2003), 'Molecular tools to study epidemiology and toxicology of fusarium head blight of cereals', *Eur J Plant Pathol*, 109, 691–703.

NICOLAISEN M, JUSTESEN A F, THRANE U, SKOUBOE P and HOLMSTRØM K (2005), 'An oligonucleotide microarray for the identification and differentiation of trichothecene producing and non-producing *Fusarium* species occurring on cereal grain', *J Microbiol Methods*, 62, 57–69.

O'CONNOR-COX E S C, MOCHABA F M, LODOLO E J, MAJARA M and AXCELL B (1997), 'Methylene blue staining: use at your own risk', *Mast Brew Ass Amer Tech Quart*, 34, 306–312.

O'SULLIVAN T F and VAUGHAN A (2005), 'Microbiological methods in brewing: past, present and future', *Proc. 30th EBC Congr.*, Prague, Nürnberg, Fachverlag Hans Carl [CD ROM], 128, 1083–1084.

PARK S, WOROBO R W and DURST R A (1999), '*Escherichia coli* O157:H7 as an emerging foodborne pathogen: A literature review', *Crit Rev Food Sci Nutr*, 39, 481–502.

PARKER M J (1989), 'The application of automated detection and enumeration of microcolonies using optical brighteners and image analysis to brewery microbiological control', *Proc. 22nd EBC Congr.*, Zürich, 545–552.

PERPÈTE P, VAN CUTSEM P, BOUTTE C, COLSON-CORBISIER A-M and COLLIN S (2001), 'Amplified fragment-length polymorphism, a new method for the analysis of brewer's yeast DNA polymorphism', *J Am Soc Brew Chem*, 59, 195–200.

PETTIPHER G L (1991), 'Preliminary evaluation of flow cytometry for the detection of yeasts in soft drinks', *Lett Appl Microbiol*, 12, 109–112.

POT B, VANDAMME P and KERSTERS K (1994), 'Analysis of electrophoretic whole-organism protein fingerprints', in Goodfellow M and O'Donnell A G, *Chemical Methods in Prokaryotic Systematics*, Chichester, John Wiley & Sons, 493–521.

POWELL C and SMART K (2003), 'A comparison of new and traditional methods for the detection of brewery contaminants', *Proc. 29th EBC Congr.*, Dublin, Nürnberg, Fachverlag Hans Carl [CD-ROM], 115, 1149–1162.

POWELL C D, QUAIN D E and SMART K A (2000), 'The impact of media composition and petite mutation on the longevity of a polyploid brewing yeast strain', *Lett Appl Microbiol*, 31, 46–51.

PREST A G, HAMMOND J R M and STEWART G S A B (1994), 'Biochemical and molecular characterization of *Obesumbacterium proteus*, a common contaminant of brewing yeasts', *Appl Environ Microbiol*, 60, 1635–1640.

PRIEST F G (1996), 'Gram-positive brewery bacteria', in Priest F G and Campbell I, *Brewing Microbiology*, 2nd edn, London, Chapman & Hall, 127–161.

PRIEST F G (2003), 'Rapid identification of microorganisms', in Priest F G and Campbell I, *Brewing Microbiology*, 3rd edn, New York Kluwer Academic/Plenum Publishers, 305–328.

PRIEST F G and CAMPBELL I (2003), *Brewing Microbiology*, 3rd edn, New York, Kluwer Academic/Plenum Publishers, 399 p.

QUAIN D (1999), 'The "new" microbiology', *Proc. 27th EBC Congr.*, Cannes, 239–248.

QUEROL A, BARRIO E and RAMÓN D (1992), 'A comparative study of different methods of yeast strain characterization', *System Appl Microbiol*, 15, 439–446.

RAINIERI S, ZAMBONELLI C and KANEKO Y (2003), Review. '*Saccharomyces sensu stricto*: Systematics, genetic diversity and evolution', *J Biosci Bioeng*, 96, 1–9.

RASSMANN K and LEIBHARD S (1999), 'Hochauflösende Klassifizierung von Hefestämmen mit genetischen Methoden. Mikrosatelliten-Typisierung von *Saccharomyces cerevisiae* und *carlsbergensis* Stämmen', *Monatsschr Brauwiss*, 52, 152–156.

REISCHER G H, LEMMENS M, FARNLEITNER A, ADLER A and MACH R L (2004), 'Quantification of *Fusarium graminearum* in infected wheat by species specific real-time PCR applying a TaqMan probe', *J Microbiol Methods*, 59, 141–146.

RICHARD G F, HENNEQUIN A, THIERRY A and DUJON B (1999), 'Trinucleotide repeats and other microsatellites in yeast', *Res Microbiol*, 150, 589–602.

RINCK M and WACKERBAUER K (1987), 'Detection and serological classification of lactic acid bacteria, especially *L. lindneri*', *Monatsschr Brauwiss*, 40, 324–327.

RODRIGUES P G, BARROS A A, RODRIGUES J A, FERREIRA A A, GONÇALVES C and HAMMOND J R M (2003), 'Yeast quality control in industrial brewing process using vitaltitration, a new method for vitality determination', *Proc. 29th EBC Congr.*, Dublin, Nürnberg, Fachverlag Hans Carl [CD-ROM], 57, 618–626.

RUSSELL I and STEWART R (2003), 'Rapid detection and identification of microbial spoilage', in Priest F G and Campbell I, *Brewing Microbiology*, 3rd edn, New York, Kluwer Academic/Plenum Publishers, 267–304.

SAKAMOTO K and KONINGS W N (2003), 'Beer spoilage bacteria and hop resistance', *Int J Food Microbiol*, 89, 105–124.

SAMI M, IKEDA M and YABUUCHI S (1994), 'Evaluation of the alkaline methylene blue staining method for yeast activity determination', *J Ferment Bioeng*, 78, 212–216.

SAMI M, YAMASHITA H, KADOKURA H, YODA K and YAMASAKI M (1997), 'A new and rapid method for determination of beer-spoilage ability of lactobacilli', *J Am Soc Brew Chem*, 55, 137–140.

SARLIN T, YLI-MATTILA T, JESTOI M, RIZZO A, PAAVANEN-HUHTALA S and HAIKARA A (2006), 'Real-time PCR for quantification of toxigenic *Fusarium* species in barley and malt, *Eur J Plant Pathol*, 114, 371–380.

SATOKARI R, JUVONEN R, MALLISON K, VON WRIGHT A and HAIKARA A (1998), 'Detection of beer spoilage bacteria *Megasphaera* and *Pectinatus* by polymerase chain reaction and colorimetric microplate hybridization', *Int J Food Microbiol*, 45, 119–127.

SATOKARI R, MATTILA-SANDHOLM T and SUIHKO M-L (2000), 'Identification of pediococci by ribotyping', *J Appl Microbiol*, 88, 260–265.

SATOKARI R M, KATAJA K and SÖDERLUND H (2005), 'Multiplexed quantification of bacterial 16S rRNA by solution hybridization with oligonucleotide probes and affinity capture', *Microbial Ecology*, 50, 120–127.

SCHWILL-MIEDANER A and EICHERT U (1998), 'Methodenentwicklung zur Beurteilung des Hygienestatus von Schankanlagen', *Monatsschr Brauwiss*, 51, 104–110.

SCHWILL-MIEDANER A, ILBERG V and SOMMER K (1996), 'Der Hygienestatus von Schankanlagen', *Brauwelt*, 136, 395–400.

SEMANCHEK J J and GOLDEN D A (1996), 'Survival of *Escherichia coli* O157:H7 during fermentation of apple cider', *J Food Prot*, 59, 1256–1259.

SIMPSON W J, HAMMOND J R M, THURSTON P A and KYRIAKIDES A L (1989), 'Brewery process control and the role of "instant" microbiological techniques', *Proc. 22nd EBC Congr.*, Brussels, Oxford, IRL Press, 663–674.

SMART K A, CHAMBERS K M, LAMBERT I, JENKINS C and SMART C A (1999), 'Use of methylene violet staining procedures to determine yeast viability and vitality', *J Am Soc Brew Chem*, 57, 18–23.

SÖDERLUND H, KATAJA K, SATOKARI R and TAKKINEN K (2004), 'Method and test kit for quantitative determination of polynucleotides in a mixture'. Int. Pat. Appl. No. PCT/FI01/00544.

SÖDERLUND H, ORESIC M, KATAJA K, SATOKARI R, PULLI T, TAKKINEN K and HOME S (2005), 'Biomarkers, a new tool for process control in malting and brewing', *Proc. 30th EBC Congr.*, Prague, Nürnberg, Fachverlag Hans Carl [CD ROM], 134, 1109–1114.

STORGÅRDS E (1997), 'Microbiological quality of draught beer is there reason for concern?', Eur. Brew. Conv. Symp. Draught Beer, Packaging & Dispense, Edinburgh, September 1996. *EBC Monograph XXV*, Nürnberg, Verlag Hans Carl, 92–103.

STORGÅRDS E (2000), 'Process hygiene control in beer production and dispensing', *VTT Publications 410*, Espoo, Finland, 105 p. + app. 66 p. ISBN 951-38-5559-7.

STORGÅRDS E and HAIKARA A (1996), 'ATP bioluminescence in the hygiene control of draught beer dispense systems', *Ferment*, 9, 352–360.

STORGÅRDS E, JUVONEN R, VANNE L and HAIKARA A (1998a), 'Detection methods in process and hygiene control', Eur. Brew. Conv. Symp. Quality Issues and HACCP, Stockholm, October 1997. *EBC Monograph XXVI*, Nürnberg, Verlag Hans Carl, 95–107.

STORGÅRDS E, POT B, VANHONACKER K, JANSSENS D, BROOMFIELD P L E, BANKS J G and SUIHKO M-L (1998b), 'Detection and identification of *Lactobacillus lindneri* from brewery environments', *J Inst Brew*, 104, 47–54.

STORGÅRDS E, TAPANI K, HARTWALL P, SALEVA R and SUIHKO M-L (2006), 'Microbial attachment and biofilm formation in brewery bottling plants', *J Am Soc Brew Chem*, 64, 8–15.

SUIHKO M-L and HAIKARA A (2001), 'Characterization of *Pectinatus* and *Megasphaera* strains by automated ribotyping', *J Inst Brew*, 107, 175–184.

SUIHKO M-L and STACKEBRANDT E (2003), 'Identification of aerobic mesophilic bacilli isolated from board and paper products containing recycled fibres', *J Appl Microbiol*, 94, 25–34.

TAKAHASHI T, NAKAKITA Y, WATARI J and SHINOTSUKA K (1999), 'New automatic rapid detection system of brewing contaminants: Application of MicroStar-RMDS-SPS (ATP-bioluminescence method) in brewery', *Cerveza Malta*, 144, 3–8.

TAKAHASHI T, NAKAKITA Y, WATARI J and SHINOTSUKA K (2000), 'Application of a bioluminescence method for the beer industry', *Biosci Biotechnol Biochem*, 64, 1032–1037.

TASCHAN H (1996), 'Mikrobiologische Untersuchung von Bieren aus Schankanlagen in der Gastronomie', *Brauwelt*, 136, 1014–1016.

THELEN K, BEIMFOHR C, BOHAK I and BACK W (2002), 'Specific rapid detection test for beer-spoilage bacteria using fluorescently labelled gene probes', *Brauwelt Int*, 20, 156–159.

THIELE F and BACK W (2005), 'Measurement of yeast vitality using a modified version of the intracellular pH measurement (ICP)', *Monatsschr Brauwiss*, 57, 2–5.

TIMMINS E M, QUAIN, D E and GOODACRE R (1998), 'Differentiation of brewing yeast strains by pyrolysis mass spectrometry and Fourier transform infrared spectroscopy', *Yeast*, 14, 885–893.

TORNAI-LEHOCZKI J and DLAUCHY D (2000), 'Delimination of brewing yeast strains using different molecular techniques', *Int J Food Microbiol*, 61, 37–45.

TSUCHIYA Y, NAKAKITA Y, NARA Y and TAKASHIO M (2003), 'Real-time PCR and melting curve analysis for rapid detection and identification of beer spoilage microorganisms', *Proc. 29th EBC Congr.*, Dublin, Nürnberg, Fachverlag Hans Carl [CD-ROM], 113, 1136–1144.

UMEK R M, LIN S W, VIELMETTER J, TERBRUEGGEN R H, IRVINE E, YU C J, KAYYEM J F, YOWANTO H, BLACKBURN G F, FARKAS D H and CHEN Y-P (2001), 'Electronic detection of nucleic acids. A versatile platform for molecular diagnostics', *J Mol Diag*, 3, 74–84.

UNKEL M (1990), 'Biologischer Schnellnachweis – ein zukunftsweisendes Analysensystem für die Qualitätssicherung', *Brauwelt*, 4, 100–107.

VAN DER VOSSEN J M B M, RAHAOU H, DE NIJS M W C M and HARTOG B J (2003), 'PCR methods for tracing and detection of yeasts in the food chain', in Boekhout T and Robert V, *Yeasts in Food, Beneficial and Detrimental Aspects*, Hamburg, Behr's Verlag, 123–138.

VAN ZANDYCKE S M, SIMAL O, GUALDONI S and SMART K A (2003), 'Determination of yeast viability using fluorophores', *J Am Soc Brew Chem*, 61, 15–22.

VANNE L and HAIKARA A (2001), 'Mycotoxins in the total chain from barley to beer', *Proc. 28th EBC Congr.*, Budapest, Nürnberg, Fachverlag Hans Carl [CD-ROM], 91, 839–848.

VAUGHAN-MARTINI A (2003), 'Reflections on the classification of yeasts for different end-users in biotechnology, ecology, and medicine', *Int Microbiol*, 6, 175–182.

VAUGHAN-MARTINI A and MARTINI A (1998), '*Saccharomyces* Meyen ex Rees', in Kurtzman C P and Fell J W, *The Yeasts, a Taxonomic Study*, 4th edn, Amsterdam, Elsevier Science, 358–371.

VIRKAJÄRVI I, VAUHKONEN T and STORGÅRDS E (2001), 'Control of microbial contamination in continuous primary fermentation by immobilized yeast', *J Am Soc Brew Chem*, 59, 63–68.

VOGESER G and DAHMEN M (2004), 'Improvement of the microbiological analysis by use of real-time PCR', *World Brewing Congress*, 24–28 July, San Diego, CA, p. 69.

VOGESER G, HEIDENREICH B and HERTEL C (2005), 'Identification of a new beer spoiling bacteria species enrichment and specific detection by PCR', *Proc. 30th EBC Congr.*, Prague, Nürnberg, Fachverlag Hans Carl [CD ROM], 133, 1107–1108.

VOS P, HOGERS R, BLEEKER M, REIJANS M, VAN DE LEE T, HORNES M, FRIJTERS A, POT J, PELEMAN J, KUIPER M and ZABEAU M (1995), 'AFLP: a new technique for DNA fingerprinting', *Nucl Acids Res*, 23, 4407–4414.

WAALWIJK C, VAN DER HEIDE R, DE VRIES I, VAN DER LEE T, SCHOEN C, COSTREL-DE

CORAINVILLE G, HäUSER-HAHN I, KASTELEIN P, KöHL J, LONNET P, DEMARQUET T and KEMA G H J (2004), 'Quantitative detection of *Fusarium* species in wheat using TaqMan', *Eur J Plant Pathol*, 110, 481–494.

WALKER S L, HAMMOND J R M, PAWLOWSKY K D M and MAUGUERET T M-J (2003), 'A PCR-ELISA assay for the detection of beer spoilage organisms', *Proc. 29th EBC Congr.*, Dublin, Nürnberg, Fachverlag Hans Carl [CD-ROM], 104, 1057–1064.

WALMSLEY R M (1994), 'DNA fingerprinting of yeast', *Ferment*, 7, 231–234.

WELLHOENER U and GEIGER E (2003), 'Definition of the physiological condition of a brewers yeast by means of enzyme activity measurements during propagation and fermentation', *Proc 30th EBC Congr.*, Dulelin/Nürnberg, Fachverlag Hans Carl [CD ROM], 55, 598–608.

WENNING M, SEILER H and SCHERER S (2002), 'Fourier-transform infrared microspectroscopy, a novel and rapid tool for identification of yeasts', *Appl Environ Microbiol*, 68, 4717–4721.

WERLEIN H-D (1998), 'Einsatzmöglichkeiten der Biolumineszenztechnologie zur Überprüfung von Wässern in der Brauerei', *Brauwelt*, 138, 1421–1423.

WHITING M, CRICHLOW M, INGLEDEW W M and ZIOLA B (1992), 'Detection of *Pediococcus* spp. in brewing yeast by a rapid immunoassay', *Appl Environ Microbiol*, 58, 713–716.

WHITING M S, GARES S L, INGLEDEW W M and ZIOLA B (1999a), 'Brewing spoilage Lactobacilli detected using monoclonal antibodies to bacterial surface antigens', *Can J Microbiol*, 45, 51–58.

WHITING M S, INGLEDEW W M, LEE S Y and ZIOLA B (1999b), 'Bacterial surface antigen-specific monoclonal antibodies used to detect beer spoilage pediococci', *Can J Microbiol*, 45, 670–677.

WHITMORE T and KEENAN E (2005), 'The selective separation and concentration of microbial contaminants from fermentation samples', *Proc. 30th EBC Congr.*, Prague, poster.

WOLD K, BLEKEN A and HAGE T (2005), 'Practical experiences on the use of TaqMan real-time PCR as a microbiological routine quality control method for fermenters at Ringnes brewery', *Proc. 30th EBC Congr.*, Prague, Nürnberg, Fachverlag Hans Carl [CD ROM], 129, 1085–1087.

YANSANJAV A, ŠVEC P, SEDLÁČEK I, HOLLEROVÁ I and NÌMEC M (2003), 'Ribotyping of lactobacilli isolated from spoiled beer', *FEMS Microbiol Lett*, 229, 141–144.

YASUI T and YODA K (1997), 'Imaging of *Lactobacillus brevis* single cells and microcolonies without a microscope by an ultrasensitive chemiluminescent enzyme immunoassay with a photon-counting television camera', *Appl Environ Microbiol*, 63, 4528–4533.

ZABEAU M (1992), 'Selective restriction fragment amplification: A general method for DNA fingerprinting', European Patent 92402629.7.

ZIOLA B, GARES S L, LORRAIN B, GEE L, INGLEDEW W M and LEE S Y (1999), 'Epitope mapping of monoclonal antibodies specific for the directly cross-linked mesodiaminopimelic acid peptidoglycan found in the anaerobic beer spoilage bacterium *Pectinatus cerevisiiphilus*', *Can J Microbiol*, 45, 779–785.

ZIOLA B, GEE L, BERG N N and LEE S Y (2000), 'Serogroups of the beer spoilage bacterium *Megasphaera cerevisiae* correlate with the molecular weight of the major EDTA-extractable surface protein', *Can J Microbiol*, 46, 95–100.

ZWIRGLMAIER K, LUDWIG W and SCHLEIFER K H (2003), 'Improved fluorescence *in situ* hybridization of individual microbial cells using polynucleotide probes: the network hypothesis', *Syst Appl Microbiol*, 26, 327–337.

20

Brewing control systems: sensory evaluation

W. J. Simpson, FlavorActiV Ltd, UK

20.1 Introduction

Breweries employ a variety of sensory control systems to assure the flavour quality of their beers. Although some have predicted the replacement of sensory analysis tools in the brewery by so-called 'electronic noses', such predictions were, and remain, premature. See Martí *et al.*. (2005) for a review of the current state of the art.

In this chapter I will review progress in the area of brewery sensory control systems and explore the need for improved tools, techniques and processes. Details of the history, underpinning science and application of such methods can be found elsewhere (Hootman, 1992; Stone and Sidel, 1993; Lawless and Heymann, 1998; Meilgaard *et al.*, 1999; Carpenter *et al.*, 2000; Simpson and Canterranne, 2001).

20.2 Brands

20.2.1 Brand sensory specifications

Brands encapsulate in consumers' minds information on the price, positioning and physico-chemical attributes of a beer. Key among such physico-chemical attributes are those relating to aroma, taste and mouthfeel.

Brewers need to have clear and relevant specifications for their products. While this can be partially achieved by means of analytical parameters such as alcohol content, pH value, colour, etc., the use of detailed sensory profiles generated by expert taste panels provides greater insight into product flavour.

Such descriptions are variously referred to as 'brand profiles' and 'brand flavour fingerprints'. For a typical pale lager beer about 40 flavour attributes are sufficient to describe the brand. Beers such as stouts and strong ales, notwithstanding their more complex flavour structure, can be described with few additional terms.

Despite the availability of such tools, some brewers struggle to define the flavour of their brands, favouring the safe ground of 'no off-flavours and taints' to the effort of cataloguing their positive flavours and precisely identifying any off-flavours and taints (house flavours) present.

In doing so, such brewers miss the opportunity to take timely improvement actions in direct response to the flavours detected in their beers. Table 20.1 lists some of the flavours found in commercial beers, together with the stage in the brewing process at which control can be exercised.

20.2.2 'Preserving' the brand

In an attempt to minimize the chance of a brand's flavour 'drifting' over time, at least one major brewing company preserves samples of its beers at ultra-low temperatures using liquid nitrogen. Samples representative of previous years' production can be assessed alongside contemporary samples, allowing potential flavour drift to be evaluated.

Table 20.1 Beer flavours and the points in the brewing process in which they can be controlled

Process point or risk area	Example flavours
Raw materials and brewhouse	Astringent, bitter, burnt, caramel, chocolate, DMS, freshly-cut grass, grainy, isovaleric, kettle hop, liquorice, malty, salty, smoky, vanilla, worty
Fermentation	Acetaldehyde, alcoholic, ethyl acetate, ethyl butyrate, ethyl hexanoate, isoamyl acetate, H_2S, sulphitic, sour, sweet
Conditioning and end-processing	Acetaldehyde, burnt rubber, caprylic, carbonation, diacetyl, ethyl lactate, H_2S, meaty, mercaptan, onion, yeasty
Packaging	None – process should be neutral to beer flavour
Beer distribution and storage	Astringent, catty, honey, leathery, lightstruck, metallic, onion, papery, powdery, tobacco, winey
Brewery hygiene	Acetaldehyde, acetic, butyric, diacetyl, DMS, ethyl acetate, H_2S, indole, lactic, mercaptan, phenolic (4-vinyl guaiacol), plastics
Taints	Alkaline, bromophenol, chlorophenol, earthy, musty

20.2.3 Links to consumers

Development of an understanding of consumer preference for different products and product categories, and an awareness of the drivers of preference, are the aims of most consumer-related sensory methodologies.

Historically, hall tests (central location tests) have been used to derive preference scores. While they can provide useful insight into consumer behaviour, that is only possible if the presentation order of samples is tightly controlled (MacFie *et al.*, 1989). Although methods to determine sample order are well established, they are not always correctly applied in market tests.

Focus groups were once the preferred vehicle for capturing information on consumers' views about product attributes: they have lately fallen out of favour. One particular difficulty with the approach is that of objectively interpreting consumers' comments. In an attempt to bring the methodology into the twenty-first century, Dransfield *et al.* (2004) have applied text clustering algorithms to aid comment interpretation.

Internal preference mapping

Internal preference mapping allows the likes and dislikes of consumers and groups of consumers for sets of products to be mapped onto a product space alongside the sensory attributes perceived in the products by an expert panel (McEwan, 1996). The technique, which can help identify opportunities for new products and consumers to which they might most appeal, has been widely applied (Schlich and McEwan, 1992; Greenhoff and MacFie, 1994; Costell *et al.*, 2000; Murray and Delahunty, 2000; Chavanne and Courcoux, 2001; Young *et al.*, 2004, 2005).

Generation of preference maps involves two different techniques and types of assessors. Firstly, hedonic assessment tools are used with consumers. Secondly, descriptive analysis tools are used with highly trained expert tasters. Statistical tools are used to analyse the two data sets before representing them within the same graphical space.

Technically, the methods to collect data from consumers and experts alike are well established. It has always been thought that the conditions under which consumer data are collected had to be tightly controlled. But recent research on hard cheeses has questioned this assumption, the location at which preference data were collected (laboratory, central location, or home) having little effect on the results (Hersleth *et al.*, 2005).

Information about beers is usually withheld from consumers in such tests for fear of bias. However, in doing so, unrepresentative preference scores might be generated since consumer choices are influenced by prior knowledge and opinion as well as product intrinsics. Perhaps the best example of such a risk in the beverage world may be the introduction of 'New Coke' in 1985. Reformulated as a result of blind consumer tests to better resemble its sweeter and less carbonated competitor (Pepsi), the reformulated product was a commercial disaster. Following a consumer revolt, the original 'Coca-Cola' formulation was reintroduced less than three months after the launch of 'New

Coke' (Hays, 2004). The brewing industry can also lay claim to a number of similar though less well-documented cases.

Two recent studies have attempted to manipulate consumer preferences prior to testing. In one study, information was given to consumers concerning the use of genetically modified yeast, organic barley and hops, and traditional brewing technology to produce beers (Caporale and Monteleone, 2004). In the other, assessors were told that one beer had been made using traditional brewing techniques, while a second had been made with a 'revolutionary new process using modified yeast and temperature controls that allow the beer to be produced within 10 hours' (Smythe and Bamforth, 2002). In both cases consumers' preferences were influenced by prior knowledge of the products. The effect was greatest in the case of naive assessors and smaller in the case of trained assessors (Smythe and Bamforth, 2002). Work on bread (Kihlberg *et al.*, 2005) also supports these findings.

Development of statistical tools for preference mapping is on-going (Monteleone *et al.*, 1998; Courcoux and Chavanne, 2001; Xiong and Meullenet, 2004), having the aim of improving the robustness of analyses, and producing more information from the data.

Ben Slama *et al.* (1998) have devised a statistical method of sample selection for preference mapping. D-optimal designs improve selection of subsets of samples from larger batches and reduce the number of samples needed to achieve specific test objectives.

External preference mapping

While internal preference mapping starts with hedonic scores generated by consumers and adds detailed sensory attribute information generated by expert tasters, external preference mapping takes detailed sensory attribute information generated by expert tasters and overlays consumer preference data within it (Schlich and McEwan, 1992; Thompson *et al.*, 2004; Young *et al.*, 2004). External preference mapping is particularly useful when establishing the ideal sensory profile for a specific group of consumers.

Despite their strengths, these techniques remain intellectually demanding. The field is dogged (or blessed, depending on your viewpoint) by complex statistical concepts and technical jargon.

Other consumer methods

Computer-aided analysis of answers to open-ended questions can be used to gain an insight into consumer perceptions. Words used by 165 consumers to describe 13 individual samples of mayonnaise were counted and the combinations of words used to describe products evaluated (ten Kleij *et al.*, 2003). Correspondence analysis was used to generate a visual map that resembled a preference map in how it represented the relationship of one product to another. A high degree of concordance was obtained between results obtained with this method and internal preference mapping.

Drinkability of beer is an important driver of commercial success, but there is little agreement on how it can best be measured. Work in Japan (Nagao *et al.*,

1999) has drawn a connection between beer drinkability and gastric emptying. Beer quality attributes perceived by consumers as negative restrict gastric emptying, minimizing the frequency and volume of urination, with consequential effects on beer intake.

An attempt to evaluate differences in drinkability of beers and establish some of the drivers of this parameter has been reported by Parker and Murray (2003) and the area has been reviewed by Sharpe *et al.* (2003).

20.3 Tasters

The competence of the tasters determines the success or failure of each and every sensory test. Sometimes it seems as if this important point is not as obvious as it appears. At least occasionally, brewers inadvisably deploy tasters who have little training to carry out critical taste evaluations in the brewery. Fortunately, tools for measurement and development of taster competence are now widely available and are an area of considerable industrial interest.

20.3.1 Competency standards

What should a brewery expect of its tasters? Simpson (2004) has proposed the following:

1. Those who qualify as novice tasters should have undergone selection and screening and completed basic training. They should be proficient in simple tasting procedures
2. Those with intermediate skills should, in addition, be capable of naming and recognizing up to six beer styles, adept at identifying three brands within each of two of these beer styles and able to recognize and name 25 beer flavours. They should be able to rate product quality using a numeric or category scale.
3. Expert assessors, in addition to all of the above, should be able to recognize more than 10 beer styles and at least 50 beer flavours.

The key competencies of an expert professional beer taster are summarized in Table 20.2.

In the past, a one million hectolitre per year brewery – not a large brewery by today's standards – might have employed several hundred people. It was easy to find people willing to taste beer; there were so many people to choose from. With modern operations and 'lean manning', the situation has changed. People are too busy to make time for 'subsidiary' tasks such as tasting and there are fewer people.

How can we overcome this difficulty? Tasting activities can be integrated into an employee's job, eliminating any conflict of duty. All brewery employees can be considered potential tasters, rather than just brewing and quality-management personnel. Part-time employees for whom tasting is their sole

Table 20.2 Key competencies of an expert professional beer taster

Competency	Assessment criteria
Ability to taste beer to a professional standard	• Can carry out difference test assessments (including triangular and duo–trio tests) • Can carry out descriptive test assessments (including Quantitative Descriptive Analysis or other quantitative descriptive techniques) • Can recognize and name at least 50 different beer flavour attributes • Can scale all attributes using one or more methods
Ability to undertake assessments of beer quality and type	• Can judge commercial acceptability (tell good beer from bad) • Can judge whether a beer is true to style • Can judge whether a beer is true to type • Can identify and detect off-flavours and taints at commercially significant concentrations
Ability to participate in a positive way to beer taste sessions	• Demonstrates interest in and enthusiasm for the activities of the taste panel • Shows sensitivity to the risk of bias in other assessors • Recognizes the risks involved in professional beer tasting and takes all reasonable steps to minimize them

responsibility can be hired. Such employees can make up what is known as a 'town' panel or 'external' panel (Hegarty *et al.*, 2001). This should not be confused with a 'consumer' panel, as unlike consumers, 'external' panellists undergo considerable training.

Although several brewing companies nowadays delegate some of their tasting activities to such panels, there remains a reluctance by some brewers to entrust assessment of beer to those not intimately involved in its production. To some extent this is rooted in a belief that if production staff have direct responsibility for tasting they will be able to react to problems more quickly than if others do the tasting. While the principle is logical, it is not consistent with industrial practice, nor with 'lean' manufacturing principles.

20.3.2 Tools to aid selection and screening of tasters

To attract, train and retain those with the best aptitude and enthusiasm for professional beer tasting, good recruitment and training practices are needed. The key points are straightforward. Potential trainees should not be recruited at random. They should be selected from a larger pool of talent. Having been selected, recruits should be screened with the aim of identifying those most likely to benefit from training. Only selected and screened individuals should be trained. Detailed training performance records should guide decisions about whether training should continue in the case of each trainee taster. Only those proven to have reached the required predefined level of competence should be deployed in professional beer tasting activities. A description of some of the approaches used follows.

Questionnaires and interviews
What types of people make good beer tasters? Firstly, they should be of legal drinking age; this varies from place to place and from time to time. They should not have a dislike of beer – while this may appear to be self evident, the opposite has been suggested (Anon, 1995). They may be regular beer drinkers, but not necessarily so. They should have an interest in flavour and food – which often translates into their not being overly cautious or restricted in their food choices. They should have good mental abilities, and concentration skills. They should be in good health and, in particular, not predisposed to, or suffering from, alcohol-related health problems or major impairments of their senses such as anosmia (smell blindness) and agueusia (taste blindness). They should not be taking medicines that might impair their sensory capabilities. They should not suffer from allergies or sensitivity to common beer components, such as sulphur dioxide.

Questionnaires used to recruit tasters should: (i) encourage and enthuse the respondent to become a trainee taster; (ii) provide a basis for their selection based on their level of self-declared interest and motivation; (iii) provide a basis for their selection based on some aspect of skill or ability; and (iv) provide a first-level screen to identify those whose health might be placed at risk.

Responses to recruitment questions should span a range, rather than being in the form of comments or yes/no answers. This allows decisions concerning candidate selection to be based on quantitative data rather than on subjective criteria.

While publications relating to selection of beer tasters are few, papers relating to selection of dairy product (Fernandez-Albalat *et al.*, 2005) and meat (Barcenas *et al.*, 2000) assessors, together with relevant international standards (ISO, 1993, 1994), provide some insight into the processes involved.

Questionnaires can also be used to estimate an individual's risk of developing alcohol-related health disorders. These include the CAGE and AUDIT screening questionnaires (Aertgeerts *et al.*, 2002; McCusker *et al.*, 2002). These are notable for their simplicity. For example, the CAGE method assesses alcohol-related risk from an individual's answers to four questions. A web-based version of the questionnaire can be found at http://mssm.edu/medicine/general-medicine/ebm/CPR/cage.html.

While such tests have not yet been widely used in recruitment of beer tasters, they may have a place in protecting brewing companies from the risk of litigation, and protecting tasters from the risk of alcohol-related health disorders (see Section 20.3.5).

Having identified suitable candidates through use of a questionnaire, an interview can be used to choose among them (Anon, 1995; Piggott and Hunter, 1999). Interviews can be used to assess 'soft' personal factors, including an ability to work as part of a group, attitude, and physical and intellectual ability. On the downside, it can be difficult to maintain objectivity in this type of selection.

Screening tests

Screening helps trainers identify those with inherent defects in their sensory acuity (including anosmia and agueusia). It also allows the likely response of candidate tasters to training to be predicted, identifying those individuals most likely to perform best after training. For this latter reason, screening should not focus on current competence. ISO8586-1 and 8586-2 (ISO, 1993, 1994) contain details of principles and practices.

For such tests to have the predictive power expected of them, the way in which the training is carried out during the screening phase must be comparable to how it will be done during the training phase. Boughton and Simpson (2000) have described one such screening procedure that meets this objective.

In contrast, the harmonized methods of the European Brewery Convention, American Society of Brewing Chemists and Brewing Congress of Japan recommend that tasters are screened with solutions of four basic tastes (sweet, salt, sour, bitter) in water. Given that many naive assessors confuse these tastes on account of having received no training in the correct use of the terms (Mackey and Jones, 1954), and that sensitivity in one sense (taste) is no predictor of ability in a second (e.g. odour) such a screening method is not only ineffective, it is in fact unhelpful.

20.3.3 Tools to help train beer tasters

Terminology

For almost 30 years the brewing industry has benefited from the availability of a terminology system and reference standards for many of the important flavours found in beer. This system defines 122 different beer flavour attributes, organized into 44 constructs in nine primary categories (Meilgaard *et al.*, 1979). Graphically, the system is represented in the form of the 'Beer Flavour Wheel' (Fig. 20.1). The terminology has been translated into a number of languages. To complement this terminology system, 27 reference standards were proposed (Meilgaard *et al.*, 1982). Since the original publication of the system, an extension to include additional mouthfeel characters has been suggested (Langstaff and Lewis, 1993). In addition, icons have been developed for many of the beer flavour attributes, which are independent of language (see Fig. 20.2).

Stabilized reference beer flavours

The use of reference standards to assist in training of assessors is not universally accepted in the field of sensory analysis. Indeed, in some methodologies, such as Quantitative Descriptive Analysis, it is expected that reference standards should be needed in the case of no more than 10% of attributes (Stone and Sidel, 1993). Attributes can be generated by tasters prior to a sensory evaluation project,

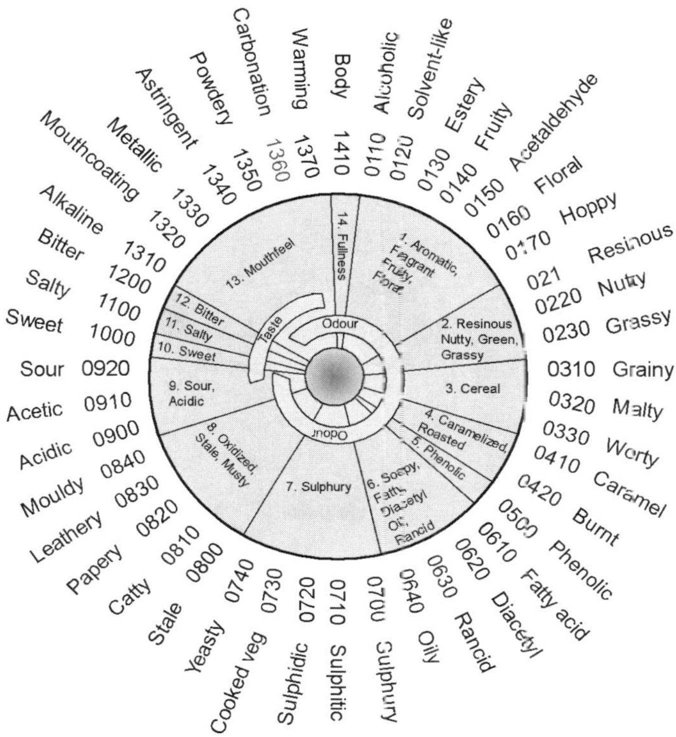

Fig. 20.1 The Beer Flavour Wheel. The wheel represents the lower levels of the harmonized terminology system of the American Society of Brewing Chemists, Master Brewers' Association of the Americas, European Brewery Convention and Brewing Congress of Japan. Adapted from Meilgaard *et al.* (1979).

either by a sequence of assessment, discussion and consensus (Stone and Sidel, 1993), or using the Repertory Grid technique (Kelly, 1955; Piggott and Watson, 1992; Jack *et al.*, 1994).

An inspection of papers relating to generation of terms, and proposals for reference standards, in the wider field of food and beverage sensory analysis makes us aware of the advantageous position we hold within the brewing industry in this respect (Murray and Delahunty, 2000; Chapman *et al.*, 2001; Drake *et al.*, 2001, 2003; McDonnell *et al.*, 2001; Jeong *et al.*, 2004; Mirarefi *et al.*, 2004; N'Kouka *et al.*, 2004; Thompson *et al.*, 2004; Young *et al.*, 2004, 2005; Verdu Jover *et al.*, 2004; Zamora and Guirao, 2004; Chollet *et al.*, 2005; Peña y Lillo *et al.*, 2005). In contrast to the pure chemical reference materials used in the brewing industry to represent attributes, a variety of poorly defined materials are often used elsewhere.

Fig. 20.2 Icons used to represent common beer flavour terms. Visual representations can be used to reinforce learning with trainee tasters through cross-modal integration and to circumvent language barriers. Reproduced by permission of FlavorActiV Limited.

While in some contexts it is desirable not to pre-judge the attributes that may best discriminate products and groups of products, in an industrial situation, the advantages of a fixed flavour vocabulary far outweigh the disadvantages since brewers must act on the results of sensory tests to correct process problems.

One factor of great importance is that of representing the attribute or construct within the product itself, rather than in isolation – on, for example, a sniff stick. Thus, flavour standards are added to and assessed in beer.

In the last decade the system of the American Society of Brewing Chemists and European Brewery Convention has been further extended by the development and use of technologies which allow such reference flavours to be stabilized. Stabilized reference beer flavour standards are now used to train more than 10 000 professional beer tasters in 700 breweries (Adam Fenton, personal communication). Standards are currently available to cover 38 beer flavour terms.

Training methodologies

Approaches used to train brewery tasters have undergone something of a revolution. This has been made possible by developments in our understanding of the determinants of success in sensory assessments and how the process of learning takes place for those involved (Peron and Allen, 1988; Bartoshuk, 2000; Chollet and Valentin, 2001; Köster *et al.*, 2002; Djordjevic *et al.*, 2004; Green and George, 2004; Labbe *et al.*, 2004, Parr *et al.*, 2004; Ballester *et al.*, 2005).

In breweries, unstructured, free-form 'learning' activities have largely been replaced by outcomes-based modular training programmes. For an example of this type of training see Boughton and Simpson (2000).

Detailed feedback during training is key (Kuesten *et al.*, 1994). Findlay *et al.* (2004) present an excellent example for the case of panellists learning to carry out descriptive assessment of wine. In their hands, tasters received immediate feedback concerning their use of scales during training. This was achieved by means of the software used to record the tasters' results. The authors concluded that (i) immediate feedback provides panellists with a strong individualized method of learning attributes and scaling of those attributes; (ii) the feedback method can be used to help panels develop and refine their own targets with respect to use of scales; and (iii) calibration of a panel's use of scales can be achieved through use of specific lexicons with reproducible attribute standards. Importantly, they concluded that, through use of appropriate feedback, panel times can be cut in half with no penalty on performance.

20.3.4 Tools to assess taster competence

Ten to 15 years ago, the competence of professional beer tasters was expressed in terms of their experience. A taster with 20 years' tasting experience was assumed to be more competent than one with several months' experience. Of course, the assumptions on which such generalizations were based are suspect.

Experience and competence in the field of professional beer tasting are not synonymous (Simpson *et al.*, 1999).

Before considering how best to assess competence, let us consider specific aspects in which taster performance can be found wanting. Poor taster performance can result from (i) confusion of flavour terms; (ii) differences in how assessors use intensity scales; (iii) differences in how they perceive one or more attributes; and (iv) inconsistencies in assessments.

Performance tests can be carried out separately from routine taste sessions. Others are based on analysis of routine test results. A useful set of guidelines for proficiency testing of assessors has been published (Lyon, 2001). These guidelines were one of the outputs of a project which was part-funded by the European Union under the Standards, Measurement and Testing Programme. The PROFISENS project involved the combined efforts of 17 research organizations and industrial companies.

Tests based on specific assessment programmes

Several 'ring analysis' or 'inter-collaborative' schemes are available to measure beer taster competence. The most widely used of these schemes currently encompasses more than 3000 professional beer tasters in 275 brewery taste panels (Adam Fenton, personal communication). The FlavorActiV™ Taster Validation Scheme uses stabilized reference beer flavour standards to generate beers with consistent sensory properties, which differ from a reference beer sample in the level of a single flavour attribute. In this scheme assessors have to identify the attribute present in each of six samples, choosing from a list of 34 possible flavours (described using the official terms of the American Society of Brewing Chemists, European Brewery Convention and Brewing Congress of Japan). The results of each test are then entered into a secure database via the internet. The resulting data can be analysed in a variety of ways (Fig. 20.3).

Tests based on analysis of routine tasting results

Many statistical procedures have been devised which allow the competence of individual tasters to be established from data collected in routine taste sessions (Cliff and Dever, 1996; Piggott and Hunter, 1999). Data can be analysed using one-way analysis of variance (ANOVA) (Naes, 1990; Naes and Solheim, 1991; Lea *et al.*, 1997). The aim of such analyses is to measure: (i) the repeatability (Mean Square Error: MSE) of the results of each assessor for each attribute (at least two replicates are needed to carry out such analyses); (ii) the degree of agreement between the mean score for each assessor for each attribute compared to the mean value for the panel (the consensus value); (iii) the degree of discrimination (p value) that an assessor achieves between closely separated levels of individual attributes; and (iv) the way in which the assessor uses the intensity scale compared to how the panel uses the scale.

The ideal assessor shows a high degree of repeatability for assessment of all attributes, together with a high degree of discrimination. They use the scale in a way that is representative of the panel as a whole. By constructing a graph of p

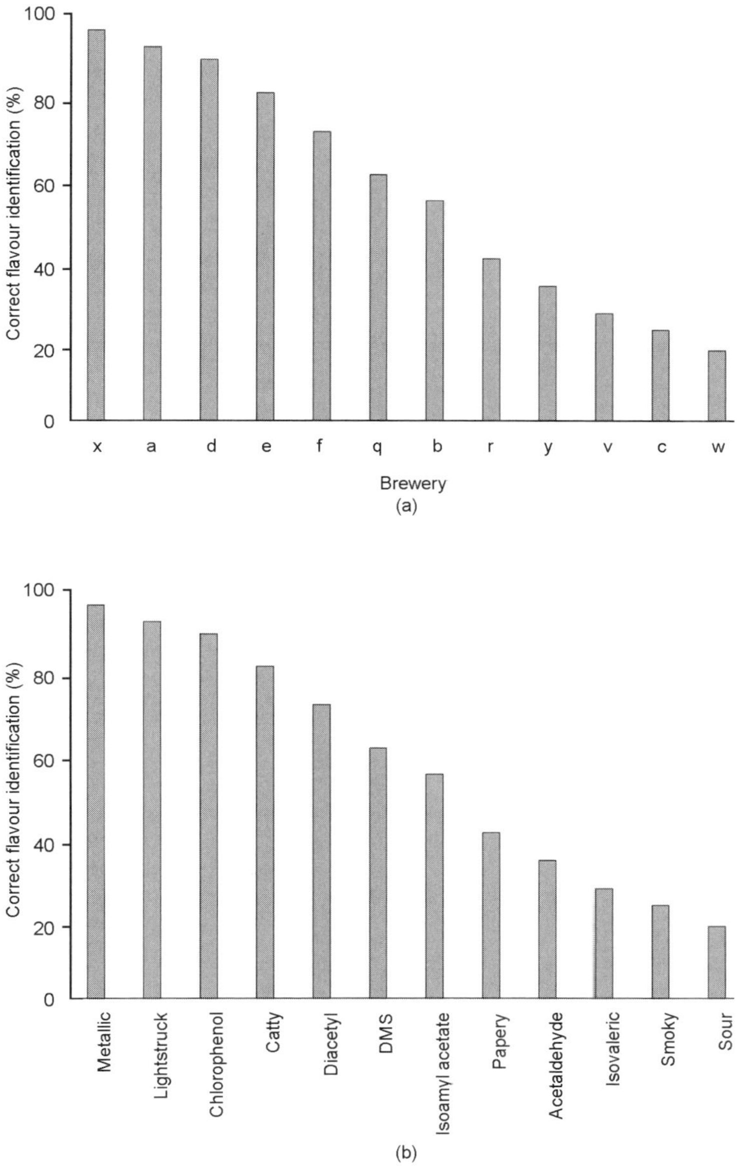

Fig. 20.3 Examples of graphs used in a web-based taster competency assessment system: (a) plot showing the relative performance of different brewery taste panels tested with the same samples; (b) plot showing the relative ease with which all of the tasters in the same brewery group are able to identify individual beer flavour attributes. Data shown are indicative only and are not intended to represent results obtained in practice. For an on-line demonstration of a web-based beer taster competence assessment system, see www.flavoractiv.com/validation.

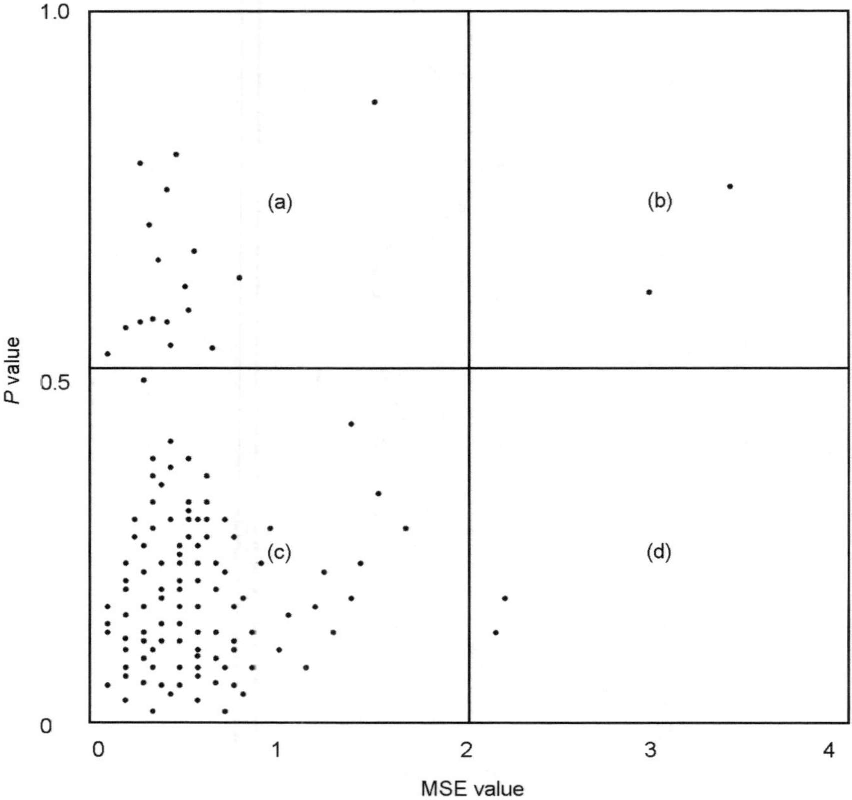

Fig. 20.4 Taste panel performance assessments from routine tasting data. Repeatability (Mean Square Error: MSE) of the results of each assessor for each attribute are plotted against the degree of discrimination (*p* value) that an assessor achieves between closely separated levels of individual attributes. Quadrants represent areas of: (a) good repeatability but poor discrimination; (b) poor repeatability and discrimination; (c) good repeatability and discrimination; (d) poor repeatability but good discrimination. Adapted from Naes (1990).

value against MSE for all attributes (Fig. 20.4), the relative performance of each assessor for each attribute can be compared (Naes, 1990). By plotting such data in the form of Control Charts, trends in performance can be identified (Gatchalian *et al.*, 1991).

In the author's view, such techniques should be part of the routine work of all brewery sensory analysts. Unfortunately, the awareness of such tools in the industry is currently low. For an illustration of how the tools can be applied to monitor the performance of a panel involved in assessment of distilled beverages, see McDonnell *et al.* (2001).

'Eggshell' plots can be used to gain further insight into assessor performance. These make use of ranking data, or continuous data converted to ranks (Naes,

1998). They have the advantage that, being based on non-parametric statistics, they do not rely on underlying assumptions concerning the distribution of the data.

Panel performance can also be assessed using multivariate statistical tools (for a review see Piggott and Hunter, 1999). These include principal component analysis, generalized procrustes analysis and partial least squares regression. Such techniques provide additional information on taster and panel performance but require more statistical expertise than is usually found in breweries.

20.3.5 Tools for monitoring taster welfare

In 2004, a Brazilian court awarded Bernd Naveke, a former master brewer and beer taster with a major brewing company, a US$2 000 000 lump sum. In doing so they overturned a 1996 judgement in which he was awarded US$30 000 plus a pension equivalent to his former salary. The taster claimed that he had developed alcoholism as a result of participating in beer tasting activities. It was also reported that he had to consume eight litres of beer each day to carry out his duties! The court concluded that the brewing company had not done enough to protect him from the risks associated with his job.

Subsequent to that ruling and, as a result of a generally increased awareness of risk among the professional beer tasting community, some breweries have established programmes to monitor taster health and keep records of tasting activities. Screening of tasters at an early stage (see Section 20.3.2 under 'Questionnaires and interviews') is an important first step.

General guidance relating to ethical and professional practices for sensory analysis of foods has recently been published by the Institute of Food Science and Technology (Anon, 2005). The Institute of Brewing (now known as the Institute of Brewing and Distilling) has proposed similar guidelines relating to sensory evaluation of beer (Anon, 1997). A guidance note from the American Society of Testing and Materials (ASTM, 2001a) describes the specific issues involved in tasting alcoholic beverages, including regulatory issues and those relating to assessor safety.

The main points for breweries to be aware of are as follows:

1. Since sensory tests involve human subjects, the scope of tests and authorization to sanction them should be defined in a written Ethical Policy.
2. Assessors should be volunteers, either through contractual agreement or on an *ad hoc* basis.
3. Organizers of taste sessions have a legal liability towards their tasters and the public.
4. Tasters should be made aware of the risks associated with consumption of alcohol.
5. Individuals for whom alcohol might be harmful, including pregnant women, should be excluded from tasting activities.
6. Sample volumes should be minimized as far as possible and volumes consumed recorded for each taste session.

7. In some jurisdictions, government agencies may have an interest in the activities related to serving and tasting of alcoholic beverages.

Routine health checks for tasters should include tests for non-specific indicators of liver function such as gamma-glutamyl transferase (GGT), Carbohydrate Deficient Transferrin (CDT), alkaline phosphatase and alanine amino transferase (Montalto and Bean, 2003; Miller and Anton, 2004).

An alternative method that (unlike blood tests) could be carried out remotely involves measurement of fatty acid ethyl esters found in hair (Wurst *et al.*, 2004). This could allow alcohol consumption patterns of individuals to be monitored routinely or used by employers to maintain a bank of evidence for later analysis in cases where self-declared fitness for tasting was called into question. Such an approach has an advantage over liver function tests in that abstinence from alcohol in the days prior to testing has little influence on the results.

Screening is particularly important given that a predisposition to alcohol-related diseases is associated with certain tasting behaviours (Duffy *et al.*, 2004). In selecting the best tasters, we may inadvertently select for a sub-group of the population for whom alcohol is more hazardous.

Ultimately genetic screening may provide the best protection against tasting-related health risks. Tests to identify at-risk individuals are already available (Tabakoof *et al.*, 2004). However, in common with other forms of genetic testing, obstacles remain to their use.

Little information is available concerning specific risks associated with beer tasting, or how they might best be managed. However, it has been reported that a wine taster who had tasted at least 20 wines a day for 10 years developed sensitive teeth (Gray *et al.*, 1998). This was found to be caused by erosion and pitting, together with loss of enamel around dental fillings. The authors showed that immersing human teeth in white wine (pH 3.3) caused marked changes in their surface within 24 hours. Kaneko *et al.* (1994) have described an approach to wine taster health management. It is timely to remind brewery managers and directors that responsibility for the health and welfare of tasters ultimately rests with them.

20.4 Assessment methods

A variety of approaches can be used to evaluate beer flavour using professional tasters (Table 20.3). Several recent publications have addressed issues of relevance to a variety of sensory assessment techniques which are either currently used or potentially could be used in breweries.

Pagès and Périnel (2003) looked at the number of samples that can be evaluated by assessors before their performance suffers. For the case of mineral water samples and descriptive profiling they concluded that the numbers of samples usually considered appropriate for assessment in a single session are often underestimated. Similar studies relating to beer assessment have not been reported.

Table 20.3 Sensory test methods and their application

Type of test	Examples	Type of assessor	Application
Difference tests/ similarity tests	Triangular test; duo–trio test	Untrained or trained assessors	Used to establish whether there is a difference between one ore more samples, or whether two or more samples are identical to each other
Descriptive tests	Flavour Profile® method; Quantitative Descriptive Analysis®; Quantitative Flavour Profiling; Trueness to Type test; Flash Profiling; Free-choice Profiling	Highly trained assessors	Used to 'measure' the flavour of one or more samples using a vocabulary of flavour terms
Preference tests	Paired comparison test; 'A/not A' test	Consumers	Used to 'measure' the degree to which one sample is preferred over another, or to assess the relative preference of several samples within a set
Scaling tests	Ranking test; Magnitude estimation test; Rank-Rating test; Difference from control test	Highly trained tasters	Used to measure the intensity of one or more flavour attributes or constructs
Drinkability tests	Volumetric consumption test	Consumers	Used to measure relative 'drinkability' of different samples
Hybrid test methods	Internal Preference Mapping; External Preference Mapping	Consumers and highly trained tasters	Used to identify relationships between different products, and to segment consumers as a result of their response to different products

The issue of how assessors should rinse their mouths between samples has been dealt with by Johnson and Vickers (2004). They used seven methods of rinsing in a study of the bitterness of cream cheese samples. They found no differences in the effectiveness of these methods with respect to their ability to control sensory adaptation or build-up of flavour, or in their ability to increase the panellists' discrimination among samples. Breweries often differ in their preferred approach, some mandating that assessors rinse with water between samples, or use unsalted crackers, and others having no fixed way of working. It seems that the choice is of no great significance.

20.4.1 Difference tests

Difference tests are used to establish whether there are differences in the flavour of two or more samples, together with a probability value that expresses the degree of confidence that can be attached to the result. Tests used in industry include the triangle (three-glass) test, the duo–trio test, the paired comparison test, and the two out of five test (Carpenter *et al.*, 2000). Detailed descriptions of these methods can be found in standard textbooks on sensory analysis (Meilgaard *et al.*, 1999; Stone and Sidel, 1993; Carpenter *et al.*, 2000) and in relevant international standards (ISO, 1983, 2004; ASTM, 1997, 2001a, 2001b).

Difference test methods are among the best established in sensory analysis, but they are also among the most widely abused. Sometimes, even the most basic details of methods are not adhered to as brewery sensory analysts try to 'save time'. The devil is the detail of such tests (O'Mahony, 1995)! Improved techniques for dealing with replicate assessments in difference testing have been published (Brockhoff and Schlich, 1998; Kunert and Meyners, 1999). This is a particularly important aspect of difference testing in the brewery environment, in which assessor numbers are often limited.

It should be appreciated that such methods are designed to detect differences among samples, not to measure similarity. In a brewery setting, difference tests are sometimes applied in situations in which 'similarity tests' are more appropriate – for example, when attempting to approve a new supplier of bottle caps. While the details of the statistical basis of this point are beyond the scope of this chapter, the reader can find a discussion of the salient points in the literature (Bi, 2005).

When establishing differences between samples, a number of authors have shown that the discriminatory abilities of assessors are greater when they are asked to choose on the basis of preference rather than on some other basis (Macrae and Geelhoed, 1992; Sauvageot and Rabier, 2004).

20.4.2 Descriptive tests

Descriptive analysis techniques are among the most widely used in the brewing industry. Several international standards (e.g. ISO, 2003) and a practical workbook (Carpenter *et al.*, 2000) describe how descriptive analysis can be carried out. The sensory analyst can choose to apply Quantitative Descriptive

Analysis (QDA), the Flavour Profile® method, Quantitative Flavour Profiling (QFP), Free Choice Profiling (FCP) or the Spectrum™ Method.

In breweries, the methods used do not typically conform to published procedures. Elements of several methods are often assimilated, with individual procedures differing in detail from one brewing company to another. For a review of this field, see Powers (1988) and Murray *et al.* (2001). For examples of how the methods have been applied in contemporary studies, see Chapman *et al.* (2001), Drake *et al.* (2001, 2003), Koussissi *et al.* (2002, 2003), Pagès and Périnel (2003), Jeong *et al.* (2004), Mirarefi *et al.* (2004), N'Kouka *et al.* (2004) and Peña y Lillo (2005).

Recently, a new method of descriptive analysis has been proposed. The 'Flash' profiling method (Delarue and Sieffermann, 2004) is intended as a rapid sensory profiling technique for industrial use. It is based on a combination of free choice profiling and comparative evaluation of product sets.

So called 'trueness-to-type' tests (Anon, 1995) are widespread in the brewing industry. Such tests allow the degree of conformity to a sensory specification of an individual beer sample to be represented as a single number (usually a percentage). The technique usually involves carrying out a descriptive analysis using a 'just-about-right' scale for each attribute, in which the level of attribute relative to that in an ideal example of the brand is rated. Off-flavours and taints are scaled using an absolute scale.

While the method has many attractive features, it is open to criticism. Unlike other methods used in the brewery taste room, it is neither based on sound scientific principles nor been subjected to peer review. Research is urgently needed in this area to avoid undermining the quality of sensory evaluation activities in breweries that use this technique.

Nielsen *et al.* (2005) have shown that panel drift can be minimized if a reference sample representative of the test product can be provided at each taste session. In their case the test sample was frozen herring. While many breweries have attempted to provide reference beers to represent ideal batches of specific brands, this is problematic. Both selection and preservation of such beers has proven difficult.

20.4.3 Attribute scaling and quality grading tests

Scaling of attribute intensity is critical to many sensory procedures. Sometimes there is a need to estimate the intensity of single characteristics, for example bitterness. In other methodologies, such as Quantitative Descriptive Analysis, the intensity of many attributes has to be estimated within a short period of time. For a review see Land and Shepherd (1988).

Attribute intensity can be estimated using a nominal scale, which consists of a series of categories that are labelled using a name or number. Or it can be estimated using an ordinal scale, which is similar to a nominal scale except for the fact that the categories lie in a specific order, allowing the observations to be ordered according to whether they have more or less of an attribute. A nine-point hedonic scale is an example of an ordinal scale, while a ranking test is an example of a nominal scale. An interval scale has properties similar to an ordinal scale, but the intervals between the levels on the scale are assumed to be equal. A continuous line scale is an example of an interval scale. Ratio scaling is a special case of an interval scale with the additional feature that the relationship of one value to another is fixed. This type of scaling has been used to estimate the relative bitterness of congeners and stereoisomers of hop-derived bitter acids found in beer (Hughes and Simpson, 1996).

An interesting advance in this area has been the development of the Rank-Rating method (Kim and O'Mahony, 1998), which tasks assessors with ranking a set of samples in order of attribute intensity before applying an intensity rating to each sample. Assessors make fewer errors when evaluating the taste intensity of salt solutions using the Rank-Rating method compared to when they use conventional scaling procedures.

The Rank-Rating method appears to facilitate good discrimination and scaling of samples regardless of the type of scale used to position the samples relative to one another (Kim *et al.*, 1998). This method has an advantage over monadic and sequential-monadic rating methods in that it forces assessors to repeatedly assess attribute intensity. This reduces errors and increases discrimination. The method has been successfully employed to study the flavour of Greek wines (Koussissi, 2003) and lager beers (Techakriengkrai *et al.*, 2004a, 2004b, 2004c).

The difference-from-control test (Sust *et al.*, 1985) has recently been approved as a recommended method by the American Society of Brewing Chemists (Thompson, 1999). In this procedure assessors receive a control sample together with one or more test samples. They are asked to rate the size of the difference between each sample and the control for the chosen attribute. Some of the time, the test sample may be a second control sample. The mean difference-from-control values for each sample and for the blind control sample are calculated and analysed by ANOVA or with a paired *t*-test. In the hands of some brewing companies, this test has proven to be a useful way of estimating the degree of difference between the same brand produced in different breweries.

Objective assessment of 'commercial quality' or 'grade' of beer remains a problem. Ultimately, the sensory specialist has to be capable of advising whether a batch of beer is likely to give rise to a consumer reaction or complaint and, in a worst case, whether it might result in the need for a product recall. Different approaches are used by brewing companies to assign a quality grade to beer. This area is one of the most active in terms of method development in the industry today.

In the simplest approach, assessors are asked to assign a quality grade to each beer sample. This may be rated in parallel to assessment of individual flavour attributes, for example when carrying out Quantitative Flavour Profiling or Quantitative Descriptive Analysis. Alternatively the quality score may be assigned in isolation, without assessment of any other attributes.

In an attempt to make quality assessment more objective, some breweries now calculate grade from descriptive data – awarding high scores for beers which match the product flavour specification, and deducting points for departures from profile and for off-flavours and taints.

An interesting contribution to the science of quality assessment has been made by Prescott *et al.* (2005) who developed a methodology to measure the 'consumer rejection threshold (CRT)' of taints and off-flavours and applied it to assessment of a common wine taint (trichloroanisole, TCA). The CRT provides a measure of the concentration at which consumers begin to reject a wine. A similar approach applied to beer-related off-flavours and taints may well prove useful.

A method which is widely used in Germany to assign beer quality ratings is the DLG-scheme (Pfenninger, 1993). In this method points are awarded for aroma, taste, body, liveliness and bitterness. Schönberger *et al.* (2004) have proposed an improvement to this method (based on a trueness-to-type method) and compared it to the standard DLG method.

20.4.4 Time-intensity (TI) methods

Measurement of attribute intensity within the timeframe of a single sensory assessment can provide a useful insight into some aspects of beer flavour and its relationship to product acceptability. Collection of such temporal sensory data is generally carried out using specialist software. While the goals of TI assessments are undoubtedly worthy, single attribute TI is particularly demanding on the assessor and dual attribute TI (Duizer *et al.*, 1997) even more so. The curves produced are also difficult to analyse (Dijksterhuis and Eilers, 1997; Eilers and Dijksterhuis, 2004). While the TI technique has been the subject of many research publications relating to beer bitterness (e.g. King and Moreau, 1996) and, recently, to sweetness (Techakriengkrai *et al.*, 2004c), the technique has not yet found application in routine analysis of beer.

20.4.5 Tests of product flavour stability

Sensory assessment of beer flavour stability remains challenging. While a full discussion of this area is beyond the scope of this chapter, suffice to say that all of the above methods can be used to assess the degree of difference between fresh and aged beers, and the nature of those differences (Meilgaard, 1989). Unfortunately, the published literature on beer flavour stability provides much evidence for the fact that methods are sometimes implemented without the required degree of rigour and control needed for their success.

Recognizing the fact that more work is needed on the flavour attributes associated with the beer ageing process, the sensory subgroup of the European Brewery Convention has proposed the concept of a 'beer flavour stability wheel' in which terms related to beer ageing are represented (Hill, 2003).

20.4.6 Best practice in the area of in-process and at-line tasting

Making sure that bad beer does not proceed any further in the brewing process than absolutely necessary is an important aim of in-process and at-line tasting.

Samples of raw materials (including malt, adjuncts, hops, water, gases, sugar syrups and processing aids), fermenting wort, maturing beer, beer from bright beer tanks, and packaged beer are tasted each day in breweries throughout the world with the aim of detecting any off-flavours and taints present. The methods used for such tasting are generally empirical and do not meet the exacting standards of the defined sensory testing procedures referred to above. Further work is needed to place at-line and in-process tasting in breweries on a more objective and professional footing.

Curt *et al.* (2001) have explored the principles of at-line assessment of samples during manufacture of dry sausage. They used a knowledge management technique – the 'M3A' method (Method for the Autonomous Analysis of Activities) – to collect and formalize the knowledge of experienced sausage factory workers. They then documented an at-line evaluation programme from that knowledge.

An interesting attempt to place at-line assessment in a wider context has been made by Barylko-Pikielna and Matuszewska (2000). Through the vehicle of a European-wide research programme, they devised a system of sensory quality management which is analogous to the HACCP (Hazard Analysis Critical Control Point) system of product and process safety management. They named this system 'SQCCP' – Sensory Quality Critical Control Point methodology. The system has five steps: (i) determination of the optimum sensory profile for the product; (ii) determination of the sensory attributes of key importance for consumer acceptance; (iii) identification of the Critical Points that affect the sensory quality of the finished product; (iv) determination of the qualitative and quantitative effects of raw materials and other ingredients on sensory quality and product acceptance; and (v) monitoring of day-to-day variation in product sensory quality against target sensory profiles, together with parallel monitoring of variation in formulation and processing factors.

20.4.7 Coupling of human assessors to modern chemicals separation technology

Modern chemicals separation technology, combined with human assessors as detectors, can help us gain an insight into beer flavour and the influence of process variables and storage on those attributes. Techniques include gas chromatography–olfactometry, the CHARM technique, and Aroma Extract Dilution Analysis (AEDA). See Dattatreya *et al.* (2002) for a review.

20.5 Assessment facilities

20.5.1 Taste rooms

Taste room designs have changed little over the last 30 years. Basic descriptions offered by international standards (ISO, 1988) provide sufficient detail for their construction. What has changed is how the taste room is used. In common with other areas of brewery operations, taste room activities have benefited from the application of process improvement tools, notably '5S' (a 'Lean' or 'Kaizen' tool) to improve standards of operation and housekeeping (George, 2003).

20.5.2 Data collection tools

Computerized data collection is now increasingly common in brewery taste rooms. There are two levels at which such computerization can be achieved. Responses obtained from assessors using paper-based forms can be made available in electronic format for further analysis and reporting. This can be achieved by scanning the forms into appropriate software. With fully computerized data collection, responses from assessors are entered directly into a desktop, notebook or hand-held computer.

Software

Software to manage the design and execution of sensory experiments is available and relatively mature. Products include FIZZ (BioSystèmes, France), Compusense 5 (Compusense, Canada), Tastel (ABT Informatique, France) and SIMS 2000 (New Jersey, USA).

20.6 Data analysis tools

20.6.1 Tests of statistical significance

Tests of statistical significance are central to the use of almost all sensory evaluation methods. Tests fall into two categories: (i) parametric tests, which assume that the data fits a predefined pattern of distribution, such as a normal distribution; and (ii) non-parametric tests, which make no assumptions about the distribution of the data. Common tests include the chi-squared test, F-test, T-test, Kramer's rank-sum test, Friedman test, and Analysis of Variance (ANOVA). A discussion of the various methods used to analyse the statistical significance of data is beyond the scope of this chapter. The reader is referred elsewhere (Stone and Sidel, 1993; Carpenter *et al.*, 2000). Basic statistical testing can be carried out using spreadsheet applications (such as Microsoft® Excel). More advanced procedures require the use of specialist software.

20.6.2 Design of Experiments (DoE)

DoE is a powerful statistical technique which is used by some breweries to improve their products and the efficiency and robustness of their processes. While publications relating to the application of DoE in beer production have been few, an example of what can be achieved can be found in a paper by Wormbs *et al.* (2004). It is likely that DoE will be increasingly used in brewery sensory applications on account of the fact that it allows a desired experimental target to be achieved with minimal effort on the part of both sensory analyst and brewery technologist.

20.6.3 Multivariate analyses

Multivariate analysis techniques can be used to explore relationships within complex data sets. Useful techniques include principal component analysis, generalized procrustes analysis, and cluster analysis. A list of software suited to multivariate analysis techniques can be found in Carpenter *et al.* (2000). Such techniques are constantly being developed. For example, two recent improvements to principal component analysis include automation of the process of identifying groups in the product space (Husson *et al.*, 2004) and detecting outliers in data sets (Nakai *et al.*, 2002).

A number of publications deal with the issue of predicting sensory perceptions from chemical and physical analyses of beers. While interesting and challenging from an academic viewpoint, this area is of considerable industrial significance as it could help reduce the need for routine tasting in breweries and in centrally located laboratories.

Multivariate models have been constructed using a variety of techniques, including multiple linear regression, partial least squares regression (Techakriengkrai *et al.*, 2004b), and neural network modelling (Kvaal and McEwan, 1996; Techakriengkrai *et al.*, 2004a). While some limited success has been obtained in the case of some attributes and constructs (Foster *et al.*, 2001) it appears that we are still some way from achieving our goal.

20.7 Data reporting and distribution tools

While the systems, tools and methods used to generate sensory data have become increasingly sophisticated and robust over the years, the same cannot be said for those used to generate reports and distribute them. While some breweries have attempted to integrate their sensory data within their Laboratory Information Management System (LIMS), this has not always led to a satisfactory outcome. To an extent this is because the structure and type of data produced by sensory analysis are somewhat different from that usually handled by LIMS. In the future, it is likely that we will see the emergence of better systems which allow data to be shared among different brewery sites in real time over the Internet. Such a system is already in use by one major brewery group.

20.8 Future trends

From a consideration of the progress described above, it is tempting to speculate on the future of sensory analysis in the brewing industry. Firstly, it is likely that sensory analysis will be regarded as increasingly important by the world's brewers, as they realize competitive advantage from such methods. To achieve this they will strive to develop tasters with the greatest possible competence, generating precise, repeatable results with the minimum expenditure of time (and money). Better tasters will mean that fewer are needed to perform the task well. They may take advantage of better selection methods, perhaps involving the application of psychometric testing in addition to currently used selection and screening tests. They will use detailed information from taster performance management systems to help tailor selection, screening and training methods to achieve the best possible results.

As brewery groups grow in size and brewers are separated by greater distances, the importance and influence of central (brewery group) panels is likely to decline, with local regional panels taking on the bulk of the responsibility for tasting. For this reason, standardization of test methods (including analysis of results), and improved methods of sample selection are likely to become increasingly significant. Predictive statistical methods and data mining tools will be used to reduce the numbers of samples routinely assessed and derive more information from the limited number of tests that are made. Assessment of identical samples on multiple sites may become increasingly common. The fields of sensory quality management and consumer testing will converge, allowing the 'voice of the customer' to be heard loud and clear within the confines of the brewery.

Distribution of sensory data is likely to change from being a batch (push) process to a continuous (pull) process, probably using secure Internet protocols. To control the risk of litigation, practices relating to the human resources side of beer tasting are likely to become increasingly specialized. On-line and at-line tests are likely to become more structured and disciplined, and better integrated into the brewery quality management system

Table 20.4 lists the key areas of competence relating to the sensory quality management activities of a modern brewery, together with the criteria against which they can be evaluated. With increased confidence in the results of sensory tests, brewers will become more proficient at diagnosing, dealing with, and ultimately preventing beer flavour problems, leveraging more than a century of research on the origin and nature of beer flavours. This is likely to give rise to significant cost savings.

20.9 Sources of further information

20.9.1 Interest groups and societies

Interest groups and societies which specialize in the area of sensory analysis include the following:

The European Sensory Network: http://www.esn-network.com;
Italian Society of Sensory Science: http://www.scienzesensoriali.it/;
The Sensometric Society: http://www.sensometric.org/;
The Sensory Nexus: http://www.sensory.org/;
ASTM Technical Committee E18: http://www.astm.org/cgi-bin/SoftCart.exe/COMMIT/COMMITTEE/E18.htm?E+mystore.

Table 20.4 Key focus areas for sensory quality management in a modern brewery

Focus area	Assessment criteria
Organization and staffing	The company should have the right level of resource available within the sensory function, and these should be organized in the most efficient way.
Brand specifications	There should be a clear view of what sort of products the company is trying to make, and Brand Flavour Fingerprints or their equivalent should be available for all brands.
Competitive position	The company should have a clear view on its competitive position from the perspective of beer flavour and consumer behaviour. The flavour of competitors' products should be understood as well as that of their own.
Inter-plant matching	If a brand is produced in more than one brewery, there should be systems in place to assure plant-to-plant consistency and to provide an objective measure of the size and nature of any differences.
Link to action	All sensory reports should lead to clear actions, with a direct link between the flavour issue highlighted and the person responsible for bringing about improvements.
Sensory testing facilities	The company should have access to sensory testing facilities appropriate to its needs and objectives.
Product release activities	No beer should leave the brewery unless it has been tasted by a group of expert tasters.
Taster competence	The level of taster competence aimed for should be clearly defined and documented, and appropriate to the company's needs and objectives.
Taster validation	The company's tasters should meet or exceed the level of competency aspired to. A robust system should be in place to allow valid measurements to be made, ideally calibrated against international benchmarks.
Training and development	A system of training and development should be in place to cover sensory professionals, tasters, and the users of information derived from sensory tests.
Reward and recognition	A system of reward and recognition should be in place to assure taster motivation, performance and attendance.
Health, safety and liability	A rigorous assessment of the risks and liabilities associated with running a sensory operation related to alcoholic beverages should be carried out and appropriate protective actions implemented.

20.9.2 Journals

Journals which specialize in sensory analysis include *Food Quality and Preference*, *Journal of Sensory Studies*, *Chemical Senses* and *Journal of the Science of Food and Agriculture*.

20.10 Acknowledgements

The author acknowledges the assistance of Valerie Simpson and Hilary Flockhart in preparing this chapter, and Frieda Dehrmann and Simon Hadman for helpful comments.

20.11 References

AERTGEERTS B, BUNTINX F, ANSOMS S and FEVERY J (2002), 'Questionnaires are better than laboratory tests to screen for current alcohol abuse or dependence in a male inpatient population', *Acta Clinica Belgica*, 57(5), 241–249.

ANON (1995), *Sensory Analysis Manual*, London, Institute of Brewing, 44 pp.

ANON (1997), *IOB Methods of Sensory Analysis Health and Safety Aspects General Policy Statement*, Addendum to *Sensory Analysis Manual*, London, Institute of Brewing (first published 1995).

ANON (2005), 'IFST: Policy statement on ethical and professional practices for the sensory analysis of foods'. Available from *http://www.ifst.org/hoptop44.htm*, issued May 2005, accessed 1 September 2005.

ASTM (1997), *Standard test method for sensory analysis – Triangle test*, ASTM E1885-97.

ASTM (2001a), *Standard guide for sensory evaluation of beverages containing alcohol*, ASTM E1879-00.

ASTM (2001b), *Standard test method for directional difference test*, ASTM E2164-01.

BALLESTER J, DACREMONT C, LE FUR Y and ETIÉVANT P (2005), 'The role of olfaction in the elaboration and use of the Chardonnay wine concept', *Food Quality and Preference*, 16, 351–359.

BARCENAS P, PEREZ ELORTONDO F J, SALMERON J and ALBISU M (2000), 'Analisis sensorial de alimentos', *Alimentaria*, 31–38.

BARTOSHUK L M (2000), 'Comparing sensory experiences across individuals: Recent psychophysical advances illuminate genetic variation in taster perception', *Chemical Senses*, 25, 447–460.

BARYLKO-PIKIELNA N and MATUSZEWSKA I (2000), 'Sensory analysis in food research, quality assurance and product development', *Acta Alimentaria*, 29(3), 255–271.

BEN SLAMA M, HEYD B, DANZART M and DUCAUZE C J (1998), 'Plans *D*-optimaux: une stratégie de réduction du nombre de produits en cartographie des préférences', *Sciences des Aliments*, 18, 471–483.

BI J (2005), 'Similarity testing in sensory and consumer research', *Food Quality and Preference*, 16, 139–149.

BOUGHTON R A and SIMPSON W J (2000), 'A ten-part program for rapid training and validation of novice beer flavour assessors', in *Proceedings of the 26th Convention of the Institute of Brewing Asia Pacific Section*, 168–173.

BROCKHOFF P B and SCHLICH P (1998), 'Handling replications in discrimination tests', *Food Quality and Preference*, 9, 303–312.

CAPORALE G and MONTELEONE E (2004), 'Influence of information about manufacturing process on beer acceptability', *Food Quality and Preference*, 15, 271–278.

CARPENTER R P, LYON D H and HASDELL T A (2000), *Guidelines for Sensory Analysis in Food Product Development and Quality Control*, 2nd edn, Gaithersburg, MD, Aspen Publishers.

CHAPMAN K W, LAWLESS H T and BOOR K J (2001), 'Quantitative descriptive analysis and principal component analysis for sensory characterization of ultrapasteurized milk', *American Dairy Science Association*, 84, 12–20.

CHAVANNE P C and COURCOUX PH (2001), 'Preference mapping using a latent class vector model', *Food Quality and Preference*, 12, 369–372.

CHOLLET S and VALENTIN D (2001), 'Impact of training on beer flavour perception and description: Are trained and untrained subjects really different?', *Journal of Sensory Studies*, 16, 601–618.

CHOLLET S, VALENTIN D and ABDI H (2005), 'Do trained assessors generalize their knowledge to new stimuli?', *Food Quality and Preference*, 16, 13–23.

CLIFF M A and DEVER M (1996), 'A proposed approach for evaluating expert wine judge performance using descriptive statistics', *Journal of Wine Research*, 7(2), 83–90.

COSTELL E, VICENTA PASTOR M, IZQUIERDO L and DURÁN L (2000), 'Relationships between acceptability and sensory attributes of peach nectars using internal preference mapping', *European Food Research Technology*, 211, 199–204.

COURCOUX PH and CHAVANNE P C (2001), 'Preference mapping using a latent class vector model', *Food Quality and Preference*, 12, 369–372.

CURT C, TRYSTRAM G and HOSSENLOPP J (2001), 'Formalisation of at-line human evaluations to monitor product changes during processing. Integration of human decision in the dry sausage ripening process', *Sciences des Aliments*, 21, 663–681.

DATTATREYA B S, KAMATH A and BHAT K K (2002), 'Developments and challenges in flavour perception and measurement – a review', *Food Reviews International*, 18(2–3), 223–242.

DELARUE J and SIEFFERMANN J-M (2004), 'Sensory mapping using Flash profile. Comparison with a conventional descriptive method for the evaluation of the flavour of fruit dairy products', *Food Quality and Preference*, 15(4), 383–392.

DIJKSTERHUIS G B and EILERS P (1997), 'Modelling time-intensity curves using prototype curves', *Food Quality and Preference*, 8, 131–140.

DJORDJEVIC J, ZATORRE R J, PETRIDES M and JONES-GOTMAN M (2004), 'The mind's nose: Effects of odor and visual imagery on odor detection', *Psychological Sciences*, 15(3), 143–148.

DRAKE M A, MCINGVALE S C, GERARD P D, CADWALLADER K R and CIVILLE G V (2001), 'Development of a descriptive language for Cheddar cheese', *Journal of Food Science*, 66(9), 1422–1427.

DRAKE M A, KARAGUL-YUCEER Y, CADWALLADER K R, CIVILLE G V and TONG P S (2003), 'Determination of the sensory attributes of dried milk powders and dairy ingredients', *Journal of Sensory Studies*, 18, 199–216.

DRANSFIELD E, MORROT G, MARTIN J F and NGAPO T M (2004), 'The application of a text clustering statistical analysis to aid the interpretation of focus group interview', *Food Quality and Preference*, 15(5), 477–488.

DUFFY V B, DAVIDSON A C, KIDD J R, KIDD K K, SPEED W C, PAKISTIS A J, REED D R, SNYDER D J and BARTOSHUK L M (2004), 'Bitter receptor gene (TAS2R38), 6-*n*-Propylthiouracil

(PROP) bitterness and alcohol intake', *Alcoholism: Clinical and Experimental Research*, 28(11), 1629–1637.

DUIZER L M, BLOOM K and FINDLAY C J (1997), 'Dual-attribute time-intensity sensory evaluation: a new method for temporal measurement of sensory perceptions', *Food Quality and Preference*, 8, 261–269.

EILERS P and DIJKSTERHUIS G B (2004), 'A parametric model for time-intensity curves', *Food Quality and Preference*, 15, 239–245.

FERNANDEZ-ALBALAT M P R, FERNANDEZ MARTINEZ M A and MENDEZ DONEGA J (2005), 'Candidates recruitment, screening and selection for a technique panel constitution', *Milchwissenschaft*, 60, 21–23.

FINDLAY C, CASTURA J, SCHLICH P and LESSCHAEVE I (2004), 'Optimizing the proficiency of wine panels trained using feedback calibration', paper presented at the 7th Sensometrics Conferences, University of California, Davis, CA, 28–30 July 2004. Available on-line from http://www.sensometric.org/lectures/findlay32.pdf, accessed 12 November 2005.

FOSTER R T III, SAMP E J and PATINO H (2001), 'Multivariate modeling of sensory and chemical data to understand staling in light beer', *Journal of the American Society of Brewing Chemists*, 59(4), 201–210.

GATCHALIAN, M M, DE LEON S Y and YANO T (1991), 'Control chart technique: a feasible approach to measurement of panelist performance in product profile development', *Journal of Sensory Studies*, 6, 239–254.

GEORGE, M L (2003), *Lean Six Sigma for Service,* New York, McGraw-Hill, 386 pp.

GRAY A, FERGUSON M M and WALL J G (1998), 'Wine tasting and dental erosion. Case report', *Australian Dental Journal*, 43, 32–34.

GREEN B G and GEORGE P (2004), ' "Thermal Taste" predicts higher responsiveness to chemical taste and flavor', *Chemical Senses*, 29(7), 617–628.

GREENHOFF K and MacFIE H J H (1994), 'Preference mapping in practice', in MacFie H J F and Thomson D M H, *Measurement of Food Preference*, Glasgow & London, Blackie Academic and Professional, 137–166.

HAYS, C L (2004), *Pop: Truth and Power at the Coca-Cola Company*, London, Hutchinson, 352 pp.

HEGARTY P, CHILVER J and THREAPLETON L (2001), 'Sensory analysis – a bridge with the consumer. Use of an external expert panel', in *Proceedings of the 28th European Brewery Convention Congress*, Nürnberg, Fachverlag Hans Carl, 89–98.

HERSLETH M, UELAND O, ALLAIN H and NÆS T (2005), 'Consumer acceptance of cheese, influence of different testing conditions', *Food Quality and Preference*, 16, 103–110.

HILL P (on behalf of the EBC Sensory Subgroup) (2003), 'Keeping the flavour wheel turning – the development of a flavour stability wheel', in *Proceedings of the European Brewery Convention, Dublin*, Nürnberg, Fachverlag Hans Carl, 791–795.

HOOTMAN R C (1992), *Manual on descriptive analysis testing for sensory evaluation*, ASTM Manual 13, Philadelphia, American Society for Testing and Materials, 52 pp.

HUGHES P S and SIMPSON W J (1996), 'Bitterness of congeners and stereoisomers of hop-derived bitter acids found in beer', *Journal of the American Society of Brewing Chemists*, 54, 234–237.

HUSSON F, BOCQUET V and PAGÈS J (2004), 'Use of confidence ellipses in a PCA applied to sensory analysis application to the comparison of monovarietal ciders', *Journal of Sensory Studies*, 19, 510–518.

ISO (1983), *Standard 5495: Sensory Analysis – Methodology – Paired Comparison Test*, Paris, International Organization for Standardization.

ISO (1988), *Standard 8589: Sensory Analysis – General Guidance for the Design of Test Rooms*, Paris, International Organization for Standardization.

ISO (1993), *Standard 8586-1: Assessors for Sensory Analysis. Part 1. Guide to the Selection, Training and Monitoring of Selected Assessors*, Paris, International Organization for Standardization.

ISO (1994), *Standard 8586-2: Assessors for Sensory Analysis. Part 2. Guide to the Selection, Training and Monitoring of Experts*, Paris, International Organization for Standardization.

ISO (2003), *Standard 13299: Sensory Analysis – Methodology – General Guidance for Establishing a Sensory Profile*, Paris, International Organization for Standardization.

ISO (2004), *Standard 4120: Sensory analysis – Methodology – Triangle Test*, Paris, International Organization for Standardization.

JACK F R, PIGGOTT J R and PATERSON A (1994), 'Use and appropriateness in cheese choice, and an evaluation of attributes influencing appropriateness', *Food Quality and Preference*, 5, 281–290.

JEONG S-Y, CHUNG S J, SUH D S, SUH B C and KIM K O (2004), 'Developing a descriptive analysis procedure for evaluating the sensory characteristics of soy sauce', *Journal of Food Science*, 69(8), 319–325.

JOHNSON E A and VICKERS Z (2004), 'The effectiveness of palate cleansing strategies for evaluating the bitterness of caffeine in cream cheese', *Food Quality and Preference*, 15(4), 311–316.

KANEKO A, RAMAIN J and PALOC R (1994), 'Dégustation et santé prévention des risques', *Revue Française d'Oenologie*, 34, 11–14.

KELLY G A (1955), *The Psychology of Personal Constructs: a Theory of Personality*, New York, Norton.

KIHLBERG I, JOHANSSON L, LANGSRUD Ø and RISVIK E (2005), 'Effects of information on liking of bread', *Food Quality and Preference*, 16(1), 25–35.

KIM K-O and O'MAHONY M (1998), 'A new approach to category scales of intensity. I: Traditional versus rank-rating', *Journal of Sensory Studies*, 13, 241–249.

KIM K-O, ENNIS D M and O'MAHONY M (1998), 'A new approach to category scales of intensity. II: Use of *d′* values', *Journal of Sensory Studies*, 13, 251–267.

KING B M and MOREAU N A (1996), 'A comparison of bitter perception in high-alcohol, low alcohol and alcohol-free beer', *Journal of the Institute of Brewing*, 102, 419–425.

KöSTER E P, DEGEL J and PIPER D (2002), 'Proactive and retroactive interference in implicit odor memory', *Chemical Senses*, 27, 191–206.

KOUSSISSI E (2003), 'Sensory character in relation to composition of Greek red wines', PhD Thesis, University of Strathclyde.

KOUSSISSI E, PATERSON A and CRISTOVAM E (2002), 'Sensory discrimination of dry red wines from Greece', *Journal of Wine Research*, 13, 165–179.

KOUSSISSI E, PATERSON A and PIGGOTT J R (2003), 'Sensory flavour discrimination of Greek dry red wines', *Journal of the Science of Food and Agriculture*, 83, 797–808.

KUESTEN C L, MCLELLAN M R and ALTMAN N (1994), 'Computerized panel training: effects of using graphic feedback on scale usage', *Journal of Sensory Studies*, 9, 413–444.

KUNERT J and MEYNERS M (1999), 'On the triangle test with replications', *Food Quality and Preference*, 12, 385–391.

KVAAL K and McEWAN J A (1996), 'Analysing complex sensory data by non-linear artificial neural networks', Volume 16 in Naes T and Risvik E (eds), *Data Handling in Science and Technology, Multivariate Analysis of Data in Sensory Science*, Amsterdam, Elsevier.

LABBE D, RYTZ A and HUGI A (2004), 'Training is a critical step to obtain reliable product profiles in a real food industry context', *Food Quality and Preference*, 15, 341–348.

LAND D G and SHEPHERD R (1988), 'Scaling and ranking methods', in Piggott J R (ed.), *Sensory Analysis of Foods*, 2nd edn, Amsterdam, Elsevier Applied Science, 155–185.

LANGSTAFF S A and LEWIS M J (1993), 'The mouthfeel of beer – a review', *Journal of the Institute of Brewing*, 99, 31–37.

LAWLESS H T and HEYMANN H (1998), *Sensory Evaluation of Food: Principles and Practices*, New York, Chapman & Hall.

LEA P, NAES T and RØDBOTTEN M (1997), *Analysis of Variance for Sensory Data*, New York, John Wiley & Sons.

LYON D H (ed.) (2001), *Guideline no. 35, International guidelines for proficiency testing in sensory analysis*, Chipping Campden, UK, CCFRA.

MACFIE H J, BRATCHELL N, GREENHOFF K and VALLIS L V (1989), 'Designs to balance the effect of order of presentation and first-order carry-over effects in hall tests', *Journal of Sensory Studies*, 4, 129–148.

MACKEY A O and JONES P (1954), 'Selection of members of a food tasting panel: Discernment of primary tastes in water solution compared with judging ability for foods', *Food Technology*, 8, 527–530.

MACRAE A W and GEELHOED H (1992), 'Preference can be more powerful than detection of oddity as a test of discriminability', *Perception & Psychophysics*, 51, 179–181.

MARTÍ M P, BOQUÉ R, BUSTO O and GUASCH J (2005), 'Electronic noses in the quality control of alcoholic beverages', *Trends in Analytical Chemistry*, 24, 57–66.

McCUSKER M T, BASQUILLE J, KHWAJA M, MURRAY-LYON I M and CATALAN J (2002), 'Hazardous and harmful drinking: a comparison of the AUDIT and CAGE screening questionnaires', *Quarterly Journal of Medicine*, 95, 591–595.

McDONNELL E, HULIN-BERTAUD S, SHEEHAN E M and DELAHUNTY C M (2001), 'Development and learning process of a sensory vocabulary for the odor evaluation of selected distilled beverages using descriptive analysis', *Journal of Sensory Studies*, 16, 425–445.

McEWAN J A (1996), 'Preference mapping for product optimization', in Næs T and Risvik E (eds), *Multivariate Analysis of Data in Sensory Science*, Amsterdam, Elsevier, 71–102.

MEILGAARD M C (1989), 'Use of the taste panel in flavour stability studies: organization, threshold, synergism, antagonism', *Louvain Brewing Letters*, 2, 3–11.

MEILGAARD M C, DALGLIESH C E and CLAPPERTON J F (1979), 'Beer flavor terminology', *Journal of the American Society of Brewing Chemists*, 37, 47–52.

MEILGAARD M C, REID D S and WYBORSKI K A (1982), 'Reference standards for beer flavor terminology system', *Journal of the American Society Brewing Chemists*, 40, 119–128.

MEILGAARD M, CIVILLE G V and CARR B T (1999), *Sensory Evaluation Techniques*, 3rd edn, Boca Raton, FL, CRC Press, 387 pp.

MILLER P M and ANTON R F (2004), 'Biochemical alcohol screening in primary health care', *Addictive Behaviors*, 29, 1427–1437.

MIRAREFI S, MENKE S D and LEE S-Y (2004), 'Sensory profiling of Chardonel wine by descriptive analysis', *Journal of Food Science*, 69(6), 211–217.

MONTALTO N J and BEAN P (2003), 'Use of contemporary biomarkers in the detection of chronic alcohol use', *Medical Science Monitor*, 9(12), 285–290.

MONTELEONE E, FREWER L, WAKELING I and MELA D J (1998), 'Individual differences in starchy food consumption: The application of preference mapping', *Food Quality and Preference*, 9(4), 211–219.

MURRAY J M and DELAHUNTY C M (2000), 'Consumer preference for Irish farmhouse and factory cheeses', *Irish Journal of Agricultural and Food Research*, 39, 433–449.

MURRAY J M, DELAHUNTY C M and BAXTER I A (2001), 'Descriptive sensory analysis: past, present and future', *Food Research International*, 34, 461–471.

NAES T (1990), 'Handling individual differences between assessors in sensory profiling', *Food Quality and Preference*, 2, 187–199.

NAES T (1998), 'Detecting individual differences among assessors and differences among replicates in sensory profiling', *Food Quality and Preference*, 9, 107–110.

NAES T and SOLHEIM R (1991), 'Detection and interpretation of variation within and between assessors in sensory profiling', *Journal of Sensory Studies*, 6, 159–177.

NAGAO Y, KODAMA H, YAMAGUCHI T, YONEZAWA T, TAGUCHI A, FUJINO S, MORIMOTO K and FUSHIKI T (1999), 'Reduced urination rate while drinking beer with an unpleasant taste and off-flavor', *Bioscience Biotechnology and Biochemistry*, 63, 468–473.

NAKAI S, AMANTEA G, NAKAI H, OGAWA M and KANAGAWA S (2002), 'Definition of outliers using unsupervised principal component similarity analysis for sensory evaluation of foods', *International Journal of Food Properties*, 5(2), 289–306.

NIELSEN D, HYLDIG G and SØRENSEN R (2005), 'An effective way to minimize drifting and monitor the performance of a sensory panel during long-term projects – A case study from a project on herring quality', *Journal of Sensory Studies* 20(1), 35–47.

N'KOUKA K D, KLEIN B P and LEE S Y (2004), 'Developing a lexicon for descriptive analysis of soymilks', *Journal of Food Science*, 69(7), 259–263.

O'MAHONY M (1995), 'Who told you the triangle test was simple?', *Food Quality and Preference*, 6, 227–238.

PAGÈS J and PÉRINEL E (2003), 'Panel performance and number of evaluations in a descriptive sensory study', *Journal of Sensory Studies*, 19, 273–291.

PARKER D K and MURRAY J (2003), 'The different roles of consumer and expert panels', in *Proceedings of the European Brewery Convention, Dublin*, Nürnberg, Fachverlag Hans Carl, 814–821.

PARR W V, WHITE K G and HEATHERBELL D A (2004), 'Exploring the nature of wine expertise: what underlies wine experts' olfactory recognition memory advantage?', *Food Quality and Preference*, 15, 411–420.

PEñA Y LILLO M, LATRILLE E, CASAUBON G, AGOSIN E, BORDEU E and MARTIN N (2005), 'Comparison between odour and aroma profiles of Chilean Pisco spirit', *Food Quality and Preference*, 16, 59–70.

PERON R M and ALLEN G L (1988), 'Attempts to train novices for beer flavor discrimination: a matter of taste', *The Journal of General Psychology*, 115, 403–418.

PFENNINGER H (1993), *Brautechnische Analysenmethoden, Vol. II*, 3rd impression, Fresing, Austria, Selbstverlag der MEBAK, 81–83.

PIGGOTT J R and HUNTER E A (1999), 'Evaluation of assessor performance in sensory analysis', *Italian Journal of Food Science*, 11, 289–303.

PIGGOTT J R and WATSON M P (1992), 'A comparison of free-choice profiling and the repertory grid method in the flavour profiling of cider', *Journal of Sensory Studies*, 7, 133–145.

POWERS J J (1988), 'Current practices and application of descriptive methods', in Piggott J R (ed.), *Sensory Analysis of Foods*, 2nd edn, Amsterdam, Elsevier Applied Science, 187–266.

PRESCOTT J, NORRIS L, KUNST M and KIM S (2005), 'Estimating a "consumer rejection threshold" for cork taint in white wine', *Food Quality and Preference*, 16, 345–349.

SAUVAGEOT F and RABIER G (2004), 'Une comparaison entre deux formes de l'épreuve par paire: la forme hédonique et la forme non hédonique', *Sciences des Aliments*, 24(4), 289–306.

SCHLICH P and MCEWAN J A (1992), 'Cartographie des préférences', *Sciences des Aliments*, 12, 339–355.

SCHÖNBERGER CH, KROTTENTHALER M and BACK W (2004), 'Formulation of an alternative tasting scheme for the sensory evaluation of beer', *Monatsschrift für Brauwissenschaft*, April, 16–24.

SHARPE R, PAWLOWSKY K and CHANDRA S (2003), 'Drinkability and drinkability mapping', in *Institute and Guild of Brewing Africa Section, Proceedings of the 9th Brewing Convention, Victoria Falls, Zambia*, South Africa, IGB Africa Section, 58–64.

SIMPSON W J (2004), 'Modern beer tasting operations', *The Brewer International*, 4, 31–34.

SIMPSON W J and CANTERRANNE E (2001), 'Sensory analysis in modern brewery operations', *Brauwelt International*, 2001(iv), 280–286.

SIMPSON W J, HADMAN S I and BOCCORH R K (1999), 'Validation of beer taster performance', in *Institute of Brewing Africa Section, Proceedings of the 7th Brewing Convention, Nairobi, Kenya*, Hout Bay, IOB, 38–45.

SMYTHE J E and BAMFORTH C W (2002), 'The effect of perceived beer history on reported preferences by sensory panels with different levels of training', *Journal of the Institute of Brewing*, 108, 34–36.

STONE H and SIDEL J L (1993), *Sensory Evaluation Practices*, New York, Academic Press, 338 pp.

SUST L, GACULA M JR, BEARD S and WASHAM R (1985), 'Degree of difference test method in sensory evaluation of heterogeneous product types', *Journal of Food Science*, 50, 511–513.

TABAKOOF B, MARINIZ L and HOFFMAN P (2004), *Genetic diagnosis of alcoholism subtype*, Patent no. WO2004067763.

TECHAKRIENGKRAI I, PATERSON A and PIGGOTT J R (2004a), 'Relationships of sweetness in lager to selected volatile congeners', *Journal of the Institute of Brewing*, 110, 360–366.

TECHAKRIENGKRAI I, PATERSON A, TAIDI B and PIGGOTT J R (2004b), 'Relationships of sensory bitterness in lager beers to iso-alpha-acid contents', *Journal of the Institute of Brewing*, 110, 51–56.

TECHAKRIENGKRAI I, PATERSON A and PIGGOTT J R (2004c), 'Time intensity parameters of sweetness perceptions in lager beers', *Journal of the Institute of Brewing*, 110, 352–359.

TEN KLEIJ F, LAU K and MUSTERS P A D (2003), 'Text analysis of open-ended survey responses: a complementary method to preference mapping', *Food Quality and Preference*, 14, 43–52.

THOMPSON J L, DRAKE M A, LOPETCHARAT K and YATES M D (2004), 'Preference mapping of commercial chocolate milks', *Journal of Food Science*, 69(9), 406–413.

THOMPSON S (1999), 'Difference-from-control sensory test (DCM)', *Journal of the American Society of Brewing Chemists*, 57, 168–171.

VERDU JOVER A J, MONTES F J L and FUENTES M M F (2004), 'Measuring perceptions of quality in food products: the case of red wine', *Food Quality and Preference*, 15, 453–469.

WORMBS G, LARSSON A, ALM J, TUNKLINT-ASPELIN C, STRINNING O, DANIELSSON E and LARSSON H (2004), 'The use of design of experiment and sensory analysis as tools for the evaluation of production methods for milk', *Chemometrics and Intelligent Laboratory Systems*, 73, 67–71.

WURST F M, ALEXSON S, WOLFERSDORF M, BECHTEL G, FORSTER S, ALLING C, ARADÓTTIR S, JACAU K, HUBER P, ALLEN J P, AUKWäRTER V and PRAGST F (2004), 'Concentration of fatty acid ethyl esters in hair of alcoholics: comparison to other biological state markers and self-reported ethanol intake', *Alcohol and Alcoholism*, 39, 33–38.

XIONG R and MEULLENET J F (2004), 'Application of multivariate adaptive regression splines (MARS) to the preference mapping of cheese sticks', *Journal of Food Science*, 69(4), 131–139.

YOUNG N D, DRAKE M, LOPETCHARAT K and McDANEIL M R (2004), 'Preference mapping of Cheddar cheese with varying maturity levels', *Journal of Dairy Science*, 87, 11–19.

YOUNG N D, SANDERS T H, DRAKE M A, OSBORNE J and CIVILLE G V (2005), 'Descriptive analysis and US consumer acceptability of peanuts from different origins', *Food Quality and Preference*, 16, 37–43.

ZAMORA M C and GUIRAO M (2004), 'Performance comparison between trained assessors and wine experts using specific sensory attributes', *Journal of Sensory Studies*, 19, 530–545.

21

The future of brewing

C. W. Bamforth, University of California, USA

21.1 Choice

The key word is 'choice' (Fig. 21.1). Various factors influence the drinker's decision of which beverage to buy.

Innate pleasure

A range of hedonic parameters determines the acceptability of a product. We might take these in impact sequence.

First the customer encounters the package itself, whether it is a small-pack container or the dispense periphery of a draught product. How is its appeal? Is its colour appealing? Is there enticing imagery? How meaningful and satisfying is the labelling? What is the integrity quotient: is the bottle scuffed, the foiling damaged, the label 'square-on' in application? Is the can dented? Is the pump prominent? Is the product 'fit for purpose': is it convenient for carrying (weight, 'hold-ability')? Is it in the appropriate container for the drinking location?

Having accessed the package (and assuming that the beer hasn't gushed), the customer is confronted by the liquid itself. Or, rather, they will be if they choose to deliver the drink into a glass. Likely he or she will first note the foam. Is there too much or too little? Does it linger? Does it lace? Is it attractive: are the

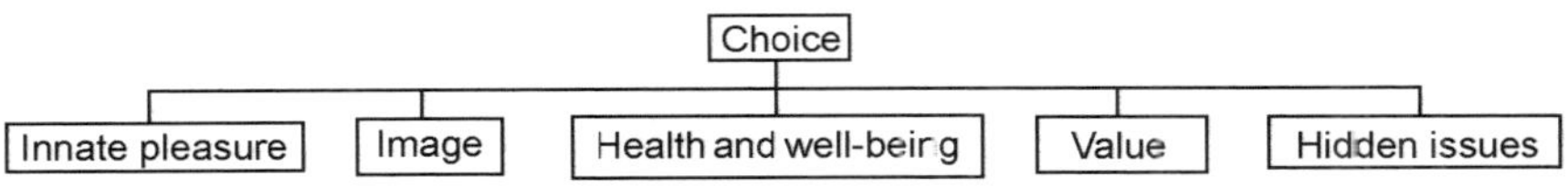

Fig. 21.1 Components of choice.

bubbles fine, white and even, or are they coarse and blotchy? Does the head match with anticipation for a beer of that style?

Is the colour of the beer appealing? What are the subtleties of its hue and shade? Does it marry with the expectation afforded by the message from the container? Is the beer bright? If not, is the 'cloud' appropriate and meeting expectation?

What does the beer smell and taste like? Does it agree with what the customer anticipates, due to either messages from the container or the appearance or from previous organoleptic experience? Is the flavour pleasing, does it linger appropriately, does it merit revisitation? Does it accompany the other sensory experiences being encountered by the drinker – environment, other foodstuffs?

In all respects, how does this beer compare with other selections available to the purchaser, including alternative beverages?

Image
Do I want to be seen with this product? Do its 'credentials' match my own? What are pressure groups telling me?

Health and well-being
What will this beer do for me? Will it meet a short-term requirement, e.g. will it slake my thirst? But will it actually do me some good? And is it likely to do me any harm?

Value
Balancing all other aspects, does this product represent good value for money? Am I getting an adequate 'bang for my buck'? Or am I investing in quality that satisfies, for instance, peer-pressure?

Hidden issues
What is lurking in the background of this product? Is the brewer using choicest raw materials? How traditional are they – have we got 'chemical beer'? How clean is the brewery wherein it was made? What are those brewers doing to the environment directly or indirectly as a result of their processes? Is there any semblance of the 'footballs stitched by child labour'-type scandal?

21.2 Impact of choice

Let us now relate these various issues to the technical needs that flow from them (Table 21.1).

21.3 Technical need drives research

In turn we can highlight from the identified technical needs where the research focus might usefully lie (Table 21.2).

Table 21.1 Technical needs arising from issues of choice

Issue of choice	Technical needs
Package appeal	• Design of suitable packages for feel, convenience, drinking location • Packages that don't jeopardise (but which might even enhance) beer quality • Procedures that allow 'clean' labelling – minimal additives • Robust and reliable packaging lines
'Assuming the beer hasn't gushed'	• No gushing risks
'If they choose to deliver the drink into a glass'	• Designing containers which enhance consumption therefrom
Foam quality	• Stable and attractive foams in the right amount and appropriate to the genre
Colour appeal	• Means for delivering consistent and appropriate colours
Bright beer	• Consistent clarity throughout shelf-life
Flavour	• Aroma and taste in keeping with other quality parameters • Consistent flavour – no surprises • Beers that complement food • Beers that complement drinking locations and events
Beer as the beverage of choice *coupled with* image	• How to 'blow other beverages out of the water' • Other beverages from a brewery
Will this beer slake my thirst?	• What are the relevant factors?
Will this beer do me good?	• Understanding and enhancing wholesomeness factors • Design of 'functional' beers to appeal to a specific dietary pressure (*cf.* light beers)
Will this beer do me harm?	• Understanding and avoiding health negative aspects
Value for money	• Cost paring • Quality enhancement
Hidden issues	• Availability of choice raw materials • Taxation legislation and opportunities • 'Greening' of malting and brewery operations – reduced emissions and wastes • Energy and water conservation

Table 21.2 Relating needs to research focus

Technical need	Research focus
Suitable packages	• Enhanced plastics – barrier properties, hand feel • Robust glass • Entirely novel containers – including other than metal, glass or plastic
Packages enhancing beer quality	• Self-temperature regulating containers • Foam-promoting packages (beyond the widget – and for beers of 'normal' carbonation and no nitrogen) • Packages (and glasses) which bind materials that jeopardise quality (flavour, foam, clarity negatives)
'Clean' labelling	• Avoidance of additives: enhancing endogenous stabilisation; fundamental appreciation of factors determining instability
Robust and reliable packaging lines	• Zero-defects packaging at high speed
No gushing risks	• Identification of gushing factors, especially of fungal origin, and rapid methods for detecting them in raw materials
Drinker-friendly packages	• 'Glass as a package' containers (i.e. take off the lid, lo! you have a glass) • Neck designs to facilitate direct drinking
Stable and attractive foams	• How to deliver robust foam not susceptible to lipid damage in trade • How to deliver exactly the desired amount of foam to satisfy the demands of different consumers • How to achieve stable (*and appealing*) foams despite fluctuations in foam-positive components
Consistent colour	• Green malt composition and kilning/roasting conditions with respect to hue and intensity of colour
Consistent clarity	• *Most of the information is in place*
Aroma and taste in keeping with other quality parameters	• The relationship between beer composition and perceived flavour (aroma, taste, texture) • How to achieve genuine quality in lower alcohol products – the relationship between alcohol and flavour • Factors determining drinkability • The psychophysics of beer drinking • The physiology of taste
Consistent flavour – no surprises	• Avoidance of flavour change in package
Beers that complement food	• Psychophysics physiology and of food/beer matches
Beers that complement drinking locations and events	• Psychophysics of beer/location interactions

Table 21.2 Continued

Technical need	Research focus
How to 'blow other beverages out of the water'	• Understanding the consumer's understanding of beer versus other drinks: how to fold the desirable elements of all into a single super-drink
Other beverages from a brewery	• Achieving versatility in appearance and taste of beverages with a minimum of process streams
Thirst factors	• Optimising beer composition in relation to thirst relief
Health positive factors	• Why exactly is alcohol in moderation beneficial? • Any genuine benefits from the soluble fibre in beer? • Vitamins in balance (avoidance of thiamine shortage) • Beers for specific types of customer – e.g. coeliac sufferers
Health negative factors	• Systems for rapidly identifying, evaluating danger and defusing issues • What, if anything, is responsible for headaches after consuming beer?
Cost paring	• Automation – in-line analytical procedures and response mechanisms • Process intensification – maximising beer production per unit of capacity
Raw material availability	• Raw material composition and physiology in relation to process performance and quality • Development of pest- and disease-resistant strains
Taxation legislation	• Maximising beer output per unit of taxation – exploiting legitimate loopholes in legislation (*cf. Happoshu*) and how to do so with no discernible jeopardising of quality
'Greening'	• Recycling (and if possible adding value to) emissions and co-products • Mechanisms for minimising co-products
Energy and water conservation	• Process intensification (e.g. stronger worts and higher pitching gravities, continuous systems) • Recycling • Low temperature processing

21.4 Global influences

Lest we forget, the brewing business is not operating in a vacuum. Several gross factors impact upon brewing, its future and the demands for research within it (Fig. 21.2).

Of natural pre-eminent significance is population: what is the size of the market? The rate of population growth since 1970 has exceeded the growth in

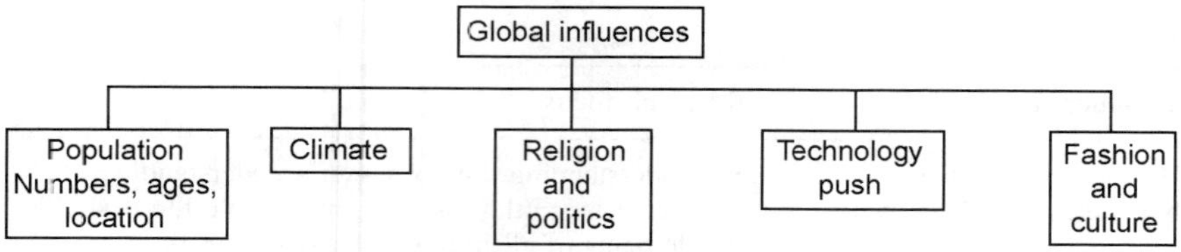

Fig. 21.2 Global influences.

total beer volumes, quite simply because much of the growth has occurred in countries without the culture or the currency to buy beer. However, the potential market will nevertheless grow significantly, particularly as we might envision increased disposable income and leisure time in many cultures. It also warrants mentioning that there will be a net ageing of the population over the coming quarter of a century. The percentage of people aged below 50 will decline, while the proportion older than this will increase substantially. Product concepts should be targeted not only at the younger element! It is incumbent upon brewers to ensure that their technical programmes are in place to satisfy the potential growth – the establishment of brewing facilities in suitable locations, knowledge of how to produce a stable product, ability to handle a diversity of raw material qualities, use of robust control mechanisms, and so on.

This presupposes that traditional raw materials will continue to be available. Recent data on global warming are alarming, presaging debate about the projected shape of agriculture in future decades (unless checked, what will this mean for crop quality and availability?). The pressure is on greenhouse gas emissions (which in brewing terms means far more for carbon dioxide generated in energy generation at all stages in production of packaged beer and its raw materials than in fermentation). And there will be no lessening in the tendency of those in agriculture to favour high value products. Hopefully those will continue to be malting barley and hops.

Political decisions on this issue will be of prime significance. They do say you shouldn't discuss religion or politics in polite society – but for a product such as beer they are seldom far away. Politicians will continue to bear major influence through the levy of taxes and the driving of legislation, such as what ingredients may or may not be employed in brewing and how packages must be labelled. For religious reasons, vast swathes of humanity do not consume alcohol. Conversely (and remarkably) there are signs that hitherto unacceptable 'recreational materials' may be legitimate competitors for a consumer's spend in future. Those herbivorous cousins, hops and marijuana, may yet meet on a level playing field.

Technology will not be denied. Current alarm about and rebellion against gene technology will not prevent its future application, provided the benefits are apparent to the consumer and any risks are sensibly eliminated. Brewing will continue to respond to technological advances in the general domain – sensor technology, communications, materials of fabrication, and so on. But we might anticipate remarkable developments in the knowledge of the human sensory systems. How might a greater appreciation of the sensory apparatus of nose and mouth help us better design a foodstuff such as beer to meet with more predictable customer satisfaction?

Last but not least, beer is as much at the mercy of fashion trends as any other beverage. In a period of less than 20 years, for instance, the British have transformed from being a nation of pub dwellers rejoicing in glasses of flat ale to a people who thrive on sparkling premium lagers drunk from novel coloured bottles or who sequester cans of beer for home consumption. No disrespect intended to anyone or to any product style, but what better example does one need of the impact of other changes in society, in this case the emergence of a wine bar generation, ingress of mega-buck advertising and the broadening of personal horizons: the world extends beyond the road to the local.

21.5 The nature of the brewing process in 2050

We might conceive of two extremes of approach to the making of beer: which will hold sway in half a century's time?

The traditionalists and the passionate will insist and hope that the shape of the process will essentially reflect that which exists today, in terms of malting of barley followed by the same unit stages of brewing, fermentation and stabilisation. They will simply be yet more efficient and controlled.

The converse vision (one that I usually blame on the logic of the chemical engineer!) is that beer production will be performed 'in a bucket'. The argument is that were an alien to land on Earth, analyse beer, decide they like it and desire to make it, he or she would not come up with prolonged malting and brewing operations as the logical route to recreating the product. Rather they would employ the most efficient and economic procedure to yield ethanol and mix in the key ingredients of flavour, foam and colour.

Much as we traditionalists may be appalled at the concept, it is impossible to argue against the *logic* of such thinking. And I have very little doubt that in 50 years' time the body of knowledge will be in place that would permit beers to be made in this way and which are indistinguishable from 'the real thing'. The advantages are various, including the elimination of agricultural vagaries and environmental challenges, reduction in food safety scares, diminution of losses, etc. The counter-arguments concern the worth of tradition and furthermore the fundamental concerns about the origins of the foods that we partake of.

Irrespective of which approach will hold sway, it happens not to matter one jot for the establishment of research priorities. An increased body of knowledge on the sensory properties of beer and the other quality dimensions of the product would be of equal value to the traditionalist or the revolutionary. It will be up to them to apply the information in the manner that they see fit.

Note: this chapter is modified from an article first published by C.W. Bamforth in the Pauls Malts Brewing Room Book, 2001–2003.

Index